Lecture Notes in Computer Science   3504

Commenced Publication in 1973
Founding and Former Series Editors:
Gerhard Goos, Juris Hartmanis, and Jan van Leeuwen

Editorial Board

David Hutchison
   Lancaster University, UK
Takeo Kanade
   Carnegie Mellon University, Pittsburgh, PA, USA
Josef Kittler
   University of Surrey, Guildford, UK
Jon M. Kleinberg
   Cornell University, Ithaca, NY, USA
Friedemann Mattern
   ETH Zurich, Switzerland
John C. Mitchell
   Stanford University, CA, USA
Moni Naor
   Weizmann Institute of Science, Rehovot, Israel
Oscar Nierstrasz
   University of Bern, Switzerland
C. Pandu Rangan
   Indian Institute of Technology, Madras, India
Bernhard Steffen
   University of Dortmund, Germany
Madhu Sudan
   Massachusetts Institute of Technology, MA, USA
Demetri Terzopoulos
   New York University, NY, USA
Doug Tygar
   University of California, Berkeley, CA, USA
Moshe Y. Vardi
   Rice University, Houston, TX, USA
Gerhard Weikum
   Max-Planck Institute of Computer Science, Saarbruecken, Germany

Alejandro F. Frangi   Petia I. Radeva
Andres Santos   Monica Hernandez (Eds.)

Functional Imaging and Modeling of the Heart

Third International Workshop, FIMH 2005
Barcelona, Spain, June 2-4, 2005
Proceedings

Volume Editors

Alejandro F. Frangi
Pompeu Fabra University, Department of Technology
Pg Circumvallacio 8, 08003 Barcelona, Spain
E-mail: alejandro.frangi@upf.edu.es

Petia Ivanova Radeva
Autonomous University of Barcelona
08193 Cerdanyola, Barcelona, Spain
E-mail: petia@cvc.uab.es

Andres Santos
Polytechnic University of Madrid, School of Telecommunications Engineering
28040 Madrid, Spain
E-mail: andres@die.upm.es

Monica Hernandez
University of Zaragoza, Ada Byron Building, D.008
1, Maria de Luna street, 50018 Zaragoza, Spain
E-mail: mhg@unizar.es

Library of Congress Control Number: 2005926936

CR Subject Classification (1998): I.4, J.3, I.6, I.2.10

ISSN   0302-9743
ISBN-10   3-540-26161-3 Springer Berlin Heidelberg New York
ISBN-13   978-3-540-26161-2 Springer Berlin Heidelberg New York

This work is subject to copyright. All rights are reserved, whether the whole or part of the material is concerned, specifically the rights of translation, reprinting, re-use of illustrations, recitation, broadcasting, reproduction on microfilms or in any other way, and storage in data banks. Duplication of this publication or parts thereof is permitted only under the provisions of the German Copyright Law of September 9, 1965, in its current version, and permission for use must always be obtained from Springer. Violations are liable to prosecution under the German Copyright Law.

Springer is a part of Springer Science+Business Media

springeronline.com

© Springer-Verlag Berlin Heidelberg 2005
Printed in Germany

Typesetting: Camera-ready by author, data conversion by Scientific Publishing Services, Chennai, India
Printed on acid-free paper   SPIN: 11494621   06/3142   5 4 3 2 1 0

Preface

The 1st and 2nd International Conferences on Functional Imaging and Modelling of the Heart (FIMH) were held in Helsinki, Finland, in November 2001, and in Lyon, France, in June 2003. These meetings were born through a fruitful scientific collaboration between France and Finland that outreached to other groups and led to the start of this biennial event. The FIMH conference was the first attempt to agglutinate researchers from several complementary but often isolated fields: cardiac imaging, signal and image processing, applied mathematics and physics, biomedical engineering and computer science, cardiology, radiology, biology, and physiology. In the first two editions, the conference received an enthusiastic acceptance by experts of all these communities. FIMH was originally started as a European event and has increasingly attracted more and more people from the US and Asia.

This edition of FIMH received the largest number of submissions so far with a result of 47 papers being accepted as either oral presentations or posters. There were a number of submissions from non-EU institutions which confirms the growing interest in this series of meetings. All papers were reviewed by up to four reviewers. The accepted contributions were organized into 8 oral sessions and 3 poster sessions complemented by a number of invited talks. This year we tried to allocate as many papers as possible as oral presentations to facilitate more active participation and to stimulate multidisciplinary discussions. Papers were organized around several tracks: anatomical modelling of the heart, electrophysiology, electro- and magnetography, modelling of the cardiac mechanics and function, cardiac motion estimation, and also a miscellaneous section. The order of presentation in these proceedings follows that of presentation at the conference.

For the communities related to this conference, it would be impossible to overlook ongoing international efforts such as Cardiome[1], which tackles the fields of cardiac imaging, and multiscale modelling and simulation of the heart. Cardiome is the first, and possibly currently the most advanced, effort within the Physiome[2] initiative that is sponsored by the International Union for Physiological Sciences (IUPS). Also related to Physiome is the UK-sponsored project Integrative Biology[3], a key part of which relates to projects on the heart. All these efforts aim at developing a vision for computational physiology where knowledge and understanding at every length and temporal scale of the different organic systems of the human body can be integrated through computational models. We believe that FIMH will contribute to several key challenges within Cardiome

---

[1] http://www.cardiome.org/
[2] http://www.physiome.org/
[3] http://www.integrativebiology.ox.ac.uk/

and, therefore, be of importance to the overall objectives of the Physiome and Integrative Biology projects.

Another important ingredient of FIMH 2005 was the participation of the European Commission through the organization of a satellite workshop entitled *Towards Virtual Physiological Human: Multilevel Modelling and Simulation of the Human Anatomy and Physiology*. This half-day workshop, jointly organized by the Directorate-General Information Society and Media in collaboration with the Institute for Prospective Technological Studies (IPTS), further motivated the conference by presenting progress to date, providing a wider perspective on modelling and simulation, and allowing the exchange of ideas about this exciting topic.

Finally, we would like to take the opportunity to thank all the authors for the outstanding contributions to FIMH, and the Program Committee and additional reviewers for their invaluable efforts in a timely review process. Last but not least, we would like to express our gratitude to all the sponsoring and organizing institutions for their support of this conference.

We hope that the attendees enjoyed the atmosphere and program of the conference and we hope to see you again at FIMH 2007.

June 2005

Alejandro F. Frangi
Petia Radeva
Andres Santos
Monica Hernandez

Acknowledgments

The members of the Program Committee deserve special thanks for their work in reviewing all the papers, and for their support in the organization of the conference.

We thank the members of the Pompeu Fabra University, the Autonomous University of Barcelona and the Polytechnic University of Madrid for their involvement and support in the organization of this conference.

Conference Supporters

Autonomous University of Barcelona, Barcelona, Spain
Pompeu Fabra University, Barcelona, Spain
Polytechnic University of Madrid, Madrid, Spain
Department d'Universitats, Recerca i Societat de la Informacio (DURSI),
Barcelona, Spain
Aragon Institute on Engineering Research (I3A), Zaragoza, Spain
Molecular and Multi-modality Medical Imaging Thematic Network (IM3)
IEEE Spain Section
IEEE Engineering in Medicine and Biology Society (EMBS)
Ministry of Education and Sciences (MEyC), Spain
Ministry of Health and Consumption (MSyC), Spain
MEDIS Medical Imaging Systems B.V.
Medtronic, Inc.
European Commission, DG-INFSO
Generalitat de Catalunya
Barcelona City

Organization

FIMH 2005 was co-organized by Pompeu Fabra University, the Autonomous University of Barcelona and the Polytechnic University of Madrid.

Conference Chairs

Alejandro Frangi, PhD	Computational Imaging Lab, Pompeu Fabra University, Barcelona, Spain
Petia Radeva, PhD	Computer Vision Center, Autonomous University of Barcelona, Barcelona, Spain
Andres Santos, PhD	ETSI Telecomunication, Polytechnic University of Madrid, Madrid, Spain

Organization Committee

Jordi Vitria, PhD	Computer Vision Center, Autonomous University of Barcelona, Barcelona, Spain
Monica Hernandez	University of Zaragoza, Zaragoza, Spain
Maria J. Ledesma, PhD	Polytechnic University of Madrid, Madrid, Spain

Coordination Committee

Adrian M. Rosolen	Pompeu Fabra University, Barcelona, Spain
Catalina Tobon	Pompeu Fabra University, Barcelona, Spain
Estanislao Oubel	Pompeu Fabra University, Barcelona, Spain
Sebastian Ordas	Pompeu Fabra University, Barcelona, Spain
Milton Hoz	Pompeu Fabra University, Barcelona, Spain
Oriol Pujol	Autonomous University of Barcelona, Spain
David Rotger	Autonomous University of Barcelona, Spain
Jaume Garcia	Autonomous University of Barcelona, Spain
Aura Hernandez	Autonomous University of Barcelona, Spain
Vanessa Escriva	Polytechnic University of Madrid, Spain
Ana Bajo	Polytechnic University of Madrid, Spain

Program Committee

A. Amini	Washington University in St. Louis, USA
T. Arts	Maastricht University, The Netherlands
L. Axel	New York University, New York, USA
N. Ayache	INRIA, Sophia-Antipolis, France
F.B. Bijnens	Catholic University of Leuven, Belgium
J.R. Cebral	George Mason University, Fairfax, USA
P. Clarysse	CREATIS, Lyon, France
P. Claus	Catholic University of Leuven, Belgium
J.L. Coatrieux	LTSI, Rennes, France
H. Delingette	INRIA, Sophia-Antipolis, France
M. Desco	Gregorio Maranon General Hospital, Madrid, Spain
M. Doblare	University of Zaragoza, Zaragoza, Spain
J. Duncan	Yale University, New Haven, USA
R. Fenici	Catholic University of Rome, Italy
A.F. Frangi	Pompeu Fabra University, Spain
M. Garreau	LTSI, Rennes, France
F. Gerritsen	Technical University Eindhoven, The Netherlands
J.M. Goicolea	Polytechnic University of Madrid, Madrid, Spain
A. Holden	University of Leeds, UK
G. Holzapfel	Graz University of Technology, Graz, Austria
P. Hunter	Auckland University, New Zealand
T. Katila	Laboratory of Biomedical Engineering, Helsinki, Finland
P. Kohl	University of Oxford, Oxford, UK
A. Laine	Columbia University, New York, USA
B. Lelieveldt	Leiden University Medical Center, Leiden, The Netherlands
S. Loncaric	University of Zagreb, Croatia
I. Magnin	CREATIS, Lyon, France
A. McCulloch	University of California, San Diego, USA
J. Montagnat	CREATIS, Lyon, France
J. Nenonen	Laboratory of Biomedical Engineering, Helsinki, Finland
W.J. Niessen	UMC, Utrecht, The Netherlands
A. Noble	Medical Vision Laboratory, Oxford, UK
P. Radeva	Computer Vision Center, Barcelona, Spain
H. Reiber	Leiden University Medical Center, Leiden, The Netherlands
N. Rougon	INT Evry, France
D. Rueckert	Imperial College of Science and Technology, London, UK
F. Sachse	University of Utah, USA
A. Santos	Polytechnic University of Madrid, Spain
B. Tilg	University for Health Informatics and Technology, Innsbruck, Austria
M.A. Viergever	UMC, Utrecht, The Netherlands

Table of Contents

Modeling of the Heart - Anatomy Extraction and Description

Multi-surface Cardiac Modelling, Segmentation, and Tracking
  Jens von Berg, Cristian Lorenz 1

Analysis of the Interdependencies Among Plaque Development,
Vessel Curvature, and Wall Shear Stress in Coronary Arteries
  *Andreas Wahle, John J. Lopez, Mark E. Olszewski,
  Sarah C. Vigmostad, Krishnan B. Chandran,
  James D. Rossen, Milan Sonka* 12

Automated Segmentation of X-ray Left Ventricular Angiograms
Using Multi-View Active Appearance Models and Dynamic
Programming
  *Elco Oost, Gerhard Koning, Milan Sonka, Johan H.C. Reiber,
  Boudewijn P.F. Lelieveldt* 23

SPASM: Segmentation of Sparse and Arbitrarily Oriented Cardiac MRI
Data Using a 3D-ASM
  *Hans C. van Assen, Mikhail G. Danilouchkine,
  Alejandro F. Frangi, Sebastián Ordás, Jos J.M. Westenberg,
  Johan H.C. Reiber, Boudewijn P.F. Lelieveldt* 33

Combining Active Appearance Models and Morphological Operators
Using a Pipeline for Automatic Myocardium Extraction
  *Bernhard Pfeifer, Friedrich Hanser, Thomas Trieb,
  Christoph Hintermüller, Michael Seger, Gerald Fischer,
  Robert Modre, Bernhard Tilg* 44

Long-Axis Cardiac MRI Contour Detection with Adaptive Virtual
Exploring Robot
  *Mark Blok, Mikhail G. Danilouchkine, Cor J. Veenman,
  Faiza Admiraal-Behloul, Emile A. Hendriks, Johan H.C. Reiber,
  Boudewijn P.F. Lelieveldt* 54

A Deterministic-Statistic Adventitia Detection in IVUS Images
  *Debora Gil, Aura Hernandez, Antoni Carol, Orial Rodriguez,
  Petia Radeva* 65

Trajectory Planning Applied to the Estimation of Cardiac Activation
Circuits
  *Lorena González, Jerónimo J. Rubio, Enrique Baeyens,
  Juan C. Fraile, Jose R. Perán* 75

A Functional Heart Model for Medical Education
  Vassilios Hurmusiadis, Chris Briscoe 85

Artificial Enlargement of a Training Set for Statistical Shape Models:
Application to Cardiac Images
  *J. Lötjönen, K. Antila, E. Lamminmäki, J. Koikkalainen, M. Lilja,
  T. Cootes* 92

Towards a Comprehensive Geometric Model of the Heart
  Cristian Lorenz, Jens von Berg 102

Automatic Cardiac 4D Segmentation Using Level Sets
  Karl D. Fritscher, Roland Pilgram, Rainer Schubert 113

Level Set Segmentation of the Fetal Heart
  *I. Dindoyal, T. Lambrou, J. Deng, C.F. Ruff, A.D. Linney,
  C.H. Rodeck, A. Todd-Pokropek* 123

Supporting the TECAB Grafting Through CT Based Analysis of
Coronary Arteries
  Stefan Wesarg 133

Electro-Physiology, Electro- and Magnetography

Clinical Validation of Machine Learning for Automatic Analysis of
Multichannel Magnetocardiography
  *Riccardo Fenici, Donatella Brisinda, Anna Maria Meloni,
  Karsten Sternickel, Peter Fenici* 143

Hypertrophy in Rat Virtual Left Ventricular Cells and Tissue
  S. Kharche, H. Zhang, R.C. Clayton, Arun V. Holden 153

Virtual Ventricular Wall: Effects of Pathophysiology and Pharmacology
on Transmural Propagation
  *Oleg V. Aslanidi, Jennifer L. Lambert, Neil T. Srinivasan,
  Arun V. Holden* 162

Electrophysiology and Tension Development in a Transmural
Heterogeneous Model of the Visible Female Left Ventricle
  Gunnar Seemann, Daniel L. Weiß, Frank B. Sachse, Olaf Dössel ... 172

Reentry Anchoring at a Pair of Pulmonary Vein Ostia
  L. Wieser, G. Fischer, F. Hintringer, S.Y. Ho, B. Tilg 183

A Method to Reconstruct Activation Wavefronts Without Isotropy
Assumptions Using a Level Sets Approach
  *Felipe Calderero, Alireza Ghodrati, Dana H. Brooks, Gilead Tadmor,
  Rob MacLeod* 195

Magnetocardiographic Imaging of Ventricular Repolarization in Rett
Syndrome
  *Donatella Brisinda, Anna Maria Meloni, Giuseppe Hayek,
  Menotti Calvani, Riccardo Fenici* 205

Insights into Electrophysiological Studies with Papillary Muscle by
Computational Models
  Frank B. Sachse, Gunnar Seemann, Bruno Taccardi 216

Induced Pacemaker Activity in Virtual Mammalian Ventricular Cells
  Wing Chiu Tong, Arun V. Holden 226

Transvenous Path Finding in Cardiac Resynchronization Therapy
  *Jean Louis Coatrieux, Alfredo I. Hernández, Philippe Mabo,
  Mireille Garreau, Pascal Haigron* 236

Methods for Identifying and Tracking Phase Singularities in
Computational Models of Re-entrant Fibrillation
  Ekaterina Zhuchkova, Richard Clayton 246

Estimating Local Apparent Conductivity with a 2-D
Electrophysiological Model of the Heart
  *Valérie Moreau-Villéger, Hervé Delingette, Maxime Sermesant,
  Hiroshi Ashikaga, Owen Faris, Elliot McVeigh, Nicholas Ayache* 256

Monodomain Simulations of Excitation and Recovery in Cardiac Blocks
with Intramural Heterogeneity
  Piero Colli Franzone, Luca F. Pavarino, Bruno Taccardi 267

Spatial Inversion of Depolarization and Repolarization Waves in Body
Surface Potential Mapping as Indicator of Old Myocardial Infarction
  *Paula Vesterinen, Helena Hänninen, Matti Stenroos,
  Petri Korhonen, Terhi Husa, Ilkka Tierala, Heikki Väänänen,
  Lauri Toivonen* 278

Dissipation of Excitation Fronts as a Mechanism of Conduction Block
in Re-entrant Waves
  Vadim N. Biktashev, Irina V. Biktasheva 283

Wavebreaks and Self-termination of Spiral Waves in a Model of Human
Atrial Tissue
  Irina V. Biktasheva, Vadim N. Biktashev, Arun V. Holden 293

Calcium Oscillations and Ectopic Beats in Virtual Ventricular Myocytes
and Tissues: Bifurcations, Autorhythmicity and Propagation
  Alan P. Benson, Arun V. Holden 304

Modeling of the Cardiac Mechanics and Functions

Left Ventricular Shear Strain in Model and Experiment: The Role of
Myofiber Orientation
  *Sander Ubbink, Peter Bovendeerd, Tammo Delhaas, Theo Arts,
  Frans van de Vosse* 314

Cardiac Function Estimation from MRI Using a Heart Model and Data
Assimilation: Advances and Difficulties
  *M. Sermesant, P. Moireau, O. Camara, J. Sainte-Marie,
  R. Andriantsimiavona, R. Cimrman, D.L.G. Hill, D. Chapelle,
  R. Razavi* 325

Assessment of Separation of Functional Components with ICA
from Dynamic Cardiac Perfusion PET Phantom Images for Volume
Extraction with Deformable Surface Models
  *Anu Juslin, Anthonin Reilhac, Margarita Magadán-Méndez,
  Edisson Albán, Jussi Tohka, Ulla Ruotsalainen* 338

Detecting and Comparing the Onset of Myocardial Activation
and Regional Motion Changes in Tagged MR for XMR-Guided RF
Ablation
  *Gerardo I. Sanchez-Ortiz, Maxime Sermesant, Kawal S. Rhode,
  Raghavendra Chandrashekara, Reza Razavi, Derek L.G. Hill,
  Daniel Rueckert* 348

Suppression of IVUS Image Rotation. A Kinematic Approach
  Misael Rosales, Petia Radeva, Oriol Rodriguez, Debora Gil 359

Computational Modeling and Simulation of Heart Ventricular
Mechanics from Tagged MRI
  Zhenhua Hu, Dimitris Metaxas, Leon Axel 369

A Realistic Anthropomorphic Numerical Model of the Beating
Heart
  *Rana Haddad, Patrick Clarysse, Maciej Orkisz, Pierre Croisille,
  Didier Revel, Isabelle E. Magnin* 384

Multi-formalism Modelling of Cardiac Tissue
  Antoine Defontaine, Alfredo Hernández, Guy Carrault 394

Analysis of Tagged Cardiac MRI Sequences
  *Aymeric Histace, Christine Cavaro-Ménard, Vincent Courboulay,
  Michel Ménard* 404

Fast Spatio-temporal Free-Form Registration of Cardiac MR Image
Sequences
  Dimitrios Perperidis, Raad Mohiaddin, Daniel Rueckert 414

Comparison of Cardiac Motion Fields from Tagged and Untagged
MR Images Using Nonrigid Registration
  *Raghavendra Chandrashekara, Raad H. Mohiaddin,
  Daniel Rueckert* 425

Tracking of LV Endocardial Surface on Real-Time Three-Dimensional
Ultrasound with Optical Flow
  *Qi Duan, Elsa D. Angelini, Susan L. Herz, Olivier Gerard,
  Pascal Allain, Christopher M. Ingrassia, Kevin D. Costa,
  Jeffrey W. Holmes, Shunichi Homma, Andrew F. Laine* 434

Cardiac Motion Estimation

Dense Myocardium Deformation Estimation for 2D Tagged MRI
  Leon Axel, Ting Chen, Tushar Manglik 446

A Surface-Volume Matching Process Using a Markov Random Field
Model for Cardiac Motion Extraction in MSCT Imaging
  *Antoine Simon, Mireille Garreau, Dominique Boulmier,
  Jean-Louis Coatrieux, Hervé Le Breton* 457

Evaluation of Two Free Form Deformation Based Motion Estimators in
Cardiac and Chest Imaging
  *Bertrand Delhay, Patrick Clarysse, Jyrki Lötjönen, Toivo Katila,
  Isabelle E. Magnin* 467

Classification of Segmental Wall Motion in Echocardiography Using
Quantified Parametric Images
  *Cinta Ruiz Dominguez, Nadjia Kachenoura, Sébastien Mulé,
  Arthur Tenenhaus, Annie Delouche, Olivier Nardi, Olivier Gérard,
  Benoît Diebold, Alain Herment, Frédérique Frouin* 477

Author Index 487

Multi-surface Cardiac Modelling, Segmentation, and Tracking

Jens von Berg and Cristian Lorenz

Philips Research Laboratories, Sector Technical Systems,
Röntgenstr. 24-26, 22335 Hamburg, Germany
Jens.von.Berg@philips.com

Abstract. Multi–slice computed tomography image series are a valuable source of information to extract shape and motion parameters of the heart. We present a method how to segment and label all main chambers (both ventricles and atria) and connected vessels (arteries and main vein trunks) from such images and to track their movement over the cardiac cycle. A framework is presented to construct a multi–surface triangular model enclosing all blood–filled cavities and the main myocardium as well as to adapt this model to unseen images, and to propagate it from phase to phase. While model construction still requires a reasonable amount of user interaction, adaptation is mostly automated, and propagation works fully automatically. The adaptation method by deformable surface models requires a set of landmarks to be manually located for one of the cardiac phases for model initialisation.

1 Introduction

The aim of our work is a comprehensive model of the geometry of the human heart contraction as well as its inter–individual variations. Such a model introducing a priori knowledge about typical properties of a beating heart will be highly beneficial in the whole chain of image–based cardiac diagnostics, as well as in many cardiac treatment procedures. The model covers landmarks, the coronary tree, and the surfaces of the large vessels [1]. The latter is the subject of the work reported here. The most valuable and practically unique source of information for the modelling process are cardiac images from clinical practice. In this paper the use of multi–slice computed tomography (MSCT) images is reported that have a voxel size of about 0.5 mm in each direction and a temporal resolution of 10 volumes per cardiac cycle. Mostly, cardiac MRI were used previously for this purpose [2, 3, 4, 5, 6]. MSCT may provide even better insight into the morphology of the human heart [7]. Extracting the relevant information from these images is hardly feasible without a priori knowledge [8]. Many approaches to cardiac segmentation were based on manually segmented images, which is a good means to both tune parameters by automated supervised learning, and to finally prove their performance in comparison to human expertise. However, manually segmenting an MSCT series means delineating each object of interest in about two thousand images. This dilemma led us to a bootstrap approach with a consecutive refinement of the model during successive analysis of new images.

2 Model Construction

The model covers the blood pool of both the left and the right heart. The blood pools of the ventricles should be distinguishable from those of the atria. All attached vessels should also be modelled, i.e. the aorta, the pulmonary artery, the vena cava, and the pulmonary vein trunks. As it is clearly visible, and diagnostically relevant, also the left myocardium should be represented in the model. Including adjacent volumetric entities required a surface modelling scheme beyond two–dimensional manifolds. In the discrete case with triangular faces this means that there are faces with more than three neighbours wherever multiple surfaces share an edge. In order to enable multi–scale / multi–resolution approaches or to just find an ideal trade–off between accuracy and complexity, a multi–resolution representation of the surface discretisation was desired.

The initial step was the construction of single basic shapes like spheres (atria), tubes (attached vessels), and opened ellipsoids (ventricles). Each one modelled an anatomical entity. These shapes were then positioned in the training image and adapted to the corresponding entities. A re-sampling closed this step to get a defined level of granularity. The third step was the most important one that combines the single basic two–dimensional manifolds to form the multi–surface model. The method used in this third step is explained in some detail below. In the resulting model, each face should be assigned a label that indicates the anatomical structure it belongs to. This information was derived by storing which of the initial basic shapes a face originates from. The basic shapes were *left atrium* a_l, *left ventricle endocardium* (inner part v_i), *left ventricle epicardium* (outer part v_o), *aorta* a, *vena cava superior* v_s, *vena cava inferior* v_l, *right atrium* a_r, *right ventricle* v_r, *pulmonary artery* a_p (right branch only), and the pulmonary vein trunks (v_1, v_2, v_3, v_4,) that drain into the left atrium.

2.1 Building a Multi-surface Model

The combination step was made by successive application of a handful of basic operations on surface meshes, starting with the basic meshes. There are volumetric set operations that consider the enclosed volume of two meshes, apply the union ($\cup$) or the difference ($\setminus$) operation on them, and yield the resulting surface mesh. A similar approach but with implicit surface models was proposed in [9]. Each of the present operations was defined as $\mathbf{B} \times \mathbf{B} \rightarrow \mathbf{B}$, where $m \in \mathbf{B}$ is a two–dimensional manifold mesh. As a further constraint on theses operations, the intersection line between both meshes had to be closed polygons. This required open basic meshes to fully overlap with their neighbours (e.g. ventricle with atrium). The join operator ($\diamond$) was defined as $\mathbf{B} \times \mathbf{B} \rightarrow \mathbf{M}$, where $\bar{m} \in \mathbf{M}$ may be a non–two–dimensional manifold mesh. The join operator just unites

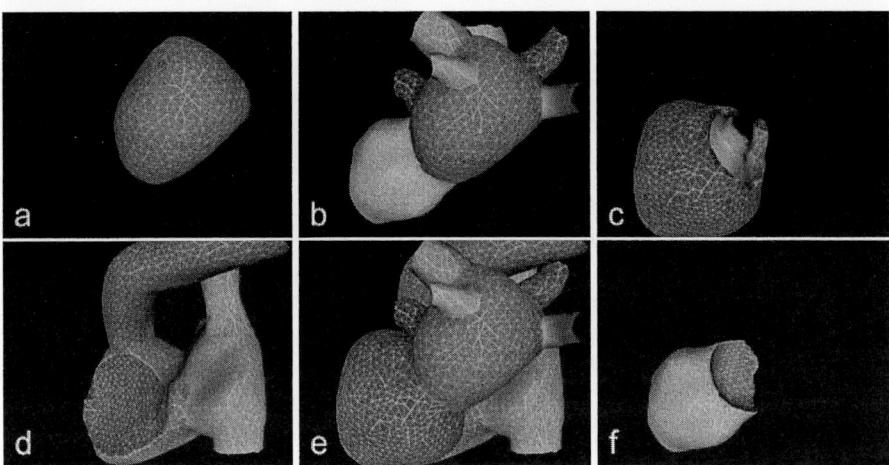

Fig. 1. A basic (a), some intermediate (b to d), the final (e), and a derived (f) sub mesh, all seen from left anterior. a: *left atrium* a_l, b: *left blood pool* p_l, c: *left myocardium* a_1, e: whole heart $\bar{h}$, f: blood pool of the *left ventricle* only p_{v_i}. The colour–coding denotes face labels. All shown meshes are just subsets of the complete multi–surface model $\bar{h}$

both sets of faces and unifies corresponding faces that occur in both meshes. The unary operator c_l ($\mathbf{M} \rightarrow \mathbf{M}$) removes all edges smaller than given by parameter l and preserves the triangles' labels. It was needed to replace auxiliary triangles created by volumetric set operations. The left blood pool $p_l \in \mathbf{B}$ was build by

$$p_l = c_l\left(v_1 \cup v_2 \cup v_3 \cup v_4 \cup a_l \cup v_i \cup a\right). \tag{1}$$

In order to construct the complete multi–surface model $\bar{h}$, first the intermediate meshes a_1 and a_2 were constructed by

$$a_1 = c_l\left(v_o \setminus p_l\right) \tag{2}$$
$$a_2 = c_l\left(v_o \cup p_l\right), \tag{3}$$

where a_1 now exactly enclosed the left myocardium. The complete left heart model $\bar{h}_l$ was built by

$$\bar{h}_l = a_1 \diamond a_2, \tag{4}$$

and the right blood pool was built by

$$p_r = c_l(a_r \cup v_r \cup a_p \cup v_l \cup v_s) \setminus a_1. \tag{5}$$

Left and right part were fused to

$$\bar{h} = \bar{h}_l \diamond p_r. \tag{6}$$

In Figure 1 some basic, intermediate, and the final mesh $\bar{h}$ are rendered from the same viewing position. The edge size was set to range between 2.5 mm and

5 mm. Also other sub meshes than those required to build the final mesh may be constructed, for instance the blood pool of the left ventricle excluding the left atrium by

$$p_{v_i} = v_i \setminus a_l. \tag{7}$$

3 Model Adaptation

For adaptation of the multi–surface model to a cardiac CT image, a shape–constrained deformable surface model approach was followed as previously described in [10, 4]. The model with given vertex positions $\hat{\mathbf{v}}$ taken from a training image both served for the initialisation of the initial mesh $\mathbf{v}^0$ and as constraint during its adaptation to $\mathbf{v}^{1\cdots n}$. The number of triangles remained unchanged in this process.

3.1 Affine Pre-registration

In order to pose the initial mesh into the image as accurately as possible, 25 anatomical landmarks were manually located both for the image the model was built from and for each target image [1]. These landmarks are mainly centre locations of chambers, valves, and ostia. A point–based affine registration [11] was applied on the two sets of landmarks. The resulting affine matrix $\mathbf{A}$ and translation vector $\mathbf{t}$ were then used to pre–register the initial mesh by $\mathbf{v}^0 = \mathbf{A}\hat{\mathbf{v}} + \mathbf{t}$.

3.2 Model Deformation

In the optimisation scheme the vertex positions of the triangular surface mesh were the parameters to be varied. Mesh deformation was done by minimizing the energy term

$$E = E_{ext} + \alpha E_{int}. \tag{8}$$

The external energy E_{ext} drives the mesh towards the surface points obtained in a surface detection step. The internal energy E_{int} restricts the flexibility by maintaining the vertex configuration of a shape model. The parameter α weights the influence of both terms. A fixed number n of such minimisation steps is performed on the mesh. The different components of the deformation algorithm are described below.

Surface Detection. Surface detection was carried out for each triangle barycentre $\mathbf{x_i}$. Within a sampling grid of points $\mathbf{c_k}$, defined in a local co-ordinate system, that point $\tilde{\mathbf{c}}_\mathbf{i}$ is chosen that maximizes the objective function

$$\tilde{\mathbf{c}}_\mathbf{i} = \mathrm{argmax}_{k=-l,\ldots,l} \left\{ F_i(\mathbf{x_i} + \mathbf{M_i c_k}) - \delta \left\| \mathbf{c_k} \right\|^2 \right\}. \tag{9}$$

$\mathbf{M_i}$ is a rotational matrix that rotates the z–axis of the local co–ordinate system to the triangle surface normal $\mathbf{n_i}$ and

$$\tilde{\mathbf{x}}_i = \mathbf{x}_i + \mathbf{M}_i \tilde{\mathbf{c}}_i \qquad (10)$$

is the new surface point for $\mathbf{x}_i$. The parameter δ controls the trade–off between feature strength and distance. The sampling grid

$$\mathbf{c_k} \equiv \mathbf{G_L} = (0,0,k\varepsilon) : k = -l, \ldots, l \qquad (11)$$

was used, that results in $(2l+1)$ equidistant sampling points along the triangle surface normal.

Feature Function. The feature function

$$F_i(\mathbf{x}) = \begin{cases} -\mathbf{n_i}^t \nabla I(\mathbf{x}) \frac{g_{max}(g_{max}+\|\nabla I(\mathbf{x})\|)}{g_{max}^2 + \|\nabla I(\mathbf{x})\|^2} & : \; I_{min} < I(\mathbf{x}) < I_{max} \\ 0 & : \; otherwise \end{cases} \qquad (12)$$

was used that projects the image gradient $\nabla I(\mathbf{x})$ onto the face normal $\mathbf{n_i}$ and damps its value so that surface points with image gradients stronger than g_{max} do not give higher response. The restriction to a dedicated intensity range may make the feature function more specific and thus makes adaptation less vulnerable to adjacent false attractors (see below).

External Energy. The external energy

$$E_{ext} = \sum_i w_i \left(\mathbf{e}_{\nabla I} \tilde{\mathbf{c}}_i\right)^2, \; w_i = \max\left\{0, F_i(\mathbf{x_i} + \mathbf{M_i}\tilde{\mathbf{c}}_i) - \delta \|\tilde{\mathbf{c}}_i\|^2\right\} \qquad (13)$$

drives each triangle barycentre $\mathbf{x}_i$ towards the detected surface point $\tilde{\mathbf{x}}_i$. $\mathbf{e}_{\nabla I}$ is the unit vector in the direction of the image gradient at the surface point $\tilde{\mathbf{x}}_i$. Since only the projection onto $\mathbf{e}_{\nabla I}$ is penalized, this allows the triangle centre to locally slide along an iso–contour. This method proved to be superior to direct attraction by the candidate in [10] in case of intermediate false attractions.

Internal Energy. The internal energy

$$E_{int} = \sum_j \sum_{k \in N(j)} \left((\hat{\mathbf{v}}_j - \hat{\mathbf{v}}_k) - s\mathbf{R}(\mathbf{v}_j - \mathbf{v}_k)\right)^2 \qquad (14)$$

preserves shape similarity of all mesh vertices $\mathbf{v}_i$ to the model vertices $\hat{\mathbf{v}}_i$. $N(j)$ is the set of neighbours of vertex j. The neighbouring vertices are those connected by a single triangle edge. The scaling factor s and the rotational matrix $\mathbf{R}$ are determined by a closed–form point–based registration method based on a singular value decomposition [11] prior to calculation of (14).

Optimisation. As only interdependences between neighbour vertices exist (14) and the energy terms are of a quadratic form, the conjugate gradient method [12] could be used for minimisation of (8) with a sparsely filled matrix.

Multi-surface Parameterisation. The labels assigned to each face of the multi–surface model may be used to parameterise interfaces between different anatomical entities specifically. However, dedicated parameter tuning was restricted to the epicardium border towards the lung parenchyma that differs in its appearance significantly from the other surfaces that enclose the blood pool. Thus, g_{max} was set to 60 $\frac{HU}{mm}$ here instead of 120 $\frac{HU}{mm}$ elsewhere. and the intensity range $(I_{min} \cdots I_{max})$ was adjusted to $350 \cdots 800$ HU instead of 1000 HU and up. The other parameters were globally set to $\alpha = \delta = 1$, $\varepsilon = 1$ mm, $l = 10$.

4 Surface Tracking

In order to capture tissue trajectories one has to find corresponding tissue landmarks in images from different cardiac phases. This was mainly done previously either by non–linear registration [13] or by active appearance models [14]. With some modifications that rather belong to the second category and that are explained below, the adaptation method presented above was also applied for a surface tracking approach that utilizes point correspondence.

Surface detection is carried out following equation (9). In order for a surface point not only to be attracted along the surface normal, a sampling grid is used that extends into all direction. A multi–icosahedron grid

$$\mathbf{c_{k=1\cdots 37}} \equiv \mathbf{G_I} = \{(0,0,0), P_2, P_4, P_8\} \quad (15)$$

was used where each P_n is a set of 12 icosahedron surface points with a radius of $n\varepsilon$ mm around the origin of the local coordinate system. Individual feature functions are required for each surface point in this case to take the local image properties into account. The feature function

$$F_i(\mathbf{x}) = \frac{2l+1}{\sum_{k=-l\ldots l} \left(I\left(\mathbf{x} + \mathbf{M_i s_k}\right) - \hat{g}_{i,k} \right)^2} \quad (16)$$

that replaces (12) thus evaluates similarity of local appearance samples to the once learnt model $\hat{\mathbf{g}}_i$. The linear sampling grid $\mathbf{s_k} = \mathbf{G_L}$ from equation (11) is taken. It is applied at each sample point. The external energy is calculated by

$$E_{ext} = \sum_i \tilde{\mathbf{c}}_i^2 \quad (17)$$

instead of (13), and the internal energy is taken from (14).

In order to demonstrate the general feasibility of surface tracking with deformable models and appearance models a simple test study was carried out. A cylindrical surface mesh was posed into a cardiac CT image to roughly fit the left myocardium. The image appearance $\hat{\mathbf{g}}$ was learnt and the mesh was rotated around its main axis by $r = \pm\frac{\pi}{16}, \pm\frac{\pi}{8}, \pm\frac{\pi}{4}$, and $\pm\frac{\pi}{2}$ afterwards in a number of trials. An adaptation with $n = 80$ iterations was performed for each rotation angle. Up to $r = \pm\frac{\pi}{4}$ the mesh successfully recovered the initial position. This simple test showed that the tracking method has a remarkable capture range, and that with the rotationally symmetric shape model, appearance alone is sufficient as driving force.

The propagation through all cardiac phases started with a phase to which the shape model was adapted successfully. From the image of this phase 1 the appearance $\hat{\mathbf{g}}_1$ was learnt at positions given by $\mathbf{v}_1$. This shape $\mathbf{v}_1$ was used as initial mesh and further on $\mathbf{v}_i^0 = \mathbf{v}_{i-1}^n$. The same holds for the shape model $\hat{\mathbf{v}}_i = \mathbf{v}_{i-1}^n$. This was repeated until all phases were processed. For each phase the initial appearance model $\hat{\mathbf{g}}_1$ was used.

5 Results

5.1 Model Construction

A multi–surface mesh with a total number of about 7,000 vertices and 13,000 triangle faces was constructed with edge lengths ranging between 2.5 mm and 5 mm. Its shape is shown in Figure 1. The basic meshes this model was constructed from, were adapted to the anatomical entities of the end–diastolic phase of the training image. The resulting multi–surface model was then adapted to a set of five other cardiac MSCT images from different hospitals but all acquired with a Philips MX8000 IDT 16–line CT scanner. The images were contrast–enhanced as they were acquired for the purpose of coronary assessment.

5.2 Pre-registration

The affine pre–registration led to a mean ($\pm$ standard deviation) residual landmark distance of 7.5 ± 4.3 mm, 7.0 ± 3.3 mm, 8.3 ± 3.5 mm, 5.6 ± 2.6 mm, and 13.0 ± 12.1 mm for the five images. The latter resulted in an unacceptable pre–registration, both visually and with respect to the subsequent adaptation result. An alternative rigid registration with an isotropic scale parameter

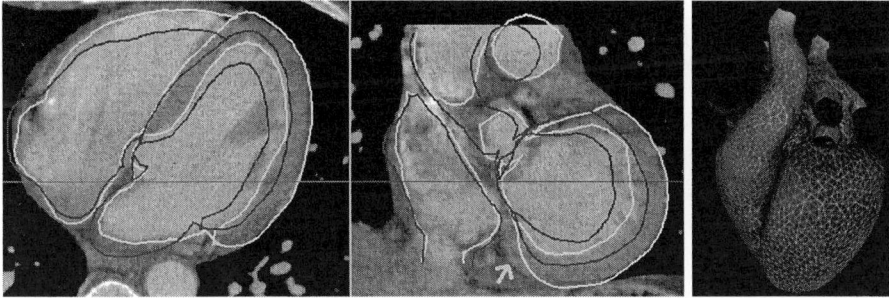

Fig. 2. Left: Pre–registration (dark mesh) and subsequent automatic adaptation (light mesh) of the multi–surface model to an unseen image. The arrow marks a local misadaptation. Right: Mean end–diastolic model of the five patients with colour–coded standard deviation (dark:0.8 mm, light:7.8 mm)

(13.0 ± 14.4 mm) resulted in an acceptable pre–registration for further processing. A typical affine pre–registered model is shown in Figure 2 in comparison to the automatic deformable adaptation based on this pre–registration.

5.3 Adaptation

The adaptation by model–based deformation significantly improved on the results of the affine pre–registration (Figure 2). Automatic adaptation with $n = 10$ iterations took about 15 seconds on a 2.6 GHz PC including real–time surface rendering. The majority of surface parts could be considered well–adapted. The reasons for remaining local mis–adaptations were mainly adaptations to false attractors e.g. of the epicardium mesh to the endocardium (see Figure 2) or to coronaries, and of the aorta mesh to the vena cava. Using the methods described in [15] manual corrections that survive subsequent automatic adaptation steps could be applied to these mis–adapted parts.

5.4 Calculating a Mean Model

The resulting individualized models were mutually registered (rigid plus isotropic scale) using a procrustes analysis of their corresponding anatomical landmarks. A mean model of the five subjects was calculated (Fig. 2).

5.5 Surface Tracking

The surface tracking method was applied to the training image sequence. The initial mesh $\mathbf{v}_1$ was the one that resulted from model construction and that was fit to the end–diastolic phase image of the training sequence. Each propagation step $\mathbf{v}_{i-1} \to \mathbf{v}_i$ was done with $n = 12$ iterations. Propagation was done for all nine images of subsequent phases and back to the initial image with $\mathbf{v}_1^0 = \mathbf{v}_9^{12}$. This allows for a comparison of the round–trip adaptation result $\mathbf{v}_1^{12}$ with the initial mesh $\mathbf{v}_1$. The mean (± standard deviation) distance of corresponding vertices between both was 1.4 ± 0.7 mm. The meshes are shown in Figure 3. The mean distance of all corresponding vertices in all phases between the forward ($\mathbf{v}_{i-1} \to \mathbf{v}_i$) and the backward ($\mathbf{v}_{i+1} \to \mathbf{v}_i$) propagation was 2.1 ± 1.3 mm. The propagation from the initial mesh ($\mathbf{v}_1 \to \mathbf{v}_i$) differs from forward propagation by

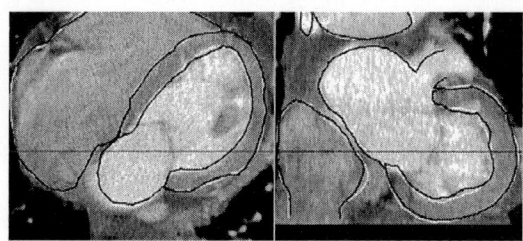

Fig. 3. Result of the consistency test: Initial mesh (white) and result of a round–trip adaptation (black) to the end–diastolic initial image

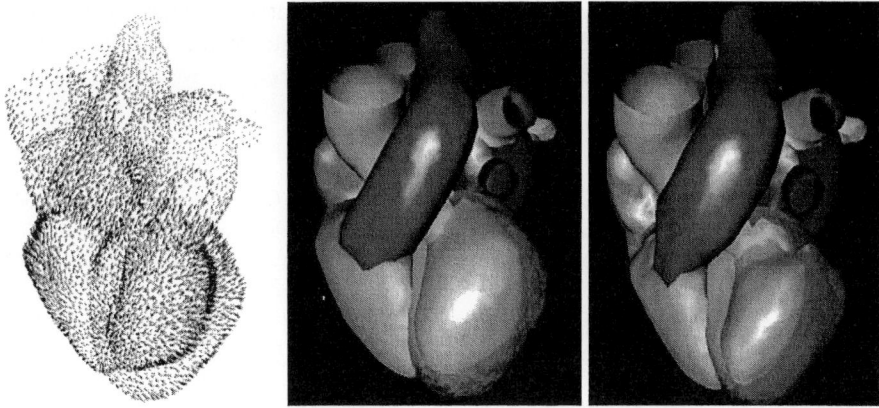

Fig. 4. Left: Trajectories of each vertex through the cardiac cycle. For visibility reasons they were scaled down by a factor 4 with respect to their initial (end–diastolic) vertex position. Centre: end–diastolic mesh. Right: end–systolic mesh

1.6 ± 1.0 mm and from backward propagation by 1.5 ± 0.9 mm. A visualisation of the moving model by a surface rendering loop gives a very natural impression of contraction (ventricles), parallel displacement (valve plane), and rather stable parts (atria). Figure 4 tries to show the results in a printed form.

6 Discussion

A method was presented that enables a widely automated construction of a multi–surface triangular mesh of cardiac chambers and vessels, mostly automatic adaptation to individual MSCT images, and automatic propagation of such an individualized model through the cardiac phases. For model construction a set of single basic shapes was adapted each to its anatomical entity. The multi–surface model resulted from their automatic combination. Some anatomical landmarks were manually located in order to pre–register this model to an unseen image by either affine or rigid registration. The subsequent deformation to fit the image boundaries was mainly gradient–based. All parameters were set explicitly during an explorative test phase resulting in a small knowledge base. Some individual surfaces of the model were parameterised specifically, which was well supported by the anatomical labels given in the multi–surface model. The surface tracking however used individual grey value profiles for each surface location learnt from the initial phase's mesh. This method was chosen in order to closely approximate the real tissue trajectories. Due to the large data volume (up to two thousand slices for a cardiac cycle) and the difficulties in manually finding reliable trajectories there was no high quality expert data available for validation. We were able to perform a capture range test and a consistency test of the method with good results. Also the animated visualisation gave a realistic impression. Only the rotational component and the twist of the left ventricle seemed to be underestimated, which we suppose to be due to the too rigid regularisation in (14).

Acknowledgements

We would like to thank our colleagues from PMS–CT Cleveland and PMS–CT Haifa for the abundance of cardiac MSCT images. We also thank our colleagues from PMS–MIT, PMS–MR, and PMS–XRD in Best for many fruitful discussions.

References

[1] Lorenz, C., von Berg, J.: Towards a comprehensive geometric model of the heart. In: Functional Imaging and Modeling of the Heart, to appear, Springer–Verlag (2005)
[2] Frangi, A.F., Rueckert, D., Schnabel, J.A., Niessen, W.J.: Automatic construction of multiple–object three–dimensional statistical shape models: Application to cardiac modeling. IEEE Trans. Med. Imag. **21** (2002) 1151–1166
[3] Lorenzo-Valdés, M., Sanchez-Ortiz, G.I., Mohiaddin, R., Rückert, D.: Atlas-based segmentation and tracking of 3D cardiac MR images using non-rigid registration. In: Proc. of MICCAI, Springer-Verlag (2002) 642–650
[4] Kaus, M.R., von Berg, J., Weese, J., Niessen, W., Pekar, V.: Automated segmentation of the left ventricle in cardiac MRI. Med. Img. Anal. **8** (2004) 245–254
[5] Lötjönen, J., Kivistö, S., Koikkalainen, J., Smutek, D., Lauerma, K.: Statistical shape model of atria, ventricles and epicardium from short– and long–axis MR images. Med. Img. Anal. **8** (2004) 371–386
[6] Pilgram, R., Fritscher, K.D., Schubert, R.: Modeling of the geometric variation and analysis of the right atrium and right ventricle motion of the human heart using PCA. In: Proc. of CARS, Elsevier (2004) 1108–1113
[7] Chen, T., Metaxas, D., Axel, L.: 3D cardiac anatomy reconstruction using high resolution CT data. In: Proc. of MICCAI, Springer–Verlag (2004) 411–418
[8] Frangi, A.F., Niessen, W.J., Viergever, M.A.: Three–dimensional modeling for functional analysis of cardiac images: A review. IEEE TMI **20** (2001) 2–25
[9] Lelieveldt, B.P.F., van der Geest, R.J., Rezaee, M.R., Bosch, J.G., Reiber: Anatomical model matching with fuzzy implicit surfaces for segmentation of thoracic volume scans. IEEE Trans. Med. Imag. **18** (1999) 218–230
[10] Weese, J., Kaus, M., Lorenz, C., Lobregt, S., Truyen, R., Pekar, V.: Shape constrained deformable models for 3D medical image segmentation. In: Proc. of IPMI, Springer–Verlag (2001) 380–387
[11] Golub, G., van Loan, C.: Matrix Computation. 3rd edn. John's Hopkins University Press, Baltimore, MD, USA (1996)
[12] Gill, P., Murray, W., Wright, M.: Practical Optimization. Academ. Press (1981)
[13] Wierzbicki, M., Drangova, M., Guiraudon, G., Peters, T.: Validation of dynamic heart models obtained using non–linear registration for virtual reality training, planning, and guidance of minimally invasive cardiac surgeries. Med. Img. Anal. **8** (2004) 387–401

[14] Mitchell, S.C., Bosch, J.G., Lelieveldt, B.P.F., van der Geest, R., Reiber, J.H.C., Sonka, M.: 3-D active appearance models: Segmentation of cardiac MR and ultrasound images. IEEE Trans. Med. Imag. **21** (2002) 1167–1178
[15] Timinger, H., Pekar, V., von Berg, J., Dietmeyer, K., Kaus, M.: Integration of interactive corrections to model–based segmentation algorithms. In: Proc. Bildverarbeitung für die Medizin, Springer–Verlag (2003) 11–15

Analysis of the Interdependencies Among Plaque Development, Vessel Curvature, and Wall Shear Stress in Coronary Arteries*

Andreas Wahle[1], John J. Lopez[4], Mark E. Olszewski[1], Sarah C. Vigmostad[2], Krishnan B. Chandran[2], James D. Rossen[3], and Milan Sonka[1]

[1] Department of Electrical and Computer Engineering,
[2] Department of Biomedical Engineering,
[3] Department of Internal Medicine,
The University of Iowa, Iowa City, IA 52242, USA
[4] Department of Medicine,
The University of Chicago, Chicago, IL 60637, USA
andreas-wahle@uiowa.edu
http://www.engineering.uiowa.edu/~awahle

Abstract. The relationships among vascular geometry, hemodynamics, and plaque development in coronary arteries are not yet well understood. This in-vivo study was based on the observation that plaque frequently develops at the inner curvature of a vessel, presumably due to a relatively lower wall shear stress. We have shown that circumferential plaque distribution depends on the vessel curvature in the majority of vessels. Consequently, we studied the correlation of plaque distribution and hemodynamics in a set of 48 vessel segments reconstructed by 3-D fusion of intravascular ultrasound and x-ray angiography. The inverse relationship between local wall shear stress and plaque thickness was significantly more pronounced ($p<0.025$) in vessel cross sections exhibiting compensatory enlargement (positive remodeling) without luminal narrowing than when the full spectrum of vessel stenosis severity was considered. Our findings confirmed that relatively lower wall shear stress is associated with increased plaque development.

1 Introduction

Coronary atherosclerosis starts at a young age and is a major cause of death in developed countries. As shown in Fig. 1(a)–(c), the intimal layer (mid gray) thickens as plaque develops. However, the lumen (light gray) initially remains unchanged due to *compensatory enlargement* as part of a remodeling process that causes the media (dark gray) to grow outward. Luminal narrowing forming an angiographically visible stenosis generally occurs after the plaque area exceeds about 40% of the cross-sectional vessel area [1]. Intravascular ultrasound

---

* Supported in part by the National Institutes of Health, grant R01 HL63373.

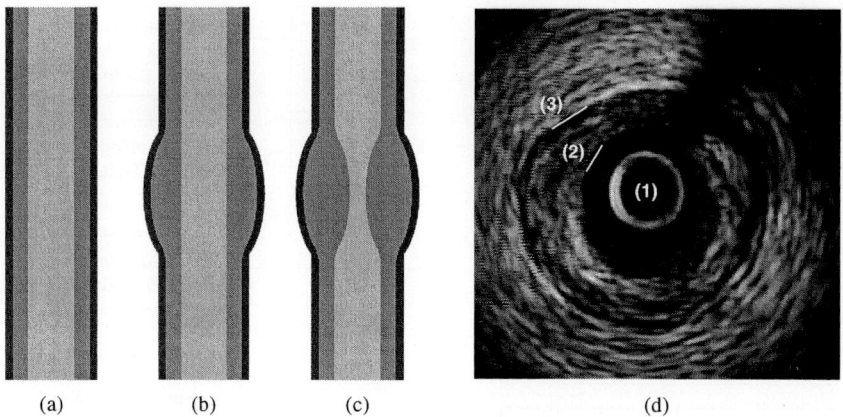

Fig. 1. Development of atherosclerotic plaque: (a) vessel without any stenosis; (b) compensatory enlargement; (c) luminal narrowing; (d) IVUS image with (1) catheter, (2) lumen/plaque, and (3) media/adventitia borders

(IVUS) is able to visualize plaque development, as shown in Fig. 1(d). Previous studies have linked plaque development with low wall shear stress [2]. Thus, the identification of areas of initially low wall shear stress and evaluation of the plaque distribution is of major interest, especially given the capabilities of IVUS to image plaque. As is typical for coronary IVUS studies, all subjects imaged had clinically indicated coronary catheterization. It is imprudent to perform IVUS imaging in patients with healthy or minimally diseased coronary vessels. Consequently, the enrolled subjects invariably suffered from advanced coronary artery disease. As such, the relationships we observed were between an already substantially altered coronary morphology and the related altered hemodynamic shear stress conditions. It has been shown that luminal narrowing deminishes the inverse relationship between plaque thickness and wall shear stress [3]. In addition to this phenomenon, we were also interested in the notion that hemodynamic shear stress plays a role in the onset of coronary disease. In contrast to wall shear stress, vascular geometry (curvature) is not changed by the course of the disease and thus can serve as a surrogate of the hemodynamic conditions prior to atherosclerotic disease development. Therefore, the relationship between vessel curvature and plaque distribution was studied as well as the relationship between wall shear stress and plaque distribution with special consideration of vascular remodeling.

2 Methods

2.1 Multi-modality Fusion

We have developed a comprehensive system that generates geometrically correct 3-D and/or 4-D (i.e., 3-D+time) reconstructions of coronary arteries and

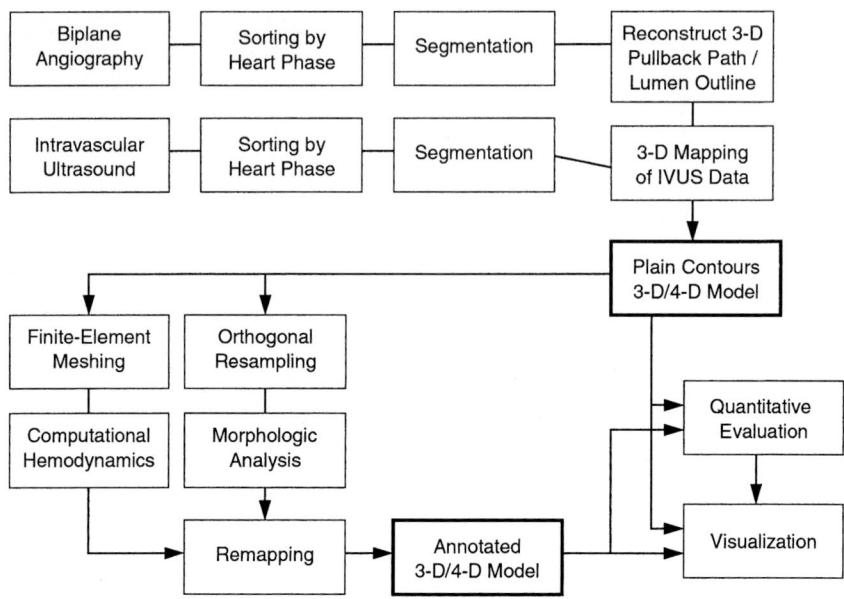

Fig. 2. Processing of the data as outlined in Section 2.1

computes corresponding quantitative indices of coronary lumen and wall morphology. The reconstructions serve as input for hemodynamic analyses and allow for interactive visualization [4, 5, 6]. A flowchart outlining the system is given in Fig. 2. In brief, the vessel geometry is obtained from biplane (or a pair of single-plane) x-ray angiographic projections, whereas the cross-sectional information is retrieved from IVUS. Thus, the resulting model accurately reflects the curvature and torsion of the vessel as well as any accumulated plaque. The angiography and IVUS data are retrospectively ECG-gated and segmented. Fusion leads to the 3-D/4-D *plain model* representing both the lumen and the vessel wall. The model consists of the lumen/plaque and media/adventitia contours oriented relative to the IVUS catheter. After tetrahedral meshing, this model is suitable for hemodynamic analyses. Following resampling orthogonal to the vessel centerline, morphologic analyses are performed. The quantitative results *annotate* the resampled contour model, which is subsequently used for visualization and further analyses. Our system utilizes conventional PC hardware and widely available software tools. Standardized storage formats for parameters and contour lists have been adopted to ensure proper interfacing between our fusion system and commercially available analysis software packages and to enhance data sharing and collaboration.

2.2 Segmentation of IVUS Image Data

While many components of the fusion system perform to full satisfaction, several challenges remain. One of them is the segmentation of the IVUS data. It is well known that IVUS images contain artifacts from various sources, thus requiring the design of cost functions that incorporate a-priori knowlegde of regional and border properties to robustly determine the optimum contours. The cost function employed in our graph-based IVUS segmentation method combines three major groups of features: (a) image data terms such as edge detectors and intensity patterns; (b) physics-based terms that distinguish different tissue types based on their Rayleigh distribution patterns [7, 8]; and (c) border probabilities based on expert tracings. A scoring system is employed to evaluate these feature classes in each image to be analyzed, and the borders are found using a multiresolution approach that mimics human vision [9].

2.3 Morphologic and Hemodynamic Indices

The reconstructed vascular model provides 3-D locations for 72 circumferential vertices on both lumen/plaque and media/adventitia contours, radially oriented with respect to the vessel centerline. This allows a straightforward determination of the plaque thickness at each location, as well as volumetric measurements over any given subsegment of the vessel [10]. In order to determine local curvature magnitude and direction, Frenet-based computational geometry was employed. To distinguish between locations of "inner" vs. "outer" curvature on the circumference of the vessel, a new scheme was introduced that weights the curvature magnitude by an index of the circumferential position of each element [11]. Blood flow through the coronary arteries was simulated using *computational fluid dynamics* (CFD) methodology. Tetrahedral meshing of the lumen using commercially available meshing software provides an unstructured grid for simulations with U^2RANS, a CFD software developed at The University of Iowa [12]. Positive and negative wall shear stress values are determined at each circumferential lumen location and mapped onto the 3-D model.

2.4 Classification of Circumferential Regions

Each of the 72 circumferential locations in each vessel cross section was categorized with respect to its relative plaque thickness (above or below average for this cross section), its location relative to the local vessel curvature (inner or outer curvature), and its wall shear stress (above or below cross-sectional average). In this way, eight different "regions" resulted. A ninth "neutral" region included those areas of curvature magnitude below a certain threshold that were eliminated from further analysis to avoid distortion of the results by noise. The following two studies correlate independently plaque distribution with curvature and wall shear stress.

3 Studies and Results

3.1 Plaque Distribution in Relation to Vessel Curvature

To verify the observation that plaque accumulation in curved vessels is biased towards the inner bend of the curvature rather than the outer bend of

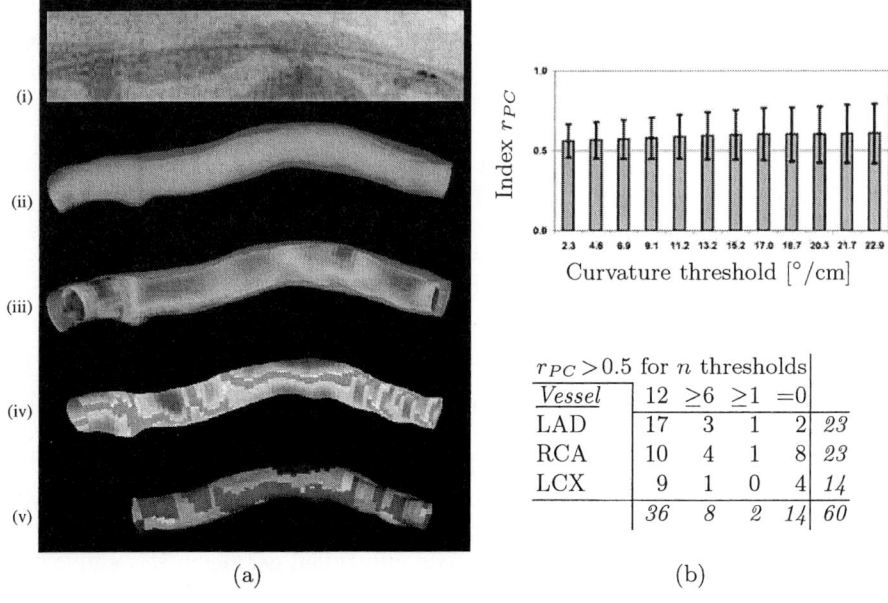

Fig. 3. Plaque thickness vs. curvature: (a)(i) angiogram of a left anterior descending artery with the IVUS catheter inserted, (ii) lumen and adventitia borders from fusion, (iii) plaque-thickness annotation, (iv) curvature-index annotation, (v) after classification into regions, with the branch segment removed from analysis; (b) results from 60 analyzed vessels, by curvature threshold and vessel, with $r_{PC}>0.5$ indicating that our guiding hypothesis was satisfied for all (12), at least half (≥ 6), at least one (≥ 1), or none ($=0$) of the curvature thresholds

the curvature, the relative amount r_{PC} of regions where inner curvature coincides with above-average plaque accumulation, or outer curvature coincides with below-average plaque accumulation, was determined in a set of 60 vessels. Preliminary results in 37 vessels and methodology were reported in [11]. The ratio r_{PC} represents a "plaque/curvature index" with a value $r_{PC}>0.5$ indicating that more plaque has accumulated along the inner curvature as compared to the outer curvature, thus supporting the hypothesis. As an example, Fig. 3(a)(iii) shows a color-coded plaque-thickness distribution in a geometrically correct 3-D representation, with red indicating high and blue indicating low plaque thickness, normalized over the entire vessel segment. As described above, a curvature index was determined for each circumferential location on the contour. Fig. 3(a)(iv) shows the color-coded curvature-index distribution, with red indicating inner curvature and blue indicating outer curvature. Four regions were defined, as depicted in Fig. 3(a)(v): R_{ai} (red), R_{ao} (magenta), R_{bi} (yellow), and R_{bo} (blue). These regions represent pairs distinguishing circumferentially considered "above-average" plaque thickness (a) from "below-average" plaque thickness (b), coinciding with either "inner curvature" (i) or "outer curvature" (o) of the vessel wall. Thus, the plaque/curvature index was defined as

$$r_{PC} = \frac{\|R_{ai} + R_{bo}\|}{\|R_{ai} + R_{bo} + R_{ao} + R_{bi}\|} \qquad (1)$$

Impact of Curvature Threshold and Vessel Type. The results are depicted in Fig. 3(b). Twelve different threshold values were empirically selected ranging from 2.31 to 22.94°/cm, resulting in 10.1–77.8% of circumferential locations being assigned to the neutral region R_n (green). The chart shows that the average r_{PC} over all 60 vessels increases steadily with increase of the curvature threshold. Thus, the more regions of low curvature are included into R_n, and therefore increasing the proportion of higher curvature regions included in the calculation of r_{PC}, the more the hypothesis was supported. The increase

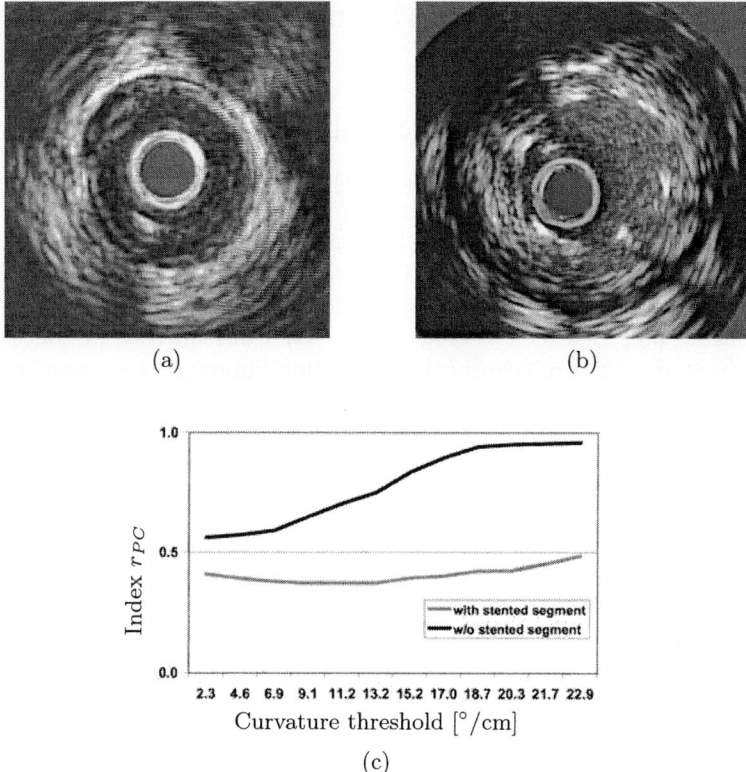

Fig. 4. (a) IVUS frame of an untreated vessel segment with slight stenosis; (b) the same vessel after stent placement at a location with heavy disease; (c) the hypothesis $r_{PC} > 0.5$ is only satisfied after exclusion of the stented segment, thus including segments with lesser disease only, as compared to the results over the entire vessel

in standard deviation of r_{PC} prompted us to categorize the results by vessel. While almost two thirds of the vessels satisfied $r_{PC}>0.5$ for *all* thresholds, the hypothesis was more strongly supported in left anterior descending (LAD) arteries (87% for all or at least half of the thresholds). Since the right coronary (RCA) and left circumflex (LCX) arteries have higher tortuosity than the LAD, the less supportive results may be caused by the more complex flow patterns that can no longer be explained by the curved-tube model.

Impact of Interventions. Stenting may have a substantial impact on the outcome of the plaque/curvature index r_{PC}. In several of the vessels analyzed, a below-threshold value of r_{PC} ($r_{PC}<0.5$) was determined when all segments were included and only branch locations were excluded. After also excluding known regions of intervention and stenting, $r_{PC}>0.5$ was reached, frequently showing the increase of r_{PC} with the increase in curvature threshold (Fig. 4). This contradicts our initial findings reported in [11] that stenting does not significantly affect the plaque/curvature index r_{PC} and indicates that substantial disease and stenting *may* have a distorting impact on the relation between vessel geometry and plaque distribution.

3.2 Plaque Distribution in Relation to Wall Shear Stress

While disease progression and stenting impact the curvature/plaque relationship to some extent, an even more substantial effect can be expected on the wall shear stress distribution. The distribution is substantially altered when the limits of positive remodeling are reached [3]. Thus, the vessel subsegments for which the area stenosis is between 10% and 40% are of specific interest (the compensatory-enlargement range identified by Glagov et al. [1]). Consequently, we concentrated on whether and how significantly the correlation improves once vessel segments of certain properties are excluded from the analysis. In this way, indirect evidence of which local conditions favor the underlying hypothesis of below-average wall shear stress inducing above-average plaque thickness was sought.

Grouping of Vessels and Segments by Disease Severity. 48 vessels (a subset of Section 3.1, since some parameters were not available for vessels received from collaborating sites) were analyzed. The data was smoothed with a moving 45°-wedge over 5 frames to limit the impact of local noise. The analyses were performed in 4 increasingly restrictive subsets of data. First, the relative amount r_{PW} of elements for which circumferentially above-average plaque thickness coincides with below-average wall shear stress (and vice versa) was determined for each vessel segment – similar to the plaque-thickness/curvature study. By replacing "inner curvature" (i) with "lower-than-average wall shear stress" (l) and "outer curvature" (o) with "higher-than-average wall shear stress" (h) in Eq. (1),

$$r_{PW} = \frac{\|R_{al} + R_{bh}\|}{\|R_{al} + R_{bh} + R_{ah} + R_{bl}\|} \qquad (2)$$

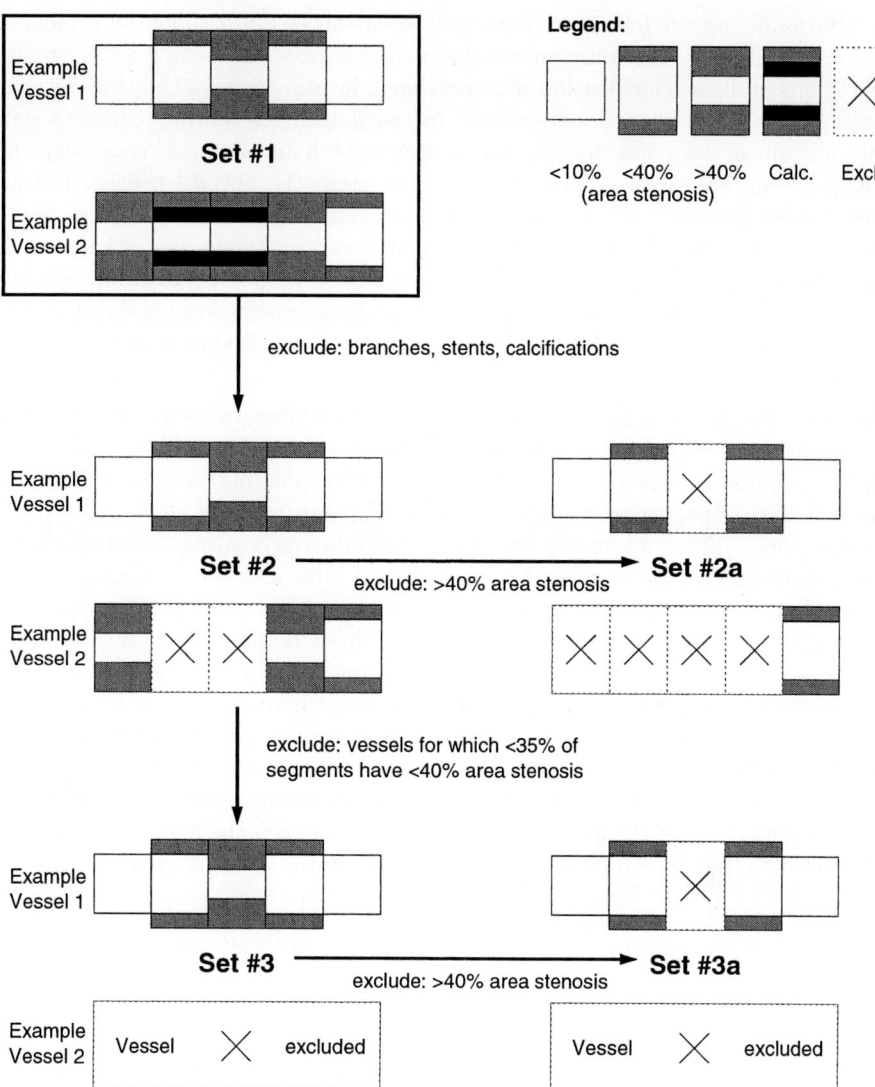

Fig. 5. Example for the definition of the sets: Vessel 1 shows only minor disease, whereas Vessel 2 is subject to advanced atherosclerosis; both form Set #1. For Vessel 1, all subsegments are retained when proceeding to Set #2, whereas two subsegments of Vessel 2 were discarded due to calcifications. All subsegments outside the 10–40% area stenosis range are removed from Set #2 to create Set #2a, thus discarding 1 subsegment from Vessel 1 and 2 subsegments from Vessel 2. Only Vessel 1 proceeds from Set #2 to Set #3, since less than 35% of Vessel 2 are within the 10–40% area stenosis range. For Vessel 1, (in analogy to the step from Set #2 to Set #2a) the center segment is discarded from Set #3 to Set #3a

results as definition for the plaque/wall-shear-stress index. This step created Set #1. Next, all vessel subsegments that included vessel branching areas, stents, or regions of dense calcification were excluded, forming Set #2. Within Set #2, percent-area stenosis was determined for each frame following Glagov's definition, which does not require the presence of a normal reference segment (plaque+wall area over cross-sectional vessel area) [1]. Set #3 consisted of all such vessles from Set #2 for which the percent-area stenosis was in the range of 10–40% in at least 35% of the non-excluded vessel segments. Set #3 consisted of 31 vessels satisfying this criterion. In each vessel, the segments of Sets #2 and #3 that were within the 10–40% range of area stenosis formed Subsets #2a and #3a. An illustration for the definition of these sets is shown in Fig. 5.

Hypothesis Test. If the hypothesis is correct and observable in regions where severe luminal narrowing is not present, the vessels in Subsets #2a and #3a should provide higher r_{PW} ratios than the corresponding vessels in Sets #2 and #3. Therefore, we determined factors g_{PW} quantifying the change $g_{PW\{2\}} = r_{PW\{2a\}}/r_{PW\{2\}}$ for all vessels and $g_{PW\{3\}} = r_{PW\{3a\}}/r_{PW\{3\}}$ for vessels with the minimum of 35% of frames within the 10–40% area-stenosis range. Note that the $g_{PW\{x\}}$ represent the differences in hypothesis validity. Consequently, $g_{PW\{x\}} > 1$ suggests a case for which the hypothesis is more strongly supported in those segments of vessel x with compensatory enlargement as compared to those with lumen narrowing. The analysis rationale is to determine: (1) whether applying the hypothesis test on the subset of segments defined in Set #2a (10–40% stenosis) increases the validity of the hypothesis compared to the Set #2; and, (2) whether applying the hypothesis test on Set #3a (10–40% stenosis in vessels with ≥35% of the wall within this range) increases the hypothesis validity compared to the Set #3 (≥35% of the wall within the 10–40% stenosis range).

Changes in Hypothesis Validity. The results can be summarized in the following table ($g_{PW\{x\}} \geq 1.01$ "increase" and $g_{PW\{x\}} \leq 0.99$ "decrease"):

	$g_{PW\{x\}}$	≥1.01	<1.01 >0.99	≤0.99	*	
Sets #2a/#2	n=48	25	3	16	4	R=0.61, p>0.75
#3a/#3	31	19	3	7	2	R=0.92, p<0.025

where for the vessels marked with ∗, either all or none of the frames were within the 10–40% area-stenosis range, therefore $g_{PW\{x\}}$ was considered undefined, and these vessels were excluded from the R and p calculations. Evidently, hypothesis validity improves and becomes statistically significant in Set #3 vs. Set #2, thus confirming our assumption. A notable cluster of 12 vessels in Set #3, having 35–63% of frames in the 10–40% area-stenosis range, shows an average 10.2% increase in hypothesis validity which is highly significant ($R=0.96$, $p<0.001$). This can be explained, in part, by the minimization of statistical noise with an even distribution of frames within vs. outside of the 10–40% area-stenosis range.

4 Discussion and Conclusions

Plaque development depends on the wall shear stress distribution, which in turn depends on the vessel geometry. The presented study demonstrated in-vivo that plaque distribution correlates with vessel curvature, and also correlates with wall shear stress in early stages of atherosclerosis. The analysis of a direct relationship between curvature and shear stress is ongoing. We have shown that, in the majority of vessels, plaque tends to form at the inner curvature of the vessel wall. These findings suggest that low wall shear stress, which is typically associated with inner vessel curvature locations, likely contributes to the *initial* formation of atherosclerotic plaque in the early stages of the disease in human coronary arteries. However, the wall shear stress distribution is altered in the *later* stages of atherosclerosis, when positive remodeling can no longer compensate for the disease and the lumen narrows. We have demonstrated that the hypothesis of above-average plaque thickness being associated with below-average wall shear stress is more strongly supported in the *early* stages of disease progression.

References

1. Glagov, S., Weisenberg, E., Zarins, C.K., Stankunavicius, R., Kolettis, G.J.: Compensatory enlargement of human atherosclerotic coronary arteries. New England Journal of Medicine **316** (1987) 1371–1375
2. Gibson, C.M., Diaz, L., Kandarpa, K., Sacks, F.M., Pasternak, R.C., et al.: Relation of vessel wall shear stress to atherosclerosis progression in human coronary arteries. Arteriosclerosis and Thrombosis **13** (1993) 310–315
3. Wentzel, J.J., Janssen, E., Vos, J., Schuurbiers, J.C.H., Krams, R., et al.: Extension of increased atherosclerotic wall thickness into high shear stress regions is associated with loss of compensatory remodeling. Circulation **108** (2003) 17–23
4. Wahle, A., Prause, G.P.M., DeJong, S.C., Sonka, M.: Geometrically correct 3-D reconstruction of intravascular ultrasound images by fusion with biplane angiography — Methods and validation. IEEE Transactions on Medical Imaging **18** (1999) 686–699
5. Olszewski, M.E., Wahle, A., Medina, R., Mitchell, S.C., Sonka, M.: Integrated system for quantitative analysis of coronary plaque via data fusion of biplane angiography and intravascular ultrasound. In Lemke, H.U., et al., eds.: Computer Assisted Radiology and Surgery, Amsterdam, Elsevier (2003) 1117–1122
6. Wahle, A., Olszewski, M.E., Sonka, M.: Interactive virtual endoscopy in coronary arteries based on multi-modality fusion. IEEE Transactions on Medical Imaging **23** (2004) 1391–1403
7. Burckhardt, C.B.: Speckle in ultrasound B-mode scans. IEEE Transactions on Sonics and Ultrasonics **25** (1978) 1–6
8. Wagner, R.F., Smith, S.W., Sandrick, J.M., Lopez, H.: Statistics of speckle in ultrasound B-scans. IEEE Transactions on Sonics and Ultrasonics **30** (1983) 156–163
9. Olszewski, M.E., Wahle, A., Mitchell, S.C., Sonka, M.: Segmentation of intravascular ultrasound images: A machine learning approach mimicking human vision. In Lemke, H.U., et al., eds.: Computer Assisted Radiology and Surgery, Amsterdam, Elsevier (2004) 1045–1049

10. Medina, R., Wahle, A., Olszewski, M.E., Sonka, M.: Three methods for accurate quantification of plaque volume in coronary arteries. International Journal of Cardiovascular Imaging **19** (2003) 301–311
11. Wahle, A., Olszewski, M.E., Vigmostad, S.C., Medina, R., Coşkun, A.Ü., et al.: Quantitative analysis of circumferential plaque distribution in human coronary arteries in relation to local vessel curvature. In: Proc. 2004 IEEE International Symposium on Biomedical Imaging, Piscataway NJ, IEEE Press (2004) 531–534
12. Lai, Y.G., Przekwas, A.J.: A finite-volume method for fluid flow simulations with moving boundaries. Computational Fluid Dynamics **2** (1994) 19–40

Automated Segmentation of X-ray Left Ventricular Angiograms Using Multi-View Active Appearance Models and Dynamic Programming

Elco Oost[1], Gerhard Koning[1], Milan Sonka[2], Johan H.C. Reiber[1], and Boudewijn P.F. Lelieveldt[1]

[1] Division of Image Processing, Department of Radiology,
1-C2S, Leiden University, Medical Center,
P.O. Box 9600, 2300 RC Leiden, the Netherlands
{c.r.oost, g.koning, j.h.c.reiber, b.p.f.lelieveldt}@lumc.nl
http://www.lkeb.nl
[2] Department of Electrical and Computer Engineering,
The University of Iowa, Iowa City IA 52242, USA
milan-sonka@uiowa.edu

Abstract. A novel approach to automated segmentation of X-ray Left Ventricular (LV) angiograms is proposed, based on Active Appearance Models (AAMs) and dynamic programming (DP). Due to combined modeling of the end-diastolic (ED) and end-systolic (ES) phase, existing correlations in shape and texture representation are exploited, resulting in a better segmentation in the ES phase. The intrinsic over-constraining by the model is compensated by a DP algorithm, in which also cardiac contraction motion features are incorporated. An elaborate evaluation of the algorithm, based on 70 paired ED-ES images, shows success rates of 100% for ED and 99% for ES, with average border positioning errors of 0.68 mm and 1.45 mm respectively. Calculated volumes were accurate and unbiased, proving the high clinical potential of our method.

1 Introduction

X-ray LV angiography is a widely applied modality for the assessment of cardiac function. In both the end-diastolic (ED) and end-systolic (ES) image frame endocardial contours are drawn around the LV manually, from which the ventricle volume in ED and ES can be estimated [1]. In addition, relevant clinical parameters such as regional wall motion and Ejection Fraction (EF) can be quantified.

Currently, several packages are available that assist the cardiologists in manually drawing contours in LV angiograms. However, due to poor image quality, drawing contours by hand is difficult, time-consuming and prone to inter- and intra-observer variability. When an expert examines an X-ray image sequence, he also inspects neighboring frames around ED and ES, to decide on correct boundary locations. This way, knowledge about contraction dynamics is used to improve the segmentation accuracy. The goal of this work is to automate the contour detection process by integrating prior knowledge about cardiac shape, appearance and motion. We aim to achieve this with the following contributions:

- A Multi-View AAM [2, 3] is employed in which statistical information of different views of the same object is modeled simultaneously. The existing correlation in shape and texture between ED and ES is exploited. The more reliable LV information present in the ED images supports the segmentation of the frequently poorly defined LV in the ES images.
- To prevent the model from locking in on local minima, we propose a novel, controlled gradient descent optimization, in which a limited number of model parameters is updated at a time. This greatly improves convergence robustness.
- Dynamic Programming is applied to compensate for over-constraining by the model and thus to attain better local border delineation. Full use of all available priors is achieved by constructing a cost function from both image features and contraction motion features.
- An elaborate evaluation of clinical efficiency of the algorithm is described based on 70 ED-ES image pairs. To our knowledge, this is the largest evaluation of an automated segmentation method for clinically realistic X-ray LV angiograms.

2 Segmentation Method

2.1 Multi-View Active Appearance Models

Active Appearance Models, introduced by Cootes [4, 5], are an extension of the well-established Active Shape Models [6], and integrate knowledge about object shape and image texture variability into the segmentation. An AAM is built by warping a complete image patch around the training shapes to the average shape. After intensity normalization to zero mean and unit variance, the shape-normalized intensity average and principal components are computed. A subsequent combined Principal Component Analysis (PCA) on the shape and intensity model parameters yields a set of components that simultaneously capture shape and texture variability. AAM matching is based on minimizing the difference between model intensities and the target image. This enables a rapid search for the correct model location, while utilizing pre-calculated derivative images for optimizing the parameters. AAMs are described in detail in [5], and an elaborate overview of medical applications is given in [7].

Typically, AAMs are applied to segmentation of single image sets, whereas in cardiac imaging, often multiple acquisitions are acquired within one patient examination, where images may depict different geometrical or functional features of the heart. Different time frames from an angiographic image sequence are examples of such interrelated views. Multi-View AAMs exploit existing shape- and intensity correlations between different images of the same heart. Potentially, this increases robustness and enforces segmentation consistency between views, yielding a better segmentation.

The Multi-View model is constructed by aligning the training shapes for different views separately, and concatenating the aligned shape vectors x_i for each of the N views. A shape vector for N frames is defined as:

$$x = \left(x_1^T, x_2^T, \ldots x_N^T\right)^T \qquad (1)$$

By applying a PCA on the sample covariance matrix of the combined shapes, a shape model is computed for all frames simultaneously. The principal model components represent shape variations, which are intrinsically coupled for all views. For the intensity model, the same applies: an image patch is warped on the average shape for view i and sampled into an intensity vector g_i, the intensity vectors for each single frame are normalized to zero mean and unit variance, and concatenated:

$$g = \left(g_1^T, g_2^T, \ldots, g_N^T\right)^T \qquad (2)$$

Analogous to the conventional AAMs, a PCA is applied to the sample covariance matrices of the concatenated intensity sample vectors. Subsequently, each training sample is expressed as a set of shape and appearance coefficients. A combined model is computed from the combined shape-intensity sample vectors. In the combined model, the shape and appearance of both views are strongly interrelated.

Like in all AAMs, estimation of the gradient matrices for computing parameter updates during image matching is performed by applying perturbations on the model and pose parameters, and measuring their effect on the residual images. In Multi-View AAMs, a disturbance in an individual model parameter yields residual images in all views simultaneously. The pose parameters however, are perturbed for each view separately to accommodate for trivial differences in object pose in each view, whereas the shape and intensity gradients are correlated for all views. During matching, the pose transformation for each view is also applied separately, whereas the model coefficients intrinsically influence multiple frames at once.

Multi-View AAMs have been successfully applied to segmentation of long-axis cardiac MR views and a pilot study on left ventricular angiograms was performed [2]. This pilot study revealed two limiting factors that needed to be addressed to make the method suitable for clinical application: sensitivity to local minima, and over-constraining of the model towards the trained data. In addition, this study showed that the exact location of the LV border can only be determined based on motion features. In the next sections, solutions to these problems are proposed.

2.2 Controlled Gradient Descent

The conventional AAM matching strategy occasionally shows difficulties in converging to the true contour positions. The gradient descent [8] in regular AAM segmentation minimizes the difference between model and true LV representation. When the model is initialized far away from the actual LV position or with a largely different scale or orientation, the model will lock in on its direct surroundings, matching to the closest local minimum. To overcome this, we developed a more controlled gradient descent, updating only a limited number of directions at a time. First all parameter updates corresponding to the specific modes of variation, are sorted to descending magnitude and the largest single parameter update is executed. If this lowers the error criterion, the proposed update is accepted, new model and pose parameters are calculated and a new parameter update vector is determined and ordered in the next iteration. In case an update proposal does not lower the error criterion, the number of updated model parameters is incremented in the next attempt. With this strategy, a large decrease of the error criterion is achieved based on one or a few parameters at a time.

2.3 Contour Refinement Using Dynamic Programming

The power of the AAM algorithm is that it is still able to come to an acceptable global segmentation in an environment with vaguely defined features. The model freedom to deform however, is limited by the modes of variation derived from the training data set. Therefore, a shape that slightly deviates from a model-generated LV contour should also be considered as valid, and a refinement of the contour is desirable. In previous work [3] an AAM contour refinement was done by applying a second AAM, in which only image intensities close to the contour were incorporated. This approach slightly improved the segmentation, but being statistically trained, it still intrinsically over-constrained the contours towards the training data.

To allow for more shape flexibility we have used a locally selective DP, in which the cost function is constructed from image and motion features, to mimic the experts routine of including knowledge of contraction dynamics. DP is well-established for contour detection in X-ray angiography [9]. Typically, angiographic DP searches for an optimal contour path through a cost matrix, where the cost function is based on a mix of first and second order image derivatives. In addition, we integrate features from a subtraction image (ES image minus ED image), from which contraction information can be extracted (Figure 1). The cost matrix C is defined as:

$$C(i,j) = \beta(\alpha_1 G_1(i,j) + (1-\alpha_1)T_1(i,j)) + (1-\beta)(\alpha_2 G_2(i,j) + (1-\alpha_2)T_2(i,j)) \quad (3)$$

where $C(i,j)$ is the cost of element in row i and column j, β is the weighing factor between the costs in the true image data and the costs in the subtraction image, $G_1(i,j)$ and $G_2(i,j)$ are the gradients of both images, $T_1(i,j)$ and $T_2(i,j)$ are the second order derivatives and α_1 and α_2 are weighing factors between the first and second order derivatives for the true image data and the subtraction image respectively. The polarity of edges in the subtraction image is defined differently for ED and ES, making the cost function locally selective for each phase. In ED the area outside the contour should be dark and the area inside the contour should be light. For ES this edge polarity is opposite. The use of these directed edges is only possible, since the Multi-View AAM already produces a reliable global segmentation in each frame.

3 Clinical Evaluation

To determine the clinical utility of our approach and to assess whether Multi-View AAM segmentation results are comparable to manual segmentation results produced by experts, experiments were executed using a data set of 70 paired ED-ES images from infarct patients. This data was used to train 14 leave-five-out Multi-View AAMs. All models were constructed retaining 100 % shape variability and 95 % intensity variability.

Automatically determined ED volume, ES volume and EF were compared with corresponding values derived from manual contours. Volumes were calculated using the area-length volume estimate [1]. Linear regression was used to determine relationships between manually traced and computer determined values. A two-tailed paired samples t-test was applied to volume measurements from manual and automatic con-

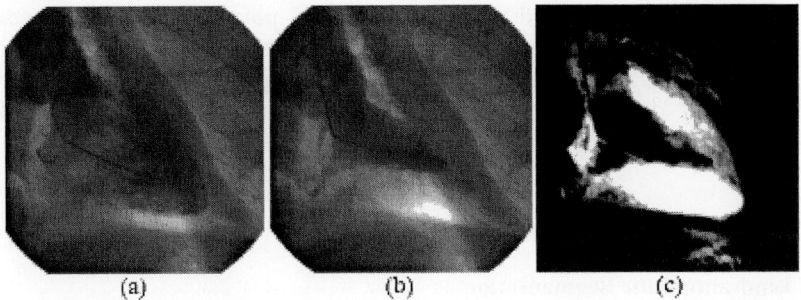

Fig. 1. Additional information can be extracted from the subtraction image (c): ES (b) minus ED (a). This example shows that the subtraction image contributes to a better definition of the mitral valve area for both ED and ES. Furthermore, the diagonal shadow is removed

tours to investigate systematic errors. A p-value smaller than 0.05 was considered significant. In addition, point to curve errors were determined, and similar to [10] we used the following equations to calculate contour errors and area errors respectively:

$$E_C = \frac{\sum_{x,y \in R_E} \{a_P(x,y) \otimes a_D(x,y)\}}{\sum_{x,y \in R_E} a_D(x,y)} \quad (4)$$

$$E_A = \frac{\left|\sum_{x,y \in R_E} a_D(x,y) - \sum_{x,y \in R_E} a_P(x,y)\right|}{\sum_{x,y \in R_E} a_D(x,y)} \quad (5)$$

with

$$a_P(x,y) = \begin{cases} 1, (x,y) \in R_P \\ 0, \text{ otherwise} \end{cases} \quad \text{and} \quad a_D(x,y) = \begin{cases} 1, (x,y) \in R_D \\ 0, \text{ otherwise} \end{cases}$$

in which R_p is the region within the automatically drawn contour, R_D is the region within the manually drawn contour, R_E is the region of evaluation and $\otimes$ denotes the logical exclusive OR operator.

The performance of our algorithms was tested by comparing obtained results with the manual contours that were used to train the 14 AAMs (expert #1 contours). Using the leave-five-out setup, none of the tested image pairs was included in the model used for segmentation. To asses the clinical relevance, calculated contours were compared with manually drawn contours of three experts. Furthermore we determined the state of automation that can be achieved, by comparing a fully automatic method with a semi-automatic approach in which for both ED and ES the endpoints of the aortic valve and the apex are predefined by a user. In the fully automatic method the model was initialized in the image center, with average scale and orientation. The benefit of the controlled gradient descent was tested by comparing it with regular AAM results.

The difficulty in interpreting LV angiograms results in a large inter- (and intra-) observer variability. For example, differences in ES volume estimation by different experts can amount to over 80 % and average ES point to curve differences can amount to 10 to 15 mm. Consequently, defining a notion of success for an automatic LV segmentation algorithm is difficult. To decide on success or failure, we have chosen not to look at quantitative numbers only, but to use them as a reference while scoring the segmentations visually.

4 Results

4.1 Semi-automatic Segmentation

The semi automatic algorithm yielded borders that agreed closely to the manual expert contours. The success rate of the algorithm is 100 % for ED and 99 % (1 outlier) for ES. After removal of the image pair with this partial failure, both ED and ES contour errors, calculated areas and calculated volumes were, to our knowledge, better then any previously reported method. Figure 2 displays representative examples of obtained contours, proving that accurate segmentation is also feasible in images with acquisition artifacts.

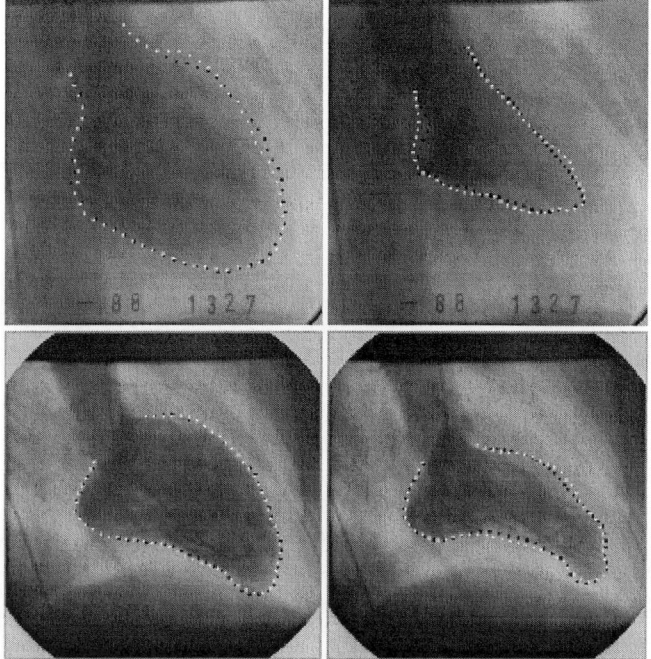

Fig. 2. Successful matches for ED (left column) and ES (right column) generated with the semi-automatic algorithm. Black dotted lines denote the manual contour, white dotted lines represent the semi-automatic contours. Semi-automatic contours correspond closely with manual contours, also when for example contrast is low (upper row)

Table 1. Point-to-curve distances (PtC), contour errors (E_C), area errors (E_A) and volume errors (E_V) for ED and ES. Six comparisons are displayed: semi-automatic model vs. expert #1, semi-automatic model vs. expert #2, semi-automatic model vs. expert #3, expert #2 vs. expert #1, expert #3 vs. expert #1 and expert #3 vs. Expert #2

ED	PtC [mm]	E_C [%]	E_A [%]	E_V [%]
semi vs #1	0.68 ± 0.37	4.13 ± 1.90	1.90 ± 1.71	3.50 ± 3.45
semi vs #2	0.74 ± 0.26	5.56 ± 2.97	2.01 ± 2.43	3.81 ± 3.23
semi vs #3	0.72 ± 0.27	5.24 ± 2.07	2.39 ± 1.93	4.04 ± 3.83
#2 vs #1	0.57 ± 0.20	4.37 ± 2.37	2.13 ± 2.54	3.36 ± 3.04
#3 vs #1	0.59 ± 0.28	4.27 ± 1.80	2.03 ± 1.65	3.34 ± 2.27
#3 vs #2	0.72 ± 0.39	5.21 ± 3.36	2.46 ± 2.98	3.87 ± 3.26

ES	PtC [mm]	E_C [%]	E_A [%]	E_V [%]
semi vs #1	1.45 ± 0.76	12.8 ± 6.30	6.42 ± 5.36	13.5 ± 11.7
semi vs #2	2.13 ± 1.73	26.4 ± 27.4	21.0 ± 28.1	38.3 ± 56.0
semi vs #3	1.77 ± 1.29	20.8 ± 18.9	14.0 ± 18.9	31.4 ± 49.7
#2 vs #1	1.23 ± 0.63	14.3 ± 8.64	11.6 ± 9.31	15.5 ± 11.6
#3 vs #1	1.05 ± 0.69	11.1 ± 7.26	7.31 ± 7.48	11.7 ± 10.9
#3 vs #2	1.25 ± 1.06	14.0 ± 13.0	9.37 ± 13.2	10.9 ± 16.5

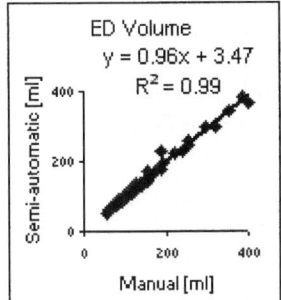 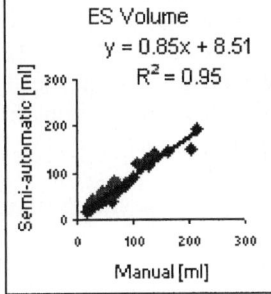 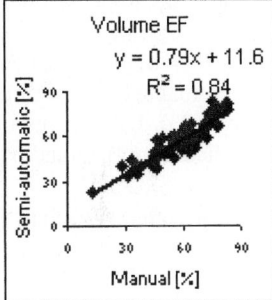

Fig. 3. Volume regression plots for ED, ES and EF for the semi-automatic algorithm

Table 2. Comparison of semi-automatic contours with 3 experts and comparing the experts mutually: relative ED volume and ES volume errors and the absolute ejection fraction error

	ED error [%]	ES error [%]	EF error [%]
semi vs #1	-1.56	-0.88	-1.20
semi vs #2	-0.86	12.79	-6.20
semi vs #3	0.30	9.54	-4.26
#2 vs #1	-0.70	-12.11	4.95
#3 vs #1	-1.85	-9.45	2.98
#3 vs #2	-1.15	3.03	-1.96

Border positioning errors were generally small. Average point to curve errors were 0.68 ± 0.37 mm for ED and 1.45 ± 0.76 mm for ES. All quantitative results (model vs. expert #1) are summarized in Table 1, together with a comparison of automatically generated contours with expert #2 and expert #3 and a mutual comparison of all three experts. For the mutual comparison of experts, only 43 samples of the original 70 paired ED-ES data were available.

Excellent correlation between volumes based on manual and semi-automatic contours was achieved, as shown in Figure 3. In a paired samples t-test differences between manually and semi-automatically calculated ED volume, ES volume and ejection fraction were found statistically insignificant ($p=0.13$, $p=0.76$ and $p=0.15$ respectively). Table 2 gives an overview of errors in ED volume, ES volume and EF. The semi-automatic algorithm compared to expert #1 gives the overall best results. Especially the differences in calculated ES volume and EF were remarkably small, smaller than any of the inter-expert differences. This indicates that the method performed within limits of inter-observer variability.

4.2 Fully Automatic Segmentation

The success rate of the fully automatic algorithm was 91 % for ED and 83 % for ES. 6 complete failures were observed, in which both ED and ES segmentation diverged, and 6 partial failures in which only ES segmentation failed. After removing these failures, point to curve errors were 0.79 ± 0.43 for ED and 1.55 ± 0.66 for ES, which is comparable to the semi-automatic results.

Linear regression is acceptable for ED ($y = 0.93x + 6.88$, $R^2 = 0.99$), ES ($y = 0.90x + 4.55$, $R^2 = 0.96$) and EF ($y = 0.84x + 9.46$, $R^2 = 0.82$). Only ED volume comparison between manual and automatic contours was statistically significant, according to a t-test ($p=0.03$). Differences in ES volume and ejection fraction were found statistically insignificant ($p=0.33$ and $p=0.72$ respectively).

4.3 Controlled Gradient Descent versus Standard AAM

To determine the effect of the controlled gradient descent, experiments were repeated while using a regular Multi-View AAM instead of the proposed gradient descent. When applying a regular Multi-View AAM in semi-automatic segmentation, performance and accuracy remained similar. Large difference in performance however occurred when applying a regular Multi-View AAM in fully automatic segmentation. The number of failures amounted to 40 % in ED and 50 % in ES segmentation: substantially worse than the controlled gradient descent matching. With a success rate of 91 % for ED and 83 % for ES the controlled gradient descent showed to be far more robust in evading local minima and converging to the desired solution.

5 Discussion and Conclusions

The semi-automatic algorithm shows a high success rate of 100 % for ED and 99 % for ES. The only failure occurred when the ES image showed an extremely slim and elongated shape. The results are based on the same data set as in [2]. Results have improved significantly compared to [2], in which a combined success rate of 87 % for both ED and ES was reported. Correlation between manually determined LV volumes and semi-automatically calculated LV volumes was good and, to our knowledge, better than any previously reported method. Especially the ES results have improved significantly, which can be mainly attributed to the combined modeling of the ED and ES phase. The correlation values shown in Figure 3 are, to our knowledge, the best values reported until now. However, correlation values ($R^2 = \{0.99; 0.95; 0.84\}$) do not match inter-observer correlations ($R^2 = \{0.99; 0.98; 0.93\}$). Due to a lack of image information, ES volumes are generally underestimated slightly. When using this method in daily clinical practice, a cardiologist will need to redraw about 20 % of the ES contour. This will increase both the ES segmentation quality and the EF calculation accuracy. To put this number in perspective, based on the similar conditions, expert #1 would on average redraw 14 % of an ES contour drawn by expert #2 and 12 % of an ES contour drawn by expert #3.

Quantitative evaluation results of the semi-automatic algorithm proved to be within boundaries of inter-observer variability. The average difference and standard deviation in comparing the semi-automatic method with expert #1 contours (the expert who produced the training contours) were comparable to values obtained when comparing different experts (Table 1 and Table 2). The ability to mimic expert drawing behavior is evident in Table 2. Differences between the semi-automatic algorithm and expert #1 are generally smaller then differences between experts.

Both the amount of failures and the quantitative results for the fully automatic algorithm were not as good as the semi-automatic approach. The major difficulty in fully automatic segmentation is the location of the three landmark points; upper aortic valve point, lower aortic valve point and apex. Errors for these landmarks are 3.8 mm, 4.1 mm and 3.0 mm respectively for ED and 4.4 mm, 3.8 mm and 6.2 mm for ES. These errors strongly influence the volume estimates from the area-length method. Still the fully automatic algorithm provided acceptable segmentation results. After removal of failures, quantitative results were comparable to results of the semi-automatic algorithm.

The benefit of controlled gradient descent became evident when large adjustments of model and pose parameters were necessary, mainly when automatic initialization was applied. The success rate improved from 50-60% for the conventional matching to 80-90% for the controlled gradient descent. It proved to be a suitable approach in which the model deformation is directed by the modes of variation that most typically describe the object of interest. The controlled gradient descent approach has proven to evade local minima by converging in smaller and more controlled steps.

The method is fast (1-2 seconds per case) and needs minimal user input: setting 6 seed points manually produces the ED and ES contours. Quantitative results demonstrate that the semi-automatic algorithm is robust and accurate, even when acquisition artifacts were present, such as poor contrast, overlapping diaphragm or strong shadows in the image.

Also, our method outperforms other recently published methods. Suzuki's neural edge detector [10], trained on 12 ED and 12 ES images, achieved average contour errors E_C of 6.2 % and 17.1 % for ED and ES respectively and average area errors E_A of 4.2 % and 11.6 % for ED and ES respectively. The semi-automatic approach presented in this paper needs a similar amount of user interaction and produces E_C values of 4.1 % and 12.8 % and E_A values of 1.9 % and 6.4 %, comparing favorably to

Suzuki's results on all indices. Moreover, more than five times as much data was used our evaluation.

In conclusion, a new algorithm for semi-automatic segmentation of the left ventricle in X-ray LV angiograms is presented. The method is a combination of a Multi-View Active Appearance Model and a locally selective dynamic programming approach and exploits knowledge about LV shape, image texture and contraction dynamics. The algorithm is capable of mimicking clinical expert drawing behavior and therefore provides excellent results. Local border accuracy is improved by a model-initialized dynamic programming step in which both image information and knowledge of contraction dynamics was integrated in the cost function. Furthermore, the robustness in fully automatic segmentation improved substantially by introducing a controlled gradient descent approach in updating the model parameters.

Acknowledgements

Financial support by the Dutch Technology Foundation STW (Project LGN 4508) is gratefully acknowledged.

References

1. H. Sandler and H.T. Dodge, The Use of Single Plane Angiocardiograms for the Calculation of Left Ventricular Volume in Man, American Heart Journal, 75(3): 325-334, 1968.
2. C.R. Oost, B.P.F. Lelieveldt, M. Üzümcü, H. Lamb, J.H.C. Reiber and M. Sonka, *Multi-View Active Appearance Models: Application to X-ray LV Angiography and Cardiac MRI*, Proceedings IPMI, LNCS Vol. 2732: 234-245, Springer, 2003.
3. E. Oost, B.P.F. Lelieveldt, G. Koning, M. Sonka and J.H.C. Reiber, *Left Ventricle Contour Detection in X-ray Angiograms using Multi-View Active Appearance Models*, Proceedings SPIE Medical Imaging, Vol. 5032: 394-404, 2003.
4. T.F. Cootes, G.J. Edwards and C.J. Taylor, *Active Appearance Models*, Proceedings European Conference on Computer Vision, Vol. 2: 484-498, Springer, 1998.
5. T.F. Cootes and C.J. Taylor, *Statistical Models of Appearance for Computer Vision*, online available at: http://www.isbe.man.ac.uk/~bim/Models/app_models.pdf, 2004.
6. T.F. Cootes, D. Cooper, C.J. Taylor and J. Graham, *Active Shape Models - Their Training and Application*, Computer Vision and Image Understanding, 61(1): 38-59, 1995.
7. M.B. Stegmann, *Generative Interpretation of Medical Images*, pp. 248, Informatics and Mathematical Modelling, Technical University of Denmark, DTU, 2004.
8. T.F. Cootes, P. Kittipanya-ngam, *Comparing Variations on the Active Appearance Model Algorithm*, Proceedings British Machine Vision Conference, Vol. 2: 837-846, 2002.
9. J.H.C. Reiber, L.R. Schiemanck, P.M.J. van der Zwet, B. Goedhart, G. Koning, M. Lammertsma, M. Danse, J.J. Gerbrands, M.J. Schalij and A.V.G. Bruschke, *State of the Art in Quantitative Coronary Arteriography as of 1996*, In: J.H.C. Reiber and E.E. van der Wall editors. Cardiovascular Imaging. Dordrecht: Kluwer Academic Publishers, 1996: 39-56.
10. K. Suzuki, I. Horiba, N. Sugie and M. Nanki, *Extraction of Left Ventricular Contours from Left Ventriculograms by means of a Neural Edge Detector*, IEEE Transactions on Medical Imaging, 23(3): 330-339, 2004.

SPASM: Segmentation of Sparse and Arbitrarily Oriented Cardiac MRI Data Using a 3D-ASM

Hans C. van Assen[1], Mikhail G. Danilouchkine[1], Alejandro F. Frangi[2],
Sebastián Ordás[2], Jos J.M. Westenberg[1], Johan H.C. Reiber[1], and
Boudewijn P.F. Lelieveldt[1]

[1] Div. of Image Processing, Dep. of Radiology, Leiden University Medical Center,
PO BOX 9600, 2300 RC, Leiden, Netherlands
b.p.f.lelieveldt@lumc.nl
http://www.lkeb.nl

[2] Computational Imaging Laboratory, Department of Technology,
Universitat Pompeu Fabra, Barcelona, Spain
http://www.cilab.upf.edu

Abstract. In this paper, a new technique (SPASM) based on a 3D-ASM is presented for automatic segmentation of cardiac MRI image data sets consisting of multiple planes with different orientations, and with large undersampled regions. SPASM was applied to sparsely sampled and radially oriented cardiac LV image data.

Performance of SPASM has been compared to results from other methods reported in literature. The accuracy of SPASM is comparable to these other methods, but SPASM uses considerably less image data.

1 Introduction

Nowadays, cardiac MRI and CT are increasingly used for cardiac functional analysis in daily clinical practice. Both modalities yield dynamic 3D image data sets. With CT, images are acquired in an axial orientation and for cardiac analysis, usually short-axis (SA) views are reconstructed from the axial image data. With MRI, images can be acquired in any spatial orientation. Commonly used orientations are short-axis and long-axis (LA) views (2-chamber and 4-chamber), and radial stacks. The SA acquisitions consist of a full stack of typically 8 to 12 (parallel) slices covering the heart from apex to base. However, there is an ongoing debate on potential improvement of functional measurements by using LA views or radially scanned long-axis (RAD) image slices, since they appear to give better volume quantification due to better definition of the apex and base [1].

For quantitative analysis of cardiac function, typically a cardiologist or radiologist manually segments the images. After segmentation, measurements of global and regional functional parameters can be performed, such as wall thickening or wall thinning, LV volume and Ejection Fraction (EF). Due to the increasing amount of data, the amount of work for manually delineating the image data has become prohibitively large, and automated segmentation is highly desired.

Recent work has shown that integration of prior knowledge into medical image segmentation methods is essential for robust performance. Many recent methods utilize a statistical shape model, and the seminal work of Cootes [2,3] on 2D Active Shape Models (ASMs)- and Active Appearance Models (AAMs) has inspired the development of 3D ASMs [4,5], 3D AAMs [6], 3D Spherical Harmonics (SPHARM) [7], 3D Statistical Deformation Models (SDMs) [8,9,10,11] and 3D medial representations (m-reps) [12]. However, all these statistical models are only applicable to densely sampled 3D volume data, because the modeling mechanism is either based on a dense volumetric registration [6,8,9,10,11] or the matching mechanism is based on a dense set of updates along the model surface [4,5,12]. Therefore they typically assume a near isotropic resolution and parallel image planes. The main goal of this work is to avoid the need for these requirements on data sampling by developing a 3D active shape model that:

- is applicable to sparsely sampled data sets without making assumptions about voxel isotropy or parallel slices.
- is extensible to other modalities without retraining the shape model

To accomplish this, we present a 3D-Active Shape Model (3D-ASM) of the cardiac left ventricle (LV). The underlying statistical shape model was based on a 3D atlas that was constructed using non-rigid registration [9,13]. Matching of the model to sparse, arbitrarily oriented image data is accomplished through a deformable mesh that enables propagation of image updates over the model surface. Independence of a trained gray level model is achieved through a Takagi-Sugeno Fuzzy Inference System (TSFIS) [14] for determining iterative model updates based on *relative* intensity differences [4].

2 Background

Active Shape Models were introduced by Cootes et al. [2,15] and consist of a statistical shape model (often referred to as Point Distribution Model (PDM)) and a matching algorithm. The PDM is trained from a population of typical examples of the target shape, and models shape variability as a linear combination of a mean shape, i.e. a mean set of (pseudo-)landmarks, and a number of eigenvariations. For an elaborate introduction to ASMs, the reader is referred to [2,15,16].

2.1 Atlas Construction

A critical issue to achieve extension of PDMs to three and more dimensions is point correspondence: the landmarks have to be placed in a consistent way over a large database of training shapes, otherwise an incorrect parameterization of the object class would result. The methodology employed to automatically achieve this point correspondence of the heart was described in detail in [9]. The general layout of the method is to align all the images of the training set to a mean atlas (Fig. 1). The transformations are a concatenation of a global rigid registration

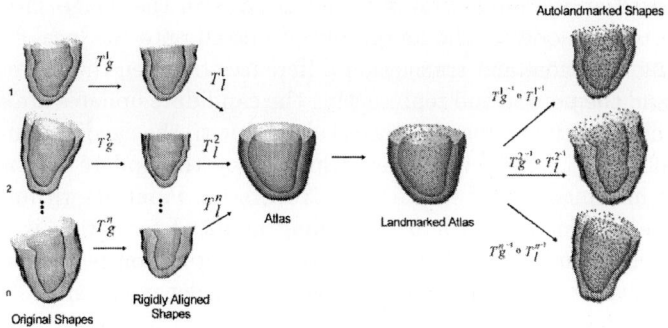

Fig. 1. Atlas construction, a set of final global (T_g) and local (T_l) transformations can take any sample shape of the training set, to the atlas coordinate system. On the left, there is landmark propagation. Once the final global and local transformations are obtained, they are inverted and used to propagate any number of arbitrarily sampled landmarks on the atlas, to the coordinate system of the original samples

with nine degrees of freedom (translation, rotation, and anisotropic scaling) and a local transformation using non-rigid registration. After registration of all samples to the mean shape, the transformations are inverted to propagate a topologically fixed point set on the atlas surface to the coordinate system of each training shape. While it is still necessary to manually segment each training image, this technique reliefs from manual landmark definition. The method can easily be set to build either 1- or 2-chamber models; in this work we have used a 1-chamber model. To build the statistical shape model, the auto-landmarked shapes are aligned using Procrustes alignment [17]. Principal Component Analysis (PCA) can then be performed on the remaining differences, which are solely shape related.

2.2 Matching Algorithm

The model described above was extended with a matching algorithm to apply it to image segmentation. A key design criterion behind this matching approach was applicability to data acquired with arbitrary image slice orientations, from different modalities (MR and CT), and even to sparsely sampled data with arbitrary image slice orientations. This implies that:

- only 2D image data may be used for updating the 3D model, to ensure applicability to arbitrarily oriented sparse data
- generation of update points is performed based on relative intensity difference to remove the dependence on training-based gray-level models.

To accomplish this, the landmark points are embedded in a surface triangular mesh. During the matching, this mesh is intersected by the image planes, generating 2D contours spanned by the intersections of the mesh triangles. To remove dependencies on image orientation or limited resolution, model update information is represented by 2D point-displacement vectors. The 2D update vectors located at the intersections of the mesh with the image slices are first propagated to the nodes of the mesh, and projected onto the local surface normals. Scaling, rotation, and translation differences between the current state of the model and the point cloud representing the candidate updates are eliminated by alignment. The current model state is aligned with the candidate model state (i.e., current model state with nodes displaced by the update vectors inferred from image information) using the Iterative Closest Point algorithm [18]. Successively, the parameter vector b controlling model deformation is calculated. An adjustment to b with respect to the previous iteration is computed, using both the candidate model state, $\hat{x}_{n+1}$, and the current model state, x_n

$$\hat{b}_{n+1} = b_n + \Delta b = b_n + \Phi^T(\hat{x}_{n+1} - x_n) \tag{1}$$

with x_n representing the aligned current state of the mesh, and b_n representing the parameter vector describing the current shape of the model within the statistical bounds. The vector $\hat{x}_{n+1}$ is the proposed model shape for the next iteration, and $\hat{b}_{n+1}$ its shape parameter vector before statistical constraints have been applied.

2.3 Update Propagation to Undersampled Surface Regions

In densely sampled data, a 3D data volume can be reconstructed that enables generation of a 3D update in each model landmark. However, in sparsely sampled data containing large undersampled regions, a (dense) 3D data volume cannot be reconstructed: interpolation between sparse image slices with different orientations (e.g., a radial stack of cardiac LA views) is non-trivial, if at all possible. In void locations, no information can be extracted from the image data to contribute to a new model instance. However, for the calculation of new model parameters, updates for the complete landmark set are required: setting updates of zero displacement would fixate the nodes to their current position, thus preventing proper model deformation.

Paulsen et al. [19] applied Gaussian smoothing of a mesh surface in combination with a Markov Random Field for restoration of point correspondences for an ear canal ASM. During the deformation of a mesh to presegmented shapes of ear canals and projection of the mesh nodes on the target shape, swapping of mesh vertices could occur. Instead of the training stage, we apply a similar method to the matching stage of SPASM. To overcome large void areas without update information, we propose a node propagation mechanism that distributes the updates from non-void update locations towards the void regions (see Fig. 2(a)). This mesh update propagation is weighted with the geodesic distance to the origin of the update using a Gaussian kernel (see Fig 2(b)):

$$w(x) = e^{-\frac{\|x-\omega\|^2}{2\sigma^2}} \tag{2}$$

where $w(x)$ is the weight at the location of the receiving node in the mesh x, ω is the source node, $\|x - \omega\|$ is the geodesic distance to the origin of the update, σ is the width of the kernel. Therefore, if multiple paths exist from source node to receiving node, only the shortest path is used. Thus, a receiving node accepts

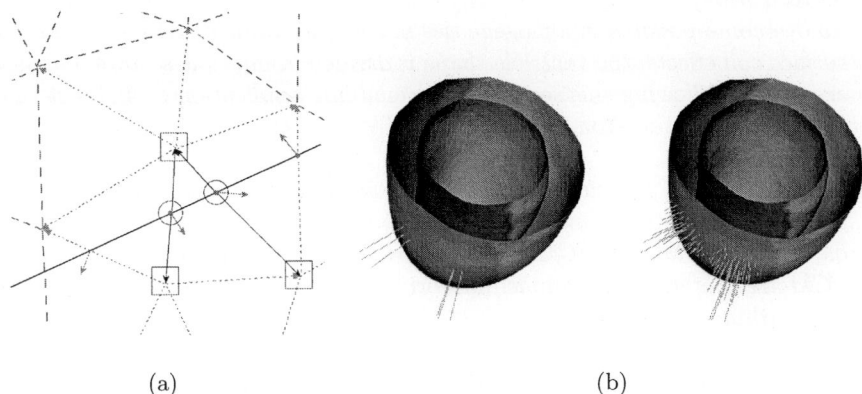

Fig. 2. (a) Propagation of single model updates at the intersection with an image plane (solid line). Propagation from the update sources surrounded with a circle is illustrated here. From a source, an update vector originates (short arrow). Updates are first propagated (longer solid arrows) to the nearest nodes in the mesh (marked with squares). Updates are further propagated to adjacent nodes weighted with a Gaussian kernel. Secondary updates (dotted lines) and tertiary updates (dashed lines) are also shown. (b) Gaussian propagation with $\sigma = 4mm$ (right) of two model updates (left)

propagated updates from any source only once. To avoid propagation updates over the entire surface, propagation is stopped when the geodesic distance exceeds a fixed threshold ($\chi \equiv 3\sigma$). After all propagations stopped, a pruning of all node updates is performed. Each node has a list of weighted contributions from source nodes, and a list of weights that were used to calculate each contribution. A total update per node is computed by summing over all contributions and normalizing with the sum of the weights.

2.4 Edge Detection Using Fuzzy Inference

The mesh structure combined with the update propagation enables applications to sparsely and arbitrarily oriented data. To apply the model to different modalities without retraining, the matching algorithm should not employ any trained intensity model to generate the updates. Instead, we have developed an update mechanism based on a Takagi-Sugeno Fuzzy Inference System (TSFIS) [14], which uses Fuzzy C-Means clustering (FCM) on the gray values of a 3D volume patch surrounding the current instance of the model (see Fig. 3(c)). This approach has been described in detail in [4], and can be summarized as follows:

1. Input
   For each intersection point between the mesh and each of the 2D images, an image patch, centered in this point, is considered. Patch size was selected such that multiple tissues were included in the patch.

2. *Sectorization*

   To overcome possible inhomogeneities in the gray value distributions due to surface coil effects, the ventricle shape is divided in multiple sectors. Patches are pooled following this sectorization, enabling application of different rule sets for different anatomical sectors the LV.

3. *Fuzzification*

   To locate tissue transitions, gray values are classified per sector based on relative intensity differences between blood, myocardium and air using standard Fuzzy C-Means (FCM) [20] clustering. In this work, the classes in the FCM are bright, dark, and medium bright, representing blood pool, air and myocardium respectively.

4. *Inference of model updates*

   For each pixel, three fuzzy membership degrees (FMDs) result from fuzzy clustering, above. Based on the FMDs, a mesh update is inferred as follows:

   (a) defuzzification for each pixel

   if (gray value is bright) then pixel is blood pool
   if (gray value is medium) then pixel is myocardium
   if (gray value is dark) then pixel is air

   However, pixels are only classified if they clearly belong to one tissue class. If a pixel does not reach a preset minimum membership degree for any tissue class (see Table 1), it is not classified and not considered for inference.

   (b) transition inference

   endocardial border: from outside to inside, find the first transition from myocardium to blood pool
   epicardial border:

   a at the septum

   from inside to outside, first transition from myocardium to blood pool

   b at the lung, anterior and posterior wall

   from inside to outside, first transition from myocardium to another tissue

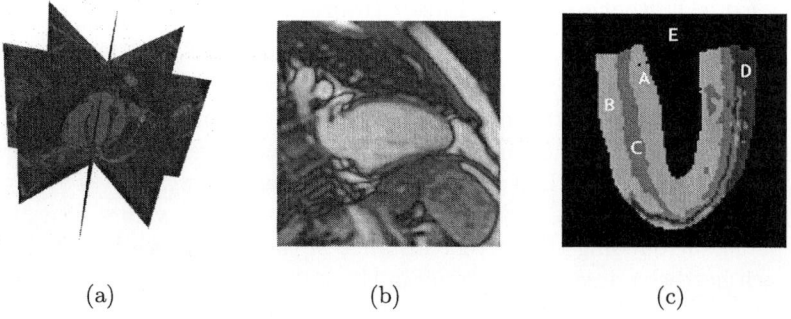

Fig. 3. (a) Radial cardiac image stack. (b) Radial slice acquired with the Turbo Field Echo (TFE) protocol. (c) Classified set of image patches. (A=LV blood pool, B=RV blood pool, C=myocardium, D=air, E=outside image patches)

Table 1. Parameters of the SPASM and their values

Defuzzification [21]		ASM	
Air cut-off proport. (See [4])	-0.20	Modes of variation	60
Blood pool mem.ship thresh.	0.20	Max variation per mode	2σ
Myocardium mem.ship thresh.	0.05	**Propagation**	
Air mem.ship thresh.	0.50	Gauss. kern. width σ (Eq. (2))	$8mm$

3 Experimental Setup

3.1 Test Data and Protocol

To test the performance of the sparse data model, a group of 15 volunteers was scanned with a Philips Gyroscan NT5 (1.5T) scanner, using the Steady State Free Procession (SSFP) and the Turbo Field Echo (TFE) protocols. For all scans and protocols, the QBody coil was used. A number of acquisitions with different slice orientations were performed during breath hold in end expiration. First, SA images were acquired, yielding a stack of typically 10-12 parallel image slices. Next, a radial scan was performed comprising four LA image slices, with inter-slice angle of 45° (see Fig. 3(a)). To avoid breathing-induced slice shifts, every slice was acquired with the TFE protocol, acquiring all four slices in the same breath hold. Image slices had a 256^2 matrix and covered a field-of-view of $300 - 400mm$, slice thickness and slice gap for the SA acquisitions were $8mm$ and $2mm$ respectively. For the RAD TFE acquisitions, the slice thickness was $8mm$. LV contours were manually drawn in all data sets. The manual contours in the radial stack were used to compensate for slice shifts in the SA volume due to differences in inspiration level. To assess inter-observer variability with respect to manual delineation, contours on all subjects were drawn by two observers.

3.2 Matching

Initialization of the model in the target data set was performed manually. Initial pose and scale were calculated from manual delineations on the image data from the SA acquisition. Due to the rotational symmetry of the model with respect to the long-axis and the sectorization, special attention was paid to initialize the model such that the myocardial sectors corresponded to the approximately correct anatomical location in the image data.

Parameter settings for the membership thresholds for the FIS used to define model updates at locations where the model is intersected by image planes were taken from previous work [21]. Best settings for the propagation parameters were determined in an exhaustive search on a computer cluster with 50 processors, using point-to-surface (P2S) error measures of the final state of the model with respect to manual segmentation as criterion for evaluation. The optimal settings for the parameters are listed in Table 1.

3.3 Quantitative Evaluation

To quantify the performance of the SPASM on the sparse radial image data, point-to-surface (P2S) error measurements were performed (see Fig. 5. Manually delineated surfaces in the SA image data were selected as gold standard. In addition, a comparison was performed between volumes of the final model states and volumes derived from the manual segmentations on the SA acquisition data.

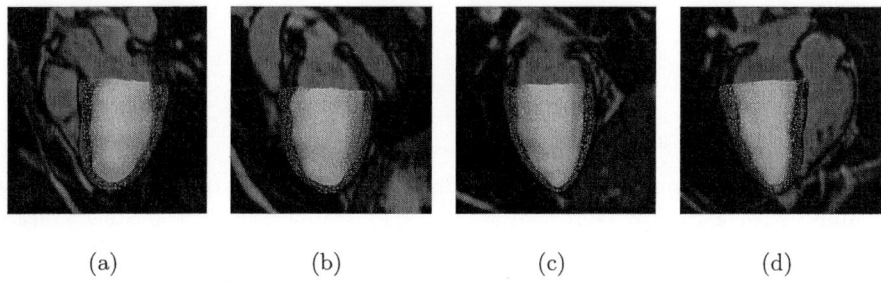

(a) (b) (c) (d)

Fig. 4. Final segmentation result of one of the subjects in the test population. (a-d) Result shown on slice 1 through 4, respectively

Fig. 5. Point-to-surface error measurement. Distances are measured from points on the automatic surface (solid) to the manually segmented surface (mesh). Note that the largest errors are made at the apical region

4 Results

Results from the tuning of the update propagation parameters are shown in Table 1. Results from the P2S evaluations between the final model instance and

---

[1] These are the best obtained results by Lötjönen et al [11] with a 4-chamber ASM based on a probabilistic atlas. Other models were built using normalized mutual information, landmark probability distribution, PCA, and ICA.

[2] Mitchell et al. compute errors of the automatic segmentation results slightly different than in this work. Mitchell et al compute (2D) distances in the image slices along lines perpendicular to the centerline between automatic and manual segmented contours. This does not guarantee shortest point-to-curve or point-to-surface distances, and may thus overestimate errors with respect to the method used in this paper and by Lötjönen et al.

Table 2. Point-to-surface distances measured per subject between manual and automatic surfaces in mm, averaged over the total population (14 subjects). Maximum distances are maximum distances per subject averaged over the total population.(all values are average ± standard deviation in mm)

	Average		Maximum	
	endocard	epicard	endocard	epicard
Inter observer	1.27 ± 0.30	1.14 ± 0.29	4.34 ± 0.88	3.93 ± 0.79
Lötjönen et al. (aut. ref.) [11][1]	2.01 ± 0.31	2.77 ± 0.49	n.a.	n.a.
Mitchell et al. [6][2]	2.75 ± 0.86	2.63 ± 0.76	n.a.	n.a.
Kaus et al. [5]	2.28 ± 0.93	2.62 ± 0.75	13.82	12.35
SPASM	2.24 ± 0.54	2.83 ± 0.78	11.1 ± 2.54	15.7 ± 5.06

Table 3. Volume regression numbers. Volumes were calculated per subject (14 subjects), separately for endocardial volume and epicardial volume. Volume calculated from SA reconstruction was taken as the reference volume (ground truth)

	Correlation coefficient (R)	
	endocardium	epicardium
Manual volume (radial image slices)	0.74	0.71
Automatic volume	0.78	0.74

manual segmentations in SA views are presented in Table 2. Correlation coefficients between manual volumes from SA views and automatic volumes from the final model instance are shown in Table 3. For comparison, the correlation coefficients between manually segmented volumes from SA views and from RAD views are presented in Table 3 as well. In the results, one subject was excluded due to a mismatch for almost all the runs in the tuning process on this subject. A final segmentation result of one of the subjects is shown on all four slices in Figure 4. Matching for $N_{max} = 100$ iterations took 727 ± 134 seconds (minimal 522 seconds, maximal 915 seconds) on a 2.8 GHz Xeon computer machine running Linux Redhat 9.

5 Discussion and Conclusion

In this paper SPASM is presented, a new technique based on a 3D-ASM, that is able to automatically segment cardiac MRI image data sets consisting of multiple planes with different orientations, and having large undersampled regions.

Because SPASM does not include a statistical gray level model, it is potentially applicable to both MRI and CT data sets without fully retraining the intensity model. For segmentation, it does not require image slices with equal orientations as present in the training data. SPASM was applied to radially oriented cardiac LV image data, which contains undersampled regions with larger sampling density at the apex than at the base.

Performance of SPASM was evaluated against manual delineations on an SA data set of the same subjects. In the SA data set, the heart can be displaced between different slice acquisitions, due to different breath hold levels. Although this displacement is minimized by acquisition during end expiration, correction of slice positions was necessary.

The maximum errors presented in Table 2 are mainly observed at the apical regions (see Fig. 5). This is due to the closed apex in SPASM, while the manual segmentation at the apex is open.

Performance of SPASM has been compared to results from other methods reported in literature [6, 11, 5] (see Table 2). The accuracy of SPASM is comparable to these other methods. However, SPASM is the only method that can be applied to a set of arbitrarily oriented and sparsely sampled image slices: it was applied to only four image slices, whereas the other models required a stack of 8-12 parallel slices, yielding comparable accuracy. Segmentation errors of all methods are substantially larger than the inter-observer variability (see Table 2). This may be caused by too rigid statistical constraints on the allowed deformation of statistical shape models in general. Further evaluation of SPASM is ongoing with respect to the minimally required sampling density, different combinations of LA and SA image slice orientations, and the sensitivity of the final segmentation results to model initialization.

References

1. T. N. Bloomer, S. Plein, A. Radjenovic, D. M. Higgins, T. R. Jones, J. P. Ridgway, and M. U. Sivananthan, "Cine mri using steady state free precession in the radial long axis orientation is a fast accurate method for obtaining volumetric data of the left ventricle," *Journal of Magnetic Resonance Imaging*, vol. 14, pp. 685–692, 2001.
2. T. F. Cootes, D. Cooper, C. J. Taylor, and J. Graham, "Active shape models - their training and application," *Computer Vision and Image Understanding*, vol. 61, no. 1, pp. 38–59, 1995.
3. T. F. Cootes, G. J. Edwards, and C. J. Taylor, "Active appearance models," *IEEE Transactions on Patern Analysis and Machine Intelligence*, vol. 23, no. 6, pp. 681–685, 2001.
4. H. C. van Assen, M. G. Danilouchkine, M. S. Dirksen, J. H. C. Reiber, and B. P. F. Lelieveldt, "A 3d-active shape model driven by fuzzy inference: Application to cardiac ct and mr," *submitted*.
5. M. R. Kaus, J. von Berg, J. Weese, W. Niessen, and V. Pekar, "Automated segmentation of the left ventricle in cardiac mri," *Medical Image Analysis*, vol. 8, no. 13, pp. 245–254, 2004.
6. S. C. Mitchell, J. G. Bosch, B. P. F. Lelieveldt, R. J. van der Geest, J. H. C. Reiber, and M. Sonka, "3d active appearance models: Segmentation of cardiac mr and ultrasound images," *IEEE Transactions on Medical Imaging*, vol. 21, no. 9, pp. 1167–1178, 2002.
7. A. Kelemen, G. Szkely, and G. Gerig, "Elastic model-based segmentation of 3-d neuroradiological data sets," *IEEE Transactions on Medical Imaging*, vol. 18, no. 10, pp. 828–839, 1999.

8. D. Rückert, A. F. Frangi, and J. A. Schnabel, "Automatic construction of 3-d statistical deformation models of the brain using nonrigid registration," *IEEE Transactions on Medical Imaging*, vol. 22, no. 8, pp. 1014–1025, 2003.
9. A. F. Frangi, D. Rueckert, J. A. Schnabel, and W. J. Niessen, "Automatic construction of multiple-object three-dimensional statistical shape models: Application to cardiac modeling," *IEEE Transactions on Medical Imaging*, vol. 21, no. 9, pp. 1151–1166, 2002.
10. R. Chandrashekara, A. Rao, G. I. Sanchez-Ortiz, R. H. Mohiaddin, and D. Rückert, "Construction of a statistical model for cardiac motion analysis using nonrigid image registration," in *Information Processing in Medical Imaging*, ser. Lecture Notes in Computer Science, C. Taylor and J. A. Noble, Eds., vol. 2732. Berlin: Springer Verlag, 2003, pp. 599–610.
11. J. Lötjönen, S. Kivistö, J. Koikkalainen, D. Smutek, and K. Lauerma, "Statistical shape model of atria, ventricles and epicardium from short- and long-axis mr images," *Medical Image Analysis*, vol. 8, pp. 371–386, 2004.
12. P. Yushkevic, P. T. Fletcher, S. Joshi, A. Thall, and S. M. Pizer, "Continuous medial representations for geometric object modeling in 2d and 3d," *Image and Vision Computing*, vol. 21, no. 1, pp. 17–27, 2003.
13. S. Ordas, L. Boisrobert, M. Bossa, M. Laucelli, M. Huguet, S. Olmos, and A. Frangi, "Grid-enabled automatic construction of a two-chamber cardiac PDM from a large database of dynamic 3D shapes," in *IEEE International Symposium of Biomedical Imaging*, 2004, pp. 416–419.
14. T. Takagi and M. Sugeno, "Fuzzy identification of systems and its applications to modeling and control," *IEEE Transactions of Systems, Man and Cybernetics*, vol. 15, no. 1, pp. 116–132, 1985.
15. T. F. Cootes, D. Cooper, C. J. Taylor, and J. Graham, "A trainable method of parametric shape description," *Image and Vision Computing*, vol. 10, no. 5, pp. 289–294, 1992.
16. T. F. Cootes and C. J. Taylor, "Statistical models of appearance for computer vision," Imaging Science and Biomedical Engineering, University of Manchester, Manchester M13 9PT, U.K., http://www.isbe.man.ac.uk/~bim/Models/app_models.pdf, Tech. Rep., March 2004.
17. J. Gower, "Generalized procrustes analysis," *Psychometrika*, vol. 40, pp. 33–50, 1975.
18. P. J. Besl and N. D. McKay, "A method for registration of 3-d shapes," *IEEE Transaction on Pattern Analysis and Machine Intelligence*, vol. 14, no. 2, pp. 239–256, 1992.
19. R. R. Paulsen and K. B. Hilger, "Shape modeling using markov random field restoration of point correspondences," in *Information Processing in Medical Imaging*, ser. Lecture Notes in Computer Science, C. Taylor and J. A. Noble, Eds., vol. 2732. Berlin: Springer Verlag, 2003, pp. 1–12.
20. J. C. Bezdek, *Pattern Recognition with Fuzzy Objective Function Algorithms*. New York: Plenum press, 1981.
21. S. Ordas, H. C. van Assen, B. P. F. Lelieveldt, and A. F. Frangi, "Segmentation performance assessment of an autolandmarked statistical shape model," *submitted*.

Combining Active Appearance Models and Morphological Operators Using a Pipeline for Automatic Myocardium Extraction

Bernhard Pfeifer[1], Friedrich Hanser[1], Thomas Trieb[2], Christoph Hintermüller[1], Michael Seger[1], Gerald Fischer[1], Robert Modre[1], and Bernhard Tilg[1]

[1] University for Health Sciences, Medical Informatics and Technology, Eduard Wallnöfer Zentrum 1, A-6060 Hall, Österreich/Austria
bernhard.pfeifer@umit.at
http://imsb.umit.at

[2] Innsbruck Medical University, Clinical Division of Diagnostic Radiology I, Anichstrasse 35, 6020 Innsbruck, Austria

Abstract. A geometrical model of the human heart is of interest in many fields of biophysics. The myocardium contains the electrical sources responsible for the generation of the body-surface ECG. An accurate geometric knowledge of these sources is crucial when dealing with the electrocardiographic forward and inverse problem. We developed a semi-automatic approach for segmenting the myocardium in order to deal with the electrocardiographic problem. The approach can be divided into two main steps. The first step extracts the atrial and ventricular blood masses by employing Active Appearance Models (AAM). The ventricular blood masses are segmented automatically after providing the positions of the apex cordis and the base of the heart. Due to the complex geometry of the atria the segmentation process of the atrial blood masses requires more information. We divided, therefore, the left and the right atrium into three divisions of appearance: the base of the heart, the lower pulmonary veins from its first up to the last appearance in the image stack, and the upper pulmonary veins. After successful extraction of the blood masses the second step involves morphologically-based operations in order to extract the myocardium either directly by detecting the myocardium in the volume block, or by reconstructing the myocardium using mean model information, in case the algorithm fails to detect the myocardium.

1 Introduction

Atrial and ventricular surface activation time imaging from body-surface ECG mapping data [8, 5, 7] may become a diagnostically powerful clinical tool for assessing cardiac arrhythmias. This cardiac source imaging technique aims to provide information in a noninvasive manner about the spread of electrical excitation in order to assist the cardiologist in developing strategies for the treatment of cardiac arrhythmias. Common cardiac arrhythmias, such as atrioventricular reentrant tachycardia, atrioventricular nodal reentrant tachycardia, or atrial fibrillation, can, in many cases, be traced back to accessory pathways, atrial or ventricular foci, e.g., from the pulmonary veins [6, 1], and reentrant circuits [13]. Identifying the site of origin of the ectopic focus or the location of an accessory pathway provides the essential information for treatment strategies, such as catheter ablation [9].

Activation time imaging from three-dimensional anatomical and body-surface ECG mapping data enables noninvasive imaging of the electrical excitation in the heart [12]. The method yields solutions to the electrocardiographic inverse problem and is based on an electrodynamic model of the patient's volume conductor and heart. The volume conductor considers a model of the electrodynamically most relevant compartments including chest, lungs, atrial and ventricular myocardium, and blood masses. A model of the heart comprises separate models for the atria and ventricles since whole heart models still resist a technical implementation with regard to the electrodynamic inverse problem. The crucial point of an atrial and ventricular model is their geometry. Geometric distances between the cardiac sources and the chest strongly influence the electrodynamic-based model and, therefore, the overall model error. The complex geometry of the atria is given by the orifices of the pulmonary veins, orifices of superior and inferior vena cava, tricuspid and mitral annuli, and right and left appendages, and this makes it more difficult, compared to the ventricle, to generate a geometrical model. It is clear that any technique that is capable of generating an atrial model will succeed also for the ventricle. Consequently, we decided to extract the ventricular blood masses using the same technique as used for the atrial blood masses, although especially the myocardium of the left ventricle could be segmented in a direct way. The reason for this decision was to get a consistent way for cardiac modeling and for incorporating the proposed technique into a segmentation pipeline with little user interaction. The main problem of constructing a realistic heart model is that the myocardium can hardly be segmented in volume data (especially the atria are a big challenge) because of the low sensitivity and resolution even for state-of-the-art medical imaging modalities like MRI and CT. We employed AAM for the extraction of blood masses and we use morphological operations to reconstruct the myocardial structure directly, in case the myocardium can be detected in the volume data, or in an indirect way, using a priori knowledge, otherwise.

The paper is organized as follows: Section 2 describes the segmentation approach and the implemented algorithms. Results of geometrical models of the atrial and ventricular myocardium are presented in section 3. The two steps of the approach are discussed in section 4, and finally, we summarize in section 5.

2 Methods

Our goal when developing this segmentation approach was to get a consistent way for cardiac modeling and for incorporating the proposed technique into a segmentation pipeline with little user interaction. We employed AAM for the extraction of blood masses because this model based technique is able to generate reliable results when segmenting the cardiac blood masses. The use of morphological operations provides a fast method for myocardium reconstruction, in case the myocardium is detectable, or estimation otherwise, and enables an easy implementation in a semiautomatic segmentation pipeline.

2.1 Blood Mass Extraction Using Active Appearance Models

In the year 1991 Craw and Cameron published one of the first appearance modeling approaches [4]. They wrapped faces to a reference shape before doing a Principal Component Analysis (PCA). In 1994 Cootes et al. introduced Statistical Models of shape and texture [2]. In 1998 Active Appearance Models were introduced [3] and since this introduction a lot of enhancements were done. For more information http://www.isbe.man.ac.at/~bim/ should be picked up.

Objects in images are represented using shapes. A shape can be described by a set of n points. Statistical methods can be applied when using shapes and, therefore, it is possible to analyze the shape differences and shape changes. Shapes can be inserted into an input or training image by searching for corresponding landmarks. Normally a human expert annotates the training sets by hand. Good landmarks are points of high curvature or junctions. Intermediate points can be used to define the boundary more precisely. The vector for representing a shape can formally be defined as

$$x = (x_1, ..., x_n, y_1, ..., y_n)^T. \tag{1}$$

If there are s training examples, then s vectors are generated by the human expert. Before applying statistical analysis on these vectors it has to be guaranteed that all shapes are in the same coordinate-frame. Therefore all shapes are aligned in a way, that the sum of distances of each shape to the mean ($D = \sum |x_i - \overline{x}|^2$) is minimized. An appearance model can represent shape and texture changes learnt in the training sets. The shape of an object is represented as a vector x and the texture as a vector g:

$$x = \overline{x} + Q_s c \tag{2}$$
$$g = \overline{g} + Q_g c \tag{3}$$

where the parameter c controls shape and texture. $\overline{x}$ is the mean shape, $\overline{g}$ is the mean texture and Q_s, Q_g are matrices describing the modes of variation (shape and texture) learnt from the training set. Generally, an AAM seeks to minimize the difference between an unseen image and one created by the appearance model.

For creating the AAM we integrated the AAM-API[1] available at http://www.imm.dtu.dk/~aam/ into our Medical Segmentation Toolkit (MST) framework. The MST framework is developed using C++ and includes some

---
[1] Copyright (c) 2000-2003 Mikkel B. Stegmann, mbs@imm.dtu.dk; This software is freely available for non-commercial use such as research and education.

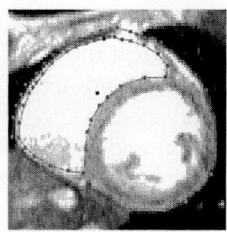

Fig. 1. Training set, annotated using 66 landmark points, of a blood mass from the right ventricle

different standard frameworks like the DCMTK framework for handling medical images (http://dicom.offs.de/dcmtk.php.en), ITK (http://www.itk.org/) for some segmentation methods, and Qt (http://www.trolltech.com/) for creating user interfaces. This toolkit enables to combine different segmentation techniques for each compartment and offers the creation of defined compartment pipelines.

When constructing a realistic cardiac model of a patient the main problem is that the myocardium can only be detected at limited locations in the volume conductor. The only structure which can be seen with sufficient accuracy are the blood masses. But also the blood masses have a big variation in shape and texture. Because of this reason a segmentation approach for this task needs a-priori information for successful and almost error-free extraction of the searched blood masses. Model based approaches like the AAM seem, therefore, to be a good choice for solving this segmentation problem. When trying to segment the atrial and ventricular blood masses from different patients the structures vary in shape and texture. With the help of AAM it is possible to figure out which are plausible variations and which are not. A new data set can, therefore, be segmented by finding the best plausible match between the model and image data.

The appearance model represents both the shape and texture variability seen in a training set. The training set consists of labelled images, where key landmark points are marked on each example object. We used 96 different training sets for establishing the right and left ventricular blood mass AAM. Every image set was annotated using 66 annotation points. 33 points were used to describe the left/right ventricular blood mass and 33 points were used to define the pericardium. Figure 1 shows an example of a training set for right ventricular blood mass extraction AAM. The pericardium (whole heart contour) is annotated because this makes it easier to initial locate the desired structure, a manner that was used for each AAM. After the preparation of the ventricular data sets, the training procedure - the processing of the principal component analysis - yielded 55 main components for the right ventricular blood mass and 56 main components for the left ventricular blood mass.

As already mentioned above, the atria show a more complex geometry and, therefore, more information is needed for the segmentation process of the atrial blood masses. In order to simplify the segmentation process we decided to divide the atria into three divisions of appearance: the base of the heart up to the left upper (LUPV) and the left lower (LLPV) pulmonary vein, the LUPV and the LLPV from its first up to its last appearance in the image stack, and from this position up to the right lower (RLPV) and upper (RUPV) pulmonary vein. We created one training set for each division. All together, we prepared 193 training sets for atrial blood mass extraction and each atrial blood mass was annotated by 66 annotation points. 33 annotation points were used to define the left/right atrial blood mass and the resting 33 annotation points were used to describe the pericardium. The creation of the appearance models yielded 56 main components for the first atrial division, 58 main components for the second division and 51 main components for the third defined atrial division. The main components of our AAMs were defined to include 97% of all shape and texture variations.

Blood Mass Search Procedure. When extracting the ventricular blood masses the user has to provide the position of the apex cordis and the base of the heart in the associated volume block. After this, the AAM approach for the left and the right ventricular blood masses is initialized and then yields the desired segmentation of the ventricular blood masses by applying the fitting procedure until convergence.

The segmentation procedure for the atrial blood masses can be described this way. First initial parameters have to be set: the base of the heart, the end of the first division of appearance and the end of the second division of appearance in the volume block have to be marked. Then the AAM need to be initialized and that means to locate the desired structures in principal. After this process the model fit approach starts and operates until the search process converges.

These steps have to be repeated for each image between the given parameter range in the volume block to extract the ventricular blood masses as well as the atrial blood masses. Because the AAM ranges are defined by the given parameters the associated AAM are used in order to extract the desired structures.

2.2 Myocardium Reconstruction/Estimation Using Morphological Operations

After blood mass extraction using the technique described in section 2.1 the labelset should be smoothed by appropriate tools. Figure 2 shows a triangulation, created by a marching cubes algorithm, of the extracted labelsets of the atrial blood masses. The extracted blood masses are the basic input for the myocardium modeling procedure. The atrial and ventricular myocardium is constructed directly, in case the myocardium can be detected by the algorithm in the volume data, or artificially otherwise by applying appropriate voxel manipulations. The method adds label voxels in the outward normal direction until the user defined wall thickness is reached. Due to given facts in human beings the atrial wall thickness is between 3 to 5mm, the right ventricular myocardium between 6 to 8mm and the left ventricular myocardium between 8 to 12mm. The necessary input parameters for the algorithm are the minimum wall thickness, the mean wall thickness, and the maximum wall thickness. The approach uses operations of mathematical morphology. In principal the dilation operation is used.

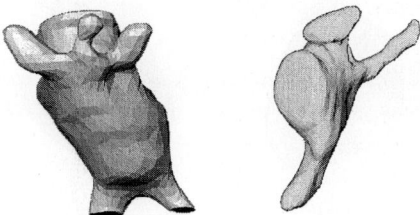

Fig. 2. Left panel shows the triangulated blood masses of the left atrium, and the right panel shows the triangulated blood masses of the right atrium

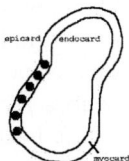

Fig. 3. On each boundary voxel of the endocardium, a virtual circle with a predefined radius rolls around the endocardium in order to reconstruct the atrial myocardium

The algorithm uses virtual circles as structuring elements with a radius range from the defined minimum wall thickness up to a maximum wall thickness. These circles roll around the blood mass boundary in order to reconstruct the myocardial structure. If the algorithm is able to determine the myocardium by probing all voxels to be element of a user defined gray value range inside the virtual circles (minimum up to maximum wall thickness) then the myocardial structure can be reconstructed directly. If the myocardial structure can not be detected the mean model information is used to reconstruct the myocardial structure. The mean model is a user defined parameter that describes the myocardium to have a standard wall thickness of 5mm for the left/right atrial myocardium, 7mm for the right ventricular myocardium and 10mm for the left ventricular myocardium, as an example for one possible parameter set.

The situation of estimation occurs predominantly when reconstructing the atrial myocardium because the atrial myocardium is almost always invisible due to its low sensitivity in the image data. This approach processes the volume stack sequentially in z direction without taking adjacent slides into account. For this reason this approach is a 2D version. In spite of the fact that this 2D version of the algorithm yields good results, the marching cubes algorithm can produce holes, especially when the segmentation of the blood masses differs too much between adjacent images or labelsets within the volumestack. Such a variation may occur because of the choosen image modalities (4mm slice thickness) and possible artefacts especially caused by motion. Although a slice thickness of 1mm is possible with new scanners such an image modality setting needs a lot of time that is not available when using the approach in a clinical application nowadays. To overcome this problem, the adjusted variant, the 3D variant, takes one slide

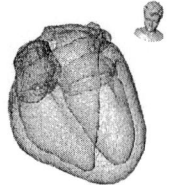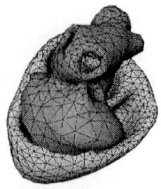

Fig. 4. Segmented ventricles and atria triangulated - for visualization - using a marching cubes algorithm

above and one slide below the initial labelset slide into account. As a main step of the 3D variant the adjacent slides are similarified by the algorithm [10]. To similarify the adjacent sets reduces the likeliness of holes when triangulating the labelset using a standard marching cubes algorithm. Because for the estimation of the electrical spread in the human heart a functional model and not an anatomical one is needed, model variations caused by similarify and smoothing operations influence the inverse solver less than having non existing structures (e.g., holes) in the model caused by above described possible situations.

3 Results

The segmented labelset is triangulated with a standard marching cubes algorithm followed by a remeshing process guaranteeing quality standards (equilaterality of triangles) that qualify for a FEM/BEM formulation used for dealing with the electrocardiographic problem. Our main problem is to get a model of the volumeconductor, on the one hand, in a very fast and efficient way to enable the estimation of the electrical excitation in a clinical application, and, on the other hand, to keep the model error as small as possible to get reliable results when trying to solve the inverse problem - and that means to find the pathological pathway in a non invasively way.

We tested our approach using volume data from eight different patients [12]. The segmentation of the left and the right ventricular blood mass needed $\mu = 148$ seconds. The segmentation of the right and the left atrial blood masses needed $\mu = 167$ seconds. The reconstruction process of the myocardial structure by using the blood masses as the main input source, needs about $\mu = 5$ seconds. For the reconstruction approach we used a *Dual Pentium Xeon* workstation with a clock frequency of 2.8 GHz and 2 GByte main memory (RAM).

Figure 4 shows a triangulated and remeshed ventricular and atrial myocardium model that qualifies for estimating the spread of electrical excitation in the patients volume conductor. Figure 5 shows a ventricular model of a female patient and the atrium with its blood masses.

To decide if the method qualifies, or with other words, if the model represents the for the estimation relevant parameters preferably close, the segmentation result of the blood masses were compared with the blood masses extracted by two

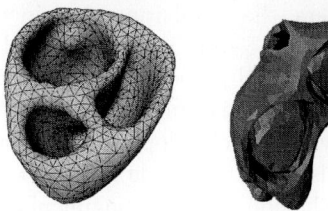

Fig. 5. Left: Reconstructed myocardium of the left and the right ventricle; Right: Atrium with blood masses

different human experts. The correlation coefficient, that was used to measure similarity, resulted in the correlation coefficient range from 0.912 to 0.931. The models of the human experts qualified as well as the automatically extracted models for the estimation of the electrical spread in the human heart.

4 Discussion

We presented a semiautomatic segmentation approach that allows to reconstruct the myocardial structure of the ventricles and the atria even if the myocardium can not be detected in the volume data. The indirect reconstruction/estimation of the myocardial structure enables the creation of a semiautomatic segmentation approach, because the main input for the myocardium extraction, the blood masses, can be seen clearly in the volume data, even if there are artefacts. This fact reduces the necessary user input and the mending process dramatically. Due to the possibility that the myocardium can be reconstructed using the blood masses, or the endocardial structure, the contribution to the model error will be sufficiently small. It is clear that it is important to have only a very small model error when trying to estimate the spread of electrical excitation in the human heart. Note that not only the segmentation task causes a model error. Also the quality of the ECG signal, the positions of the electrodes and other interferences cause an increase of the mean model error. So it is important to hold all these sources of errors down because only if the sum of all errors is low the mean model error has an acceptable value.

It seems to be imperative, when trying to reconstruct the myocardium, to extract the blood masses primary and then to use an indirect technique using the blood masses as a-priori information for the myocardial extraction. The reason for this strategy was that the myocardium can be detected only at limited locations in the medical image data. This myocardium visibility problem occurs because of low image resolutions, artefacts (caused by motion of the patient and/or the heartbeat) and image modality settings. Our cine gated short axis scan image data were acquired using a slice thickness of $4mm$. This slice thickness requires an indirect myocardium extraction approach because the thickness of the atrial myocardium is about $4mm$ and, therefore, it is almost always invisible in the image data. Although it is possible to generate slides with a thickness of $1mm$ with modern scanners, the acquisition time increases. When using the $4mm$ slice thickness the MRI procedure can be finished in an acceptable time span (about 30 to 45 min). The MRI procedure consists of the preparation of the patient and the acquisition of the axial and the cine-gated short-axis scan. The cardiologist usually starts with the electrophysiology study (EPS) after a two hour intermission. During this break the whole volume conductor model has to be generated in order to enable the non invasive imaging of cardiac electrophysiology (NICE) [11] approach in the catheter laboratory.

5 Summary

The usage of AAM allow the extraction of ventricular and atrial blood masses in a very efficient way, which means that the run time behavior and the quality of the labelsets are in an excellent ratio. Only the annotation or rather the learning procedure of the appearance models need a lot of experience and time. The extraction result of the appearance models, the blood masses of the ventricles and the atria, can be directly used as input by the myocardium extractor. This technique allows to reconstruct the myocardium directly, when the structure can be detected in the volume data, or indirectly by using a-priori information, when the myocardium can not be identified in the volume data. The big advantage of this step is that only a few parameters have to be set and that the algorithm reconstructs the desired structure in a very efficient way.

The approach yields ventricular and atrial models that qualify, according to our experience, for cardiac source imaging. Thanks to the reduction of user interaction, the fast structure detection and the fast reconstruction of the cardiac model, this approach can be used in clinical applications.

The cardiac models were implemented many times in the construction of the patient's volume conductor model needed for solving the electrocardiographic inverse problem. The construction of this volume conductor segmentation pipeline coupled with our inverse solver can provide essential information to the cardiologist, in order to develop treatment strategies like catheter ablation. To get such information in a non invasive manner can help reducing costs, time, and it also reduces the remaining risk for the patient that arises during every invasive treatment.

The combination of AAM and morphological operators allows to create a segmentation pipeline with little user interaction and to reconstruct the desired structure even if not detectable in the volume data.

Acknowledgments

This research study was funded by the START Y144 program granted by the Austrian Federal Ministry of Education, Science and Culture (bm:bwk) in collaboration with the Austrian Science Fund (FWF).

References

1. S. A. Chen, M. H. Hsieh, C. T. Tai, C. F. Tsai, V. S. Prakash, W. C. Yu, T. L. Hsu, Y. A. Ding, and M. S. Chang, *Initiation of atrial fibrillation by ectopic beats originating from the pulmonary veins: electrophysiological responses, and effects of radiofrequency ablation.*, Circulation **100** (1999), 1879–1886.
2. T. Cootes and C. Taylor, *Modelling object appearance using grey-level surface*, 5th British Machine Vision Conference (BMVA 1994), 419–428.
3. _____, *Active appearance models*, 5th European Conference on Computer Vision (Springer 1998), 484–498.
4. I. Craw and P. Cameron, *Parameterising images for recoginition and reconstruction*, 2nd British Machine Vision Conference (London, Springer 1998), 367–370.
5. F. Greensite, *The mathematical basis for imaging cardiac electrical function*, Crit. Rev. Biomed. Eng. **22** (1994), 347–399.
6. M. Haissaguerre, P. Jais, D. C. Shah, A. Takahashi, M. Hocini, G. Quiniou, S. Garrigue, A. Le Mouroux, P. Le Metayer, and J. Clementy, *Spontaneous initiation of atrial fibrillation by ectopic beats originating from the pulmonary veins.*, N. Engl. J. Med. **9** (1998), 659–666.
7. G. Huiskamp and F. Greensite, *A new method for myocardial activation time imaging*, IEEE Trans. Biomed. End. **44** (1997), 433–446.
8. R. Modre, B. Tilg, G. Fischer, and P. Wach, *An iterative algorithm for myocardial activation time imaging*, Comput. Methods Programs Biomed. **64** (2001), 1–7.
9. F. Morady, *Radio-frequency ablation as treatment for cardiac arrhythmias.*, N. Engl. J. Med. **340** (1999), 534–544.
10. B. Pfeifer, F. Hanser, C. Hintermüller, R. Modre-Osprian, G. Fischer, M. Seger, C. Kremser, and B. Tilg, *Atrial myocardium model extraction*, Medical Imaging: Visualization, Image-Guided Procedures, and Display; Proceedings SPIE 2004 **5367** (2004), 320–331.
11. R. Modre R, M. Seger, G. Fischer, F. Hanser, B. Pfeifer, Hintermüller, and B. Tilg, *Nice: Noninvasive imaging of cardiac electrophysiology*, The International Conference on Inverse Problems: Modeling and Simulation. Fethiye, Turkey (2004).
12. B. Tilg, R. Fischer, R. Modre, F. Hanser, B. Messnarz, M. Schocke, C. Kremser, T. Berger, F. Hintringer, and F.X. Roithinger, *Model-based imaging of cardiac electrical excitation in humans*, IEEE Trans. Med. Imag. (2002), no. 21, 1031–1039.
13. E. J. Vigmond, R. Ruckdeschel, and N. A. Trayanova, *Reentry in a morphologically realistic atria*, J Cardiovasc Electrophysiol **12** (2001), no. 9, 1046–1054.

Long-Axis Cardiac MRI Contour Detection with Adaptive Virtual Exploring Robot

Mark Blok[1,2], Mikhail G. Danilouchkine[1], Cor J. Veenman[2],
Faiza Admiraal-Behloul[1], Emile A. Hendriks[2], Johan H. C. Reiber[1],
and Boudewijn P. F. Lelieveldt[1]

[1] Leiden University Medical Center, 2300 RC Leiden, The Netherlands
[2] Delft University of Technology, 2628 CD Delft, The Netherlands
mark_blok@wanadoo.nl

Abstract. This paper describes a method for automatic contour detection in long-axis cardiac MRI using an adaptive virtual exploring robot. The robot is a simulated trained virtual autonomous tri-cycle that is initially positioned in a binary representation of the left ventricle (LV) and finds the contours during navigation through the ventricle. The method incorporates global and local prior shape knowledge of the LV in order to adapt the navigational parameters. Together with kinematic constraints, the robot is able to avoid concave regions such as papillary muscles and navigate through narrow corridors such as the apex. Validation was performed on in-vivo multiphase long-axis cardiac MRI images of 11 subjects. Results showed good correlation between the quantitative parameters, computed from manual and automatic segmentation: for end-diastolic volume (EDV) r=0.91, for end-systolic volume (ESV) r=0.93, ejection fraction (EF) r=0.77, and LV mass (LVM) r=0.80.

1 Introduction

Over the last decade cardiovascular MRI imaging has become the clinical standard for the functional assessment of the human cardiovascular system. A typical MRI study consists of a large amount of data and, therefore, automated analysis of acquired data is desirable in the daily clinical practice.

For modelling and extracting myocardium borders a large number of techniques have already been proposed. For example, Active Contour Model (ACM) is a deformable contour, which is widely employed for tracing the cardiac borders [1]. Deformation is governed by the internal and external energies. The internal energy assures the contour's smoothness imposing constraints on its shape. The external energy attracts the contour to the object's boundaries. ACM comes short in segmentation of long-axis cardiac images. The contour has to be rigid enough to prevent its deformation outside the myocardial borders, where the boundaries are not well defined (e.g. boundary between the myocardium and papillary muscles or liver), and it must be flexible enough to provide reliable segmentation of regions with high curvature (apex). A technique, based on Active Appearance Models (AAM) [2] overcomes the aforementioned drawbacks of

ACM by incorporating knowledge about the myocardial boundaries. This knowledge is formalized in terms of the statistical properties of the average heart's shape and appearance (gray-level representation) as well as the main modes of variation found in a training set. The AAM model placed in a new image is deformed to minimize the difference between the model and object of interest within the image. The deformation is restricted to variations found in the training set. AAM's were used for extraction of the cardiac borders at end-diastole and end-systole and showed good correlation between automatically delineated and expert contours [3]. However AAM's can be successfully used to match an object with the statistically plausible shape, while it fails to recognize the object of interest that shows large deviation from shapes in the training set.

An novel method for myocardial border detection based on a virtual exploring robot was introduced in [4]. The robot is represented as a tricycle with a steering front wheel. It can automatically navigate through an environment (myocardium). Using the frontal and lateral range sensors the robot is able to detect the coordinates of obstacles (myocardial borders) and to plan a safe path towards the target avoiding these obstacles. The robot was used for short-axis contour extraction in MRI data sets and showed promising results.

In this paper we propose a new system for automatic delineation of long-axis contours for both two- and four-chamber orientations using a modified version of the robot. The navigational environment is constructed with an improved segmentation procedure based on both intensity and spatial information to reduce misclassification of pixels. The path planning is made more robust allowing the navigation in narrow regions with high curvature. The robot is made adaptive with respect to its length, speed and maximum turning angle, depending on the local LV geometry.

2 Methods

The global outline of the system is as follows. The multiphase images in two- or four-chamber view are automatically segmented by fitting finite gaussian mixture into the combined image histogram and applying spatial regularization in terms of Markov Random Field (MRF). The classified pixels are subsequently recombined to yield binary images consisting of the allowed navigational space or obstacles. For each binary image the initial start position and orientation for the robot as well as the target are determined. The navigational environment along with the robot's initial position and orientation provides the input for the next step. The myocardium is divided into four different segments. The robot is initialized once at the segment boundaries and cruises through the navigational environment until the end of the segment is reached. Depending on the local LV geometry the robot adapts its navigational parameters to prevent from getting stuck and to avoid concave regions such as the papillary muscles, collect information about the endocardial (endo) and epicardial (epi) boundary points. Finally, the detected edge segments are used to reconstruct the endo- and epi-contours.

2.1 Creation of Navigational Environment

The robot is designed to navigate through a binary environment. Therefore, the gray-level images must be converted into binary images consisting of allowed navigational space (myocardium) and obstacle (blood pool and air).

To automatically segment the image, the pixels are grouped into classes by modelling the pixel intensity distributions with three gaussian mixtures using Expectation-Maximization (EM) [5, 6] with greedy search heuristics [7]. During the expectation step the pixels are assigned to a class with the highest conditional probability based on previous estimates of the distribution parameters. In the maximization step the distribution parameters are updated to maximize the log-likelihood. These two steps are repeated until convergence.

With the estimated pixel intensity distributions the images can be segmented into blood, myocardium and air. However statistical segmentation based only on the pixel intensity may lead to misclassification due to the presence of noise. Therefore, the context of the pixel's neighborhood is taken into account and pixel intensity information is augmented with spatial regularization using MRF [8]. The amount of spatial regularization is limited to 8-connected neighborhood and pairwise interaction between two neighboring pixels. The spatial mixture model yields the final classification of the image and is schematically represented in Fig. 1.

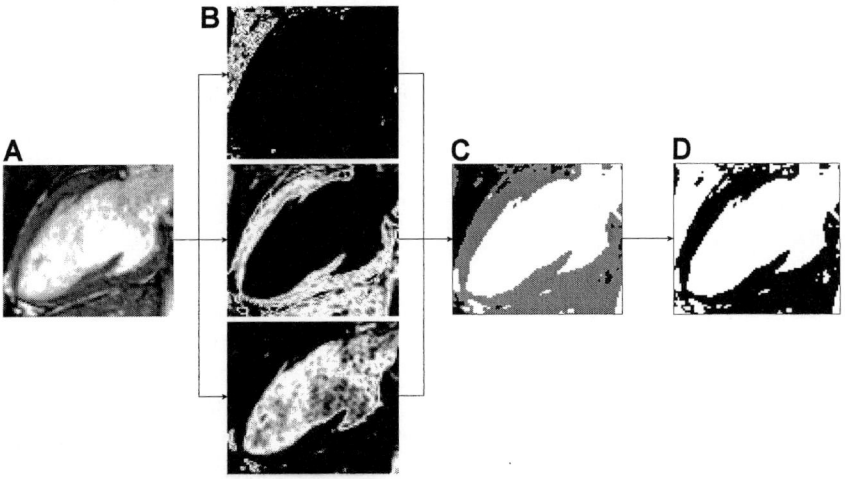

Fig. 1. Creation of navigational environment: (A) input image; (B) probability density maps for air (top), myocardium (middle), blood (bottom); (C) statistical segmentation with spatial regularization; (D) navigational environment - white corresponds to obstacle space, black - to allowed navigational space

2.2 Estimation of Initial Navigational Parameters

For the autonomous navigation in the allowed space the robot needs the starting position, starting orientation, and target. Based on the assumption that the

LV resembles a truncated ellipse at end-diastole, fitting an ellipse [9] into the myocardium gives a good approximation of the spatial orientation of the LV. Using the elliptic model the important anatomical landmarks can be localized and division of the heart into four non-overlapping segments can be done (Fig.2). The apex is located by finding the intersection point between the long axis of the ellipse and myocardium and yields the end point of the apical segments. The boundary between the basal and midventricular segments is put at the intersection of the myocardium with the line that is parallel to the ellipse short-axis and passes though the ellipse focal point, thus serves as the starting point for the robot navigation. The target is located in the ellipse center and the start orientations are the direction of the tangent line of the closest point on the ellipse. From the starting position the robot cruises through the allowed space, bounces from the endocardial LV border in attempt to reach the target and proceeds further until the end of the segments. It is initialized in each segment and navigates twice towards the mitral valve points and twice to the apex.

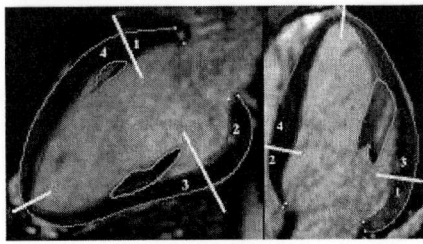

Fig. 2. Division of the LV into four segments. Segments 1 and 2 cover to the basal part, while segments 3 and 4 represent the midventricular and apical parts

2.3 Adaptive Robot Navigation

The robot is a tricycle with a front steering wheel as is illustrated in Fig.3A. The position and orientation of the robot are characterized by $p(x, y, \theta)$, where x and y are the coordinates of the front wheel, θ is the orientation of the robot with respect to the x-axis of the coordinate system, l is the length between the front and rear wheel axis, ϕ is the orientation of the front wheel. The robot moves with a constant speed v. Its motion obeys the following kinematics equations:

$$\dot{x} = v \cos(\phi) \cos(\theta); \quad \dot{y} = v \cos(\phi) \sin(\theta); \quad \dot{\theta} = \frac{v}{l} \sin(\phi) \qquad (1)$$

The robot is subject to the non-holonomic constraints: it can only move along a direction perpendicular to its rear wheel axis and its maximum front wheel angle is upper bounded.

To navigate through the allowed space towards the target the robot is equipped with range sensors of a limited length (Fig.3B). The frontal sensors have their origin at the front wheel and cast rays at the different angles with

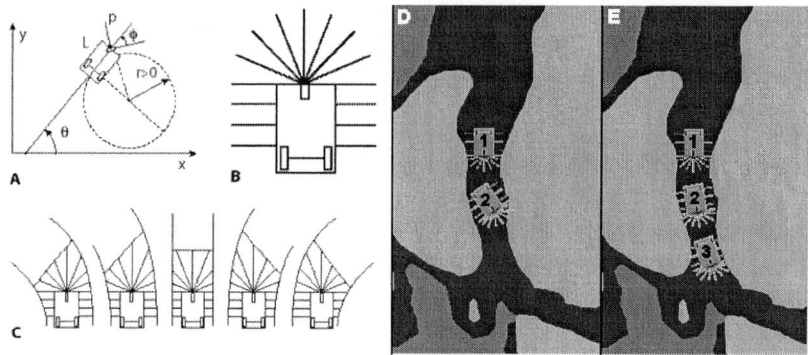

Fig. 3. The virtual mobile robot as a tricycle with the steering front wheel (A); Mounted frontal and lateral sensors (B); Navigational corridors corresponding to the different orientation of the front wheel (C); Robot navigation (D and E)

respect to the robot axis - a line connecting the front wheel and the middle point of the rear wheel axis. These sensors are primarily used to scan the environment in the vicinity of the current robot position, to detect the obstacles and to plan further movements towards the target. The lateral sensors, which are mounted on the left and right sides of the robot and perpendicular to the robot axis, are intended for inner and outer myocardial border detection.

The concept of the pre-computed corridors turns the robot navigation into a computationally efficient procedure. Depending on the orientation of the front wheel and using the kinematic equations (1) a trajectory of the robot or a corridor can be calculated. Provided that the corridor has a limited width the distances to its boundary for each range sensors can be computed and saved (Fig.3C). This procedure is repeated for each orientation of the front wheel. These pre-calculated distances are used during the navigation of the robot. While navigating, the robot scans the surrounding environment before moving forward. The range sensors give the information about the distances to the obstacle at the current robot's positions. These distances are compared to the ones of the pre-computed corridors and a decision about the corridor safety is made. The corridor is assumed to be safe if the distances to the obstacle at the current positions are bigger than the distances to the pre-computed corridor boundaries for each range sensor. Among possible safe corridors the one leading closest to the target is taken [4].

This approach to path planning was successfully applied for myocardial border delineation in short-axis images and is not robust enough for contour detection in the long-axis views. The LV geometry in two- and four-chamber projections consists of the regions with the variable myocardial width thickness and high curvature. Moreover, the information obtained from the discrete sensory system is not sufficient to reconstruct a precise topological structure of the allowed navigational space for such regions. This situation is illustrated in Fig.3D. From the initial position, marked as one, the robot moves to position two following the "closest safe corridor" strategy. However, being in position two the robot cannot enter the narrowing region in front of it without bumping into the obstacle and, therefore, cannot advance forward.

To solve the problem mentioned above the navigational strategy has been modified to assure its robustness in two- and four-chamber views. The basic idea is to find the safest path with a "look-ahead" procedure. The robot is allowed to explore the surroundings by advancing a number of steps in forward direction from its current position and choosing the longest and safest path leading closest to the target. After such a path has been found, the robot advances one step ahead. Fig.3E shows the path planning procedure using the "look-ahead" strategy. From position one the robot moves forward to position two, from which a safe move to position three is guaranteed due to better path planning. Therefore, the navigation through highly curved regions with the variable myocardial thickness is made more robust.

To account for the complex LV shape in two- and four-chamber views, the robot's navigational parameters are made regionally dependent. Three different regions, namely basal, midventricular, and apical, are commonly addressed in the medical literature and their geometric properties can be summarized as follows:

- Apical: high curvature with significantly changing myocardial thickness;
- Midventricular: low curvature with concave regions such as the papillary muscles;
- Basal: medium curvature of a constant myocardial thickness.

We exploit this knowledge to deduce estimates for the navigational parameters:

- Apical: The robot has to be highly mobile, which is guaranteed by its short length and slow speed.
- Midventricular: The robot is made long and fast enough to avoid the concave regions.
- Basal: The robot has a medium velocity and is made long enough to stop at the mitral valve points without a possibility of turning around.

The robustness of the robot navigation with respect to the navigational parameters has been tested in a pilot study performed on a dataset acquired from several subjects. The initial guesses for the parameters were chosen by taking into account the aforementioned considerations and analyzing the global geometric properties such as the size and maximal curvature of each cardiac segment. The final parameters, shown in Tab.1, were derived from the initial guesses by brute-force optimization.

Having been safely initialized inside the allowed navigational space, the robot autonomously explores the structure of the each myocardial segment. In attempt to reach the target, which set in the middle of the LV cavity, the robot bounces from the obstacle, formed by the LV blood pool, and proceeds further along the endocardial border. As the target is located inside the obstacle space and could not be possibly reached, the robot eventually arrives to the end of the segment and stops. During the trip the robot uses the lateral sensors to detect

Table 1. Robot's navigational parameters

Parameter	Basal Segment	Midventricular Segment	Apical Segment
Speed (mm/step)	4.17	6.59	2.78
Length (mm)	13.9	13.9	6.95
Number of corridors	15	21	19
Sensor length endo (mm)	6.95	2.78	6.95
Max steps looked ahead	1	2	2

the presence or absence of the myocardial borders (i.e. the transactions between the allowed navigational space and obstacle), and memories the coordinates of the candidate border points. This navigational procedure is repeated for all four segments, resulting in complete exploration of the left ventricle in a two- or four-chamber view.

2.4 Contour Reconstruction

The final step in automated contour detection is collecting the contour segments found by the robot and merge those together in a single contour. Reconstruction of the endo-contour is relatively straightforward. The papillary muscles are already removed due to the kinematic constraints of the robot and short lateral sensors used for endo-cardiac border detection. Therefore, connecting the detected points is sufficient to reconstruct the endo-contour. Reconstruction of the epi-contour is more challenging (Fig.4). Firstly, the outliers (falsely detected boundaries) has to be removed. To achieve this, goal prior knowledge about the myocardial thickness is used. The reconstructed endo-contour provides the reference to approximate the wall thickness. The distance between each epicardial candidate point and the reference is measured. All points, for which the calculated wall thickness is larger than a predefined threshold of 30mm, are deleted. Secondly, an additional step is required to approximate possibly missing segments caused by the absence of the myocardial border in the regions where the

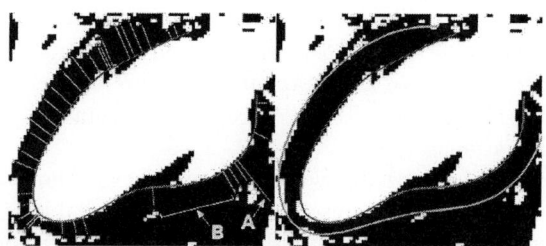

Fig. 4. Reconstruction of endo- and epi-contours (left). Prior knowledge about the myocardial wall thickness is used to remove outliers (A). The missing segments (B) are interpolated using non-uniform cubic splines. Reconstructed contours (right)

heart is adjacent to the organs with the same grey-level intensities (i.e. liver). To restore the missing information interpolation using the non-uniform cubic splines [10] is performed.

3 Results

In-vivo cardiac long-axis images were obtained at Leiden University Medical Center from 11 subjects using a Philips Gyroscan Intera 1.5T MRI scanner. Balanced-FFE protocol with prospective VCG and respiratory triggering was utilized to acquire breath-hold cardiac images in two-chamber and four-chamber views. Thirty phases provided the complete coverage of the cardiac cycle resulting in a total of 660 cardiac datasets. The field of view and slice thickness were equal to 350 mm and 8 mm, respectively. The reconstruction matrix of 256x256 was used. The total acquisition time did not exceed 10 minutes.

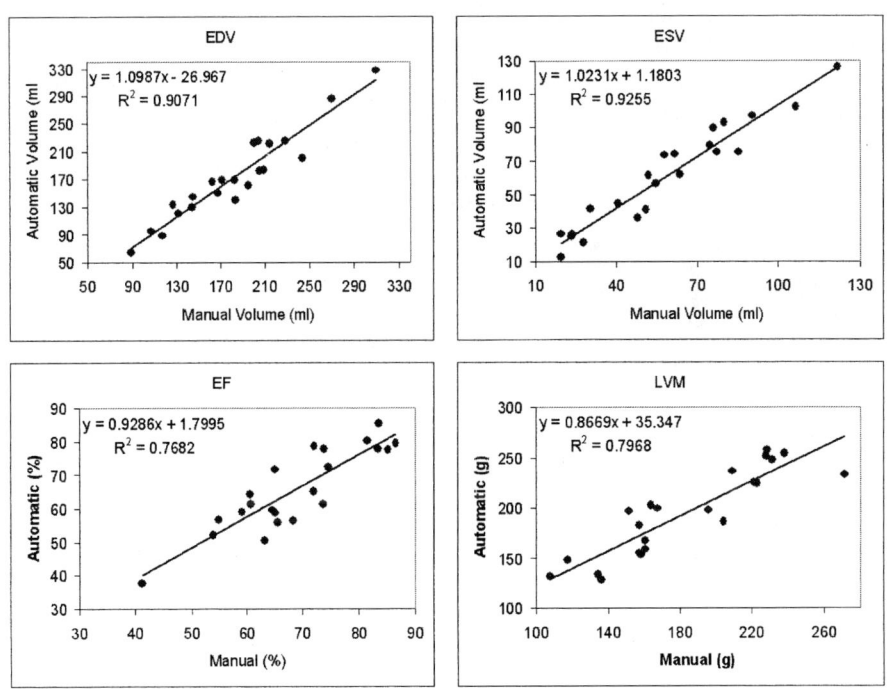

Fig. 5. Results of the statistical comparison between global LV function for manually and automatically segmented images

To assess the algorithm's performance the global LV function was computed for manually and automatically segmented images using commonly used area-length methods. The CMR measurements from two- and four-chamber views

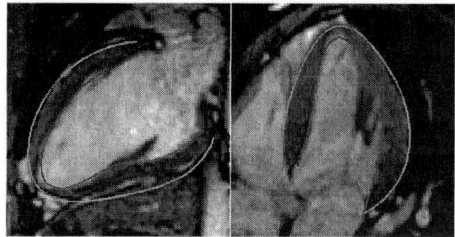

Fig. 6. Automatically detected myocardial endocardial (red) and epicardial (green) boundaries for two- (left) and four-chamber (right) view images

were polled together. The regression analysis was used to estimate the strength and direction of a linear relationship between manual and automatic measurements and graphically summarized in Fig.5. The paired t-test revealed statistically indistinguishable differences at 5% significance level between the manual and automatic segmentation for all parameters (two-chamber: EDV p=0.26; ESV p=0.09; EF p=0.1; four-chamber: EDV p=0.17; ESV p=0.34; LVM p=0.16) but two-chamber LVM (p=0.01) and four-chamber EF (p=0.04). The results of automatic segmentation are shown in Fig.6.

4 Discussion and Conclusions

Creation of the navigational environment required the classification of the input image into three profoundly distinct classes: air, blood, and myocardium. Although this assumption may not be necessary true for cardiac images where other anatomical structures, such as myocardial fat, are present. Nevertheless, due to the use of EM algorithm with greedy search heuristics this requirement can be easily incorporated by adding one extra class into the statistical segmentation scheme. However decision about the number of distinguishable classes presented in input images remains a challenging problem.

To take into account the complex LV geometry, three different sets of the navigational parameters were utilized depending on the LV region. However, the requirements for the robot navigation in the basal and midventricular regions can be combined together, resulting in only two sets of the navigational parameters. Further simplification of our method may be achieved by letting the robot navigate in only two segments (i.e. posterior and anterior myocardial walls starting from the mitral points towards the apex). Implementation of the aforementioned improvements would require accurate tracing of the mitral points in all phases because of the prominent cardiac contraction.

A better path planning procedure based on the "look-ahead" strategy resulted in a more robust navigation of the robot. A number of different paths is tried to determine the safest route, before the robot advances only one step forward. This results in increased computational demands and a slightly lower performance. An alternative approach would involve the map-based navigation using the simulated topological maps of the navigational environment [11]. In this case the neighborhood around the robot's current positions would be matched against templates of the environment from the database and the precalculated path for the chosen template can be undertaken. However, it remains debatable whether the map-based navigation will be more computationally efficient.

Our validation study was carried out only on a group of healthy subjects. Some assumptions (i.e. the predefined myocardial wall thickness in the contour reconstruction phase) may not be valid for abnormal hypertrophic hearts. Hence, further validation of our method in patients is desirable.

In this paper an unorthodox method for the myocardial border detection in long-axis views using the adaptive exploring robot was presented. Using this approach a reliable and consistent segmentation of the myocardial boundary can be achieved. A clinical validation on a group of healthy subjects showed good agreement between the global LV function computed from manually and automatically segmented images.

Acknowledgement

Support of this research by the grant from Dutch Foundation for Technical Sciences (STW Project LPG 5651) is greatly appreciated. We express our gratitude to Dr. B. P. F. Lelieveldt for proof reading the manuscript.

References

1. A. Yezzi, S. Kichenassamy, A. Kumar, P. Olver, and A. Tannenbaum, "A geometric snake model for segmentation in medical imagery," *IEEE Trans. on Pattern Analysis and Machine Intelligence*, vol. 6, no. 2, pp. 199–209, 1997.
2. T. F. Cootes, G.J.Edwards, and C. J. Taylor, "Active appearance models," *IEEE Trans. on Pattern Analysis and Machine Intelligence*, vol. 23, no. 6, pp. 681–685, 2001.
3. C. R. Oost, B. P. F. Lelieveldt, M. Uzumcu, H. Lamb, J. H. C. Reiber, and M. Sonka, "Multi-view active appearance models: Application to x-ray lv angiography and cardiac mri," in *LNCS 2732*, 2003, pp. 234–245.
4. F. Admiraal-Behloul, B. P. F. Lelieveldt, L. Ferrarini, H. Olofsen, R. J. van der Geest, and J. H. C. Reiber, "A virtual exploring mobile robot for left ventricle contour tracking," in *Proc. IJCNN*, 2004, vol. 1, pp. 333–338.
5. A. P. Dempster, N. M. Laird, and D. B. Rubin, "Maximum likelihood from incomplete data via the em algorithm," *Journal of the Royal Statistical Society (B)*, vol. 39, no. 1, pp. 1–38, 1977.
6. X. Ye and J. A. Noble, "High resolution segmentation of mr images of mouse heart chambers based on a partial-pixel effect and em algorithm," in *Proc. ISBI*, 2002, pp. 257– 260.
7. J. J. Verbeek, N. Vlassis, and B. Krose, "Efficient greedy learning of gaussian mixture models," *Neural Computation*, vol. 15, no. 2, pp. 469–485, 2003.
8. S. Geman and D. Geman, "Stochastic relaxation, gibbs distributions, and the bayesian restoration of images," *IEEE Trans. On Pattern Analysis and Machine Intelligence*, vol. 6, no. 6, pp. 721–741, 1984.

9. V. Vezhnevets, "Method for localization of human faces in color based face detectors and trackers," in *Proc. ICDIPCES*, 2002, pp. 51–56.
10. M. Unser, "Splines - a perfect fit for signal/image processing," *IEEE Signal Processing Magazine*, vol. 16, no. 6, pp. 22–38, 1999.
11. G. N. DeSouza and A. C. Kak, "Vision for mobile robot navigation: A survey," *IEEE Trans. on Pattern Analysis and Machine Intelligence*, vol. 24, no. 2, pp. 237–267, 2002.

A Deterministic-Statistic Adventitia Detection in IVUS Images

Debora Gil[1], Aura Hernandez[1], Antoni Carol[2], Oriol Rodriguez[2], and Petia Radeva[1]

[1] Computer Vision Center, Universitat Autonoma de Barcelona,
Bellaterra, Barcelona, Spain
debora@cvc.uab.es
http://www.cvc.uab.es/ debora/
[2] Hospital Universitari Germans Trias i Pujol, Badalona, Spain

Abstract. Plaque analysis in IVUS planes needs accurate intima and adventitia models. Large variety in adventitia descriptors difficulties its detection and motivates using a classification strategy for selecting points on the structure. Whatever the set of descriptors used, the selection stage suffers from fake responses due to noise and uncompleted true curves. In order to smooth background noise while strengthening responses, we apply a restricted anisotropic filter that homogenizes grey levels along the image significant structures. Candidate points are extracted by means of a simple semi supervised adaptive classification of the filtered image response to edge and calcium detectors. The final model is obtained by interpolating the former line segments with an anisotropic contour closing technique based on functional extension principles.

1 Introduction

IVUS clinical interest feeds development of image processing techniques addressing detection of arterial structures [1], [2], such as lumen/intima segmentation or plaque characterization. However, although adventitia modelling is crucial for a reliable plaque quantification, the topic has been hardly approached. Regardless of low quality in IVUS images, adventitia detection adds the difficulty of a large variety of descriptors, which include image edges points of maximum variance (calcium) and tissue region segmentation. Deterministic strategies presented in previous works on adventitia detection exclusively basing on contour extraction are not reliable enough and need of either manual intervention [6] or laborious special treatment of sequences [7]. We argue that a robust adventitia segmenting algorithm should rely on learning strategies.

In this paper we address adventitia detection in two stages: a statistical extraction of points laying on the adventitia and a deterministic recovery of a closed model of the extracted points. At the first step, we define the quantities that best characterize the adventitia, that is, in the framework of classification, we should determine the optimal feature space of image descriptors. In such representation space, the adventitia should lie on a region isolated from other image structures response, so that the problem of point selection reduces to determining the borders of such regions. Within this framework, there are several point selection strategies. On one side, we have statistical approaches [13] searching for a criterion to discriminate the target object representation in the feature space. On the other side, we apply a deterministic criterion of image smoothness to choose pixels achieving extreme values of the functions (filters) that determine the feature space. Still, even in this case, thresholding values should take into account the probability distribution of the image response to the describing filters. Therefore, whatever the decision criterion we adopt, the selection step nature is essentially statistical. Because the selected set of points is prone to be unconnected, contour completion is a compulsory second step. Usual techniques rely on deterministic principles: active models (parametric [4], geodesic [5] or region-based [14]) solve an energy minimizing problem and contour closing techniques [8] base on interpolation/functional extension methods.

The deterministic-statistical strategy for adventitia detection we propose is the following. For a better handling of the classifying problem, our feature space reduce to adventitia and calcium detectors, the latter to discard sectors with ambiguous information. In order to enhance significant structures while removing noise and texture response, we use a Restricted Anisotropic Diffusion [9] (RAD). For adventitia points selection, we search for the feature space partition (thresholds) achieving the best classification rate for a training set. For segment closing we suggest using an Anisotropic Contour Closing (ACC) [8] that bases on image local geometry for curve segment interpolation. Parametric B-spline snakes yield the final compact explicit model.

The topics are presented as follows. In Section 2 we thoroughly describe the way adventitia points are selected. Explanations about the main detection steps are given in Section 3. Section 4 is devoted to validation of the method and Section 5 to conclusions and further research.

2 A Deterministic Statistical Strategy

There are two main points in the segmentation process:

2.1 Statistical Selection of Adventitia Points

Since in an IVUS plane, the adventitia is a circular-wise structure (fig.1 (a)), we work in polar coordinates (see Section 3.1 for details). Let $AdvPol(i,j)$ denote images in polar form (fig.1 (b)) with radius $i = 1, \ldots, R_{max}$, and angle $j = 1, \ldots, 360$. The selection stage summarizes in the next steps:

Set of Descriptors. The feature space for adventitia detection we propose reduces to the following two characteristics:

1. **Horizontal Edges (X)**
   Since in the coordinate system chosen (fig. 1(b)), the adventitia layer is an horizontal dark line, horizontal edges constitute our main descriptor (see fig.

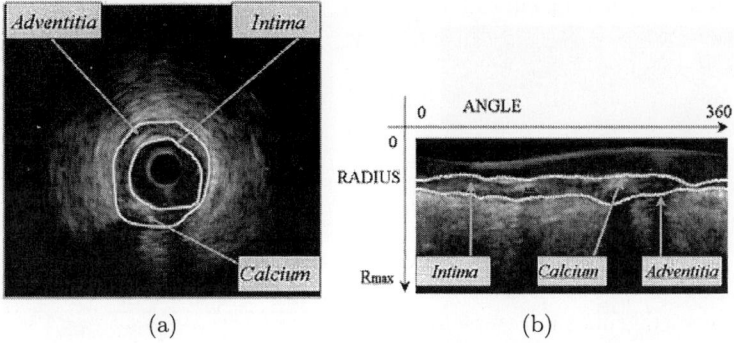

Fig. 1. IVUS images in cartesian (a) and polar (b) coordinates

2). Edges are computed by convolving the image with the y-partial derivative of a 2 dimensional gaussian kernel of variance ρ:

$$X = e_y(i,j) = g_y * AdvPol \quad \text{for} \quad g_y(x,y) = -\frac{y}{2\pi\rho^4}e^{-(x^2+y^2)/(2*\rho^2)}$$

Although intima and adventitia correspond to negative edges (fig.2), with a suitable (intima removing) strategy [10], this detector achieves optimal accuracy in the absence of calcium. Because at angles of calcium the adventitia does not appear and the detection is misled towards the intima, we discard those sectors. We base on calcium outstanding brightness to detect it by means of:

2. **Radial Standard Deviation (Y)**

   Striking brightness corresponds to an outlier of the pixel gray value in the radial distribution. We measure it by means of the difference between the pixel gray value and the radial mean. For each pixel (i,j), we define it as:

$$\sigma(i,j) = (AdvPol(i,j) - \nu(j))^2$$

where $\nu(j)$ is the radial (i.e. column-wise) mean of the polar image:

$$\nu(j) = \frac{1}{R_{max}} \sum_{i=1}^{i=R_{max}} AdvPol(i,j)$$

Point clouds in fig. 2(b) show the feature space corresponding to the images in fig. 2(a). Adventitia corresponds to large negative X values an a small Y negative range, while calcium yields in the extreme positive values of the pair (X,Y).

Statistical Thresholding. In a classification framework, determining the threshold values of the pair (X,Y) that characterize each structure reduces to finding a partition of the feature space separating adventitia and calcium from other vessel structures. Supervised techniques learn regions enclosing most of the

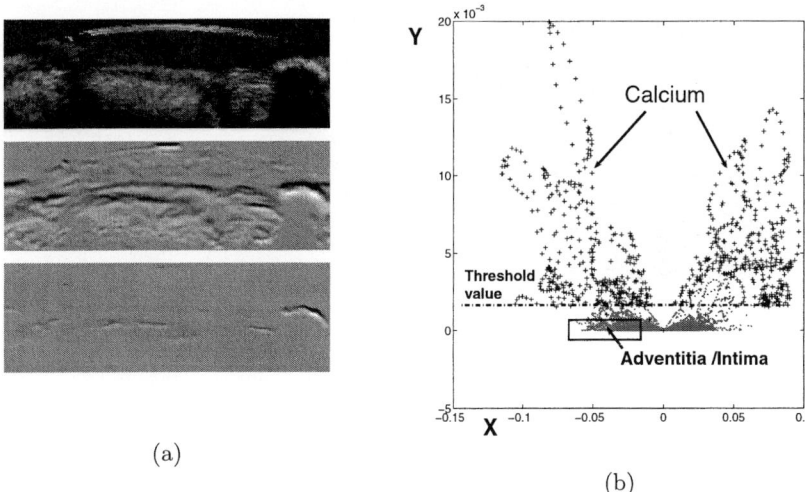

Fig. 2. Feature Space of Vessel Structures: X and Y responses, (a), and (X,Y) plane,(b)

training set, while 'ad-hoc' unsupervised clustering bases on the class instances structure given in a particular image. Although classical strategies exclusively follow either a supervised or an unsupervised approach, we adopt an adaptive criterion since mixed approaches [3] have proven to work better in IVUS images.

Because, in the feature space proposed, points of calcium correspond to extreme values, a supervised approach based on the Mahalanobis distance would work fine. However, by their spatial distribution, we have further reduced the decision criterion to choosing the threshold for Y values achieving the best compromise between true and fake classifications. On the other hand, if we consider all training images as a whole, adventitia points response presents a within class variability significant enough as to discard a fixed supervised criterion. By using a gaussian mixture [13] to model the training set density function, we have a misclassification error of 47.09% of fake detections for a test set. An unsupervised clustering is not sensible either since low dimensionality of the feature space introduces an overlapping between adventitia and other structures. What we propose is using an image sensitive classification based on searching for radial outliers in X negative values. That is, the thresholding value corresponds to the 5/6% percentile of the X values along each angle (columns in the polar image). This simple image adaptive criterion drops misclassification to 42.18% false positives corresponding to points on the intima layer.

2.2 A Restricted Diffusion Determined by Image Geometry

In order to smooth textures and strengthen response to the describing functions given in (2.1), we evolve the polar image under the following structure preserving filtering:

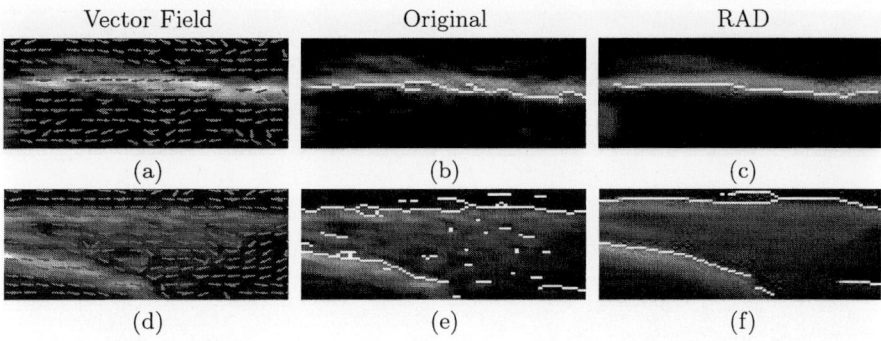

Fig. 3. RAD smoothing for calcium (1st row) and adventitia (2nd row)

Restricted Anisotropic Diffusion (RAD). Most filtering techniques based on image gray level modification [11] use the heat diffusion equation:

$$u_t(x,y,t) = \text{div}(J\nabla u)$$

The time dependent function u is the family of smoothed images and J is a 2-dimensional metric (i.e. an ellipse) that locally describes the way gray levels distribute. The diffusion tensor J is thoroughly described by means of its eigenvectors (ξ, $\eta = \xi^\perp$) and eigenvalues (λ_1, λ_2). If the latter ones are strictly positive, gray values spread on the whole image plane and the family u converges to a constant image. On the other side if we degenerate J (i.e. we admit null eigenvalues), then the final image [9] is a collection of curves of uniform gray level. Smoothing effects depend on the suitable choice of the eigenvector of positive eigenvalue. Let us consider a metric $\tilde{J}$ with eigenvalues $\lambda_1 = 1$ and $\lambda_2 = 0$, and ξ the eigenvector of minimum eigenvalue of the Structure Tensor [12]. If u_0 is the image to be denoised, then the Restricted Heat Diffusion we suggest is given by:

$$u_t = \text{div}(\tilde{J}\nabla u) \text{ with } u(x,y,0) = u_0(x,y) \quad (1)$$

Figure 3 illustrates the way restricted diffusion works. Around the image significant structures (calcium in fig.3(a)), ξ represents the tangent space to a closed model of such structures. Meanwhile at noisy areas (textured tissue in fig.3(d)), it is an irregular vector with random orientation. The result is that gray levels homogenize along image regular level sets and solutions to (1) converge to a smooth image that enhances the main features of the original image, in the sense that their response to standard detectors is uniform. Figure 3 shows the improvement of calcium (first row) and edges (second row) responses after applying RAD. Background spurious edges due to noise in fig.3(e) have been removed, in a similar fashion a gaussian smoothing would do, while edges corresponding to the vessel adventitia and calcium are continuous closed curves in the RAD images of fig.3(c),(f).

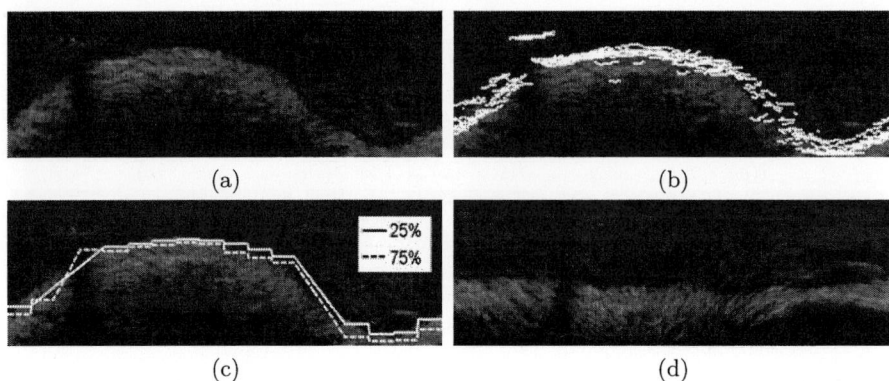

Fig. 4. Straighten adventitia layer procedure

Anisotropic Contour Closing (ACC). Heat diffusion has also the property of smoothly extending a function defined on a curve in the plane, provided that boundary conditions are changed to Dirichlet [15]. By using restricted heat operators this property can be used to complete unconnected contours [8] as follows. Let γ be the set of points to connect, χ_γ its characteristic function (a mask) and define $\tilde{J}$ as in RAD, then the extension process given by:

$$u_t = \text{div}(\tilde{J}\nabla u) \text{ with } u_{|\gamma} = u_0 \qquad (2)$$

converges to a closed model of γ. Intuitively, we are integrating the vector field ξ, that is, we are interpolating the unconnected curve segments along it. This fact not only ensures convergence to a closed model, but also yields closures more accurate than other interpolating techniques (such as geodesic snakes [5]).

3 Adventitia Modelling Steps

The characterization strategy of Section 2 serves to model the adventitia layer in the following three step procedure.

3.1 Polar Coordinates Origin

Polar coordinates with a fixed origin at the center of the cartesian image present two main artifacts produced by cardiac movement and the artery geometry. In cartesian coordinates, heart movement induces a translation followed by a rotation. This motion converts into an angular translation (rotation) and a radial dynamic wave (translation). The latter is a main artifact for the set of descriptors given in Section 2.1 and it is removed by taking as origin of coordinates the mass center of the image (fig. 4(a)). In such coordinates, the adventitia still presents a slight static wavy shape because image mass centers do not coincide with geometric centers. We correct this deformation by means of a set of points extracted using the strategy described in Section 2. The impact of noise is minimized by considering the average of the energy e_y in the sequence temporal direction (fig. 4(b)). In order to endow further continuity to the extracted edges, we use the statistical distribution of their position in angular sectors of the cartesian image. For each sector we only consider edge points within the central percentile computed for a given number of frames (fig. 4(c)). The mass center of the cartesian transform of the former radial values serves as geometric center of the adventitia layer and is the origin of our polar transform. Fig. 4 (d) shows the final polar coordinates.

3.2 Adventitia Selection

The classification of the filtered images given by RAD yields a calcium and adventitia masks. Small structures in the adventitia image are removed by applying a length filtering, so that only segments of length above the 75% percentile are kept. In order to remove intima points, we consider that an edge connected component is on the adventitia layer if it corresponds to an edge of maximum radius in a longitudinal cut of the sequence.

3.3 Adventitia Closing

We split interpolation of the selected curve segments into computation of an implicit closed representation and explicit encoding with parametric B-splines.

For adventitia completion we will use ACC with the Structure Tensor defining the vector ξ computed over the edge map used in the selection step. In order to obtain models as accurate as possible, the vector ξ is weighted by the coherence of the Structure Tensor:

$$coh = \frac{(\lambda_1 + \lambda_2)^2}{(\lambda_1 - \lambda_2)^2}$$

where $\lambda_1 \geq \lambda_2$ are the eigenvalues of the tensor. At regions where ξ is a continuous vector, λ_2 is closed to zero, so coh is maximum, meanwhile, at noisy areas, since ξ is randomly oriented, λ_1 compares to λ_2 and $coh \sim 0$. In this manner we avoid wrong interpolations at side branches and sensor shadows sectors.

The final model discards angles presenting response to calcium and uses B-splines to smoothly interpolate the adventitia at side branches and calcium sectors.

4 Results

Objective quantitative validation of the method has been based on the following assessment protocol. A total number of 3300 frames extracted from 9 different patients, including 4 sequences with calcium, have been analyzed. The measures used to quantify accuracy of the automated detections are the mean and maximum distance error (in mm) and area differences (in percentages) between our model and an expert manual segmentation. The sequences have been manually segmented by 3 different physicians every 10 frames in order to analyze inter-observer variation. Figure 5 shows adventitia snake models for soft plaque (fig.5 (a), (d)), calcium (fig.5 (b), (e)) and at a side branch (fig.5 (c), (f)).

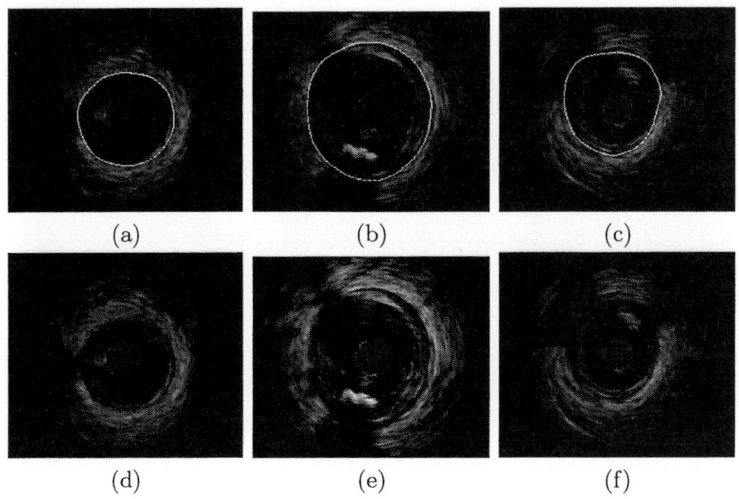

Fig. 5. Segmentations (a), (b), (c) for different plaque (d), (e) and at a side branch (f)

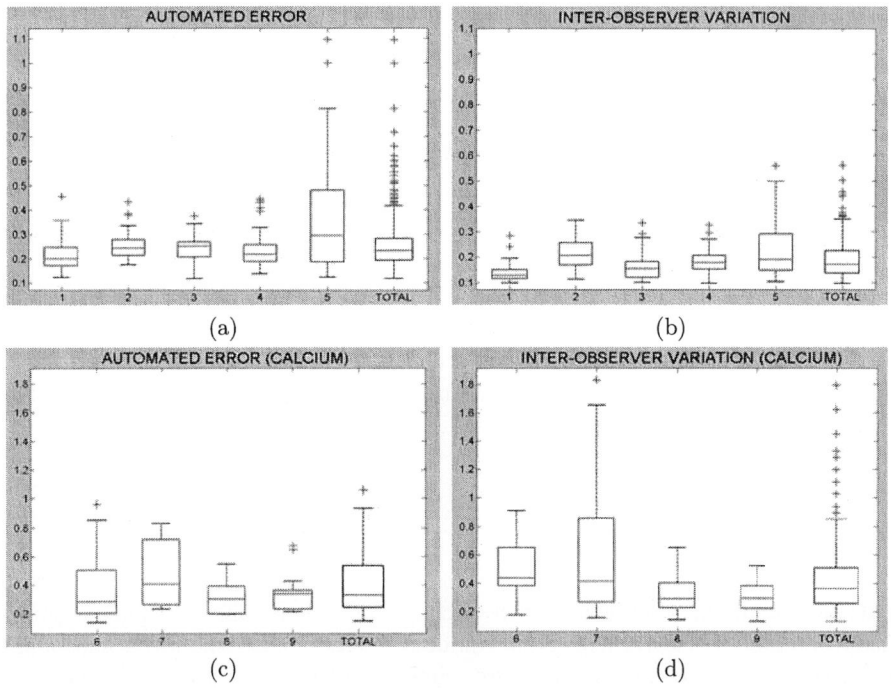

Fig. 6. Whisker Boxes for Automated Error and Inter-Observer Variation

4.1 Statistical Measurements

Figure 6 shows whisker boxes for mean distance absolute errors and inter-observer variations for soft plaque (1st row) and calcium segments (2nd row). They summarize the statistics for each patient and for the total population at the last box on the right. An analysis of the whisker boxes reflects robustness of segmentations: the smaller the boxes are, the more reliable the method is. First note that, lack of reliable information at large angular sectors, significantly increases errors variability in calcified segments (fig.6(c), (d)), especially for manual segmentations, due to the subjectivity of manually traced curves. Still, our strategy is highly stable as, in most cases, graphics present a smaller variability than manual models. Only subject 5 has a large variability, but, comparing, with manual errors (fig.6(b)), we observe that this subject is also the one presenting the largest box. Average relative and absolute errors in distances and percentage of area difference for the total number of patients (excluding the outlier case 5) are summarized in table 1. Mean distances compare to inter-observer variability and maximums, although above it, are less than 1% of the vessel radius.

Table 1. Statistics on Errors

	Max. Dist. (mm)		Mean Dist. (mm)		Area Dif. (%)
	Abs. Error	Rel. Error (%)	Abs. Error	Rel. Error (%)	
INT-OBS.	0.560 ± 0.326	0.5 ± 0.28	0.284 ± 0.222	0.247 ± 0.203	8.294 ± 3.914
AUT.	0.655 ± 0.349	0.619 ± 0.4	0.273 ± 0.131	0.243 ± 0.120	10.287 ± 4.369

5 Conclusions

Using an integrated approach of statistic classification and anisotropic filtering to detect the adventitia layer presented in this paper is a new trend in medical imaging with a straightforward clinical application to plaque area and vessel diameter measurements. The strategy proposed combines statistical classification and deterministic energy based techniques into a two step algorithm. On a first stage, a set of adventitia and calcium descriptors are proposed as a feature space. A supervised analysis of such 2-dimensional space serves to determine those regions enclosing target points. Feature extraction is optimized by applying a response regulating restricted diffusion operator to polar IVUS images. The second step involves computation of a closed model of the selected curve segments. An anisotropic contour closing is used for obtaining an implicit representation that captures all geometric features.

Statistics show that automated errors are comparable to inter-observer variability as far as adventitia can be detected by means of the proposed descriptors. Since accuracy exclusively relies on such features, our future research will focus on adding some a priori knowledge on vessel tissue.

References

1. McInerney, T.,Terzopoulos, D.: Deformable models in medical images analyis: a survey, Medical Image Analysis, **1** (1996) 91–108
2. Zhang, X., Sonka, M.: Tissue characterization in intravascular ultrasound images, IEEE Trans. on Medical Imaging, **17** (1998) 889–899
3. Pujol, O., Radeva, P.: Supervised Texture classification for Intravascular Tisue Characterization, Handbook of Medical Imaging, Kluwer Academic/Plenum Pub. (2004)
4. M.Kass, A.Witkin and D.Terzopoulos: Snakes: Active Contour Models, Int. Journal of Computer Vision, **1** (1987) 321–331
5. Caselles, V., Kimmel, R., Sapiro, G.: Geodesic Active Contours, Int. J. Comp. Vision, **22 (1)** (1997) 61–79
6. Klingensmith, JD., Shekhar, R., Vince, DG.: Evaluation of three-dimensional segmentation algorithms for the identification of luminal and Medial-adventitial borders in intravascular ultrasound, IEEE Trans. on Med. Imag., **19(10)**, (2000) 996–1011
7. von Birgelen, C., de Vrey, E. A., Mintz, G. S., Nicosia, A., Bruining, N., Li, W., Slager, C. J. , Roelandt, J. R. T. C., Serruys, P. W. and de Feyter, P. J.: ECG-gated three-dimensional intravascular ultrasound: Feasibility and reproducibility of the automated analysis of coronary lumen and atherosclerotic plaque dimensions in humans, Circulation **96**, (1997) 2944–2952
8. Gil, D., Radeva, P.: Extending Anisotropic Operators to Recover Smooth Shapes, Comp. Vis. Imag. Unders., (in press)
9. Gil, D.: Geometric Differential Operators for Shape Modelling, PhD Tesis, Universitat Autonoma de Barcelona, (2004) (available at http://www.cvc.uab.es/ debora/)
10. Hernandez, A., Gil, D., Radeva, P., E. Nofrerias,: Anisotropic Processing of Image Structures for Adventitia Detection in IVUS Images, Proc CiC, (2004)
11. Weickert, J.:A Review of Nonlinear Diffusion Filtering, Scale-Space Theory in Computer Vision, Lecture Notes in Comp. Science, Springer-Verlag, (1997)
12. Jähne, B.: Spatio-temporal image processing, Lect. Notes in Comp. Science, Springer, (1993)
13. Duda, R., Hart, P.: Pattern Classification, Wiley-Interscience, (2001)
14. Paragios, N., Deriche., R.: Geodesic Active Contours for Supervised Texture Segmentation, Proc. of Comp. Vis. and Pat. Rec. **2**, (1999), 422–427
15. Evans, L.C.: Partial Differential Equations, Berkeley Math. Lect. Notes (1993)

Trajectory Planning Applied to the Estimation of Cardiac Activation Circuits

Lorena González[1], Jerónimo J. Rubio[2], Enrique Baeyens[3], Juan C. Fraile[3], and Jose R. Perán[3]

[1] CARTIF, División de Ingeniería Biomédica,
Parque Tecnológico de Boecillo Parc. 205, Boecillo, Valladolid, Spain
lorgon@cartif.es
[2] Instituto de Ciencias del Corazon,
Hospital Clínico Universitario, Valladolid, Spain
jrubio@vitanet.nu
[3] Universidad de Valladolid,
Depto. de Ingeniería de Sistemas y Automática, Valladolid, Spain
{enrbae, jcfraile, peran}@eis.uva.es

Abstract. A procedure for helping the professional in electrophysiology in performing catheter ablation as a definitive treatment of certain types of arrythmia is presented here. This procedure uses trajectory planning techniques that have been developed in the robotics field. Starting off from signals obtained in an electrophysiological study of a patient, an electrical model of the heart with zones of different propagation properties is generated. Trajectory planning techniques are used to obtain the qualitative behavior of the heart under different types of arrythmia. A good point for ablation is computed as one that interrupts the trajectory that is sustaining the arrythmia.

1 Introduction

The upheavals of the heart rate are cause of 50% of the cardiac related deaths [1]. In addition, most of them are produced by sudden death. There are more than 300,000 sudden deaths per year only in the US and 90% are caused by cardiac arrhythmia. Cardiac arrhythmia is any alteration of the heart rate including changes of the cardiac characteristics or inadequate variations of the heart frequency. Unfortunately, the available therapeutic arsenal for cardiac arrhythmia is still relatively limited [2].

The evidence that arrhythmia needs an anatomo-electrophysiological substratum (forced conduction circuits) for its maintenance motivated the development of catheter ablation techniques. This is the only form of definitive treatment. Its advantages are efficiency, safeness, practically null mortality, low cost and almost absence of counterindications [3].

Notwithstanding, this technique has several fundamental limitations. The location of isthmuses or zones of forced conduction requires a long time electrophysiological study (EPS) that has to be performed by an experienced professional. The success of the treatment depends on the electrophysiologist's skills. In some cases, the EEF stimulation to find these zones can represent a potential danger for the patient's life. On the other hand, presently there is no tool in the market that locates the optimal place for performing ablation.

The research presented here is motivated by this necessity and aims to the improvement of the conventional cardiac arrythmia treatments by catheter ablation.

This paper presents the outline of a new procedure for estimation of cardiac activation trajectories and location of the optimal point for ablation in a virtual map previously developed and adjusted from EPS signals.

2 Frame of Application: Electrophysiological Studies

Electrophysiological studies are carried out in a cardiac catheterization laboratory. They are based in obtaining intracavitary electrograms, with the purpose of studying its cardiac activation sequence in basal conditions, during different arrhythmias as a response to a programmed heart stimulation. The indications for the accomplishment of these studies are in a constant evolution [4].

The general procedure consists of introducing electrocatheters through the vessels of the leg and carry them to the heart. The guidance is accomplished by means of fluoroscopic control. An electrocatheter registers the electrical impulses of the heart allowing to obtain a map of the electrical conduction system.

The correspondence of cardiac signals is the technique by which the signals gathered from the multiple locations of the heart are drawn as a function of time in an integrated way [5]. It requires the determination of the local activation time for each electrode and the creation of activation maps providing space models of the activation sequence. It is used to unveil the arrythmia mechanisms, its prognosis and to delimit the structures implied in its maintenance with the purpose of eliminating it (or at least modifying it) by ablation. Therefore, this technique tries to locate the origin of the arrhythmia, *i.e.* the point that has the precocious electrical activity. Recent advances in this field are new correspondence systems that do not need fluoroscopy to guide catheters [2].

The treatment of cardiac arrhythmias has evolved quickly during the last decade [6]. At the beginning of eighties, the development of invasive electrophysiology techniques as ablation revolutionized the treatment of many types of arrythmia. Ablation consists of producing a controlled injury in the vital zone for the initiation and/or the maintenance of the arrhythmia. The objective is to burn fibers and consequently suppress the electrical conduction in that zone. The controlled injury produces that an essential part of the electrical circuit responsible for the maintenance of the arrhythmia is eliminated and this avoids the initiation and/or the sustainment of it. The injury area produced, depends on the size of the electrode, the time and power of application, and the type of tissue. The development of these techniques allowed to introduce the only really curative treatment for many types of arrythmia. This has been one of the greatest electrophysiology advances.

The application of this treatment could be greatly improved with the aid of a tool able to locate the circuits without inducing a tachycardia. This tool should also minimize the registration time and locate the best point for ablation avoiding the detection of false places that are not essential part of the circuit. A computer tool supporting these features would be very valuable for the electrophysiologist and would make this treatment more reliable, simple, efficient and economic, with a minimal risk for the patient.

3   Hypothesis

The proposed hypothesis is that a model based on cardiac potential maps can be developed, where different conduction properties are given to distinct zones according to signals obtained in conventional EPS. Using this model, it is possible to apply trajectory planning ideas developed in the field of robotics [7] in order to, first elaborate a procedure that simulates the feasible propagation pathways from one point to another on the surface model; and second, search for the optimal point that would interrupt some specific trajectories. This point is selected by establishing previous conditions according to the mechanisms that originate or maintain the arrhythmia and also, according to the morphological characteristics of the involved conduction areas.

4   Methodology and Implementation

In order to validate the hypothesis, the following steps are proposed:

- Development of a basic cardiac conduction model running on a PC that integrates cardiac geometry information about origin, characteristics and propagation of the associated electrical signal from EPS.
- Development of a procedure to locate the forced conduction circuits in the model under certain given propagation conditions. The procedure would exploit the type of arrhythmia represented by the model and would apply trajectory planning techniques.
- Development of a procedure to search the points that interrupt the trajectories fulfilling the pre-established specifications.

In order to obtain this purpose, recent results about data processing, experimental modelling and system identification will be applied. These techniques will be combined with trajectory planning methods from the field of robotics and theoretical advances in heart electrophysiology.

A software tool called SCIRun/BioPSE [8] is being tested for the development of the models and simulation of the conduction features. SCIRun/BioPSE is a shareware (MIT license) scientific program, developed by The Scientific Computing and Imaging Institute (SCI Institute) of the University of Utah, that allows a modular and interactive development, error debugging and execution of scientific computations on a great scale. By using this computational tool, a data flow programming model is designed and tested by simulation. Geometric, mathematical and bioelectrical information are integrated in the model that also allows for automatic parameter adjustment, contour conditions, as well as fitting the discretization level necessary to obtain a suitable numerical solution. By comparison with the "off-line" procedures that are usually employed for this type of simulations, SCIRun allows an interactive handling of the design and simulation phases. Also, it avoids the excessive use of memory, that is one of the problems of data flow standard implementations, and improves the the computacional efficiency. In addition it allows the visualization of scalar, vectorial and tensor fields. This tool has being used to solve medical problems related to bioelectric fields and has been selected as a suitable tool for our purpose.

5 Basic Conduction Model

The purpose is the development of a base whose elements will be basic maps of propagation. Each map will be generated on a 3D geometric heart model (elaborated by means of the finite elements method) and a tensor of parameters is associated to each cell of the model. This tensor would gathers, at least, the three electrical properties of cardiac fibers:

- Automatism or the capacity to generate impulses that can propagate through the tissues. Sinus node cells and also atrio-ventricular ones are fundamentally automatic.
- Excitability or the capacity that has any cardiac cell to respond to an effective impulse. Contractile cells only respond to propagated impulses from an automatic structure. Once excited, every cell requires of a time to recover its excitability (refractory period).
- Conductivity or the capacity to propagate the impulses. This propagation takes place by an electrical phenomenon that crosses the cellular membrane and all the cardiac structures. The normal speed of conduction varies for the different cardiac structures (atrium, 1–2 m/s; atrio-ventricular node, 0.05 m/s; Purkinje system, 1.5–4 m/s; ventricle, 0.3 m/s).

These properties will be assigned to a tensor according to the signals gathered from EEF. Starting off from these signals is possible to build a map of cardiac potentials and relate the characteristics of this map to the properties associated to the tensors at each cell of the model.

In [9] a three-dimensional atlas of the human heart is given. It is based on the image data obtained by tomography, accessible magnetic resonance and cryosection in the Visible Human Project. This heart atlas offers great possibilities for analysis using computer vision techniques. The underlying cardiac model has been complemented with the addition of a temporal dimension for simulation of the excitation. For this purpose, an algorithm based on second generation of cellular automata has been implemented. It is adapted to the kinetic of the cardiac tissue excitation. This system demonstrates to be a right method for visualizing and researching the cardiac excitation.

During the last years, the resolution of the inverse problem in the field of cardiology has acquired a greater importance [10]. This problem consists of obtaining the bioelectrical image projection. The projection of the electrocardiographic image (ECG) uses an applied inverse solution to the electrical voltages registered in the surface of the thorax, and/or the actual characteristics of the cardiac source that produces the surface distributions.

6 Location of the Circuits

Our proposal here is to apply the trajectory planning techniques used in robotics to solve this step. Trajectory speed is making reference to a path associated to a kinematic profile.

The theoretical formalization of the planning problem has been widely studied by Latombe [7]. Many particular planners are found in the literature, for example those implemented by Farvejon and Tournassoud [11] or Kondo [12].

Most of the abovementioned results are applicable to static environments because the traditional algorithms for movement planning in deformable spaces are designed to work in spaces where the obstacles are rigid. This restriction is important because it limits the complexity of the model. A widely accepted method is that of L. Kavraki *et al.* [13]. There are also studies of movement planning for dynamic environments [14], but in order to validate our hypothesis a simpler model is preferred. The methodology will be later improved by using deformable space models.

A plan is a set of actions that allows an agent (in our case it will be a stimulus) to go from an initial state to a final state. Thus, the plan will be defined by searching trajectories or propagation paths in the cardiac map. The basic elements to formulate the plan will be the states (*e.g.* stimulus position, initial state), and the operators. The following elements are considered to be given:

- States:
  - position (X, Y, Z) of the cell in the map,
  - propagation tensor associated to each cell.
- Operators (or actions):
  - Movement to some neighboring cell: $X \pm 1, Y \pm 1, Z \pm 1$
  - Propagation conditions determined according to the type of arrhythmia subject of study.

Initial considerations are:

- A stimulus will be modeled as a point and a tail (with time-varying properties depending on the refractory time).
- A static environment with known obstacles (nonconduction zones).

If a set of cells or a road-map, free of obstacles, obtained previously by means of Voronoi diagrams is considered, an approximate diagram derived from the first one can be used. This auxiliary diagram considers a maximum distance to the obstacles corresponding to the limitations given by the sensors (catheter).

Applying potential field techniques [7] the following analogy could be made:

- The stimulus is a particle with electrical charge.
- The free space is considered a potential field.
- The obstacles have an electrical charge of the same sign than the stimulus.
- The goal has an electrical charge of opposed sign to the stimulus.

The differential potential field, U, is constructed adding the goal field U_g, and the obstacles field, U_o:

$$U(q) = U_g(q) + \sum U_o(q)$$

From this differential potential field, U, an artificial force field, F is obtained as:

$$F = -\nabla U(q)$$

Once derived the force field, the stimulus movement is based on the local force. A robust scheme is obtained and it has implicit a plan for any point of the space. The potential functions of the goal (parabolic attractor), center (parabolic repulsor) and obstacles (exponential potential barriers) has to be modeled. Later, the potential for each point of the free space can be computed and the forces are obtained by potencial derivation. The main advantages of this technique are the following:

- Trajectories can be generated from the force field in real time.
- Generated trajectories are smooth.
- It allows direct connection of the planning phase with the control phase.

7 Location of the Interruption Point

After obtaining the plan, a procedure for searching spheres of radius R that interrupt the abnormal circuits of propagation is developed. Criteria to decide their location will be previously established by an electrophysiologist and implemented on the computer in order to be automatically detected or even interactively selected. Certain fundamental premises are considered in order to formulate the problem:

- The spheres must have a minimum surface.
- The natural propagation path cannot be interrupted under any circumstances.
- Particular conditions for each type of arrhythmia are teaken into account.

An algorithm to estimate the risk of a wrong or false interruption will be implemented in order to avoid *bystanders* or local minima. The algorithm evaluates the convenience of one greater sphere in a place strategically better placed. Strategies used in the trajectory planning can also be applied to avoid *tramps*, as backtracking or wall pursuit (in our case, null conduction zones pursuit), and so on.

8 Study Case

Let consider the simple 2D heart model with an accesory pathway depicted in figure 1 that represent a patient suffering Wolff-Parkinson-White Syndrome (WPW). Certain properties as automatism, excitability, conductivity and speed are given to each cell and are also graphically shown in figure 1.

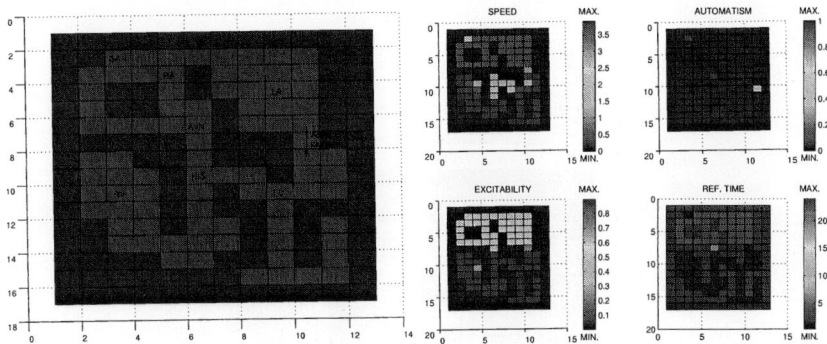

Fig. 1. Simple 2D model of a patient with WPW and properties assigned to each cell in the model

WPW is a form of supraventricular tachycardia characterized by the presence of extra pathways called accessory pathways in addition to the normal conduction ones. This is graphically shown in figure 2. The impulses travel through the extra pathway (shortcut) as well as through the normal AV-HIS Purkinje system.

The simulation of the propagation during only one beat in the model permits to observe that stimulus travels through different pathways, as can be checked in figure 3.

Points A and B represent places where two impulses collide so that they cannot continue the propagation in that direction. In that case they do not travel around the heart in a circular pattern. The collision at point B will generate a signal characterized by the *delta* wave in the ECG. The most the ventricle is depolarized by the accessory pathway, the greater *delta* wave is.

However, when multiple beats are simulated, the cell with high automatism capacity in the left ventricule of the model (see figure 1) could originate an impulse. If the neighboring cells are capable of responding to this impulse then it could occur the situation depicted at left in figure 4. Stimulous could travel very quickly through the heart in a circular pattern, causing the heart to beat unusually fast. Sinus node (SN) is inhibited and the circular pattern is sustained. Under such circumstances, a re-entry tachycardia is observed in the ECG.

Determining the optimal place for ablation is easy in this case. At right in figure 4 it can be seen the precocious activity that has been generated in point B. Nevertheless, eliminating this point is not the solution because after some time,

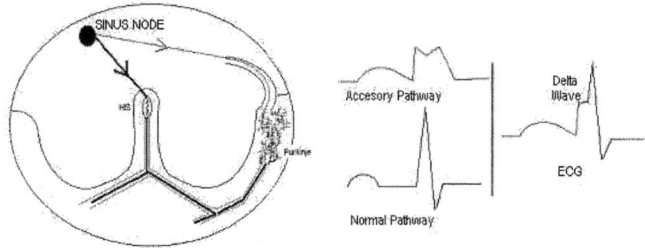

Fig. 2. Propagation in a WPW patient

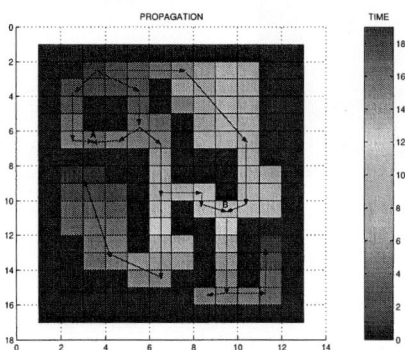

Fig. 3. Propagation of a signal characterized by the *delta* wave in the ECG

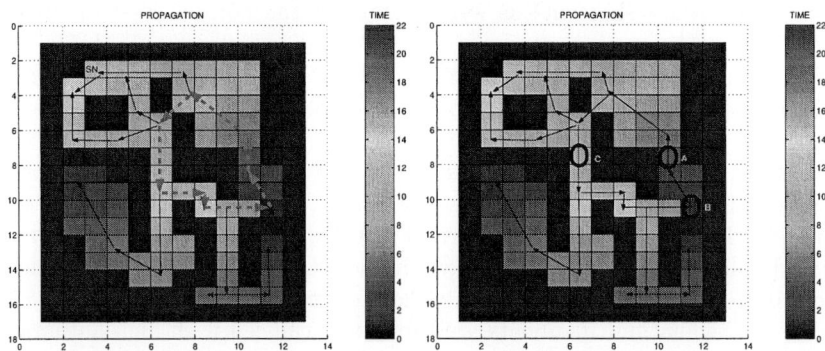

Fig. 4. Propagation in a re-entry tachycardia in a WPW patient and searching for the place for ablation

it could appear another pulse generator point in the ventricule and cause another re-entry tachycardia. This could be easily checked by simulation of the model but changing automatism properties of a contiguous cell to point B and changing cell B speed properties to zero. The elimination of other cells of the circuit will modify the trajectory but not open the circuit. Therefore, the candidate places for ablation are points A and C. Point A is not a good alternative, because it would interrupt the normal propagation pathway. As a consequence, point C is the best candidate.

Finally, it is convenient to remark that the models used here are very simple and the conclusions obtained are very promising and good from a qualitative point of view. However, more detailed models have to be developed and tested for a precise prediction of arrythmia mechanisms and for the reliable determination of ablation points. Speeds and accuracy in the location will depend on some factors (*i.e.* complexity of the model, software and hardware implementation, other systems involved, etc.).

9 Conclusions

Some deficiencies and lacks in the therapeutic arsenal used for conventional procedures of definitive treatment of patients with cardiac arrhythmias have been detected and exposed here. The essential steps to develop a procedure that carries out a significant contribution to the progress and the qualitative improvement of catheter ablation treatments have been studied and reported. The specific techniques to be used at each of the steps previously mentioned have been described. Also, a tool to be used for the implementation of the procedure, according to our criteria has been selected.

Finally, a simple example of a simulated model of a heart suffering Wolff-Parkinson-White Syndrome (WPW) has been studied using the proposed methodology. Properties as automatism, excitability, conductivity and speed are given to the model. Using this model, some propagation trajectories that validates the behavior of a WPW patient have been obtained and the optimal suggested point to apply the radiofrequency ablation has been computed.

The preliminary results obtained are quite promising, at least from a qualitative point of view, although more detailed models have to be developed and tested for prediction of particular arrhythmia mechanisms and for the consistent determination of ablation points. The validation of all the established hypotheses in this formal proposal strongly depends on the multidisciplinar cooperation between the medical and the engineering teams.

The final goal is to integrate the new procedure in the same computers that are used now for catheter mapping and make interactive use of it during the interventions.

Acknowledgments

We would to acknowledge the medical personnel of the Cardiac Electrophysiology Laboratory of the Clinical Hospital of Valladolid for their unvaluable efforts and collaboration.

References

1. Myerberg, R.J., Kessier, K.M., Kimura, S., Basset, A.L., Cox, M.M., Castellanos, A.: Life-threatening ventricular arrhythmias: The link between epidemiology and pathophysiology. In Zipes and Jalife (eds.): Cardiac electrophysiology. From cell to bedside, W.B. Saunders Company (1995)
2. Rezk, G.: Instrumentación para Electrofisiología Intervensionista. Monografías del XIII Seminario de Ingeniería Biomédica 2004, Facultades de Medicina e Ingeniería, Universidad de la República Oriental del Uruguay, Junio (2004)
3. Singer, I.: Interventional electrophysiology. Willians & Wilkins, (1997)
4. ACC/AHA guidelines for clinical intracardiac electrophysiological and catheter ablation procedures. Circulation **92** (1995) 675–691
5. Berbari, E.J., Landr, P., Geselowitz, D.B., et al.: The methodology of cardiac mapping. In Shenasa M., Borggrefe M., Breithhardt G. (eds): Cardiac mapping, Mount Kisco, New York, Futura Publishing (1993) 11–34.
6. Helguera, M., Pinski, S., de Elizalde, G., Corrado, G., Schargrodsky, H., Bazzino, O.: Ablación por radiofrecuencia de arritmias cardiacas. Servicio de Electrofisiología y Dispositivos Implantables. Departamento de Cardiología. Hospital Italiano de Buenos Aires. Argentina.
7. Latombe, J.C.: Robot motion Planning, Kluwer Academic Publishers (1991)
8. The Scientific and Computing Institute (SCI), SCIRun User Guide, para SCIRun/BioPSE version 1.22.0, University of Utah, www.sci.utah.edu (2004)
9. Freudenberg, J., Schiemann, T., Tiede, U., Hohne, K.H.: Simulation of cardiac excitation patterns in a three-dimensional anatomical heart atlas. Comput. Biol. Med. **30**(4) (2000) 191–205
10. Johnson, C.R., Mohr, M. Rüde, U., Samsonov, A., Zyp, K.: Multilevel methods for bioelectric field inverse problems. In Barth, T.J., Chan, T.F., Haimes, R., (eds.): Lecture Notes in Computational Science and Engineering-Multiscale and Multiresolution Methods: Theory and Applications. Springer-Verlag, Heidelberg (2001) 331–346
11. B. Faverjon and P. Tournassoud: A local based method for path planning of manipulators with a high number of degrees of freedom. IEEE Int. Conf. on Robotics and Automation, pages 1152-1159, 1987.
12. K. Kondo: Motion Planning with Six Degrees of Freedom by Multistrategic Bidirectional Heuristic Free-Space Enumeration. IEEE Transactions on Robotics and Automation, vol. 7, no. 3, 267-277, 1991.
13. Kavraki, L., Svestka, P., Latombe, J.-C., Overmars, M.H.,: Probabilistic roadmaps for path planning in high-dimensional configuration spaces. IEEE Trans. Rob. Autom. **12** (1996) 566-580
14. Berg, J.P.v.d., Overmars, M.H.: Roadmap-based motion planning in dynamic environments. Institute of information and computing sciences, Utrecht University, Technical Report UU-CS-2004-020, www.cs.uu.nl (2004)

A Functional Heart Model for Medical Education

Vassilios Hurmusiadis and Chris Briscoe

Primal Pictures Ltd, London, UK
`vassili@primalpictures.com`
`http://www.primalpictures.com`

We developed a 3D computer graphic model of functional anatomy of the human heart. The model provides visually correct anatomical and functional detail suitable for medical education. We reconstructed 3D surface models of the human heart based on segmentation obtained from the Visible Human image datasets. We developed a fiber based muscle action model specially adapted for the myocardium. Each muscle fiber is equipped with contractile and elastic elements and is used as a local shape deformation guide. The timing of fiber contraction activation is driven by patient specific action potential excitation patterns. As a first step we have visualized the function of a healthy heart. We are now planning to visualize a range of cardiac conditions and dysfunctions.

1 Anatomic Model

The reconstruction of a 3D surface model of the human heart was based on the Visible Human Project datasets [1]. Segmentation was extracted from the axial anatomical cross-section images of the Visible Male and Female datasets in the thoracic region (see figure 1). A male and female heart models were reconstructed. The male model, coming from a 39 year old healthy person, was used to visualize the function of a healthy heart. The female heart model, coming from a 59 year old person with enlarged heart, will be used for the visualization of heart failure.

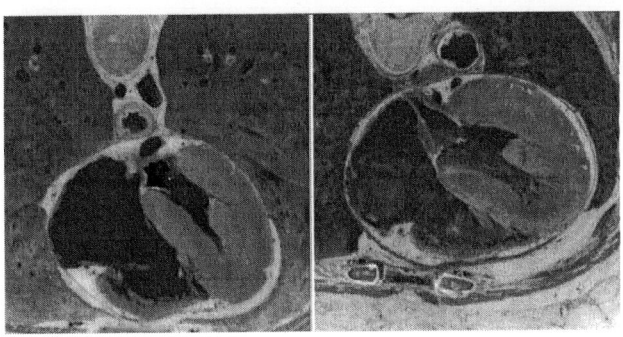

Fig. 1. Axial anatomical cross-section images from the Visible Male (left) and Visible Female (right)

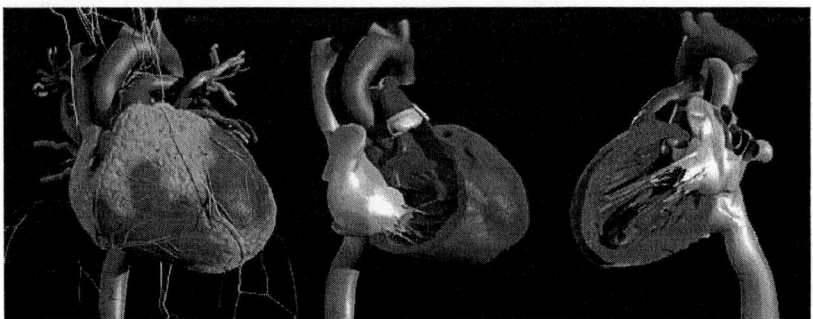

Fig. 2. Views of the reconstructed 3D female heart model

At every stage of development particular attention was paid to the functionality of each part of the model. This was done to facilitate the integration of the graphic model into the functional model.

The surface model includes detailed inner and outer wall structures on all four chambers and valves. Also structures such as the trabeculae carneae, the papillary muscles and all the main cardiac veins, arteries and fatty tissue have been modeled (see figure 2).

2 Mechanical Contraction

We developed a fiber based muscle action model specially adapted for the human heart myocardium [2]. In our model a muscle is represented by a set of fibers, which run through the muscle body.

Each fiber is equipped with contractile and elastic elements connected in parallel [3] (see figure 3). Each fiber line acts as a local shape deformation guide for the surrounding muscle tissue. By activating each fiber we can accurately specify the level of contraction and volume preservation of a muscle.

The modeled fibers in and around the ventricles were made to follow approximate heart muscle fiber orientation data obtained from diffusion tensor MRI [4] (see figure 4).

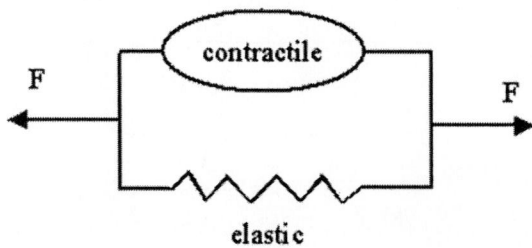

Fig. 3. Muscle action model with parallel contractile and elastic elements

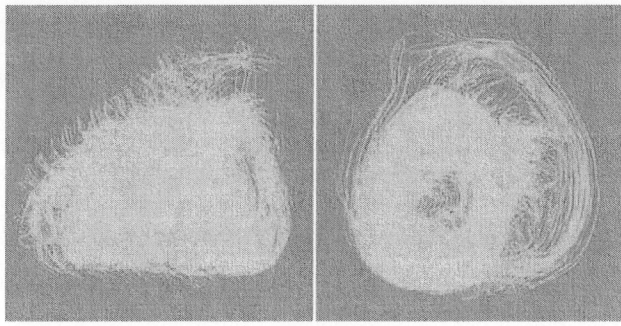

Fig. 4. Muscle fiber orientation in both ventricles

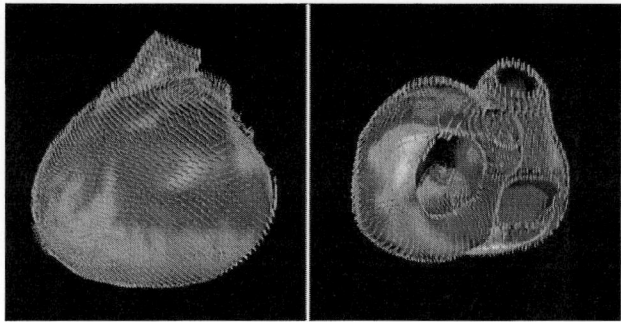

Fig. 5. Muscle fiber orientation mapped onto the 3D model

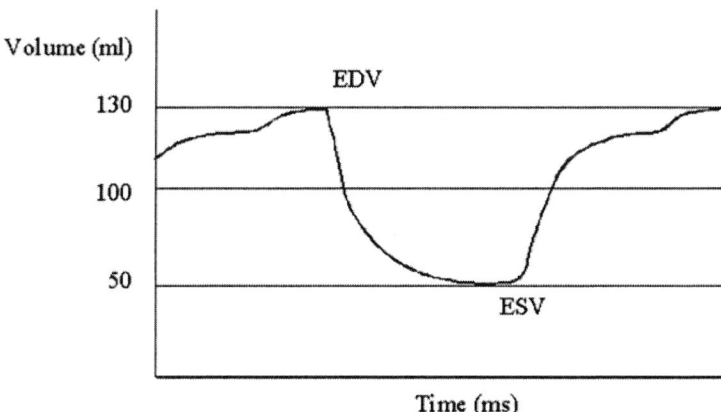

Fig. 6. Graph of the inner volume of left ventricle during a normal heartbeat cycle

Fiber orientations were geometrically mapped onto the inner and outer surfaces of the ventricle models. Thus the surface of each ventricle was equipped with the fiber based deformation system (see figure 5). Fiber orientation in the atria was derived from anatomical morphology studies [5] and by cadaveric observation.

The mechanical model was equipped with an inner volume calculation algorithm. An experimental relationship between the amount of fiber contraction and the inner volume in a heart chamber (i.e. the left ventricle) was derived as follows: the amount of fiber contraction was step increased and the resulting inner chamber volume was simultaneously calculated. By reversing this experimentally derived relationship a simple volume graph such as the one in figure 6 was used to determine the amount of mechanical contraction during a complete heartbeat cycle (see figure 7).

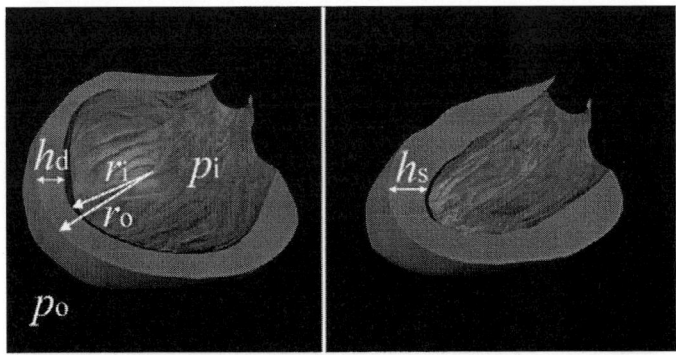

Fig. 7. Wall thickness in the left ventricle in end diastole h_d (left) and end systole h_s (right)

The ventricular wall thickness was calculated by locally approximating the chamber in question (i.e. the left ventricle in figure 7) with the shape of a spherical membrane with thick walls. In this case, the law of Laplace (see equation 1) relates the inner and outer pressures p_i and p_o and radii r_i and r_o with the membrane stress T [3]:

$$T = p_i r_i^2 - p_o r_o^2 / (r_i + r_o). \tag{1}$$

$$T = h(\sigma). \tag{2}$$

The inner pressure was made to follow a pressure graph of a normal heartbeat cycle while the outer pressure was kept constant and approximately equal to the atmospheric pressure. Equation (1) was combined with an experimentally acquired stress (T)/strain ($h(\sigma)$) function of cardiac muscle [3] (see equation 2). The wall thickness was calculated by combining equations (1) and (2) and solving for h.

All four main valves (bi-cuspid, tri-cuspid, aortic and pulmonary) were modeled. Fiber based rigging enabled their opening and closing function, synchronized with the heartbeat cycle (see figure 8).

Cardiac vessels (coronary arteries and veins) were subjected to forced displacement by the underlying muscle contraction, while their volumes were kept constant (see figure 9).

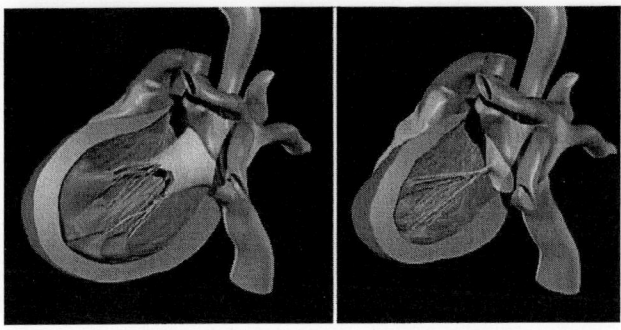

Fig. 8. Left ventricle in cross-section and bi-cuspid (mitral) valve during diastole (left) and systole (right)

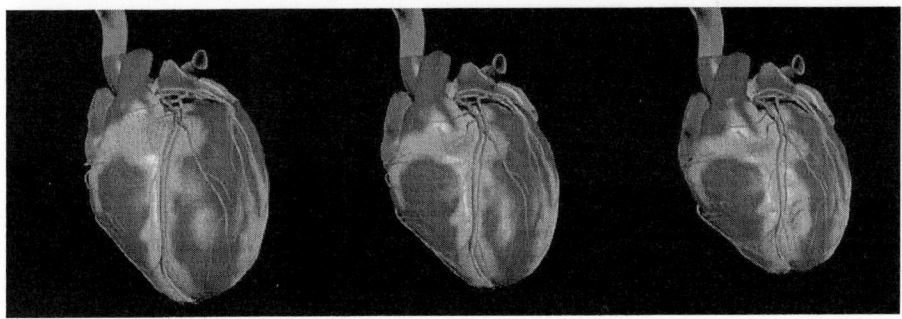

Fig. 9. Cardiac vessels during heartbeat from end diastole left to end systole right

3 Electrical Excitation

The timing of fiber activation can be driven by manually designed contraction/time graphs. All movement in atria, ventricles and valves can be independently driven by such graphs. By modifying the contraction/time graphs we can visualize normal heart beat as well as various arrhythmic conditions.

Patient specific action potential maps were also used to drive the timing of the mechanical model. In figure 10, electrical propagation maps were acquired using the Ensite catheter (from Endocardial Solutions Inc.) inside a patient's right atrium.

In a cardiac cell mechanical contraction occurs after the cell has been electrically stimulated. Each peak of electrical stimulation is followed by a single contraction peak with an approximate delay of 150 ms (see figure 11).

The above mentioned patient specific maps were geometrically projected onto the 3D surface of our modeled atrium (see figure 12). The mechanical contraction of the right atrium was activated by patient specific electrical data. Muscle contraction at each point on the atrium peaked approximately 100-150 ms after the arrival of maximum action potential on the same point.

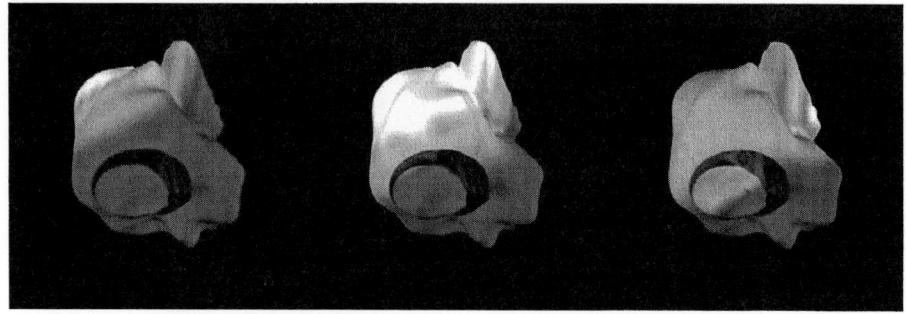

Fig. 10. Patient right atrium geometry with recorded electrical activation pattern

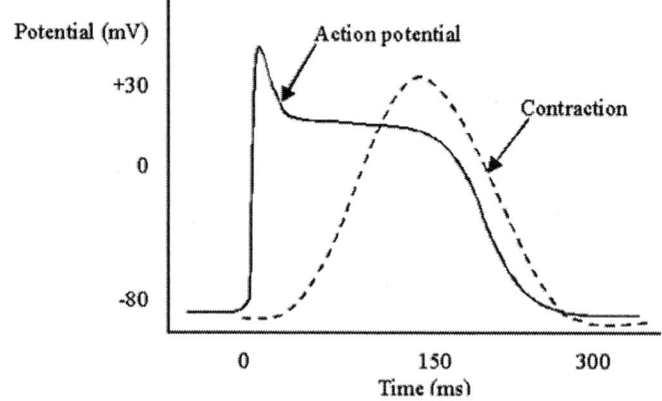

Fig. 11. Graph of action potential and contraction in a single cardiac cell (contraction graph out of scale)

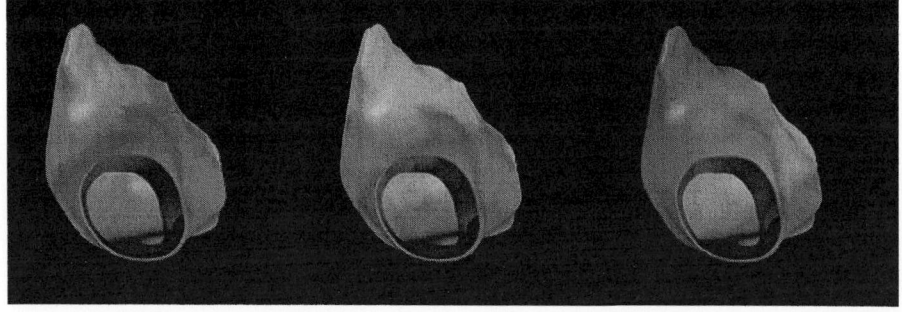

Fig. 12. 3D heart model right atrium with mapped electrical activation pattern

4 Conclusion and Future Work

A heart function visualization model was developed. A fiber based muscle action model was combined with inner volume/time and pressure/time graphs in order to achieve a visually correct contraction cycle. Manually derived and patient specific electrical data were used to activate the model. As a first step we visualized the function of a healthy heartbeat in 4 dimensions. As a further step we are planning to visualize a range of cardiac conditions and dysfunctions such as myocardial infarction, atrial fibrillation, bradycardia, tachycardia, the sick sinus syndrome, valve dysfunction, etc. Also the transition between arrhythmic and normal cycles will be addressed. This research is supported by a SMART Exceptional Award grant from the Department of Trade and Industry, UK.

References

1. Visible Human Project. National Library of medicine, Bethesda, Maryland USA (http://www.nlm.nih.gov)
2. Hurmusiadis, V., Barrick, S., Briscoe, C.: Visualization of Muscle Function for Medical Education. Medicine Meets Virtual Reality 12, IOS Press (2004)
3. Fung, Y. C.: Biomechanics: Mechanical Properties of Living Tissues. 2Ns edn. Springer-Verlag, New York (1993)
4. Zhukov, L., Barr, A.H.: Heart-Muscle Fiber Reconstruction from Diffusion Tensor MRI. Proceedings of Visualization, IEEE, Vis03 (2003)
5. Ho, S.Y., Anderson, R.H., Sanchez-Quintana, D.: Atrial Structure and Fibres: Morphologic Bases of Atrial Conduction. Cardiovascular Research 54, Elsevier (2002)

Artificial Enlargement of a Training Set for Statistical Shape Models: Application to Cardiac Images

J. Lötjönen[1], K. Antila[1], E. Lamminmäki[1], J. Koikkalainen[2], M. Lilja[2], and T. Cootes[3]

[1] VTT Information Technology, P.O.B. 1206, FIN-33101 Tampere, Finland
{Jyrki.Lotjonen@vtt.fi}
[2] Laboratory of Biomedical Engineering, Helsinki University of Technology, P.O.B. 2200, FIN-02015 HUT, Finland
[3] Division of Imaging Science and Biomedical Engineering, University of Manchester, U.K

Abstract. Different methods were evaluated to enlarge artificially a training set which is used to build a statistical shape model. In this work, the shape model was built from MR data of 25 subjects and it consisted of ventricles, atria and epicardium. The method adding smooth non-rigid deformations to original training set examples produced the best results. The results indicated also that artificial deformation modes model better an unseen object than an equal number of standard PCA modes generated from original data.

1 Introduction

Segmentation is known to be one of the most difficult problems in image analysis. Several reasons explain the difficulty, such as noise in images, image inhomogeneities, partial volume effect, complex and cluttered scenes and low visibility of edges between objects. Deformable model-based methods provide one approach to overcome partially the problem. In these methods, an *a priori* model is non-rigidly registered to the object of interest in the image by optimizing a cost function. To find the optimal non-rigid transformation is an ill-posed problem; hence some form of constraints are required. One possibility is to use physical-based models, such as viscous fluid or elastic models [1, 2]. Statistical shape models is another option. In these methods, an *a priori* model is allowed to deform only in a way consistent with the information captured from a training set.

The most popular approach for modeling the shape changes is the point distribution model, also referred to as active shape model (ASM) [3]. It defines a mean model and its typical deformation modes on the basis of a training set using principal component analysis (PCA). The deformation modes are the eigenvectors of the covariance matrix determined for corresponding points in different examples of the training set.

Statistical shape models suffer from two commonly known problems especially in medical applications. First, building a statistical shape model is laborious as the point correspondences need to be defined between the training set examples. However, automatic procedures have been recently proposed to overcome this problem [4, 5, 6]. Second, as the building process is time consuming and enough data are not always available, only a small set of examples are often used to construct the model. Since the maximum number of deformation modes can not exceed the number of examples in the training set minus one, the eigenvectors obtained span only a small subspace which can not represent the full range of shape variation present in real medical objects. For example, $10-30$ subjects have been used to construct a 3D statistical shape model of the brain and heart [5,7,8,9]. As reported in [9], the size of the training set was considered to be the most important reason for a relatively high segmentation error. This work concentrates on the modeling of the heart.

The segmentation accuracy depends on the ability 1) of the model to represent an unseen object, and 2) of the segmentation algorithm to define correctly the model parameters. This work concentrates on the first point. The objectives of this work are two-fold:

- To define an appropriate method for enlarging the training set artificially and efficiently.
- To estimate the relation between the size of the training set and the ability of the model to represent an unseen object, as applied to cardiac data.

The latter objective is closely related to commonly known bias-variance trade-off. In other words, if the number of the deformation modes is too small, the model is over-constrained. However, choosing too many deformation modes based on a large training set leads to overfitting.

2 Methods

2.1 Statistical Shape Models

In statistical shape models, new examples of the shape, $\mathbf{x} = [x_1, \ldots, x_n]^T$, that are specific to the studied object, are generated using a linear combination

$$\mathbf{x} = \bar{\mathbf{x}} + \mathbf{\Phi b}, \qquad (1)$$

where $\bar{\mathbf{x}} = [\bar{x}_1, \ldots, \bar{x}_n]^T$ is a reference shape, typically a mean shape constructed from a training set, $\mathbf{\Phi} = [\phi_1, \ldots, \phi_m]$ is a matrix consisting of the modes of shape variation, ϕ_i, and $\mathbf{b} = [b_1, \ldots, b_m]^T$ is a weight vector.

In ASM [3], the object is represented by a point set. The training set is first affinely aligned, and the mean shape is calculated:

$$\bar{\mathbf{x}} = \frac{1}{N} \sum_{i=1}^{N} \mathbf{x}_i, \qquad (2)$$

where N is the size of the training set. Next, PCA is applied to the variations of the training set, i.e., the eigenvectors and eigenvalues of the covariance matrix

$$\mathbf{\Sigma} = \frac{1}{N-1} \sum_{i=1}^{N} (\mathbf{x}_i - \bar{\mathbf{x}})(\mathbf{x}_i - \bar{\mathbf{x}})^T \qquad (3)$$

are calculated. The eigenvectors of the covariance matrix describe the ways in which the shapes vary, and the corresponding eigenvalues explain the variance of the data projected onto each eigenvector.

If the number of points in a training set example is denoted by P and the dimensionality by D ($D = 2$ in 2D and $D = 3$ in 3D), the maximum number of deformation modes is $min(N-1, D \cdot P)$.

The model instance $\mathbf{x}'$ representing an unseen object $\mathbf{x}$ is

$$\mathbf{x}' = \bar{\mathbf{x}} + \mathbf{\Phi} \mathbf{b}', \tag{4}$$

where the weights $\mathbf{b}'$ are computed from

$$\mathbf{b}' = \mathbf{\Phi}^{-1}(\mathbf{x} - \bar{\mathbf{x}}). \tag{5}$$

In this work, we model five objects: 2 atria, 2 ventricles and epicardium. Contours of these objects are catenated to one vector, i.e. $\mathbf{x}_i = (x_{i1}, y_{i1}, x_{i2}, .., y_{iP})$ in 2D, i.e. each mode contained deformations for all objects.

2.2   Materials

Our dataset consisted of cardiac short- and long-axis magnetic resonance images acquired from 25 healthy subjects. The mean shape and its variation were modeled as described in detail in [9]. The procedure is shortly summarized.

The atria, ventricles and epicardium were manually segmented by fitting a triangulated surface model simultaneously to the short- and long-axis images. Thereafter, one subject was considered as a reference volume to which all other subjects were aligned using translation, rotation and isotropic scaling. The normalized mutual information (NMI) was used as a similarity measure. Segmented volumes, where each object was represented by one gray-scale value, were used in registration. Next, the reference volume was non-rigidly registered to the aligned volumes using a non-rigid registration based on a deformation sphere technique [10]. In the deformation sphere technique, smooth deformations are applied to voxels inside a sphere in such a way that the NMI is maximized. The location of the sphere is randomly chosen from the surfaces of ventricles, atria and epicardium, and it is varied during the iteration.

The nodes of the triangulated surface model of the reference subject, obtained from the manual segmentation, were considered as semi-landmarks. Semi-landmarks were used because only a few anatomical landmarks can be located from the heart in the MR images. Propagating the semi-landmarks, using the non-rigid transformations defined above, a set of corresponding semi-landmarks was achieved for each training set subject. The mean shape and its variance was then computed by applying Eqs. 2 and 3. In addition, the mean gray-scale short- and long-axis volumes were computed.

To reduce the bias of the mean shape towards the selected reference subject, and to give a better *a priori* estimate in the non-rigid registration, the preceding procedure was repeated by using the mean shape as a reference model.

In this work, two datasets were used: 1) a set of 2D contours from one long-axis image of each subject (Fig. 1a), and 2) a set of 3D triangulated surfaces from each subject (Fig. 1b). The number of data points was $P = 219$ in 2D and $P = 2086$ in

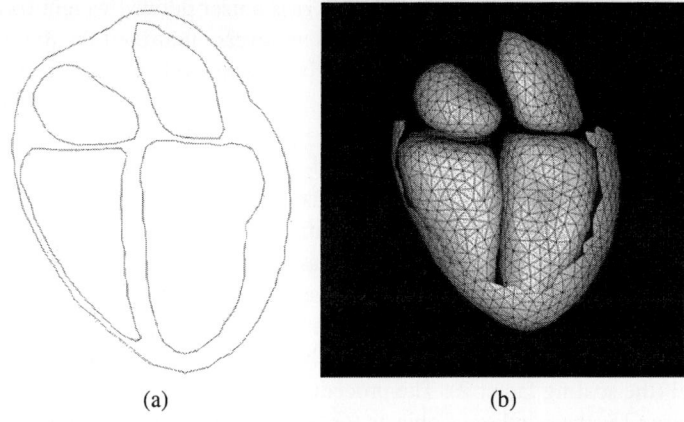

(a) (b)

Fig. 1. A set of a) 2D contours and b) 3D surfaces of atria, ventricles and epicardium from one subject

3D. The number of deformation modes to represent any arbitrary contour or surface is $D \cdot P = 2 \cdot 219 = 438$ or $3 \cdot 2086 = 6258$, respectively. Because the number of subjects is $N = 24$ (24 instead of 25 because of cross-validation) in this work, the maximum number of deformation modes is, however, 23 ($N - 1 < D \cdot P$) for the standard PCA.

2.3 Techniques to Increase the Size of Training Set

Techniques to enlarge artificially the training set have been widely studied [11, 12, 13, 14, 15]. Next, the techniques tested in this work are summarized.

The values in parentheses in the text below are related to user-defined parameters and indicate the values that produced the lowest mean point-to-point error (Section 2.4). These values were used in Section 3.

Standard PCA. The procedure described in Section 2.1 was followed producing $N-1$ deformation modes.

PCA & FEM. Cootes et al. [11] combined the standard PCA and finite element method (FEM) in shape modeling. FEMs take an instance of a shape and treat it as if it was made of flexible material. Modal analysis gives a set of linear deformations of the shape, such as bend, shear and pinch, equivalent to the modes of vibration of the original shape. These modes can be used in shape modeling:

$$\mathbf{x} = \mathbf{x}_i + \mathbf{\Omega}_i \mathbf{u}, \qquad (6)$$

where $\mathbf{x}_i$ is an example of shape (as in Eq. 2), $\mathbf{\Omega}_i$ is a matrix consisting of eigenvectors computed for the stiffness matrix of the shape i, and $\mathbf{u}$ is a weight vector (as $\mathbf{b}$ in Eq. 1).

The deformation modes are the eigenvectors of the matrix

$$S = \mathbf{\Sigma} + \alpha(\frac{1}{N}\sum_{i=1}^{N}\mathbf{\Omega}_i \mathbf{\Lambda}_i \mathbf{\Omega}_i^T), \qquad (7)$$

where Σ is the covariance matrix from Eq. 3, α is a user defined weight ($\alpha = 0.5$), N is the number of chosen example shapes, Λ is an inverse matrix of the diagonal matrix consisting of eigenvalues of the stiffness matrix. The procedure produces $D \cdot P$ deformation modes.

Adaptive Focus. In the standard PCA method, each point has an equal weight in computing the covariance matrix. Shen and Davatzikos [12, 13] proposed a method called adaptive focus where objects having a low spatial variance or objects with a high confidence were spatially scaled in order to increase the variance. Due to scaling, the deformations of the scaled object became more emphasized and better represented in the most important deformation modes (high eigenvalues because of high variance). In this work, each object was scaled separately in the co-ordinate system of the reference model (the scaling factor 2). The procedure produces $(O + 1)N - 1$ deformation modes where O is the number of objects ($O = 5$) and N is the size of the training set. The factor $O + 1$ is used instead of O because the original examples are also included in addition to scaled ones.

Non-rigid Scaling. In this approach, the adaptive focus technique was extended to non-rigid but smooth deformations. The contours were scaled inside of a deformation sphere. The scaling factor, $s = s(x, y, z)$, of a point (x, y, z) is computed from

$$s(x,y,z) = \frac{e^{-2\frac{(x-c_x)^2+(y-c_y)^2+(z-c_z)^2}{r^2}} - e^{-2}}{1.0 - e^{-2}} S + 1, \qquad (8)$$

where (c_x, c_y, c_z) and r are the location and the radius of the sphere ($r = 50$ mm), respectively, and S is the user specified scaling factor ($S = 1$). The sphere is randomly located to L locations on the contours ($L = 100$), and the original contour points are deformed at each location. The origin during the scaling is in the center of the sphere. The number of deformation modes produced is $(L + 1)N - 1$.

Non-rigid Movement. Another strategy, very similar to the non-rigid scaling technique, was also tested. The displacement vector, $\mathbf{v}(x, y, z)$, for any point inside the sphere is computed from

$$\mathbf{v}(x,y,z) = \frac{e^{-2\frac{(x-c_x)^2+(y-c_y)^2+(z-c_z)^2}{r^2}} - e^{-2}}{1.0 - e^{-2}} \mathbf{V}, \qquad (9)$$

where $\mathbf{V}$ is a random vector and other parameters as in Eq. 8. The length of the vector $\mathbf{V}$ was chosen from a uniform distribution ([0 25] mm).

Fourier. The approach adopted in this work is closely related to the hierarchical method proposed in [14] where the data were divided into different frequency and spatial location bands using wavelets, and PCA was performed for each band separately. In this work, the data were decomposed only in frequency bands. The deviations of the training set examples from the mean were transformed into the frequency space using Fourier transformation and the data were band-pass filtered into B ($B = 18$) separate frequency bands. New artificial training set examples were then generated by restoring each band separately into the shape space with inverse Fourier transformation. This procedure produces $(B+1)N-1$ deformation modes, when also original examples are included in the training set.

Noising. In this approach, Gaussian noise is added to each point which makes data less correlated. The displacement of a point was chosen from a uniform distribution ($[-2\ 2]$ mm) in each direction. From each training set example, L ($L=100$) noisy contours were generated leading to $(L+1)N-1$ deformation modes. Alternatively, the data could be made more uncorrelated by replacing the covariance matrix Σ by $\Sigma + \alpha \mathbf{I}$, where α is a weight factor and $\mathbf{I}$ a unit matrix.

2.4 Evaluation

As mentioned above, the segmentation accuracy does not depend only on the model properties but also details of the optimization method and image characteristics affect the result. In addition, normally no real gold standard exists for evaluating the accuracy of the segmentation. The automatic segmentation result is usually compared with the manual one which is commonly known to contain errors. Warfield et al. [16] recently proposed a solution to this problem. In our work, two methods were used in evaluation: 1) the model was fitted directly to a training set example using Eq. 5 (model-to-shape fit), and 2) the model was fitted iteratively to image data (model-to-image fit). The former method measures the ability of the model to represent an unseen object while the latter method includes all error sources in segmentation.

Two error measures are defined. Point-to-point (PP) error is computed as an average Euclidean distance between the corresponding points in $\mathbf{x}'$ and $\mathbf{x}$ (Eq. 4). The PP error can not normally be used in segmentation, because the point correspondences are not known. Therefore, point-to-curve/surface (PCS) error is defined: 1) search the shortest Euclidean distance from each point of $\mathbf{x}'$ to the contour or surface defined by $\mathbf{x}$, and 2) take an average of these distances. In other words, the PCS error omits the error in the tangential direction of the contour or surface, and produces lower values than the PP error. Both PP and PCS errors were used to measure the model-to-shape fit while only the PCS error was used with the model-to-image fit.

Cross-validation was used: each training set example was once regarded as a target, the shape model was built using the remaining training set, and the target was represented by the shape model.

3 Results

Model-to-Shape Fit. The 2D PP error (in [mm]) for different techniques is represented in Fig. 2. A non-parametric Wilcoxon Signed Ranks Test was used to detect statistically significant differences between the shape models when the number of deformation modes was 23. Statistical significance was considered to be obtained for p-value $p < 0.01$. The non-rigid movement technique produced the best result: as compared with the non-rigid scaling ($p < 0.01$) and with the other methods ($p < 0.0001$). This

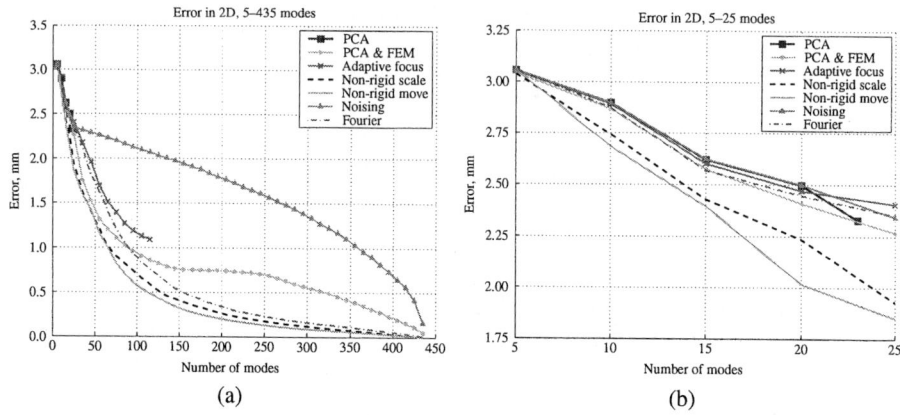

Fig. 2. a) The 2D PP error in function of deformation modes for different techniques. b) The same curves only up to 25 modes

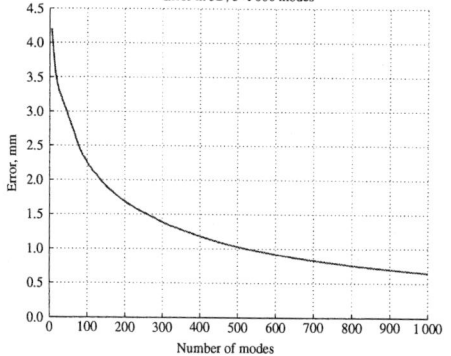

Method	PCS [mm]	NMI
PCA23	2.09 ± 0.43	1.086 ± 0.011
NRM23	2.04 ± 0.38	1.094 ± 0.012
NRM100	2.00 ± 0.34	1.102 ± 0.013
NRM250	2.04 ± 0.36	1.103 ± 0.014

Fig. 3. The 3D PP error in function of deformation modes for the non-rigid movement technique. Table shows segmentation error and the corresponding NMI value for the standard PCA and non-rigid movement technique (NRM) with different number of deformation modes

indicates that partly artificially generated modes perform better than the modes derived from the original training set using the standard PCA. The curves show also that the PP error decreases very slowly after $100 - 150$ modes. In addition, the noising technique is clearly the worst approach. Fig. 3 shows the corresponding curve for the non-rigid movement technique as applied to 3D surfaces.

Model-to-Image Fit. Preliminary 3D segmentation results are also provided. The following method was used to optimize the weights of the deformation modes. For each data point ($P = 2086$), two gray scale profiles normal to the surface were generated: one from the mean short-axis and one from the mean long-axis data sets. The NMI was

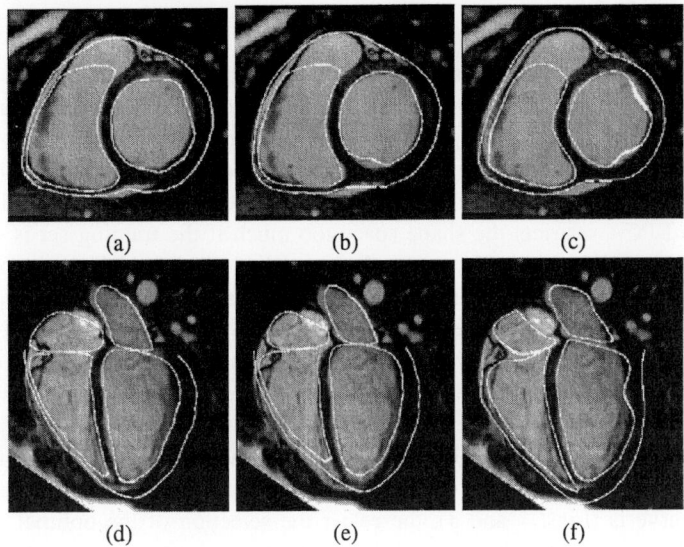

Fig. 4. Segmentation result superimposed on MR slices of one subject. A short-axis view showing the result for a) the standard PCA (PCS error 1.95 mm), b) the non-rigid movement technique (PCS error 1.80 mm, 250 modes) and c) manual segmentation. The images d), e) and f) show the corresponding results in a long-axis view

maximized between the profiles from the model and the profiles from same locations in the target data. All points from all profiles and from the both data sets were used in computing the NMI. In this work, the length of the profile was 21 points and the data sets were quantized to 32 gray values. Conjugate gradient method was used to optimize the weights of the deformation modes.

The segmentation results have been shown in Fig.3. No statistically significant differences were found between PCS values. However, when the NMI values were compared, the artificial modes were found superior compared with the standard PCA ($p < 0.00001$) and 100 or 250 artificial modes produced better results than 23 artificial modes ($p < 0.00001$). The improvement in NMI but not in the segmentation error indicates the existence of errors in our manual segmentation results used as a gold standard. The difference in the NMI values between 100 and 250 modes was not statistical significant. The segmentation result of one subject is visualized in Fig. 4.

The 3-D segmentation errors of cardiac structures, produced by automatic tools not based on the PCA-based approaches, have been recently reported to be around $2 - 3$ mm [9, 17]. The PCS error was about 55 % and 46 % of the PP error with our 2D and 3D data, respectively. This means that 23 deformation modes (3.6 mm PP error) have enough degrees of freedom to reach 1.7 mm PCS error. As our segmentation results indicated, the real segmentation error is higher: the PCS error was 2.09 mm for 23 modes defined by the non-rigid movement technique. For 250 modes, the PCS error was 2.04 mm while the minimum PCS error, due to limited degrees of freedom in the model, is only 0.7 mm (46 % of about 1.5 mm for 250 modes in Fig. 3).

4 Discussion

Different techniques to increase artificially the number of deformation modes were studied. The technique based on non-rigid movements produced the best results, also in statistical sense. The advantages of the method were that 1) the PP error decreased fastest as the size of the training set was increased, and 2) the PP error was lower than using the standard PCA with an equal number of modes. The latter point indicates that the standard PCA restricts the shape space too much if the training set is small, and introducing artificial variation improves the generality of the model. On the other hand, the artificially enlarged training set has also drawbacks. The model may become physically implausible and unrealistic segmentation results can be produced for low-quality and complex image data. In addition, statistical shape models can be used to detect different abnormalities from images if the training set is defined from healthy volunteers. With artificial training set this property is lost.

In order to avoid overfitting of the model, a common habit is to select only modes that explain, for example, 99 % of the variance in the training set. The more L-shaped the error curve is (Figs. 2 and 3), the easier the selection of the optimal number of modes is. The optimal number of modes is attained in the cross-section of vertical and horizontal parts of the L-curve because the error decrease per an added mode is small after that point [18]. In our cardiac cases, the cross-section point was approximately at 130 modes in 2D and at 250 modes in 3D as the non-rigid movement technique was used. In other words, the model does not improve considerably by adding more data to the training set after these limits. Although these limits have been derived from a partly artificially generated training set, we believe that the shape of curves for real data would be approximately similar, and the limits computed from artificial data provide a rough estimate of the optimal size of the training set also for real data.

The results indicated that although the model with extended training set is capable to represent more accurately unseen objects, the segmentation accuracy does not improve equally. Two reasons explain the relatively small improvement in the segmentation accuracy: 1) a local maximum of NMI has been found because the conjugate gradient method is not a global optimization technique, and 2) the manual segmentation used as a gold standard contains errors. The relative contribution of these error sources should be studied in future.

Acknowledgements

This research was partly supported by the National Technology Agency, Finland.

References

1. Christensen, G.E., Miller, M.I., Vannier, M.W., Grenander, U.: Individualizing neuroanatomical atlases using massively parallel computer. IEEE Computer **29** (1996) 32–38
2. Wang, Y., Staib, L.H.: Pysical model-based non-rigid registration incorporating statistical shape information. Med. Image Anal. **4** (2000) 7–20

3. Cootes, T.F., Taylor, C.J., Cooper, D.H., Graham, J.: Active shape models - their training and application. Computer Vision and Image Understanding **61** (1995) 38–59
4. Davies, R., Twining, C., Cootes, T., Waterton, J., Taylor, C.: A minimum description length approach to statistical shape modeling. IEEE Trans. Med. Imag. **21** (2002) 525–537
5. Frangi, A., Rueckert, D., Schnabel, J., Niessen, W.: Automatic construction of multiple-object three-dimensional statistical shape models: Applications to cardiac modeling. IEEE Trans. Med. Imag. **21** (2002) 1151–1166
6. Kaus, M., Pekar, V., Lorenz, C., Truyen, R., Lobregt, S., Weese, J.: Automated 3-D PDM construction from segmented images using deformable models. IEEE Trans. Med. Imag. **22** (2003) 1005–1013
7. van Ent, D., de Munck, J., Kaas, A.: A fast method to derive realistic BEM models for E/MEG source reconstruction. IEEE Trans. Biomed. Eng. **48** (2001) 414–423
8. Rueckert, D., Frangi, A., Schnabel, J.: Automatic construction of 3-D statistical deformation models of the brain using nonrigid registration. IEEE Trans. Med. Imag. **22** (2003) 1014–1025
9. Lötjönen, J., Kivistö, S., Koikkalainen, J., Smutek, D., Lauerma, K.: Statistical shape model of atria, ventricles and epicardium from short- and long-axis MR images. Med. Image Anal. **8** (2004) 371–386
10. Lötjönen, J., Mäkelä, T.: Elastic matching using a deformation sphere. In Niessen, W.J., Viergever, M.A., eds.: Lecture Notes in Computer Science 2208: Medical Image Computing and Computer-Assisted Intervention - MICCAI 2001, Springer (2001) 541–548
11. Cootes, T.F., Taylor, C.J.: Combining point distribution models with shape models based on finite element analysis. Image and Vision Computing **13** (1995) 403–409
12. Shen, D., Davatzikos, C.: An adaptive-focus statistical shape model for segmentation and shape modeling of 3-D brain structures. IEEE Trans. Med. Imag. **20** (2001) 257–270
13. Shen, D., Davatzikos, C.: An adaptive-focus deformable model using statistical and geometric information. IEEE Trans. Patt. Anal. Mach. Intell. **22** (2000) 906–913
14. Davatzikos, C., Tao, X., Shen, D.: Hierarchical active shape models using the wavelet transform. IEEE Trans. Med. Imag. **22** (2003) 414–423
15. de Bruijne, M., van Ginneken, B., Viergever, M., Niessen, W.: Adapting active shape models for 3d segmentation of tubular structures in medical images. In Taylor, C., Noble, A., eds.: Lecture Notes in Computer Science 2732: Information Processing in Medical Imaging, Springer (2003) 136–147
16. Warfield, S.K., Zou, K.H., Wells, W.M.: Simultaneous truth and performance level estimation (STAPLE): An algorithm for the validation of image segmentation. IEEE Trans. Med. Imag. **23** (2004) 903–921
17. Lorenzo-Valdés, M., Sanchez-Ortiz, G.I., Elkington, A.G., Mohiaddin, R.H., Rueckert, D.: Segmentation of 4D cardiac MR images using a probabilistic atlas and the EM algorithm. Med. Image Anal. **8** (2004) 255–265
18. Horn, J.L.: A rationale and test for the number of factors in factor analysis. Psychometrika **30** (1965) 179–186

Towards a Comprehensive Geometric Model of the Heart

Cristian Lorenz and Jens von Berg

Philips Research Laboratories Hamburg,
Röntgenstrasse 24-26, 22335 Hamburg, Germany
Cristian.Lorenz@Philips.com

Abstract. Domain knowledge about the geometrical properties of cardiac structures is an important ingredient for the segmentation of those structures in medical images or for the simulation of cardiac physiology. So far, a strong focus was put on the left ventricle due to its importance for the general pumping performance of the heart and related functional indices. However, other cardiac structures are of similar importance, e.g. the coronary arteries with respect to diagnosis and treatment of arteriosclerosis or the left atrium with respect to the treatment of atrial fibrillation. In this paper we describe the generation of a comprehensive geometric cardiac model including the four cardiac chambers and the trunks of the connected vasculature, as well as the coronary arteries and a set of cardiac landmarks. A mean geometric model has been built. A general process to add inter-individual and temporal variability is proposed and will be added in a second stage.

1 Introduction

The use of cardiac domain knowledge in terms of geometrical models of the heart has been reported in many articles (see [1] for a review). The main focus so far, was on the left ventricle and the related cardiac function and wall motion analysis. Recently, motion analysis has also been performed on the right ventricle [2] and atrium [18] and modeling approaches started to include both ventricles [3-8] or even all 4 cardiac chambers [9,17]. Other publications deal with the geometrical properties of the coronary arteries [10-12]. In clinical practice, two trends are currently gaining importance. First of all there is a strong trend towards automation. Limited budgets in terms of money and time call for "zero-click" procedures for cardiac analysis such as functional values or coronary artery assessment. A comprehensive image based cardiac diagnosis session, revealing all important parameters and producing all relevant image renderings needs to be finished in about 10 to 15 min. The second trend is about accomplishing a synoptic representation of the cardiac aspects of the patient: How is the stenosed coronary artery related to the damaged myocardial tissue? Does the wall motion artifact support the myocardial perfusion findings? A key issue arising from both trends is the extensive use of cardiac domain knowledge i.e. the use of cardiac models. A third trend actually comes from the scientific desire to understand and simulate the heart from first principles. Here, in the end, we need to include all relevant structures and the related properties into one model: The coronary arteries supplying the myocardium with oxygen, the myocardium contracting and performing a pump-action, the resulting blood flow in turn supplies oxygenated blood to the coronary arteries. Each of the three trends benefits from or even requires a comprehensive representation of all important cardiac structures in one model. In this paper the generation of such a comprehensive geometric heart model is described.

In addition to the information about shape and appearance of the object itself (as e.g. used in a model based segmentation approach), the model can provide information for proper initialization of position and pose of the object, e.g. by use of geometrical relations between the object of interest and other cardiac structures. For example the position of either manually marked or automatically detected landmarks can be used to estimate an initial spatial transformation to place the cardiac model into the image space. The landmarks used for the procedure must be part of the comprehensive model but they need not be part of the object of interest. Another possibility is a sequence of segmentation or adaptation procedures, each one being initialized with the result of the previous one. The result of an adaptation of a surface model to the left ventricle of the heart can be used to initialize the segmentation of the coronary arteries by transforming the coronary artery model into the image space and thereby restricting the search space for the subsequent coronary artery segmentation.

2 A Multi-component Model

The generation of a multi-component model raises several issues:

- **The combination of geometrical information form several sources**
  In our case, the mean geometrical model for the coronary arteries was taken from the literature [13,14], the cardiac surfaces and cardiac landmarks originate from multi-slice CT data which provides high resolution data of the complete heart (but with limited temporal resolution) and we intend to improve the motion model using cardiac MRI data which provides a better temporal but anisotropic spatial resolution and covers usually only the left and right ventricles. The information from all these sources needs to be combined.
- **The representation of consistent variability, avoiding conflicting deformation of the individual structures**
  The geometrical information provided by a comprehensive multi-object model may contain geometric entities of different representation. Some surfaces are perhaps represented as triangular meshes, others as spline surface patches, others may be represented using implicit functions. In addition to the surfaces, there may be vessel representations using centerlines and radius values etc. In this case, where we have different geometric parameterizations, the standard Eigen-value decomposition of the covariance matrix of the shape samples [15] cannot be used any longer. In addition, the available geometry samples may contain different sub-sets of the object set contained in the projected model. We think that this aspect calls for a distinct representation of the shape geometries and their variability. More specifically we propose to generate a mean comprehensive geometry model and separate deformation models, dealing with

inter-individual deformation and temporal deformation and defining each full-space deformations.
- **The representation of the topological, geometrical, anatomical, and physiological relations between sub-structures**
For higher levels of reasoning and user interaction, a complex anatomical model needs to be augmented with information beyond pure geometry. Anatomical nomenclature and its associated relations and hierarchies will e.g. help to display processing results adequately to the clinical user or to request user input, or to inhibit penetration or intersection of certain structures by other structures, during model adaptation.

3 General Structure of the Model

The model includes a definition of a set of cardiac landmarks and their mean locations, the mean geometry of the coronary arteries (centerlines and radii), the mean surfaces of the four cardiac chambers, and the connected vascular trunks, i.e. trunks of the vena cava, the pulmonary arteries, the pulmonary veins and the aorta. The mean geometries correspond to the end-diastolic cardiac phase. In addition to the mean geometries the model will be extended to include typical deformation patterns for inter-individual deformation and for temporal deformation. The deformation patterns are expressed as smooth full space transformations, independent of the geometric structures. All geometric entities feature an anatomical label. A nomenclature table allows the lookup of the respective anatomical name of the structure. Relation tables provide information about the relation of anatomical items. Currently the relations "is-part-of", "is-child-of", and "is-connected-to" are covered. To facilitate user interaction pictograms can be added to the model. Currently a pictogram of the coronary arteries derived from the one proposed by the American Heart Association (AHA) [16] is provided. The model is intended to support mainly image processing applications. In order to do so, the pure information (e.g. geometries, variability, meta-information) covered by the model is associated with application independent model related functionality. It covers basic individualization functionality (e.g. landmark based model deformation or model to image registration), meta-information related functionality (e.g. retrieve a list of related structures for a given structure of interest), and basic user interaction (e.g. rendering of image data together with geometric entities or pictogram based user-input). The model related functionality has been implemented in Java, model persistence is achieved by serialization of the model object entities to XML files.

4 Model Generation

4.1 The Coronary Artery Model

As the basis for the coronary artery model we used measurements from J. T. Dodge et al. about the location [13] and diameters [14] of human coronary arteries as reconstructed from bi-planar angiograms. In addition to the publicized values, J. T. Dodge kindly made available an updated and enlarged list of values. Dodge distinguishes 32 coronary artery segments. Each segment is trisected in a proximal, mid, and distal section and the center-point of each section is measured, giving in total 96 points defining the coronary artery tree. Basis for the measurements are 37 patients, categorized into three coronary supply types: right dominant, balanced and left dominant. The original point data is given in a spherical coordinate system. Based on this data, we constructed a coronary artery tree model. Since the point set provided by Dodge does not include the start and end point of each segment, we recovered the branching points by linear extrapolation and intersection with the parent segments. An interesting result of the measurements performed by Dodge, is the consistency of coronary artery location across the three supply types. The main property that differs depending on the supply type, is the tree topology, i.e. the connectivity between coronary artery segments. As a result, the arteries at the lower "back-side" of the heart are sometimes fed by right coronary artery and sometimes by the left circumflex coronary artery, but they stay mainly in place. Figure 1 shows a rendering of the resulting coronary artery model and the corresponding pictogram derived from the one recommended by the AHA [16]. The model was evaluated on multi-slice CT angiography (MSCTA) images [12]. The evaluation was restricted to the three main coronary arteries (left anterior descending, circumflex, and right coronary artery) being manually drawn in the images. The smaller branches could not be imaged with the necessary constant visibility over patient samples and were therefore left out. It could be shown that using an affine adaptation scheme, a mean residual distance between adapted model and sample lines of 2.7 mm could be achieved [12].

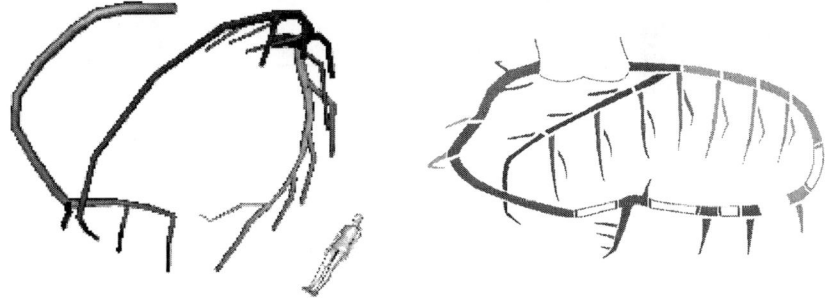

Fig. 1. Left: Coronary artery model, derived from the Measurements of Dodge et al. [13,14]. Right: Pictogram adopted from the AHA recommendation [16] using a coherent color scheme. The whitish colored coronary artery segments of the pictogram depict the variable portion depending on the supply type

4.2 Adding Landmarks to the Model

Cardiac landmarks are usually not of direct interest in cardiac diagnosis or treatment planning. However, they can serve as reference points that can be used to register image-data to image-data or model to image-data. Landmark positions may originate from user input or from automated detection algorithms. We defined a set of 25 landmarks (see figure 2). The landmarks were manually defined in 20 end-diastolic cardiac CTA datasets in order to create a mean landmark model. In addition to the landmarks, the three main coronary arteries were defined in the CTA datasets, in order to allow a registration between landmark model and coronary artery model. In order to calculate mean landmark positions, the landmark sets need to be transformed into a common reference coordinate system. We performed a Procrustes analysis [15] to find the optimal transformations for all shape samples given the allowed transformation class (similarity transformation). In order to transform the mean landmark model into the coordinate system of the coronary artery model, the transformations resulting from the Procrustes analysis are applied to the manually delineated coronary artery centerlines of the samples. By a subsequent match of the resulting bunch of coronary arteries to the coronary artery model, and applying the resulting transformation to the landmark model, we achieve a combined coronary artery and landmark model (Fig. 2).

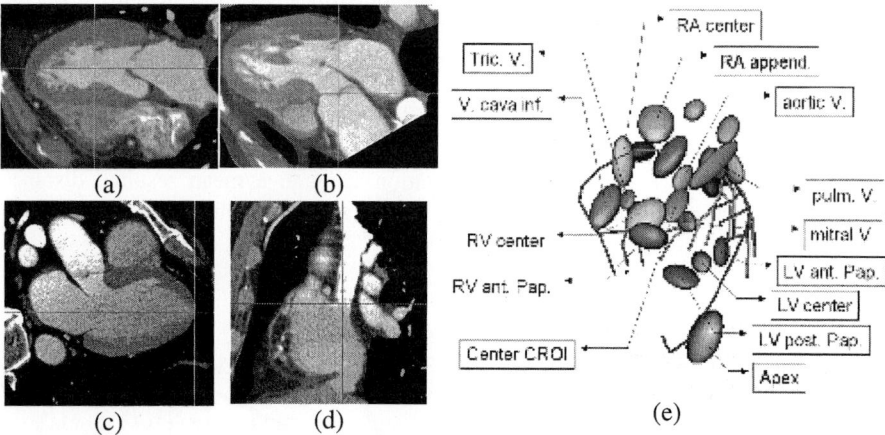

Fig. 2. Cardiac landmarks. The landmark set includes the overall center of the heart, the center-points of the four cardiac chambers, the four valve centers, apex, center of left anterior, left posterior, and right anterior papillary muscle, center points of left and right atrial appendage, left and right coronary ostium, bifurcation point of left anterior descending and circumflex coronary artery, the four ostia of the pulmonary veins, ostia of vena cava superior and inferior, and the ostium of the coronary sinus

Figures a-d: Some landmark examples, (a) center left ventricle, (b) aortic valve, (c) apex, (d) onset of vena cava superior. Figure (e) shows the error ellipsoids (directional std. deviation) centered at the landmark positions with the registered coronary artery model

4.3 Adding Cardiac Surfaces to the Model

With cardiac surfaces we mean the endo- and epicardium of the cardiac chambers and the walls of the connected vascular trunks. On the basis of state of the art 3D image material such as CT or MRI images, endo- and epicardium can often only be distinguised for the left ventricle. Therefore, for the time being, the right ventricle and the left and right atria are modeled with one surface each, representing endo- *and* epicardium. The following structures are included in the model: Left and right ventricle, left and right atrium, trunks of vena cava superior, vena cava inferior, pulmonary artery, pulmonary veins, and aorta. The surfaces are represented as a set of connected triangular meshes. A labeling scheme allows identifying for each triangle the corresponding cardiac structure.

The standard procedure to generate an anatomical surface model starts with the (usually interactive) segmentation and labeling of a learning set of data. In a second step either a mean label image is generated [3] and subsequently triangulated, or one label image is triangulated and the resulting mesh is adapted to the other label images [19]. For the generation of our cardiac surface model we tried to circumvent the necessity of a set of segmented datasets and chose for a bootstrap method working directly on un-processed 3D images. The main reason for this choice is to avoid the extremely time consuming procedure of manual or semi-automated segmentation of all the required cardiac structures. Our method makes use of available mesh generation and manipulation functionality [20] as well as active surface adaptation procedures for 3D image segmentation [21]. The procedure is somewhat similar to the one described in [19], but circumvents the use of labeled images. It consists of four main steps:

1. All cardiac structures of interest are independently interactively segmented in one high-quality, "normal" CTA image (root image). The segmentation is performed using an active shape procedure [21] starting from a simple, e.g. ellipsoidal or tubular shape. The segmentation of the individual structures is iteratively improved until a sufficient segmentation quality is reached. Each iteration consists of an automatic active surface based surface to image adaptation and a subsequent interactive correction at locations of insufficient match. The interactive corrections are mesh deformation operations working on an adjustable influence range [20]. The result of this step is a set of closed surfaces, each resembling one cardiac structure, i.e. one for the left ventricle, one for the left atrium etc.
2. Next, the set of surfaces from part one are merged to create one connected and labeled surface mesh. It requires the successive application of a handful of basic operations on surface meshes such as volumetric operations (union or difference operation applied to two closed surfaces), intersection and cut operations, and mesh refinement operations (e.g. in order to remove small triangles or to change the resolution of the triangulation) [22]. As long as the structures that need to be merged are overlapping, the merging operation can be performed automatically, given a set of closed surfaces and the desired triangle size. In case of non-overlapping structures that still need to be connected, some handcrafting is required. The result of this step is one connected and labeled mesh covering all input structures. The mesh resembles the shape of the structures as given in the input image.
3. Then, the mesh resulting from part two is adapted to a learning set of images. For initialization, a similarity transformation is applied to the vertices of the mesh resulting from part two. The transformation is estimated [23] on the basis of a set of cardiac landmarks defined in the root image and the image under

consideration. After initialization the mesh is semi-automatically adapted to the image similarly to the procedure in part two. The active surface method used during the adaptation procedure contains a shape term that minimizes triangle edge ratio differences between the model shape and the adapted shape. This leads to a predominant conservation of point correspondences [19]. The result of this step is a learning set of corresponding sample meshes.
4. Finally, based on the learning set of corresponding meshes from part three, a mean model and deformation modes can be extracted. The averaging can either be performed in the coordinate system of the landmark model or in a coordinate system resulting from a Procrustes analysis. According to our experience, the landmark based registration scheme works sufficiently well, a rigorous analysis of the influence of the registration scheme needs still to be performed.

The procedure described above is clearly biased by the selection of the root image. In order to reduce the influence, steps three and four may be iterated, similar to an iterative Procrustes procedures. The approximate time consumption of the above procedure is as follows. Step one requires about 5 min per structure. For all structures of the cardiac surface model this sums up to about one hour. The merging of structures in part two works largely automatically for nicely overlapping structures. Together with the remaining handcrafting step two requires again about one hour. The mesh adaptation to the set of learning samples in step three requires about 5 min. for the landmark definition and another 10 min. for the semi-automated adaptation procedure, summing up to 15 min. per learning sample. Thus, the construction of a model from 20 samples requires about 7 hours.

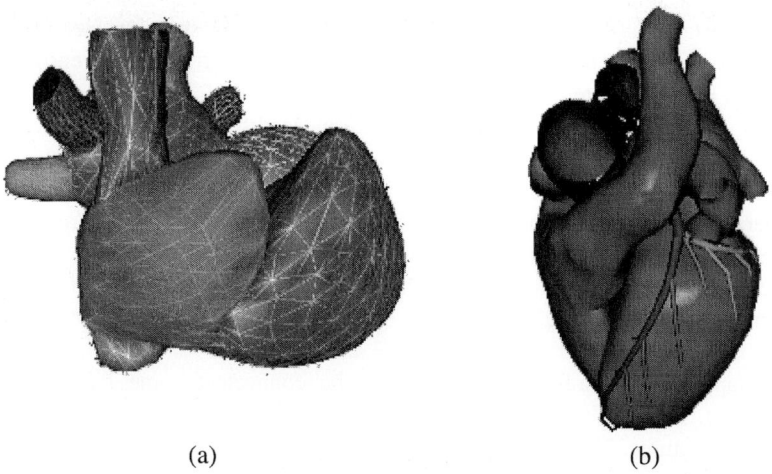

Fig. 3. (a) Triangular meshes representing the cardiac surfaces. (b) Registered surface and coronary artery model

4.4 Adding Variability to the Model

As pointed out in section 2, we follow the idea of separating the representation of (mean) shape and the representation of deformation. The advantage of this approach is firstly that it intrinsically enables a coherent deformation of the model, independent of the representation and parameterization of the geometry of the model parts. Secondly, it allows using deformation fields that originate from other sources, e.g. from elastic registration procedures or from tagged or phase contrast MRI images. The approach is related to methods that use elastic registration during the model construction or adaptation phase [24-26]. In order to realize this approach we need a common scheme to represent deformation. The steps to achieve this are sketched as follows. We assume that a deformation measurement consists of a set of deformation

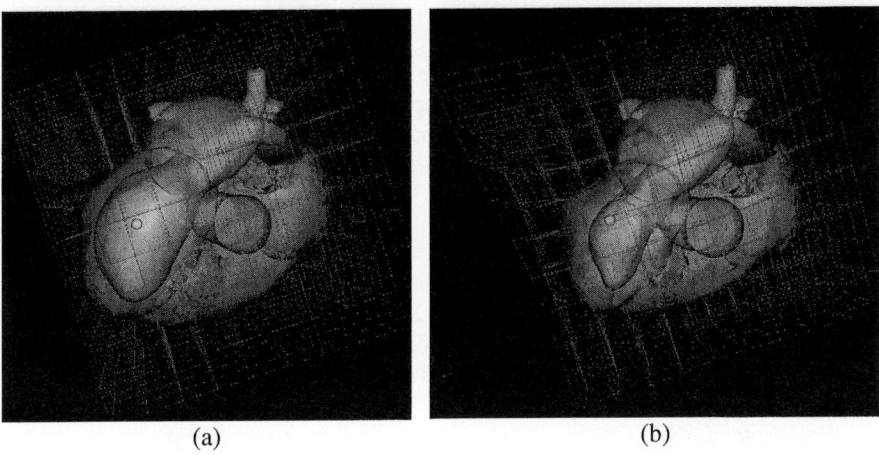

(a)   (b)

Fig. 4. Result of the adaptation of the cardiac surface model to a multi-phase CTA dataset. (a) depicts the end-diastolic and (b) the end-systolic heart phase. Based on the motion of the mesh vertices a full-space deformation field as been interpolated using a thin-plate-spline approach, depicted by the blue grid-lines

vectors in a given coordinate system. This could be a set of corresponding vertices of a surface mesh propagated through a time series of images, or corresponding tag-line crossings of a tagged MRI image series, or the motion of a grid of control points derived from an elastic registration. In a first step the deformation vectors are transformed into the model coordinate system based on the registration of a sub-set of geometrical model-items that can be delineated in the source images (e.g. a set of cardiac landmarks). In the second step a smooth deformation field is interpolated from the (potentially sparse) input vectors, e.g. by a thin-plate-spline (TPS) interpolation approach [27], resulting in a smooth full space deformation field. In a third step, the deformation field now defined everywhere is sampled in a standardized way, e.g. on a Cartesian grid. Based on the representation resulting from step three, the deformation fields can be averaged or further statistically analyzed. Figure 4 shows a deformation derived from surface tracking through the cardiac cycle. The end-diastolic (4a) and end-systolic (4b) shape of the cardiac surface model as adapted to a multi-phase cardiac CTA image is shown. The blue grid-lines visible in the images indicate the spatial deformation as derived from a TPS interpolation.

5 Conclusions and Future Work

The generation of a comprehensive geometrical model of the human heart has been described. Currently available is a mean model of the cardiac structures comprising the surfaces of the cardiac chambers and trunks of the connected vasculature, the coronary arteries and a set of 25 landmarks. The model is based on published data on the coronary arteries and on 20 multi-slice CT datasets. We advocate a distinct representation of the mean model geometry and model variability. A scheme to represent and add inter-individual and temporal variability to the model has been proposed. Current activities focus on enlarging the set of learning samples and on motion field extraction based on multi-phase cardiac CT and cine cardiac MRI image data. The next step will be the model application in the context of automated detection, segmentation and tracking of cardiac structures.

Acknowledgements

We would like to thank our colleagues from PMS-CT Cleveland, PMS-CT Haifa, PMS-MIT Best, PMS-MR Best, and PMS-XRD Best for the abundance of cardiac image material and many fruitful discussions. We also would like to thank J. T. Dodge for kindly making available updated position measurements of the coronary arteries.

References

1. A. F. Frangi, W. J. Niessen, and M. A. Viergever, "Three-Dimensional Modeling for Functional Analysis of Cardiac Images: A Review", IEEE Trans Med Imaging, vol. 20, no. 1, pp.2-25, Jan. 2001.
2. I. Haber, D. N. Metaxas, and L. Axel, "Three-dimensional motion reconstruction and analysis of the right ventricle using tagged MIR, Medical ImageAnalysis, vol.4, pp. 335-355, 2000.
3. A. F. Frangi, D. Rueckert, J. A. Schnabel, and W. J. Niessen, "Automatic Construction of Multiple-Object Three-Dimensional Statistical Shape Models: Application to Cardiac Modeling", IEEE Trans Med Imaging, vol. 21, no. 9, pp.1151-1166, Sep. 2002.
4. K. Park, D. N. Metaxas, and L. Axel, "LV-RV Shape Modeling Based on a Blended Parameterized Model", MICCAI 2002, LNCS 2488, pp. 753-761, 2002.
5. S. C. Mitchell, B. P. F. Lelieveldt, R. J. van der Geest, H. G. Bosch, J. H. C. Reiber and M. Sonka, "Multistage Hybrid Active Appearance Model Matching: Segmentation of the Left and Right Ventricles in Cardiac MR Images, IEEE TMI, vol. 20, no. 5, May 2001.

6. M. Lorenzo-Valdés, G. I. Sanchez-Ortiz, R. Mohiaddin, and D. Rueckert, "Atlas-Based Segmentation and Tracking of 3D Cardiac MR Images Using Non-rigid Registration", MICCAI 2002, LNCS 2488, pp. 642-650, 2002.
7. M. Lorenzo-Valdés, G. I. Sanchez-Ortiz, R. Mohiaddin, and D. Rueckert, "Segmentation fo 4D Cardiac MR Images Using a Probabilistic Atlas and the EM Algorithm", MICCAI 2003, LNCS 2878, pp. 440-450, 2003.
8. R. F. Schulte, G. B. Sands, F. B. Sachse, O. Dössel, and A. J. Pullan, "Creation of a human heart model and its customisation using ultrasound images", Biomedizinische Technik vol. 46, no. 2, 2001.
9. W. P. Segars, D. S. Lalush, and B. M. W. Tsui, "A Realistic Spline-Based Dynamic Heart Phantom", IEEE Trans nuclear science, vol. 46, no. 3, pp.503-506, June 1999.
10. N. P. Smith, A. J. Pullan, and P. J. Hunter, " Generation of an anatomically based geometric coronary model", Ann Biomed Eng. vol. 28, no. 1, pp. 14-25, 2000.
11. D. Sherknies, and J. Meunier, "A numerical 3D coronary tree model", Proc. 16th int. Congress and Exhibition Computer Assisted Radiology and Surgery (CARS'02), Springer, pp. 814-818, 2002.
12. C. Lorenz, J. von Berg, T. Bülow, S. Renisch, S. Wergandt, "Modeling the coronary artery tree", Shape Modeling Applications, 2004. Proceedings, pp. 354- 399, 2004.
13. J.T. Dodge, B.G. Brown, E.L. Bolson, and H.T. Dodge, "Intrathoracic spatial location of specified coronary segments on the normal human heart. Applications in quantitative arteriography, assessment of regional risk and contraction, and anatomic display", Circulation 78, pp. 1167-1180, 1988.
14. J.T. Dodge, B.G. Brown, E.L. Bolson, and H.T. Dodge, "Lumen diameter of normal human coronary arteries. Influence of age, sex, anatomic variation, and left ventricular hypertrophy or dilation", Circulation 86. pp. 232-246, 1992.
15. I. L. Dryden and K. V. Mardia, "Statistical Shape Analysis", Wiley, Chichester, 1998.
16. P.J. Scanlon, D.P. Faxon, "ACC/AHA Guidelines for Coronary Angiography", Journal of the American College of Cardiology, vol. 33, no. 6, pp. 1756-1824, 1999.
17. J. Lötjönen, S. Kivistö, J. Koikkalainen, D. Smutek, and K. Lauerma, "Statistical shape model of atria, ventricles and epicardium from short- and long-axis MR images", Medical Image Analysis, vol. 8, no. 3, pp. 371-386, Sep. 2004.
18. R. Pilgram, K. D. Fritscher, R. Schubert, "Modeling of the geometric variation and analysis of the right atrium and right ventricle motion of the human heart using PCA", Proc. 18th int. Congress and Exhibition Computer Assisted Radiology and Surgery (CARS'04), Springer, pp. 1108-1113, 2004.
19. M. R. Kaus, V. Pekar, C. Lorenz, R. Truyen, S. Lobregt, and J. Weese, "Automated 3-D PDM construction from segmented images using deformable models", IEEE TMI, vol. 22, no. 8, pp. 1005-13, Aug. 2003.
20. H. Timinger, V. Pekar, J. von Berg, K. Dietmayer, and M. Kaus, "Integration of Interactive Corrections to Model-Based Segmentation Algorithms", Bildverarbeitung für die Medizin Algorithmen, Systeme, Anwendungen. Proceedings des Workshops vom 9. - 11. März 2003 in Erlangen, Springer, 2003.
21. M.R. Kaus, J. von Berg, J. Weese, W. Niessen, and V. Pekar, "Automated segmentation of the left ventricle in cardiac MRI", Medical Image Analysis, vol. 8, no. 3, pp. 245-254, Sep. 2004.
22. J. von Berg, C. Lorenz, "Multi-Surface Cardiac Modeling, Segmentation, and Tracking", Functional Imaging and Modeling of the Heart, 2005. submitted
23. B. K. P. Horn, "Closed-form solution of absolute orientation using unit quaterions", Journal of the optical society of America A, Vol. 4, no. 4, pp. 629-642, 1987.

24. J. Lötjönen, D. Smutek, S. Kivistö, K. Lauerma, "Tracking Atria and Ventricles Simultaneously from Cardiac Short- and Long-Axis MR Images, MICCAI 2003, LNCS 2878, pp. 467-474, 2003.
25. M. Wierzbicki, M. Drangova, G. Guiraudon, ajnd T. Peters, "Validation of dynamic heart models obtained using non-linear registration for virtual reality training, planning, and guidance of minimally invasive cardiac surgeries, Medical Image Analysis, vol. 8, pp. 387-401, 2004.
26. A. Rao, G. I. Sanchez-Ortiz, R. Chandrashekara, M. Lorenzo-Valdés, R. Mohiaddin, and D. Rueckert, "Comparison of Cardiac Motion Across Subjects Using Non-rigid Registration, MICCAI 2002, LNCS 2488, pp. 722-729, 2002.
27. F. L. Bookstein, "Principal Warps: Thin-Plate Splines and the Decomposition of Deformations", IEEE PAMI, vol. 11, no. 6, pp. 567-585, 1989.

Automatic Cardiac 4D Segmentation Using Level Sets

Karl D. Fritscher, Roland Pilgram, and Rainer Schubert

Institute for Biomedical Image Analysis, University for Health Sciences,
Medical Informatics and Technology, Hall in Tirol, Austria
Karl.Fritscher@umit.at

Abstract. For the analysis of shape variations of the heart and the cardiac motion in a clinical environment it is necessary to segment a large amount of data in order to be able to build statistically significant models. Therefore it has been the aim of this project to find and develop methods that allow the creation of a fully automatic segmentation pipeline for the segmentation of endocardium and myocardium in ECG-triggered MRI images. For this purpose a combination of a number of image processing techniques, from the fields of segmentation, modeling and image registration have been used and extended to create a segmentation pipeline that reduces the need for supplementary manual correction of the segmented labels to a minimum.

1 Introduction

The analysis of shape and shape variations of organs and anatomical structures in general has become an important field of medical image processing. Detailed shape analysis gives the possibility to identify typical variations among healthy individuals in order to be able to distinguish them from pathological variations and improve the early diagnosis of diseases, which result in pathological variations of shape. Since the heart is a dynamic organ, not only the analysis of the cardiac shape, but also the analysis of the cardiac motion is a major topic in medical image analysis. For this purpose, the heart has to be segmented not only at one particular time, but during one cardiac cycle, which is typically consisting of 15-20 images using ECG-triggered MRI images. Due to the fact that such large amounts of data are needed in order to perform analysis of shape and shape variations, it has been the objective of this project to develop a pipeline that is providing methods, which allow fully automated and at the same time robust and effective segmentation of cardiac MRI-images.

In the last decade, deformable models [1] emerged as a well established method for medical image segmentation. Beside of parametric deformable models [1] also known as "snakes", introduced by Kass and Terzopoulos, geometric deformable models based on level sets [2, 3] became one of the most used methods in medical segmentation pipelines. The fact, that geometric deformable models can easily handle topological changes and are easily expandable from two to three dimensions made them a frequent choice for a number of extensions to the geometric deformable models originally introduced by Sethian and Osher [2, 3].

One characteristic of deformable models is that the segmentation process has to be started by providing an initial surface, which will be deformed and adapted to the image data by minimizing an energy functional. One possibility to generate such an initial surface, which is ideally already a good approximation of the structure to be segmented is to use user defined seed points and take them as a basis for e.g. fast marching segmentation [4]. The disadvantage of using these methods is, that they need user interaction to set the seed points and the resulting initial surface is very dependent on the location of the seed points. Therefore a better solution is to use a pre-defined initial surface to start the level set segmentation. One possibility to fulfill this task is to create a common shape template the from a number of segmented data sets by using principal component analysis [5]. The term common shape template is used, since we are only using the mean shape for initialization and not to guide the segmentation process, where the whole common shape model – including the principal components – would be used.

Having an initial surface, this surface has to be positioned - ideally - as near as possible to the boundaries of the structure to be segmented. For this purpose a registration of the dataset to be segmented, with the common shape template has to be performed. This can be done by registering the individual dataset and an atlas, containing the common shape template and a grayscale image that has once been aligned to the common shape template. A good choice for performing this task is to use mutual information metric [6].

For the segmentation of the whole cardiac cycle the segmented label from one point in time of the cardiac cycle can be used as initial template for the next point in time. The initial template for the myocardium segmentation is generated by creating distance maps of the segmented endocardium to produce initial templates for the myocardium. The final myocardium segmentation is again performed using level set segmentation.

Summing up, the objective of this project was to generate a pipeline for automatic segmentation of the endocardium and myocardium for a whole cardiac cycle, by using a common shape template for the initialization of the segmentation. For this purpose two main tasks had to be fulfilled

1. Building a common shape template of the four chambers of the heart in order to initialize the segmentation process.
2. Generating a segmentation pipeline that uses this template for initialization and succeeds in automatically generating labels of the cardiac endocardium and myocardium, which need no or minimal manual correction.

2 Methods

2.1 Geodesic Snakes/Geometric Deformable Models

Although parametric deformable models are quite intuitive to implement, they are also having some weaknesses, which are partially limiting the usability of this type of models: Firstly, realizing topologically adaptive parametric models, means to do some major modifications of the parametric deformable models, since any change in topology need new parameterization. During the evolution of a contour in the segmentation process, interfaces may change connectivity and split, thereby undergoing a topological transformation which is often very difficult to follow using traditional approaches. Moreover adapting parametric models to 3 or 4 dimensions is a very challenging task and requires computationally expensive methods [7].

In order to overcome these problems, geometric deformable models have been introduced in the field of image analysis by Caselles and Malladi [8, 9]. They are based on curve evolution theory and level set methods [4].

Being more independent from initialization than parametric deformable models, level sets are also designed to handle problems in which the evolving interfaces can develop sharp corners and cusps and change topology. Hence in order to provide a method, which is on the one hand capable of handling topological changes and on the other hand allow the usage of statistical shape models to guide the segmentation process in the future, geometric deformable models have been preferred to parametric deformable models in the course of this project.

In this project geometric deformable models have been used in two different concepts: Boundary driven geometric deformable models for the endocardium segmentation and region-competition snakes for myocardium segmentation.

Boundary-Driven Geometric Deformable Models. As posted in [10], geometric deformable models are defined as the zero level set of an implicit function ϕ, defined on the entire image. The evolution of the surface is defined via partial differential equation on the implicit function ϕ. Following the approach used by Caselles et al. [9] we are using the following formula

$$\frac{\delta \phi}{\delta t} = c(x)(\kappa + V_0)|\nabla(\phi)| + \beta(\nabla(P) \cdot \nabla(\phi)). \tag{1}$$

$\beta(\nabla(P) \cdot \nabla(\phi))$ is the projection of an attractive force vector to the surface. P is the gradient of a potential field, given as

$$P(x, y, z) = |(\nabla(G_\sigma * I(x, y, z)))|. \tag{2}$$

β denotes the strength of the attractive force and κ is the curvature dependent speed. $c(x)$ is the stopping term based on the image gradient and V_0 is a constant.

The curvature dependent stopping term adds some robustness concerning leakage through object boundaries and prevents the evolving contour from leaking through small gaps.

Region-Competition Snakes. In contrast to boundary driven snakes, geometric deformable models can also be governed by local probabilities that determine if the snake is inside or outside of the structure to be segmented. In this implementation of geometric deformable models, the propagation term is controlled in a way, that it shrinks, when the boundary encloses parts of the background and grows, when the boundary is inside the wanted regions [11].

In our implementation, based on the itkTresholdSegmentationLevelSetImageFilter of the Insight Segmentation and Registration Toolkit (ITK) [12] a speed term (feature image) with positive values inside an intensity window (between a low and high threshold) and negative values outside that intensity window is constructed. The evolving level set front will lock onto regions that are at the edges of the intensity window. In detail the feature image is calculated as follows (L...lower threshold, U...upper threshold).

$$f(x) = \begin{cases} g(x) - L & \text{if } g(x) < (U - L)/2 + L \\ U - g(x) & \text{otherwise} \end{cases} \quad . \tag{3}$$

In our application the thresholds can be calculated by calculating the mean grey value and standard deviation of the pixels, which are at the position of the template image in the original grayscale image. The thresholds are set by taking the mean grey value of the template region ± 1 standard deviation.

Furthermore, a Laplacian calculation on the image to the threshold-based speed term can be added. The Laplacian term causes the evolving surface to be more strongly attracted to image edges.

Identically to boundary driven snakes, an additional curvature based smoothing term adds robustness concerning leakage through object boundaries.

2.2 Model Building

Signed Distance Maps. For the purpose of building models of already segmented label data, we were choosing distance maps as a representation of shape following the approach of Leventon et al. [5]. A curve C which should be represented is embedded as the zero level set of a higher dimensional surface u, whose height is sampled at regular intervals. Each sample encodes the distance to the nearest point on the curve, with negative values inside the curve. The unsigned surface u is defined as

$$|u(x)| = \min_{q} \|C(q) - x\| \ . \tag{4}$$

Distance maps have the property, that the gradient magnitude of the image is constant across the image and equal to one. The direction of the gradient is equal to the outward normal of the nearest point on the curve C. From any point x in space the nearest point on the curve can be computed by

$$x - u(x)\nabla u(x) \ . \tag{5}$$

A distance map provides the propagation of the boundary information without loss of fidelity and the redundancy of information over a region in space provides stability in many types of computation.

Alignment of Distance Maps. In order to rigidly align the distance map representations of the individual labels, we were using mutual information (MI) independently introduced by Viola and Wells [6].

Given two variables U and V, mutual information is defined as

$$\text{MI}(U,V) = H(U) + H(V) - H(U,V). \tag{6}$$

Already applied to a wide range of applications for multi modality registration, MI turned out to be also very useful for the global alignment of distance functions and provided very reasonable results for the alignment of our signed distance maps.

Principal Component Analysis on Signed Distance Maps. Having a training set of signed distance maps, Principal Component Analysis can be used to derive a shape model [13]. A mean surface can be computed by taking the mean of the signed distance functions. The matrix of eigen-vectors and the diagonal matrix of corresponding eigen-values is computed from the co-variance matrix using Single Value Decomposition.

An estimate of a novel shape, u, can be represented by k principal components in a k-dimensional vector of coefficients, α:

$$\alpha = U_k^T(u-\mu). \quad (7)$$

U_k is a matrix consisting of the first k columns of the matrix of eigen-vectors U, which is used to project a surface into the eigen-space. Given the coefficients α, an estimate of the shape u is reconstructed from U_k and μ:

$$\tilde{u} = U_k \alpha + \mu. \quad (8)$$

Since distance transforms do not form a linear vector space, $\tilde{u}$ will in general not be a true distance function. However, the surfaces still have the properties of smoothness and local dependence, which is sufficient for our purposes [5].

2.3 Model to Image Registration

In order to register the common shape template to a new image, we are also using mutual information by rigidly aligning the grey-scale image, on whose segmented label data all other label data sets have been registered, and the image to be segmented.

For this purpose a multi-resolution registration approach has been used. This means that the images are registered in an iterative process, using different resolutions of the images. This fact adds robustness to the registration process and increases speed and accuracy.

3 Results

In the course of this project, we have been developing a C++ software-pipeline for fully automatic segmentation of 4D heart MRI datasets. This pipeline is implementing the methods described above, by using and extending some of the functionality of the Insight Segmentation and Registration Toolkit (ITK) and the Visualization Toolkit (VTK) [14].

As a first step we had to create a common shape template out of 10 segmented heart datasets. Since we also wanted to have the possibility to segment the four chambers of the heart separately, we have not only been creating a common shape template of the whole endocardium, but also of each chamber of the heart. For this purpose we created distance maps of the datasets, and rigidly aligned them by using mutual information. Fig. 2 is showing the results of the registration of the whole endocardium and of the left ventricle each with 5 datasets.

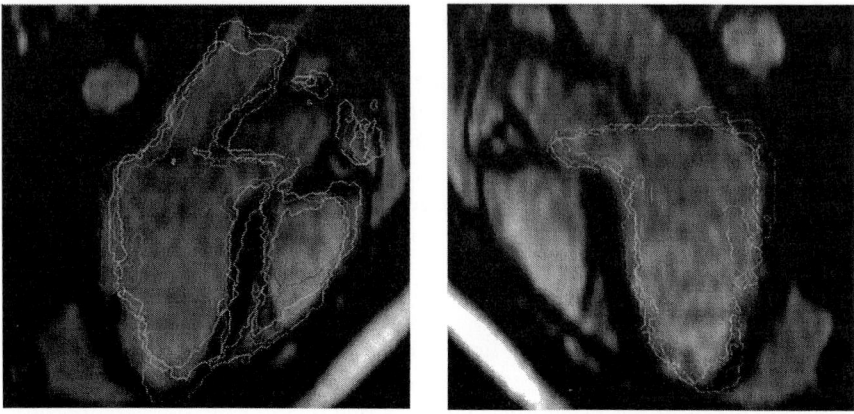

Fig. 1. Result of the rigid alignment of 5 labeled heart datasets. Left: Registered endocardium, Right: Registered left ventricles

Of course rigid alignment does not provide perfect correspondence, however, for the task of building a model/template to initialize the segmentation process and using signed distance maps, which are robust to slight misalignment as a representation of shape, we did not necessarily need perfect correspondence.

Using the methods described in section 2.2 we were calculating the common shape template and its principal components. Note, that the main variations of the model represented by the principal components are not involved in the segmentation process up to now, however, this might be part of our future work. Moreover, at this point of time the principal components are an additional important criterion to evaluate the validity of the model for our purposes. Fig. 2 is showing the common shape templates of the 4 chambers of the heart.

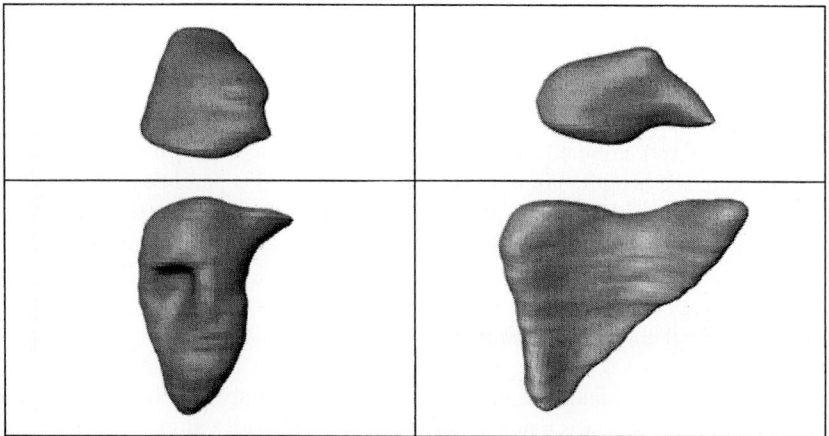

Fig. 2. Common shape templates of the four heart chambers. Top row: left atrium (left), right atrium (right), bottom row: left ventricle (left), right ventricle (right)

Having these initial templates the image to be segmented has been registered to the grayscale image containing the label data on which the distance maps have been registered, using mutual information. Another possibility would have been to directly register the new image to the distance map of the common shape template, however, in this case, the results turned out to be less robust and less correct, than in the first case.

The active geodesic level set segmentation process itself is started by using the common shape templates as initial templates and setting predefined parameters for the level set algorithm. As stopping criteria a threshold for the amount of change of the zero level set between two segmentation iterations has been used. Additionally a maximum number of iterations has been set. Due to the fact, that the segmentation is initialized very near to the object boundaries, the propagation scaling (~balloon force) can be set rather low. This brings the advantage that leaking through boundaries is less likely and the geometric deformable model is guided by the advection force, pulling the contour to edges in the image and the curvature term, preventing the contour from leaking and resulting in more robust convergence.

Fig. 3 is showing the initial templates and the results of endocardium segmentation. Note that the borders between atria and ventricles have not been manually corrected.

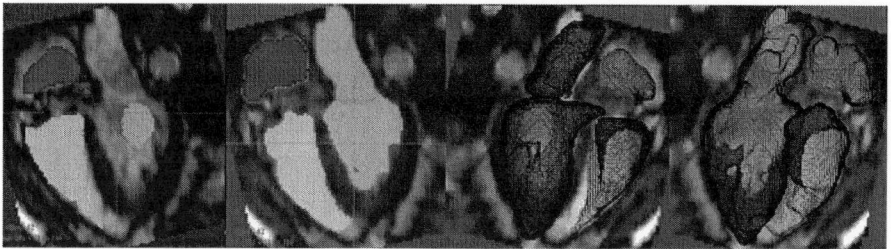

Fig. 3. Endocardium segmentation. Comparison of initial state and segmentation result of a endocardium segmentation in 2D (first two images) and 3D (third and fourth image)

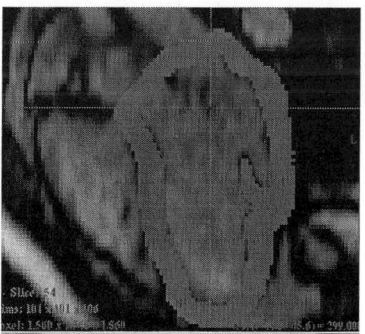 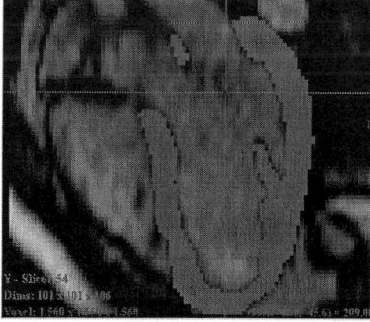

Fig. 4. Example for a template of the myocardium segmentation of the left ventricle generated via distance maps (left) and the final segmentation result for the left ventricle myocardium (right)

For the segmentation of the cardiac cycle, the segmentation results of the initial point in time have been used as initial templates [15]. For this purpose, the same transformation as for the first image has to be applied to the whole dataset of the cardiac cycle. An initial template for the myocardium segmentation can be generated by thresholding a signed distance map of the endocardium labels, for each chamber. Using this as an initial template, the thresholds for the region-competition snakes are computed by calculating the mean grey value of the pixels covered by the label. The thresholds are set by adding ±1 standard deviation to this mean value.

In this work the correctness of the segmentation has been evaluated for two different heart datasets: One dataset of the ten datasets, which have been used for the model building process (1) and one new dataset (2). In order to evaluate the correctness of the segmentation for the datasets, the generated labels have been compared with the labels after manual correction of the segmentation results, considering this as the gold standard. For endocardium and myocardium segmentation a similarity index for three different points in time of the cardiac cycle has been calculated by using

$$S = \frac{2|A \cap B|}{|A| + |B|}. \tag{9}$$

A and B are the non-zero pixels in the first and second input images. Operator $|\cdot|$ represents the size of a set and $\cap$ represents the intersection of two sets.

Table 1 is showing the results for the similarity indexes:

Table 2 is showing the undirected Hausdorff distances, comparing the automatically segmented endocardium and the manually corrected labels.

Table 1. Similarity indexes for automatic segmentation results and manually corrected segmentation results for two heart data sets (0, 60 and 110 ms after the R-peak in the ECG)

Similarity index	000 ms		060 ms		110 ms	
	Heart 1	Heart 2	Heart 1	Heart 2	Heart 1	Heart 2
Endocardium	0.980	0.992	0.970	0.980	0.971	0.976
Left ventricle	0.991	0.991	0.998	0.992	0.998	0.976
Right Ventricle	0.946	0.988	0.933	0.976	0.927	0.918
Left atrium	0.975	0.984	0.988	0.976	0.979	0.930
Right atrium	0.986	0.990	0.989	0.961	0.948	0.953
Myocardium	0.950	0.964	0.952	0.958	0.940	0.936

Table 2. Hausdorff distances in mm between automatic segmentation results and manually corrected segmentation results for two heart data sets (0, 60 and 110 ms after the R-peak in the ECG)

Hausdorff distance	000 ms		060 ms		110 ms	
	Heart 1	Heart 2	Heart 1	Heart 2	Heart 1	Heart 2
Endocardium	2.3	2.0	2.7	2.5	2.7	2.5
Myocardium	5.1	4.8	5.3	5.2	5.8	6.1

4 Discussion

Using mutual information to register the distance maps of the individual labels and the grayscale images to the atlas resulted in a precise rigid alignment and provided very satisfying results. Calculating a mean model resulted in a meaningful and feasible common shape template, which proved to be an adequate tool for the initialization of the level set segmentation process. Using geodesic level sets for the segmentation of the endocardium turned out to be an adequate choice and resulted in good segmentation results compared to the gold standard. Note that no manual correction has been performed between the segmentation of the different phases of the cardiac cycle. Performing minimal manual correction - especially the correction of the valve plane level - after segmenting the first point in time of the cardiac cycle would of course mean another improvement of the segmentation results.

The usage of ITK and VTK to implement the segmentation pipeline turned out to be an adequate choice for programming the software pipeline.

A detailed and comprehensive evaluation of the presented pipeline and an extension of the pipeline, which is using a modified template created out of more datasets and using other forms of shape representation in order to implement knowledge about shape variations in the segmentation process itself is currently under development.

Acknowledgement

We would like to thank the Department of Cardiology, Head: Univ. Prof. Dr. O. Pachinger and the Department of Radiology I, Head: Univ. Prof. Dr. W. Jaschke of the University Clinic Innsbruck for the acquisition of the MRI data. This project has been supported by the "Forschungsförderungsfonds für die gewerbliche Wirtschaft" (FFF- Austria).

References

[1] M. Kass, A. Witkin, D. Terzopoulos, "Snakes: active contour models", *International Journal on Computer Vision*, vol. 1, pp. 321-331, 1987.
[2] Sethian, J. A., "Curvature and evolution of fronts", *Commun. Math. Phys.*, vol. 101, 1985.
[3] J. Osher, J. A. Sethian, "Fronts propagating with curvature-dependent speed: algorithms based on Hamilton-Jacobi formulations", *Journal of Computational Physics*, vol. 79, pp. 12-49, 1988.
[4] Sethian, J. A., "Level Set Methods and Fast Marching Methods: Evolving Interfaces in Computational Geometry, Fluid Mechanics, Computer Vision and Material Science", 2nd Edition ed, Cambridge University Press, 1999.
[5] Leventon, M., "Statistical Models for Medical Image Analysis", in *Artificial Intelligence Lab.*: MIT, 2000.
[6] P. Viola, W M. Wells, "Alignment of maximization of mutual information", *International Journal on Computer Vision*, vol. 22, pp. 61-97, 1997.

[7] T. McInerney, D. Terzopoulos, "T-snakes: Topology adaptive snakes", *Medical Image Analysis*, vol. 4, pp. 73-91, 2000.
[8] R. Malladi, J. A. Sethian, B. C. Vemuri, "Shape modeling with front propagation: a level set approach", *IEEE TPAMI*, vol. 17, pp. 158-175, 1995.
[9] V. Caselles, R. Kimmel, G. Sapiro, "A geometric model for active contours", *Numerische Mathematik*, vol. 66, 1993.
[10] R. Goldenberg, R. Kimmel, R. Rivlin, E. Rudzsky, "Fast Geodesic Active Contours", *IEEE Transactions Imag. Proc.*, vol. 10, pp. 1476-1475, 2001.
[11] S. C. Zhu, A. Yuille, "Region Competition: Unifying Snakes, Region Growing, and Bayes/MDL for Multiband Image Segmentation", *IEEE Transactions on Pattern Analysis and machine Intelligence*, vol. 18, 1996.
[12] http://www.itk.org.
[13] M. Leventon, E. Grimson, O. Faugeras, "Statistical Shape Influence in Geodesic Active Contours", *Computer Vision and Pattern Recognition*, vol. 1, pp. 316-323, 2000.
[14] http://www.vtk.org.
[15] K. D. Fritscher, R. Schubert, "A software framework for pre-processing and level set segmentation of medical image data" presented at SPIE Medical Imaging, San Diego, 2005.

Level Set Segmentation of the Fetal Heart

I. Dindoyal[a], T. Lambrou[a], J. Deng[a,b], C.F. Ruff[c], A.D. Linney[a], C.H. Rodeck[b], and A. Todd-Pokropek[a]

[a] Departments of Medical Physics and Bioengineering,
[b] Obstetrics and Gynecology, University College London (UCL),
[c] UCL Hospitals NHS Trust (UCLH), UK
{i.dindoyal, t.lambrou, jdeng, cfr, alf, atoddpok}@medphys.ucl.ac.uk

Abstract. Segmentation of the fetal heart can facilitate the 3D assessment of the cardiac function and structure. Ultrasound acquisition typically results in dropout artifacts of the chamber walls. This paper presents a level set deformable model to simultaneously segment all four cardiac chambers using region based information. The segmented boundaries are automatically penalized from intersecting at walls with signal dropout. Root mean square errors of the perpendicular distances between the algorithm's delineation and manual tracings are within 7 pixels (<2mm) in 2D and under 3 voxels (<4.5mm) in 3D. The ejection fraction was determined from the 3D dataset. Future work will include further testing on additional datasets and validation on a phantom.

1 Introduction

Congenital heart disease affects about 8 in every 1000 births [1] and its signs can be diagnosed with prenatal echocardiography [2]. As with the adult heart, functional volume estimation of the left ventricle provides quantitative information about the state of the myocardium. However, in the fetus the blood flow in both sides of the heart is allowed to mix and so both ventricles are important for clinical assessment. One important application of fetal cardiac segmentation is for measurement of the absolute size of the chambers. This can be used for evaluation of the function of the heart, compromised either by cardiac malformations or by non-cardiac diseases such as immuno-haemolysis. In this condition the maternal immune system can kill fetal blood cells and so the fetal heart grows larger to compensate.

The prenatal heart has very thin chamber boundaries particularly in the areas consisting of the atrial septum, the membranous segment of the ventricular septum, and the valvular leaflets. Often the resolution of the ultrasound beam perpendicular to its axis is insufficient to resolve these structures and so these walls suffer from signal dropout and appear as holes in the endocardium. These dropouts can also be misleading for clinical diagnosis since the fetal heart contains septal holes which normally close at birth. Artefacts such as these complicate the automated functional volume quantification of each chamber for determining useful cardiac indices such as ejection fraction. In some cases it is difficult for fetal cardiologists to manually trace the endocardiac structures because of the missing image greyscale information.

Automated volume quantification in fetal cardiology is relatively new since it is difficult to acquire datasets without significant shadowing. Fetal body and cardiac motion artefacts are most noticeable when using slice-reconstruction 3D methods. Recent advances in volumetric acquisition have allowed the fetal heart to be imaged in 3D with considerably reduced motion artifacts [3], [4].

In the past Navaux and co-authors have published their work on segmentation of the 2D fetal heart by classification via neural networks [5], [6]. There has been relatively little use of deformable models to segment fetal cardiac data – only two papers in the literature currently exist: Lassige et al [7] used a level set snake to measure the size of the septal defects in echocardiographic images. This snake had a constant speed term that frequently overshot boundaries. In 2003 Dindoyal and co-workers presented an explicit 2D Gradient Vector Flow (GVF) snake algorithm with rigid body motion constraints to segment and track ventricles in 2D motion-gated fetal cardiac data [8]. Recently Esh-Broder et al [3] have collected over 20 3D fetal heart datasets. In this study the ejection fraction was estimated from the manual segmentation as well as comparison of both left and right ventricular volumes.

Section 2 outlines the proposed method to automatically segment the fetal cardiac chambers of two echocardiographic datasets after placement of manual seed points; one dataset was acquired from conventional 2D ultrasound slices and the other by Live 3D. The next section presents a selection of the images segmented with manual tracings for comparison as well as measurement of the accuracy. We then conclude the work and present further directions for study.

2 Method

2.1 Data Acquisition

3D acquisition of the fetal heart by serial slices was carried out by an online motion gated method pioneered in our group [9] using paired Acuson scanners (25 frames per second and a square pixel spacing of 0.26mm). True 3D acquisition of the fetal heart was performed with the Live 3D ultrasound scanner from Phillips [4]. This imaging system is capable of capturing about 24 volumes per second and can output a resampled cubic voxel resolution of $1.47mm^3$ for the penetration depth required. Although the 2D images from the paired scanners were stacked in 3D with motion gating, there was parts of the volume with noticeable motion artifacts that caused misalignment between slices. For this reason the volume dataset from the Acuson scanners was treated as separate 2D images. The Live 3D dataset did not suffer from this problem due to the volumetric acquisition method. Motion gating was unnecessary for this dataset since it was acquired with the probe kept immobile during no apparent fetal body movement.

2.2 Level Set Deformable Model

The level set method is defined implicitly compared to many adaptations of the snake model first introduced by Kass in 1988 [10] which track explicit markers. Level set methods can behave like the explicit case by chopping the level set function at the zero level (refer to Fig. 1). These implicit models have attractive properties in image segmentation such as automatic interpolation of the propagating front for irregular shaped boundaries and the ability to handle topological changes with ease.

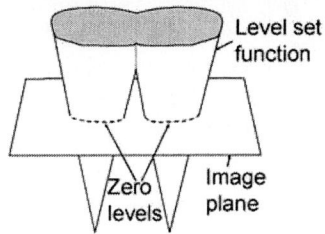

Fig. 1. Illustration of the level set function and zero levels for a 2D image

The generic level set equation for image processing can be written in the form

$$\dot{\varphi} = F|\nabla \varphi| \, . \tag{1}$$

where φ is the level set function and F is a problem dependent speed function.

The level set used in this paper was solved using a first order iterative scheme because of its low computational complexity

$$\varphi_{n+1} = \varphi_n + \Delta N F \varphi_n \, . \tag{2}$$

where n is the iteration number and ΔN is the timestep. Upwind differencing schemes were used where appropriate to maintain numerical stability around the propagating front as well as a small timestep. A form of narrow banding was used to speed up the level set propagation and to prevent nucleation of new fronts. The front is tracked on each iteration and its intersection with the edge of the narrow band can be predicted. When this occurs a new band is grown from the current zero level front by isotropic diffusion.

The front was manually initialized as a circle or sphere in each chamber. Each chamber contained a different snake which was stored in separate memory space to the others. This was implemented to prevent the level set merging of neighboring fronts. The distance transform for each snake was defined as a cone with negative values inside the front and positive elsewhere. This can be computed very quickly for such simple geometry in a single pass by computing the distance between voxel positions from the radius of the front. For non primitive initializations it may be necessary to use more general efficient distance transforms such as chamfering. The usual criterion of normalizing the distance transform was enforced.

Sarti et al 2002 [11] developed a level set algorithm with mean curvature and edge flow diffusion properties to segment datasets with missing boundaries. In this paper a new term was added to this evolution equation proposed by Sarti to incorporate region growing based on local deviations from the interior and exterior regions using the image part of the Mumford Shar (MS) functional. This term is useful in images without clear boundaries [12]. In the implementation proposed here the MS force is heavily penalized by curvature and inter-snake collision detection to reduce inter-chamber leakage. This is shown in equation (3) where Sarti's geometric model for boundary completion is enclosed in curly braces.

$$\dot{\varphi} = \left\{ \alpha g \nabla \cdot \left[\frac{\nabla \varphi}{|\varphi|} \right] + \beta (\nabla g) \cdot (\nabla \varphi) \right\} \|\nabla \varphi\| \\ + \left(\lambda_1 [I - \mu_i]^2 - \lambda_2 [I - \mu_o]^2 \right) \exp\left(-\kappa \nabla \cdot \left[\frac{\nabla \varphi}{|\varphi|} \right] \right) (1 - \xi) \|\nabla \varphi\|. \quad (3)$$

Where $g = (1+|\nabla GF|)^{-1}$ is an edge detector that returns a value between 0 and 1, with G denoting Gaussian filtering and F is the image. In the implementation for this paper image prefiltering to reduce noise was unnecessary for the volumetric data since the images were already at very low spatial resolution, but was necessary for the 2D sliced data due to the high speckle content.

In equation (3) φ is the level set function, I is the current voxel intensity under investigation. μ_i and μ_o are the means of the internal and outside regions of the dataset defined by the level set front. ξ is a function that tests if any of the enclosed regions from individual snakes overlap. If there is overlap ξ returns 1 and 0 otherwise. β is a function to penalize edge advection in the presence of local edges and is defined as $\beta = \exp(-\gamma|\nabla GF|)$. In Sarti's original formulation β is a unitary constant. The factors α, γ, λ_1, λ_2, κ are empirically determined weighting coefficients for the respective terms.

In Sarti's equation the first term is standard mean curvature flow weighted by an edge stopping coefficient. It serves to regularize the curve where the data is sparse and propagation can be further reduced by the presence of edges. The advection term drives the front towards image edges that have been defined from a pre-computed edge diffusion field. The main weakness of this term is the presence of many edges at various strengths as is often found in sonography. Edge flow by advection is heavily dependent on the quality of the edgemap and so may fail to propagate the front towards the edges sufficiently to overcome the mean curvature flow.

The proposed term aims to provide some expansion or contraction forces dependent on the local tissue type in the absence of a strong edge field, e.g. when the front is in homogeneous regions. Unlike the constant advection term in Lassige's algorithm [7] this force can propagate the front in either direction according to the position of the boundaries and so would be less prone to overshoot. The MS factor models the foreground and background of the image and tries to minimize its energy by separating these two regions. The foreground was estimated from a small circle/sphere placed inside the chamber prior to evolution and the background was assumed to be the remainder of the dataset. Since the appropriate λ_1 and λ_2 could potentially vary significantly between datasets, the images were normalized to reduce the dependence on these coefficients.

The exponential factor contains a second mean curvature component and its presence is mostly required where there is extensive shadowing to the chamber. The collision detection component is heavily penalizing and tends to stop two intersecting fronts immediately upon contact so that a steady wall is formed where the two interfaces meet. Although open valves cause blood from atria and ventricles to mix cardiac function in clinical use is measured by treating each chamber in isolation.

From preliminary experiments it was discovered that for the collision to occur at the right place (where part of the chamber wall has suffered signal dropout due to the beam resolution); the two snakes should be started from as close to the centers of their respective chambers as possible. This prevents one snake from invading the adjacent chamber due to its arrival at the missing boundary first.

3 Results and Discussion

Fig. 2 illustrates the effect of the added term to Sarti's equation. Without the presence of a clear edgemap from the data Sarti's snake fails to propagate appreciably towards the desired boundary. To overcome this restriction we used the bidirectional MS term in conjunction with Sarti's algorithm which yielded a closer segmentation to the expert's delineation. We used manual expert tracings by a fetal cardiologist as a gold standard. Full interactive segmentation of the images proved to be both challenging and tedious in particular for areas with partial volume artifacts of the papillary muscles and the missing atrial septum.

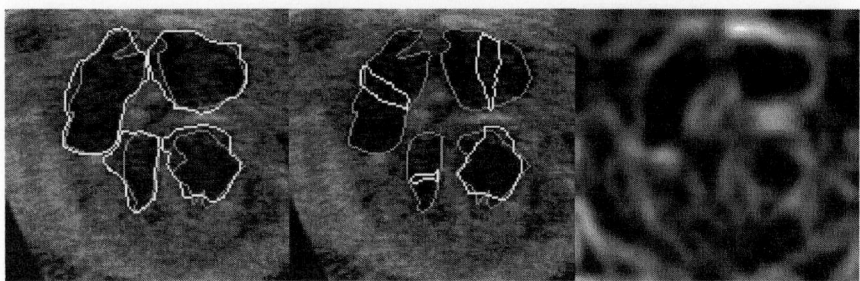

Fig. 2. Segmentation by the algorithm proposed in this paper (*left*) and segmentation by Sarti's algorithm (*middle*). The white contours are automatically generated and grey denotes manual tracings. Atria appear at the top of the image and ventricles at the bottom. The *right* image shows the edgemap

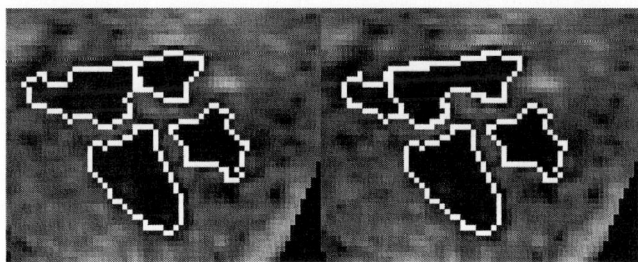

Fig. 3. Effect of the collision penalization term (enabled in *left* image and disabled in *right* image)

Fig. 3 shows the effect of the collision penalization term. The left image shows the atrial boundary reconstructed at approximately the location of the true boundary and the right image illustrates overgrowth of the fronts.

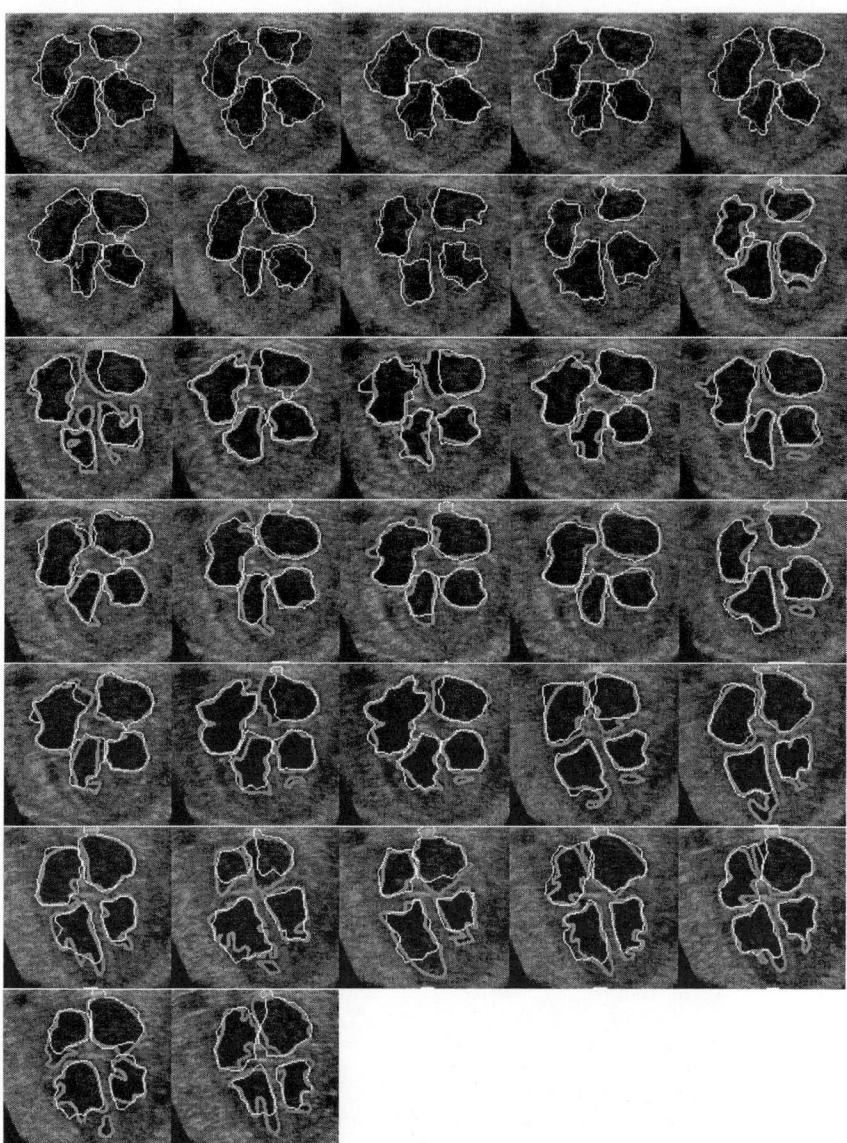

Fig. 4. Segmentation results of the algorithm on the 2D slice data (*white*) superimposed on manual tracings (*grey*)

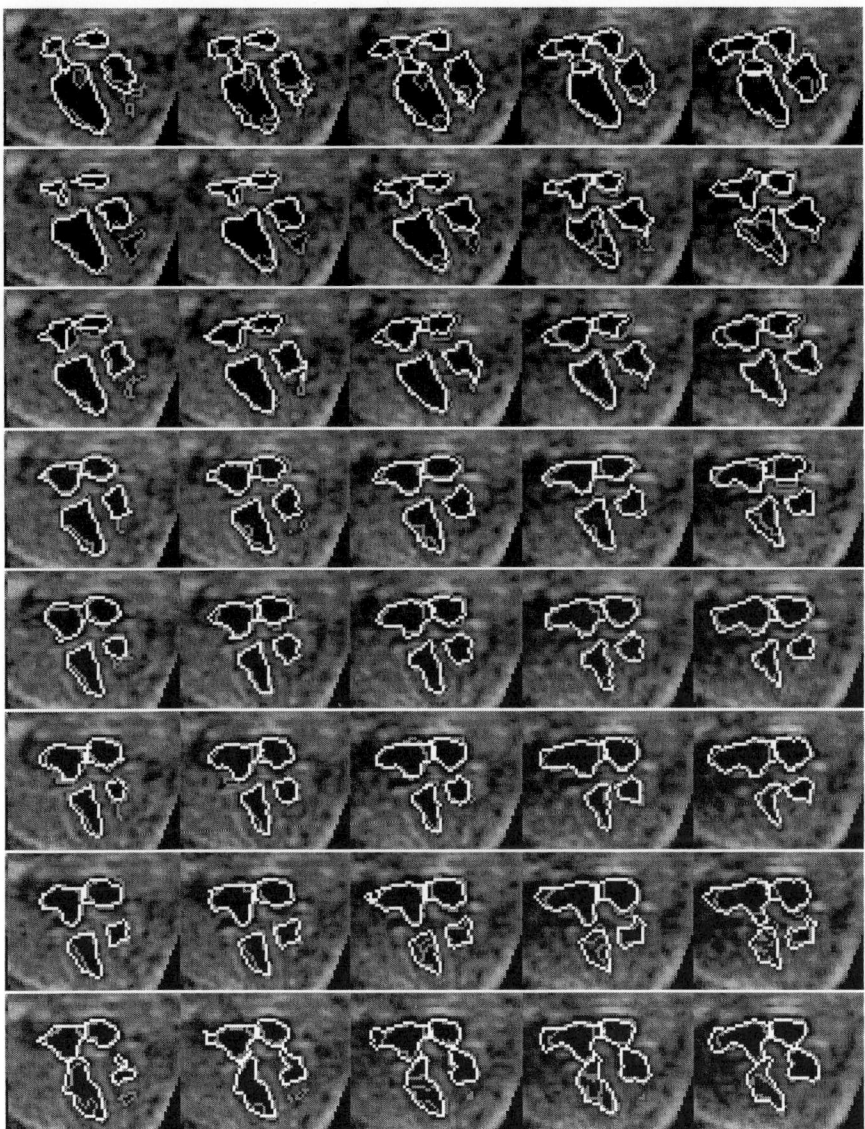

Fig. 5. 3D segmentation on the Phillips Live 3D data (only central slices are shown for clarity). Space varies *horizontally* and time *vertically* in this figure. *White* contours were generated by the algorithm and *grey* ones are manual tracings

The algorithm was applied to 32 2D images from the Acuson scanner and a 3D dataset from the Phillips Live 3D scanner. Inter-operator variability of manual segmentation was not measured for comparison with the automated delination in this paper. It was assumed that the repeatability of manual tracing to within an error of 2 voxels would be sufficient for clinical use. To access the accuracy of the algorithm root mean square (rms) errors were computed from point-wise distances between the automatic boundaries and the manual tracings. The segmentation results in 2D can be seen in Fig. 4 and 3D analysis is displayed in Fig. 5.

Frequency analysis of the rms errors is shown in the bar charts in Fig. 6. The rms errors are within 7 pixels (<2mm) in 2D for the chambers excluding the right atrium. In Fig. 6 the scale was truncated to empathize the distribution of 2D rms errors excluding outliers of the right atrium. The outliers stretch out to 15 pixels. Fig. 4 shows several examples where the snake is attracted to regions of echo enhancement in the ultrasound image above the right atrium. The algorithm appears to be influenced by strong intensity inhomogeneity artifacts and this pulls the contour away from the desired endocardial boundaries.

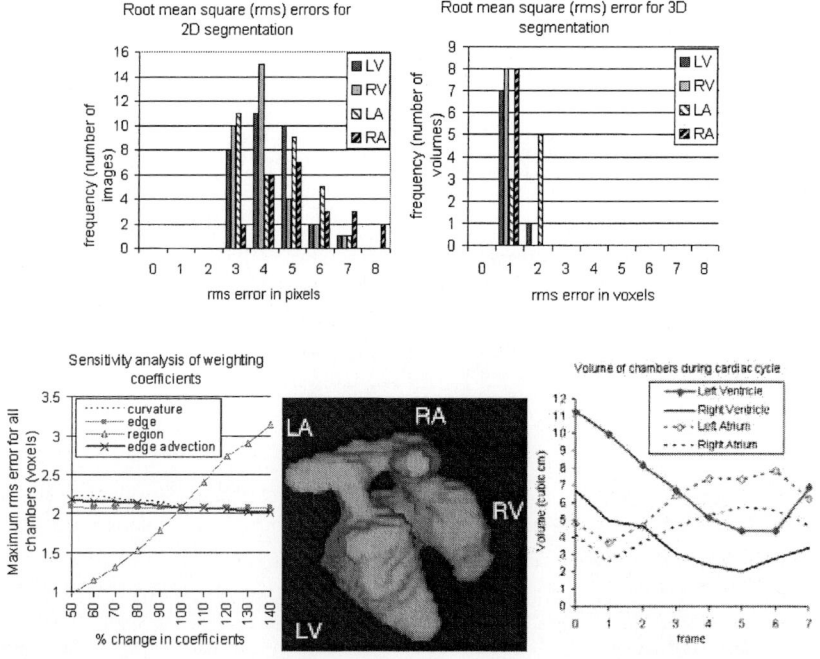

Fig. 6. *Top row*: rms errors of segmentation for each chamber. *Bottom left*: sensitivity analysis of the weighting coefficients in the 3D segmentation. *Bottom middle*: example volume rendering of segmented Live 3D dataset (during ventricular diastole and atrial systole, posterior view). *Bottom right*: volume-time curves for all frames

In 3D the rms errors were between 1-2 voxels (4.5mm). The graphs show that the spread of rms errors is greater in the 2D case although the overall errors from the 3D algorithm are larger when expressed in mm units. We attribute the cause of this to the lower spatial resolution in direct volumetric acquisition which increases the partial volume effect. Strong echo enhancement effects would further increase the spatial rms errors but these were absent in this dataset.

In many of the 2D segmented images the front stopped short of the desired boundary and this shows up in the high rms errors (refer to Fig. 6). Whilst high curvature penalization was partially responsible, premature stopping of the level set front was also due to contributions from the type of image forces used. The MS term models unchanging mean intensities both inside and outside of a small sphere or circular seed placed inside the chamber. If the mean was updated as the snake evolved the front could come to rest closer to the boundary. The edge flow term requires a diffusion equation to be applied to an edge map and so broadens edges. The Gaussian prefiltering used a large kernel of width 9 pixels and this could contribute to the edge broadening.

To assess the dependence of the segmentation on the weighting coefficients, these parameters were varied independently within a range of 50-140% of their appropriate values for this application. The test was arbitrarily run on the 1st frame of the Live 3D dataset and for this range of parameter values, rms errors in the segmentation remained below 3.5 voxels for all chambers. Fig. 6. shows that the MS region term is most strongly affected by choice of weighting coefficients.

The left and right ventricular ejection fractions were computed as 61% and 59% respectively from the automatic segmentation (frames 0 and 6 were identified as the necessary cardiac time points to perform this calculation). These values are comparable to the ranges found in a study by Esh-Broder [3] (57.5±14.6% and 54±11.2% respectively), who also measured a non-significant variation between the left and right ventricles. A volume-time graph of all four cardiac chambers is shown in Fig. 6 as well as a volume rendering of the level set front.

4 Conclusion

To our knowledge this is the first time non manual segmentation techniques have been applied on 3D prenatal heart data to measure volumes and cardiac indices. The automated method provides a segmentation that is far quicker and more repeatable than manual tracings; but has problems in delineating fine intra-cavity structures and is strongly affected by enhancement of echogenic regions. The 2D data appears to be easier to automatically segment than the new Live 3D possibly due to the smaller pixel size which allows better definition of thin walls and intra-cavity structures. However, fetal cardiac volumetric imaging simplifies the gating process and will be the method of choice for acquiring future datasets.

Future work will involve validation of the accuracy of the automated volume determination from a phantom as well as further testing on additional datasets.

Acknowledgements

We are grateful for the volume rendering provided by Daren McDonald and stimulating discussions about optimization strategies with Dr Robin Richards. This work was supported by EPSRC (GR/N14248/01) and MRC (D2025/31) under the

Interdisciplinary Research Consortium scheme - "From Medical Images and Signals to Clinical Information" (MIAS IRC). Dr Jing Deng is supported by MRC (G108/516).

References

1. Mitchell SC, Korones SB, Berendes HW. Congenital heart disease in 56,109 births. Incidence and natural history. Circulation, Vol. 43. (1971) 323-332
2. Copel JA, Gianluigi P, Green J, Hobbins JC, Kleinman CS. Fetal echocardiographic screening for congenital heart disease: The importance of the four-chamber view. British Journal of Obstetrics and Gynaecology, Vol. 157. (1987) 648-655
3. Esh-Broder E, Ushakov FB, Imbar T, Yagel S. Application of free-hand three-dimensional echocardiography in the evaluation of fetal cardiac ejection fraction: a preliminary study. Ultrasound in Obstetrics & Gynecology, Vol. 23. (2004) 546-551
4. Deng J. Terminology of three-dimensional and four-dimensional ultrasound imaging of the fetal heart and other moving parts. Ultrasound in Obstetrics & Gynecology, Vol. 22. (2003) 336-334
5. Piccoli L, Dahmer A, Scharcanski J, Navaux POA. Fetal echocardiographic image segmentation using neural networks. In. Image Processing and its Applications 1999, Seventh International Conference. (1999)
6. Siqueira ML, Scharcanski J, Navaux POA. Echocardiographic image sequence segmentation and analysis using self-organizing maps. Journal of VLSI Signal Processing, Vol. 32. (2002) 135-145
7. Lassige TA, Benkeser PJ, Fyfe D, Sharma S. Comparison of septal defects in 2D and 3D echocardiography using active contour models. Computerized Medical Imaging and Graphics, Vol. 24. (2000) 377-388
8. Dindoyal I, Lambrou T, Deng J, Ruff CF, Linney AD, Todd-Pokropek A. An active contour model to segment foetal cardiac ultrasound data. In. Medical Image Understanding and Analysis. 03 Jul 10, University of Sheffield, UK. (2003)
9. Deng J, Ruff CF, Linney AD, Lees WR, Hanson MA, Rodeck CH. Simultaneous use of two ultrasound scanners for motion-gated three-dimensional fetal echocardiography. Ultrasound in Medicine and Biology, Vol. 26. (2000) 1021-1032
10. Kass M, Witkin A, Terzopoulos D. Snakes: Active Contour Models. International Journal of Computer Vision, Vol. 1. (1988) 321-331
11. Sarti A. Subjective surfaces: a geometric model for boundary completion. International Journal of Computer Vision, Vol. 46. (2002) 201-221
12. Gibou, F. and Fedkiw, R. A fast hybrid k-means level set algorithm for segmentation. Stanford Technical Report. (2002)

Supporting the TECAB Grafting Through CT Based Analysis of Coronary Arteries

Stefan Wesarg

FhG-IGD, Dept. Cognitive Computing & Medical Imaging, Darmstadt, Germany
stefan.wesarg@igd.fraunhofer.de

Abstract. Calcified coronary arteries can cause severe cardiac problems and may provoke an infarction of the heart's wall. An established treatment method is the bypass operation. The usage of a telemanipulation system allows for the execution of that operation as a totally endoscopic coronary artery bypass (TECAB) grafting. This relatively new method narrows the surgeon's view and does not permit the palpation of the vessel in order to detect calcifications (hard plaques).

A planning based on contrast enhanced, cardiac CT data sets can compensate for that problem. This work presents analysis methods for coronary arteries. Hard plaques are detected using a tracking-based vessel segmentation technique. In addition, the vessel's neighborhood is analyzed in order to decide whether it is surrounded by tissue or fat, or if it is freely accessible for the surgeon's instruments. Furthermore, well adapted methods for the visualization of these analysis results are presented.

Keywords: Coronary arteries, vessel segmentation, calcification detection, minimally invasive surgery, cardiac imaging, computed tomography.

1 Introduction

In the developed nations, malfunctions of the cardiovascular system are widespread. Often an obstruction of coronary arteries causes severe risks for the patient's health. The herewith related coronary artery disease (CAD) can hinder the blood-flow towards the areas of the myocardium close to the heart's apex. In the worst case this can provoke an infarction of the heart's wall.

The bypass grafting is an established procedure for the treatment of obstructed coronary arteries. Conventionally, it is executed as open-chest surgery. There, the surgeon can directly look onto the artery and detect hard plaques by palpation of the vessel. New medical devices like 'telemanipulators' change the way that such operations are executed (see fig. 1). They allow for a special form of minimally invasive surgery – the totally endoscopic coronary artery bypass (TECAB) grafting [1, 2].

There, the chest is no longer opened completely, and the intervention can be done on a beating heart. The surgical instruments and an endoscope are inserted into the patient via small ports and controled by the surgeon from a

 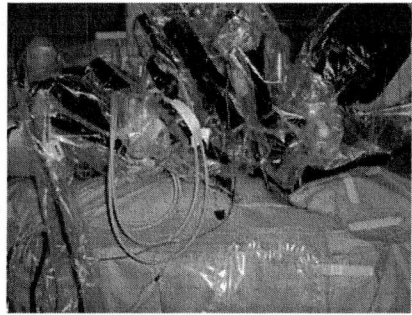

Fig. 1. A telemanipulator based system for minimally invasive surgery (*daVinci*, Intuitive Surgical): The surgeon controls the instruments from a console that is providing a stereoscopic view on the operation area (left). The instruments and the endoscope are attached to the arms of the system, and inserted into the patient via ports (right)

console that is providing a stereoscopic view on the operation area. That type of the bypass grafting results in less trauma for the patients, a faster recreation and a much lower infection risk. On the other hand, the TECAB grafting requires more experience of the surgeon due to the limited view during the operation. Finally, only a good planning of the intervention can lead to an optimal result.

An analysis of the coronary artery can compensate for the limitations due to the minimally invasive character of the operation. Considering the fact that a huge amount of time is used for accessing the artery that is often hidden behind fat or muscle tissue, it would be desireable to have this information already prior to the operation. Also, the position of hard plaques should be known before in order to decide where the bypass should be attached to the vessel.

In the past, coronary artery calcium has often been detected based on electron-beam computed tomography (EBCT) data. Several publications describe methods for quantifying the calcium [3,4], others focus on the reproducibility of the calcium scoring [5]. The reason for the usage of EBCT was the much shorter acquisition time of EBCT compared to conventional computed tomography (CT) of these days.

But the advent of multi-slice computed tomography (MSCT) scanners in combination with a still increased rotation speed now allows for high-quality cardiac imaging based on conventional CT. The simultaneous acquisition of projection data in 16 or more detector rows makes it possible to acquire the whole heart during one single breath-hold [6]. This opens new horizons for the employment of MSCT for cardiac imaging and analysis; especially in the domain of non-invasive coronary angiography [7]. CT based analysis of the coronary arteries is still a relative new technique. Most of the approaches described in the literature are limited to the detection of stenoses and an analysis of 2D image data [8,9]. To the knowledge of the author there has not been published any work regarding true 3D analysis of coronary arteries based on MSCT coronary angiography yet.

The 'gold standard' for the analysis of the coronary artery tree is still conventional angiography. However, this imaging modality is not used for preparing a TECAB grafting. The reason for that is the fact that angiography is an invasive modality that is not suitable for the examination of bypass patients. Furthermore, angiography does not provide any information about the tissue surrounding the coronary artery tree.

There are several publications dealing with the planning of a TECAB grafting using a telemanipulation system. However, the authors mainly focus on the optimization of the port placement for avoiding collisions of the instruments while assuring the reachability of the heart and the coronary artery the surgeon is focusing on [10,11].

The here presented work introduces methods for localizing calcifications in coronary arteries and inspecting the tissue in their neighborhood aiming on supporting the TECAB grafting in cardiac surgery.

2 Material and Methods

It is desireable to present the image data used for the planning of the intervention in 3D, rather than as conventional 2D slices. However, a 3D view of the image data based on direct volume rendering does neither allow a good perception of coronary arteries and calcium therin, nor the constitution of the surrounding tissue can easily be determined. Therefore, a vessel analysis based on the result of its segmentation is introduced, and well adapted methods for the presentation and exploration of the analysis results are presented.

For the acquisition of our test data a contrast agent has been used to enhance the visibility of the cardiovascular structures. The projection data has been acquired with a multislice CT scanner (SIEMENS *Somatom Sensation 16*) and reconstructed based on the simultaneously recorded ECG data. This resulted in high-quality image data without severe artifacts. The data sets consisted of nearly cubic voxels with a size of about 0.5 mm for each direction.

2.1 Segmentation of Coronary Arteries

We used our own tracking-based vessel segmentation technique that has been developed for the reliable extraction of coronary arteries from high-resolution CT data sets [12]. Vessels, that are containing a contrast agent, are relative homogeneous and show a high contrast with respect to the surrounding tissue. This allows for a detection technique we call the 'corkscrew algorithm', and that is truely working in 3D. Thus, a connection between the user defined start and end point following a helical – or corkscrew-shaped – path is searched. It provides in the first step an estimation for the centerline, that is afterwards corrected iteratively by detecting the voxels that belong to the vessel's border. The algorithm's output is a set of points defining the centerline and another one representing the border of the artery. For more details see reference [12].

2.2 Hard Plaque Detection

The output of the vessel segmentation presented in the preceding section allows for the subsequent calcification detection. Each computed point of the centerline has a corresponding set of points representing the vessel's border in perpendicular direction to the centerline segment. As a consequence, the diameter of the coronary artery can easily be computed, resulting in a diameter function for the segmented coronary artery.

The employed vessel extraction approach excludes calcifications from the segmentation result. Hence, calcified regions are expected to lower the mean diameter, since the corresponding border points lie 'in front' of them. Based on the generated diameter function and the image data a three-step analysis is performed (see fig. 2):

1. From the diameter function calcification candidates are extracted by selecting those centerline points with a corresponding diameter below a certain threshold. (This could be for instance the mean value of that function.)
2. Afterwards, these candidate points' neighborhoods are searched through whether voxels with high gray values are present. Calcifications are assumed to be 20 % to 30 % brighter than the vessel's lumen that is filled with a contrast agent. Only those candidate points possessing a neighborhood that fulfills that brightness condition are kept.
3. In a last step, the remaining candidate points are analyzed whether they are close to the same calcification. If several candidate points form a group sharing the same calcification, the position laying in the middle between the first and the last point of the group is stored.

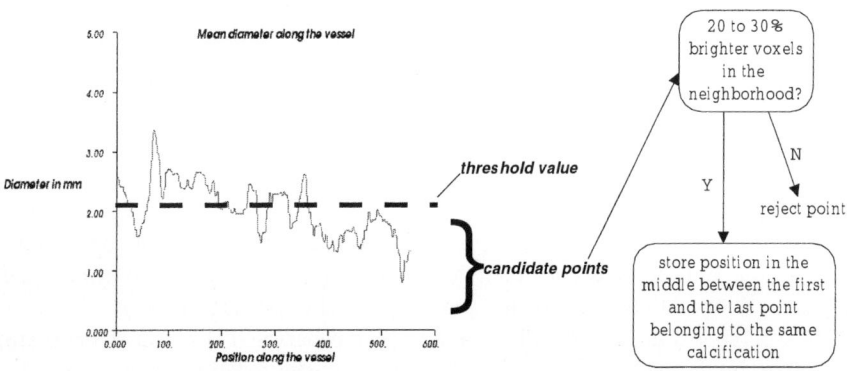

Fig. 2. The analysis for localizing calcifications in coronary arteries: A set of candidate points is selected based on an automatically computed threshold. These points are further analyzed whether bright gray values are in their neighborhood. At the end, the remaining points are decimated again for assuring that only one of them belonging to the same single calcification is stored

The result of that analysis is a set of points with on one hand a position related to a relative low diameter of the coronary artery and on the other hand a neighborhood containing voxels with a high gray value. These two conditions are expected to reliably localize calcifications.

2.3 Inspection of the Artery's Neighborhood

Coronary arteries are often not freely accessible for the surgeon's instruments, since they are surrounded by muscle tissue or fat. The preparation of a TECAB grafting requires the coagulation of that matter in order to isolate the vessel. This procedure is very time-consuming, since it has to be done very prudently in order to not hurting the artery. That process and the TECAB grafting in general could be speed up if there was a possibility for classifying and quantifying the tissue in the vessel's neighborhood.

We propose an analysis method that is based on our vessel segmentation described above. For each set of border points belonging to the same centerline point a set of rays starting from the centerline and passing through the corresponding border points is considered. Along these rays, starting behind the border point, gray value samples from the image data are taken. These values are converted into Hounsfield units (HU) (a task that can easily be done for CT data, since the necessary information is stored in the header of the DICOM [1] data). Based on the obtained HUs the tissue can be roughly classified into air (HU: ≈ -1000), fat (HU: -220 to -20), and muscle tissue (HU: 20 to 50). The length of the rays as well as the sampling rate for the gray value acquisition can be selected by the user.

2.4 Visualization of the Analysis Results

In this section we describe the methods that we have developed for the presentation of the analysis results. They have been designed to be well adapted to the planning of a TECAB grafting. All visualization tasks are done using the freely available toolkit VTK[2].

The 'natural' way of presenting the generated diameter function is an x-y plot (see left part of fig. 2). For a direct visualization of the diameter function in the volume rendered view, we implemented a filter that creates a tube around the generated centerline. This tube's cross-section dimension varies the same way the diameter function does. In addition, it is colored based on the diameter values. A red color signifies a low diameter, whereas a blue color stands for a large one (see fig. 3).

For visualizing the results of the hard plaque detection multiple cone-shaped pointers are used for indicating the detected calcifications directly in the 3D view. In addition, their positions are given in a list box control. By clicking

---

[1] The DICOM standard (Digital Imaging and Communications in Medicine) (http://medical.nema.org/)
[2] The Visualization Toolkit by Kitware, Inc. (http://www.vtk.org)

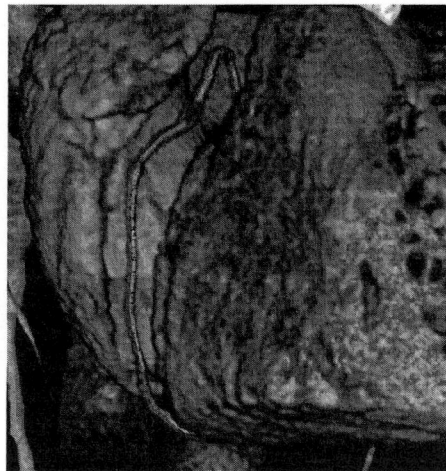

Fig. 3. Visualization of the diameter function directly in the 3D rendered volume of a cardiac MSCT data set: A tube around the computed vessel's centerline is shown. It varies in diameter corresponding to the diameter function. In addition, this variation is color-coded using a linear rainbow-based transition from red to blue. (The artery itself can be perceived as shadow around the generated tube)

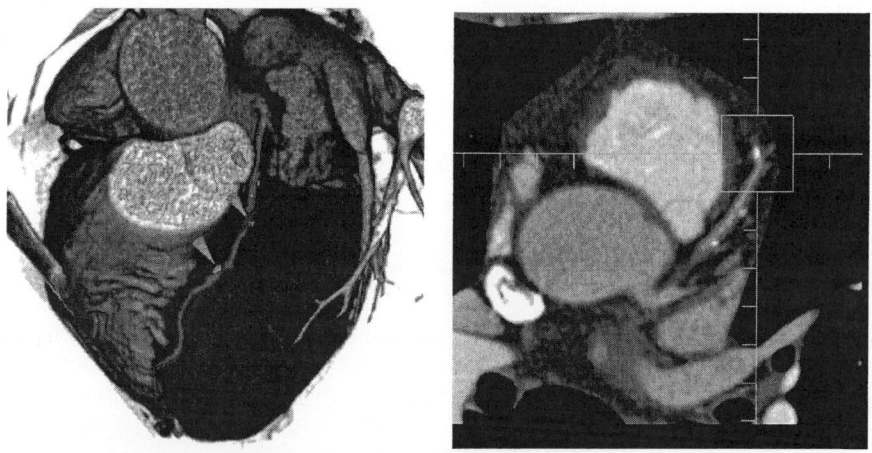

Fig. 4. Detection of multiple hard plaques in a coronary artery: All three calcifications present in the LAD (= left anterior descending) have been detected and are indicated by a pointer (left). Selecting one of them from a list box control adjusts the 2D views and highlights the position of the calcification – axial view shown as example (right)

on one of these entries the corresponding 2D views are shown, highlighting the calcification position (see fig. 4).

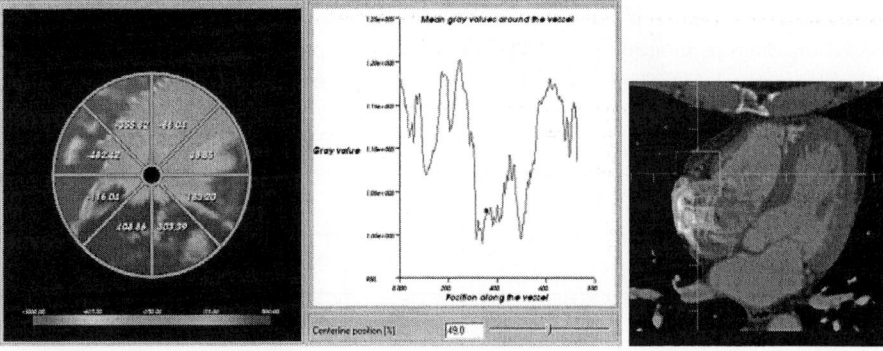

Fig. 5. Visualization of the coronary artery's neighborhood: A polar plot shows the color coded HUs (red: air, greenish cyan: fat, blueish cyan: muscle, intense blue: contrast filled right atrium and ventricle) for a selected centerline point. The mean gray values of the artery's neighborhood are shown in an x-y plot (left). Comparison with the corresponding original axial slice shows a strong coincidence with the output of the tissue classification. The small red circle is the outer boundary of the segmented artery. The region's size shown in the polar plot equals approximately the circle through the square's corners (right)

A special visualization method for the inspection of the vessel's neighborhood has been developed. It consists of two outputs. The first one displays the HUs of that neighborhood for each centerline point as color coded values in a polar plot. There, the correct relative dimensions of the vessel's mean diameter and the selected ray length are used. A slider control allows for navigating along the centerline and inspecting the coronary artery's neighborhood. An x-y plot displays the mean gray value of the neighborhood for every centerline point.

In addition, the correspondence between this special visualization method and the conventional way of displaying medical image data can easily be verified. For each position along the centerline the 2D views can be aligned according to the x (sagittal), y (coronal), and z (axial) position, and the corresponding point in such a 2D slice is highlighted (see fig. 5). In the 3D view that position is indicated by a small pointer.

3 Results

The proposed analysis techniques have been tested with CT data from 7 patients. For all of them the left anterior descending (LAD), the left circumflex (LCX), and the right coronary artery (RCA) have been segmented using the corkscrew algorithm [12]. The segmentation did not always extract the whole artery due to imaging artifacts that interupted the continuous run of the vessel. In these cases only that part that has been segmented could be further analyzed.

Table 1. The results for the automatic detection of hard plaques in coronary arteries based on their segmentation: The data sets of seven patients have been used, and in all of them the LAD, the LCX, and the RCA have been inspected

Data set	Hard Plaques	LAD	LCX	RCA
1	Visible	0	0	0
	Detected	0	0	0
2	Visible	3	1	1
	Detected	3	1	1
3	Visible	5	2	5
	Detected	6	2	5
4	Visible	1	0	0
	Detected	1	0	0
5	Visible	3	0	0
	Detected	3	0	0
6	Visible	6	0	1
	Detected	6	0	1
7	Visible	4	0	2
	Detected	5	0	2

The color mapping of the computed artery's diameter on a tube around the vessel's centerline allowed for a perception of the change of that parameter along the artery without any difficulty. This way of representing the diameter function is more convenient than providing only its x-y plot.

Table 1 shows the results for the automatic detection of hard plaques. The above introduced automatic technique detected reliably the present calcifications. All of the hard plaques that have been found during a preceding manual inspection of the axial slices have been tracked. Two of them have been indicated twice due to their large size that led the algorithm to an overestimation of the number of calcifications in these cases (data sets 3 and 7). Those segments that did not contain any visible hard plaque have been classified by our technique to be calcification-free, i. e., our tests resulted in a false-positive value of 0.

The method for inspecting the artery's neighborhood has been tested with the same datasets where the LAD, the LCX, and the RCA have been segmented. There, only a retrospective analysis could be done since the data came from patients whose treatment was already finished. Consequently, we only checked whether the color-coded display of the HUs for the surrounding tissue corresponded to what could be perceived in the conventionally displayed image data. For a run of the artery in axial direction this could simply be done by comparing the colored polar-plot of the analysis window with the axial slice (see fig. 5). For more 'difficult' directions of the vessel the other two 2D views as well as the 3D view have been used. In all of the cases this visual evaluation showed a perfect correspondence: Air was displayed red, fat in greenish cyan, and muscle tissue in blueish cyan. In addition, hard plaques and contrast agent filled cavities could easily perceived since they appeared as intense blue areas.

4 Discussion

The presented, yet preliminary, analysis results using the segmentation output of our tracking-based approach [12] turned out to be robust. The detection of hard plaques delivered in all cases the correct number and positions of the calcifications that have been found by a visual inspection of the image slices. In two cases a single but widespread hard plaque has been detected twice. But, this can not be considered as a serious drawback of our technique. Very important for a clinical use of the automatic hard plaque detection is the false-positive value. Here, the tests resulted in a value of 0 – no false indication of non-existing hard plaque. However, the criteria that the hard plaque detection is based on are of a rather qualitative nature, and an extended, clinical evaluation of our technique is needed (see the end of this section).

The highlighting of the calcifications' positions also in the conventional 2D slices establishs a relationship between the newly introduced technique and the manual inspection of the slices. Thus, radiologists will hopefully accept the automatic technique without reservation, since they can still verify the analysis' output in their habitual way.

Stenoses that may be introduced by calcifications can easily be found when inspecting the vessel's diameter. The proposed method of generating a colored tube that varies in diameter the same way the artery does makes it easy to localize areas of small vessel diameter in the 3D view of the volume. There is no need any longer to mentally map the x-y plot to the image data.

Our method for analyzing the neighborhood of the artery is an innovative approach for providing essential information for the planning of a TECAB grafting. The HU based color coded visualization of the surrounding tissue and air makes it easy for the surgeon to estimate the amount of tissue that has to be removed in order to access the artery.

As a limitation of this work one might consider the fact that the presented methods have been tested only with data from seven patients. We are aware of this, and consequently a clinical study together with our partners from the University Hospital Frankfurt has been started. This study is also aiming on the comparison of our CT based analysis methods with the current 'gold standard' – conventional angiography.

The inspection of the vessel's neighborhood could be done only as a retrospective study of patients whose treatment has already been finished. The aforementioned clinical study will also determine the expected improvements for the TECAB grafting in terms of speeding up the dissection of the artery.

Finally, the here presented methods for visualizing the analysis results are an improvement over existing approaches. But, they are not an optimal solution yet. Therefore, future work will focus on a simulation of the limitations the surgeon is experiencing through the fixed angle view of the used endoscope. In addition, an augmentation of the endoscopic view provided by the telemanipulation system's console with the analysis results would be a possible and usefull extension of our proposed approach.

Acknowledgements

We want to thank the clinic for thoracic, cardiological and thorax vasculum surgery of the University Hospital Frankfurt for providing us the CT data sets.

References

1. Loulmet, D., Carpentier, A., d'Attellis, N., Berrebi, A., Cardon, C., Ponzio, O., Aupecle, B., Relland, J.Y.: Endoscopic Coronary Artery Bypass Grafting with the Aid of Robotic Assisted Instruments. J Thorac Cardiovasc Surg **118** (1999) 4–10
2. Dogan, S., Aybek, T., Andreßen, E., Byhahn, C., Mierdl, S., Westphal, K., Matheis, G., Moritz, A., Wimmer-Greinecker, G.: Totally Endoscopic Coronary Artery Bypass Grafting on Cardiopulmonary Bypass with Robotically Enhanced Telemanipulation: Report of forty-five Cases. J Thorac Cardiovasc Surg **123** (2002) 1125–1131
3. Agatston, A.S., Janowitz, W.R., Hildner, F.J., Zusmer, N.R., Viamonte, M., Detrano, R.: Quantification of Coronary Artery Calcium Using Ultrafast Computed Tomography. J Am Coll Cardiol **15** (1990) 827–832
4. Becker, C.R., Knez, A., Jakobs, T.F., Aydemir, S., Becker, A., Schoepf, U.J., Bruening, R., Haberl, R., Reiser, M.F.: Detection and Quantification of Coronary Artery Calcification with Electron-Beam and Conventional CT. European Radiology **9** (1999) 620–624
5. Callister, T.Q., Cooil, B., Raya, S.P., Lippolis, N.J., Russo, D.J., Raggi, P.: Coronary Artery Disease: Improved Reproducibility of Calcium Scoring with an Electron-Beam CT Volumetric Method. Radiology **208** (1998) 807–814
6. Ohnesorge, B., Flohr, T.: Non-Invasive Cardiac Imaging with Fast Multi-Slice Cardio CT. electromedica 68, Siemens AG, Medical Engineering (2000)
7. Traversi, E., Bertoli, G., Barazzoni, G., Baldi, M., Tramarin, R.: Non-Invasive coronary angiography with multislice computed tomography. Technology, methods, preliminary experience and prospects. Ital Heart J **5** (2004) 89–98
8. Gerber, T.C., Kuzo, R.S., Karstaedt, N., Lane, G.E., Morin, R.L., Sheedy, P.F., Safford, R.E., Blackshear, J.L., Pietan, J.H.: Current Results and New Developments of Coronary Angiography With Use of Contrast-Enhanced Computed Tomography of the Heart. Mayo Clin Proc **77** (2002) 55–71
9. Dewey, M., Schnapauff, D., Laule, M., Lembcke, A., Borges, A.C., Rutsch, W., Hamm, B., Rogalla, P.: Multislice CT Coronary Angiography: Evaluation of an Automatic Vessel Detection Tool. Fortschr Röntgenstr **176** (2004) 478–483
10. Cannon, J.W., Stoll, J.A., Selha, S.D., Dupont, P.E., Howe, R.D., Torchiana, D.F.: Port Placement Planning in Robot-Assisted Coronary Artery Bypass. IEEE Trans. on Robotics and Automation **19** (2003)
11. Coste-Maniere, E., Adhami, L., Severac-Bastide, R., Lobontiu, A., Salisbury, J.K., Boissonnat, J.D., Swarup, N., Guthart, G., Mousseaux, E., Carpentier, A.: Optimized Port Placement for the Totally Endoscopic Coronary Artery Bypass Grafting using the da Vinci Robotic System. In Rus, D., Singh, S., eds.: Experimental Robotics VII. Volume 271 of Lecture Notes in Control and Information Sciences., Springer-Verlag Heidelberg (2001) 199–208
12. Wesarg, S., Firle, E.A.: Segmentation of Vessels: The Corkscrew Algorithm. In Robert L. Galloway, J.E., ed.: Medical Imaging Symposium 2004. Volume 5370 of Proc. of SPIE. (2004) 1609–1620

Clinical Validation of Machine Learning for Automatic Analysis of Multichannel Magnetocardiography[1]

Riccardo Fenici, Donatella Brisinda, Anna Maria Meloni,
Karsten Sternickel, and Peter Fenici

Clinical Physiology - Biomagnetism Research Center,
Catholic University of Sacred Heart, Largo A. Gemelli 8, 00168 Rome, Italy
feniciri@rm.unicatt.it

Abstract. Magnetocardiographic (MCG) mapping measures magnetic fields generated by the electrophysiological activity of the heart. Quantitative analysis of MCG ventricular repolarization (VR) parameters may be useful to detect myocardial ischemia in patients with apparently normal ECG. However, manual calculation of MCG VR is time consuming and can be dependent on the examiner's experience. Alternatively, the use of machine learning (ML) has been proposed recently to automate the interpretation of MCG recordings and to minimize human interference with the analysis. The aim of this study was to validate the predictive value of ML techniques in comparison with interactive, computer-aided, MCG analysis.

ML testing was done on a set of 140 randomly analysed MCG recordings from 74 subjects: 41 patients with ischemic heart disease (IHD) (group 1), 32 of them untreated (group 2), and 33 subjects without any evidence of cardiac disease (group 3). For each case at least 2 MCG datasets, recorded in different sessions, were analysed.

Two ML techniques combined identified abnormal VR in 25 IHD patients (group 1) and excluded VR abnormalities in 28 controls (group 3) providing 75% sensitivity, 85% specificity, 83% positive predictive value, 78% negative predictive value, 80% predictive accuracy This result was for the most part in agreement, but statistically better than that obtained with interactive analysis.

This study confirms that ML, applied on MCG recording at rest, has a predictive accuracy of 80% in detecting electrophysiological alterations associated with untreated IHD. Further work is needed to test the ML capability to differentiate VR alterations due to IHD from those due to non-ischemic cardiomyopathies.

1 Introduction

Magnetocardiographic mapping measures magnetic fields generated by the electrophysiological activity of the heart, and is a promising imaging technology developed for the rapid, non-invasive detection of ventricular repolarization abnormalities. MCG data are usually mapped, simultaneously or sequentially, from

---

[1] Partially supported by MIUR grants # 9906571299_001, 2001064829_001 and by the National Science Foundation, SBIR phase II award #0349580.

33-60 locations above the frontal torso, using superconducting quantum interference devices (SQUIDs).

Previous research[1] has shown that, compared to standard ECG, multichannel MCG provides non-invasive evaluation of cardiac electrogenesis, with similar investigation time, but higher spatial and temporal resolution.

The diagnostic potential of MCG mapping ranges from three-dimensional electroanatomical localization of arrhythmias, to the identification of VR abnormalities in patients with myocardial ischemia and non-diagnostic ECG[1,2].

The analysis of VR from MCG mapping can be done visually and/or quantitatively. Quantitative VR parameters can be calculated from the ST interval and/or the T wave[3-10]. Interactive computer-aided analysis of MCG parameters, especially of the ST interval, can be influenced by low signal to-noise ratio (SNR) and by the examiner's experience. Therefore, automatic analysis procedures are needed to speed-up the procedure and to minimize human input.

The aim of this study was to validate automatic classification of Magnetocardiograms using a Machine Learning (ML) approach, developed under the NSF SBIR phase I grant #0232215 and described by Szymanski et al[11]. The performance was compared to computer-aided interactive analysis of MCG mapping, independently performed by two expert cardiologists.

As the ST-segment has usually a low SNR in magnetocardiograms, whereas the T-wave is most likely to show primary abnormalities due to ischemia and has a high SNR, ML was applied to the magnetic field data of the T-wave only.

2 Methods

2.1 Instrumentation and Data Pre-processing

MCG mapping was performed at rest in supine position, with a 36-channel MCG system (*CardioMag Imaging Inc., USA*)[12] based on DC-SQUID sensors coupled to second order gradiometers (baseline: 50-70 mm) with pick-up coils diameter of 19 mm and sensor-to-sensor spacing of 40 mm. The distance between the measuring sensors, kept at liquid helium temperature and arranged in a horizontal plane, and the flat bottom surface of the cryostat is 19 mm^2. With a built-in automatic electronic noise suppression system (ENSS), the instrumentation reaches a sensitivity of about 20 fT/Hz½ at 1 Hz, with balance stability of gradiometers better than 0.01%.

All MCG signals and one reference 12-lead ECG were simultaneously recorded for 90 seconds, at a sampling rate of 1 kHz, in the bandwidth from DC to 100 Hz.

All recordings were performed without electromagnetic shielding, in a room fully equipped for cardiac catheterization and intensive care. Digital low pass filter at 20 Hz was used before ML was applied. To eliminate stochastic noise components, all signals were averaged. For automatic classification, data from a time window between the J point and T peak were used.

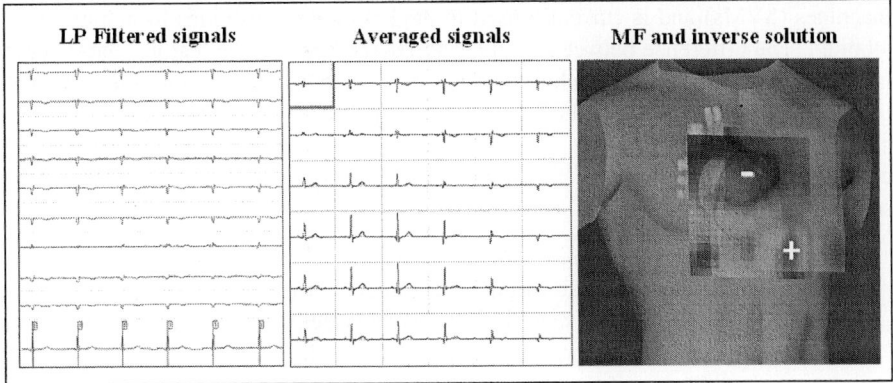

Fig. 1. MCG signal processing

2.2 Signal Processing

To eliminate stochastic noise components, all signals were averaged using the maximum of the R peak as a trigger point (Figure 1).

VR was analyzed according to specific preset parameters, and two ML scores automatically calculated for each subject resulting in an MCG classification of either normal or abnormal (Figure 2).

The tool used for ML is called Direct Kernel partial least squares (**DK-PLS**). Partial least squares (PLS) are one of the standard analysis methods in QSAR and chemo metrics[14]. Kernel PLS (K-PLS) is a recently developed nonlinear version of PLS, introduced by Rosipal and Trejo[15]. K-PLS is functionally equivalent to support vector

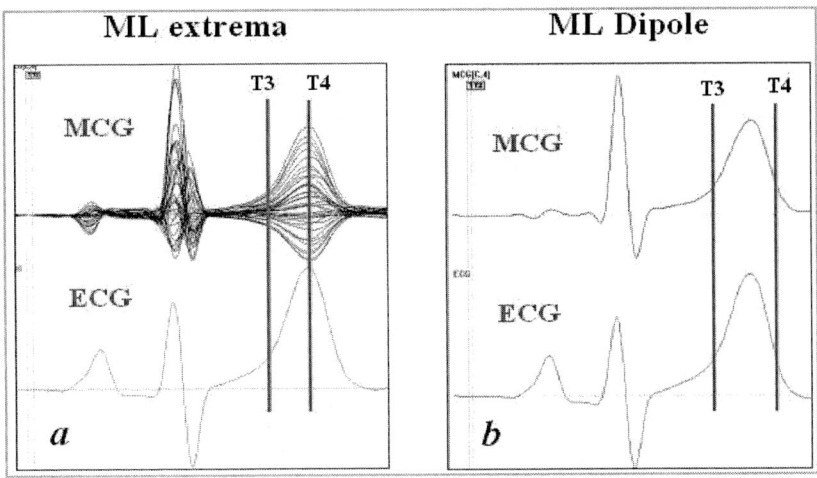

Fig. 2. Time intervals (indicated by T3 and T4 bars) from which the two ML scores were calculated

machines (SVMs) and is currently used to predict binding affinities to human serum albumin. The difference between K-PLS and DK-PLS is that the feature (data) kernel matrix is used in K methods while this matrix is replaced by the (non-linear) kernel-transformed matrix in DK methods[16]. DK-PLS reached a convincing performance in a preliminary preclinical test[11]. The algorithm was trained on data from 73 cases considering MCG patterns of ischemic and non-ischemic patients. Two diagnostic scores were calculated:

1) The "**ML extrema**" score, based on wavelet transformed MCG patterns of the upslope of the T-wave as shown in Figure 2a (abnormal if > 50).
2) The "**ML Dipole**" score, based on parameters delivered by the solution of an inverse problem. This approach assumes that the electrical processes in the heart during repolarization can be approximated by a so-called Effective Magnetic Dipole (EMD), (see Figure 2b, abnormal if > 34).

2.3 Validation

To validate ML automatic analysis, two expert cardiologists independently performed **interactive computer-aided analysis** on the same data sets. The interactive analysis of MCG mapping was based on:

- The **T-wave "extrema" Magnetic Field (MF) dynamics analysis**", which calculates cardiac magnetic field parameters, in a moving time window of 30 msec duration during the T- wave. Said time window starts at MF strength of 1/3 of that at the Tpeak and ends at the Tpeak. For each millisecond a color contour plot is calculated from the MF and displayed as shown in figure 1. In each map two points are marked indicating the extreme values of the magnetic field. The point indicating the location of the maximal magnetic field strength is labeled "+" ("+ pole"), and the point indicating the location of the minimal magnetic field strength is labeled "-" ("- pole"). Parameters calculated within this time interval are:

  1) Change of <u>angle</u> between + pole and - pole (abnormal if > 45°);
  2) Change of <u>distance</u> between + pole and - pole (abnormal if > 20 mm);
  3) <u>Ratio</u> between the strength of + pole and - pole (abnormal if > 0.3)[10];

- The **Quantitative Dipole score (Q score)**, also based on analysis of EMD parameters calculated at 20 points of the T-wave in the same T3-T4 interval used for the ML Dipole (Figure 2 b), (abnormal if > 0) [8,9].

- The **magnetic field gradient (MFG) orientation** (angle), computed at two time-intervals: 1) the integral of the second quarter from the J-point to the T-wave apex, representing ST-segment, and 2) the T-wave apex[3]. The MF α angle was then calculated as the angle between the direction of the largest gradient and the patient's right-left line. The α angle values were considered normal when in the range between 0-90° (Figure 3).

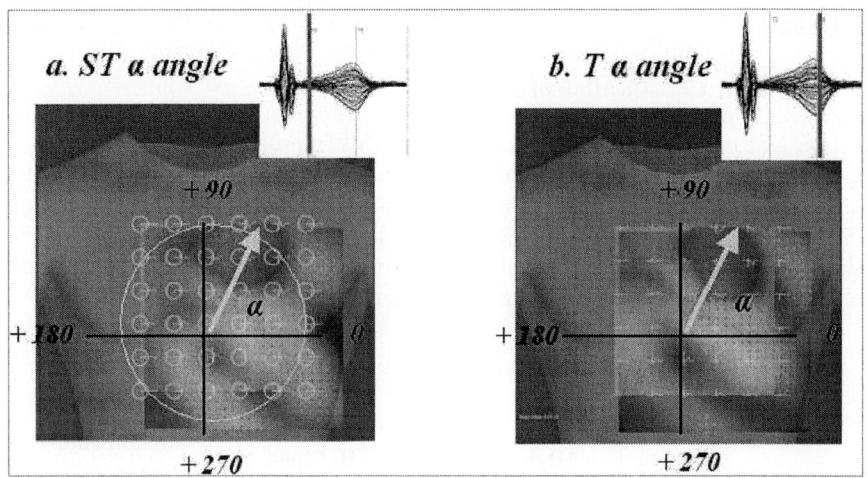

Fig. 3. Example of normal MFG orientation (α angle), at the second quarter of the ST interval (a) and at the apex of the T-wave (b)

2.4 Patients

All MCG studies were performed, after written informed consent, mainly on outpatients, as an additional simultaneous procedure during ECG control.

ML testing was done on a set of 140 randomly analysed MCG recordings belonging to **74 subjects**:

41 patients (Group 1), 26 males and 15 females; 26 with previous MI and 22 with stable class 1 or class 2 angina. Patients were classified as ischemic based on clinical criteria, and on results of exercise ECG testing, nuclear stress testing and/or coronary angiography (CA). CA was available in 31 (29 abnormal). Nuclear stress testing was available in 33 (abnormal in 30). In 9 patients MCG was performed after CA and successful therapy with PTCA. In the 10 patients without CA, nuclear stress testing was abnormal.

All patients were chest pain free at the time of testing, and 27 (67.5%) had a normal or non-specific 12-lead ECG. As 9 patients were studied only after CA, the 32 patients who were studied with MCG before CA were also analysed as a separate group **(Group 2)**.

33 subjects, without any evidence of cardiac disease at clinical history, normal physical examination and echocardiography, were included as normal controls **(Group 3)**. The mean age of the investigated subjects was 64.2 ± 9.9 years for group 1 versus 44.4 ± 9.3 years for group 3 ($p<0.0005$).

For each case at least 2 MCG datasets, recorded in different sessions were analysed.

2.5 Statistics

Data are reported as mean ± S.D. Statistical analysis was performed with the unpaired two-tails Student t-test, to evaluate the significance of differences among males and females parameters. A value of $p < 0.05$ was considered significant.

3 Results

3.1 Automatic Classification of MCG

ML classification of MCG mapping was highly reproducible. In Group 1, the combination of two ML scores, obtained by considering "pathological" any patient with at least one of the two scores abnormal, gave: 61% sensitivity, 85% specificity, 83% positive predictive value, 64% negative predictive value, and 72% predictive accuracy (Table 1). However, if only patients of Group 2 were considered (Table 2), the predictive accuracy of the combined ML scores increased to 80%.

Table 1. ML results of 41 IHD patients (Group 1) vs 33 Normals (Group 3)

	ML extrema	ML Dipole	Combination of 2 ML scores
41 IHD patients	46,4 ± 36,1	46 ± 32,3	-
33 Normals	12,9 ± 17,2	8,8 ± 16,3	-
p value	*< 0.001*	*< 0,001*	-
Sensitivity	41,4	54	61
Specificity	94	88	85
Positive PV	89	85	83
Negative PV	56	60	64
Predictive Accuracy	**65**	**69**	**72**

Table 2. ML results of 32 IHD patients (Group 2) vs 33 Normals (Group 3)

	ML extrema	ML Dipole	Combination of the 2 ML scores
32 IHD patients	53,6 ± 36,8	53,5 ± 31	-
33 Normals	12,9 ± 17,2	8,8 ± 16,3	-
p value	*< 0.001*	*< 0,001*	-
Sensitivity	47	63	75
Specificity	94	88	85
Positive PV	88	83	83
Negative PV	65	71	78
Predictive Accuracy	**71**	**75**	**80**

3.2 Validation by Comparison with Interactive Quantitative Analysis

For comparison the predictive values of computer-aided interactive estimate of VR parameters (T-wave extrema MF dynamics analysis, Q score analysis and MFG orientation) were calculated.

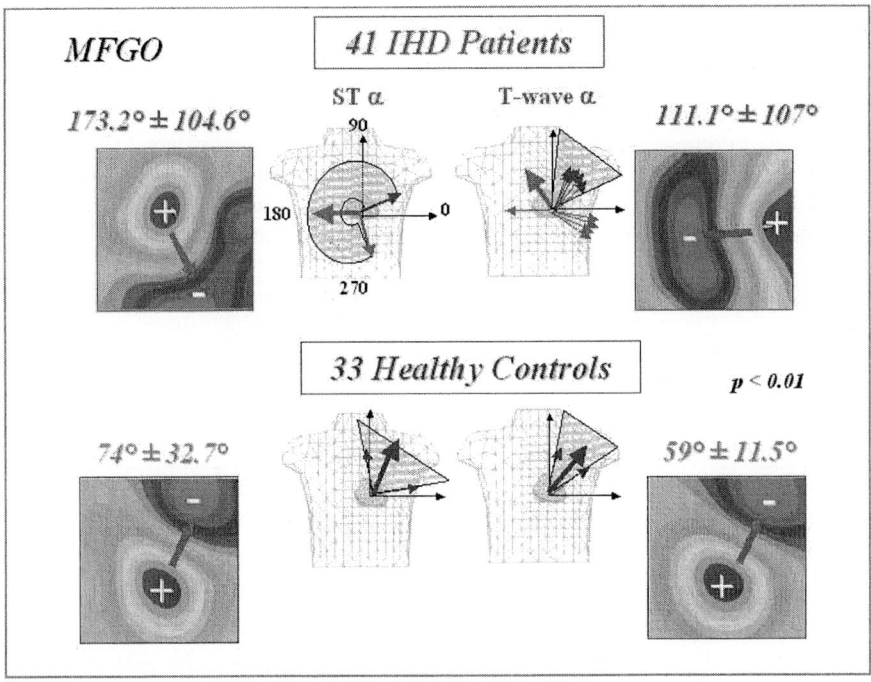

Fig. 4. Examples of typical MF distribution and of average values of MF field gradient orientation (α angles), computed during the ST interval and at the T-wave peak, are shown for Group 1 patients and for controls (Group 3)

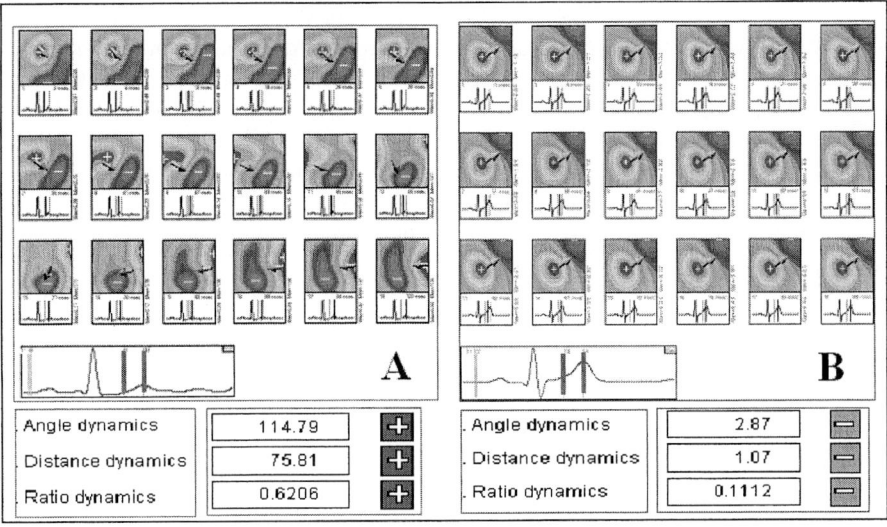

Fig. 5. Example of interactive computer-aided analysis of the MF dynamics during the ascending phase of the T-wave (vertical bars on the ECG). In A, abnormal pattern of an IHD patient. In B, a normal subject is shown for comparison

An example of typical MF distribution during the ST interval and at the T-wave peak in IHD patients and in normal controls is shown in Figure 4, where the average values of MF gradient orientation (α angles) are also included.

An example of MF dynamics analysis is shown in Figure 5.

Interactive computer-aided quantitative estimate of VR parameters (Table 3) was in good agreement with the results of automatic classification, although none of the calculated parameter reached the same predictive accuracy obtained with a combination of the two ML scores, especially in Group 2 patients.

Table 3. Interactive computer-aided analysis

%	41 IHD patients (Group 1)					32 IHD patients w/o PTCA (Group 2)				
	ST α angle	T-wave extrema *			Q score	ST α angle	T-wave Extrema *			Q score
Sensitivity	69	22	39	46	56	81	25	47	50	66
Specificity	70	100	91	79	79	70	100	91	79	79
PPV	74	100	84	73	77	72	100	83	70	75
NPV	64	51	55	59	59	79	58	64	62	70
Pred Acc	69	57	62	61	66	75	63	69	65	72

* T-Wave extrema parameters: Change of <u>angle</u> between + pole and - pole; Change of <u>distance</u> between + pole and – pole; <u>Ratio</u> between the strength of + pole and - pole (see page 4). Pred Acc: Predictive accuracy.

4 Conclusions

The possibility of accurate, rapid, and no risk diagnosis of ischemia in an emergency room setting may have a great impact on health care. Truly ischemic patients would benefit from a significant reduction of time for diagnosis while in non-ischemic subjects unnecessary admissions and more invasive testing could be avoided.

This study was performed in an unshielded hospital room fully equipped for intensive cardiac care and interventional cardiology. The MCG data and mapping quality was sufficiently high to detect ventricular repolarization abnormalities in IHD patients.

Automatic classification of rest MCG recording provided quick detection of electrophysiological alterations associated with ischemic heart diseases, with sensitivity ranging between 60 and 70%, specificity of about 85% and predictive accuracy higher than 70%, thus better than that of rest ECG, which was 50 % in Group 1 patients (Table1). Interestingly, when patients successfully treated with PTCA before MCG mapping were excluded from the statistic evaluation (Group 2), the sensitivity, specificity, and predictive accuracy improved to 75%, 85%, and 80%, respectively (Table 2). Thus, although the number in investigated patients is limited, our results confirm that, as for patients with acute chest pain and normal or non-specific 12-lead ECG and normal troponin[8-10] magnetocardiographic imaging is a promising alternative with the capability of detecting repolarization abnormalities at rest in patients presenting with class 1 and 2 angina, in the absence of significant ECG alteration. The predictive accuracy of the ML method was comparable with that obtained blindly with interactive computer-aided analysis by two expert cardiologists, or even better in untreated patients (Group 2) (Tables 2–3).

In order to improve the predictive accuracy of the method one could incorporate so-called domain knowledge into the machine learning process. Information about the patient, e.g. history, risk factors, results from other tests, could be considered as additional parameters if available. An interesting challenge will be the automatic differentiation of magnetocardiographic abnormalities due to different cardiac diseases by solving a non-ordinal multi-class classification problem[13].

References

1. Tavarozzi I, Comani S, Del Gatta C, Di Luzio S, Romani GL, Gallina S, et al. Magnetocardiography: current status and perspectives. Part II: Clinical applications. *Italian Heart J*. 3(2): 151-165, 2002.
2. Fenici R, Brisinda D, Meloni AM, Fenici P. First 36-channel System for Clinical Magnetocardiography in Unshielded Hospital Laboratory for Cardiac Electrophysiology. *International Journal of Bioelectromagnetism*. 5(1): 80-83, 2003.
3. Hänninen H, Takala P, Mäkijärvi M, Montonen J, Korhonen P, Oikarinen L, et al. Detection of exercise induced myocardial ischemia by multichannel magnetocardiography in patients with single vessel coronary artery disease. *A.N.E.* 5(2): 147-157, 2000.
4. Tsukada K, Miyashita T, Kandori A, Mitsui T, Terada Y, Sato M, et al. An iso-integral mapping technique using magnetocardiogram and its possible use for diagnosis of ischemic heart disease. *Int J Card Imaging*. 16: 55-66, 2000.
5. Brisinda D, Meloni AM, Fenici. First 36-Channel Magnetocardiographic Study of CAD Patients in an Unshielded Laboratory for Interventional and Intensive Cardiac Care. *Lecture Notes in Computer Science*. 2674: 122-131, 2003.
6. Hailer B, Chaicovsky I, Auth-Eisernitz S, Schafer H, Steinberg F, Grönemeyer DH. Magnetocardiography in coronary artery disease with a new system in unshielded setting. *Clin Cardiol*. 26(10):465-471, 2003.
7. Kanzaki H, Nakatani S, Kandori A, Tsukada K, Miyatake K. A new screening method to diagnose coronary artery disease using multichannel magnetocardiogram and simple exercise. *Basic Res Cardiol*. 98(2): 124-132, 2003.
8. Steinberg BA, Roguin A, Allen E, Wahl DR, Smith CR, St. John M, et al. Reproducibility and Interpretation of Magneto-Cardio-Gram Maps in Detecting Ischemia. The 53rd Annual Scientific Sessions of the American College of Cardiology, New Orleans, LA, March, 2004. *J Am Coll Cardiol Supplement*. 43:149A, 2004.
9. Tolstrup K, Madsen B, Brisinda D, Meloni AM, Siegel R, Smars PA. Fenici R. Resting Magnetocardiography Accurately Detects Myocardial Ischemia in Chest Pain Patients with normal or non-specific ECG Findings. Abstract N° 3440, *Circulation Supplement* 26 October 2004. 110 (17): III-743, 2004.
10. Park JW, Reichert U, Maleck M, Klabes J, Schafer J, Jung F. Sensitivity and predictivity of magnetocardiography for the diagnosis of ischemic heart disease in patients with acute chest pain: preliminary results of Hoyerswerda Registry Study. *Critical Pathways in Cardiology*. 1:253-254, 2002.

11. Szymanski B, Embrechts M, Sternickel K, Naenna T, Bragaspathi R. Use of Machine Learning for Classification of Magnetocardiograms. *Proceedings of the 2003 IEEE Conference on Systems, Man, and Cybernetics, SMC 2003, October 5-8, Washington, D. C.*, 1400-1406, 2003.
12. CardioMag Imaging Inc (CMI). Schenectady, USA. 36-channel system 2436 (Alfa version).
13. Sternickel K, Tralshawala N, Bakharev A, et al. Unshielded Measurements of Cardiac Electric Activity Using Magnetocardiography. *International Journal of Bioelectromagnetism.* 4:189-190, 2002.
14. Wold S, Sjöström, Eriksson L. PLS-Regression: a Basic Tool of Chemometrics, *Chemometrics and Intelligent Laboratory Systems.* 58:109-130, 2001.
15. Rosipal R, Trejo LJ. Kernel Partial Least Squares Regression in Reproducing Kernel Hilbert Spaces. *Journal of Machine Learning Research.* 2, 97-128, 2001.
16. Embrechts M, Szymanski B, Sternickel K. A Brief Introduction to Scientific Data Mining: Direct Kernel Methods as a Fusion of Soft and Hard Computing" in "Computationally Intelligent Hybrid Systems: The Fusion of Soft Computing and Hard Computing", Seppo Ovasko, Ed. *IEEE Press*, October 2004.

Hypertrophy in Rat Virtual Left Ventricular Cells and Tissue

S. Kharche[1], H. Zhang[2], R.C. Clayton[3], and Arun V. Holden[1]

[1] Computational Biology Laboratory, School of Biomedical Sciences,
University of Leeds, Leeds LS2 9JT, UK
[2] Department of Physics, UMIST, Manchester, UK
[3] Department of Computing, University of Sheffield, Sheffield, UK
S.Kharche@leeds.ac.uk
http://cbiol.leeds.ac.uk

Abstract. Left ventricular hypertrophy induces remodeling of various ion channels and prolongs depolarization of the ventricles. We modified a model of electrical activity of rat ventricular cell by incorporating available experimental data. Hypertrophy was modeled by incorporating experimental data of changes in sodium (I_{Na}), hyperpolarizing (I_f), outward transient potassium (I_{to}) and T-type calcium currents channel kinetics (I_{CaT}), cell size and Ca^{2+} handling. In 1D simulations, a continuous increase in action potential duration (APD) and corresponding decrease in conduction velocity (CV) with subsequent beats was observed, resulting in conduction block at low values of stimulus intervals (SI), for which the simulated action potential (AP) restitution of the cell models has negative slope.

1 Introduction

Cardiac hypertrophy is a response to long-term pathologic (*e.g.* hypertension) or physiologic (*e.g.* exercise) hemodynamic overload accompanied by changes in energy substrate utilization, or advancement of age by cellular hypertrophy. Hypertrophy is a major cause of cardiac disease.

Hypertrophy induces an increased action potential duration and $[Ca^{2+}]_i$ transient duration. It reduces the $[Ca^{2+}]_i$ transient amplitude. The electrical restitution properties of cells often dictate spatial behaviours and their instabilities. The relationship between APD of successive pulses APD_n and their corresponding diastolic intervals DI_n is given by

$$APD_{n+1} = f(SI - DI_n) \tag{1}$$

where f is the restitution relationship for single cell. An equilibrium point of equation (**1**) corresponds to a periodic response. This equilibrium point, and consequently the APD, will become unstable if $|\frac{df}{dDI_n}| > 1$. The case for $\frac{df}{dDI_n} > 1$ leads to alternans instability. The case when $\frac{df}{dDI_n} < -1$ leads to a new type of instability. Idealised models have been shown to exhibit a monotonically increasing APD rather than alternans behaviour when $\frac{df}{dDI_n} < -1$ [1] that leads to conduction block in 1D virtual strand.

2 Methods

A computer model of rat left ventricular cell was constructed by modifying the model of Pandit *et al.* [2]. The main modifications included were as following.

For the normal cell model, we incorporated experimental data on the kinetics and conductances of L-type Ca^{2+} channel [3], transient outward current [4], and the Na^+ current [5, 6]. The calcium handling mechanism was taken from [7].

To simulate hypertrophy [8], we incorporated data of hypertrophy induced down-regulation of transient outward current (a decrease in conductance by 35 %) [4], up-regulation of Na^+ current (increase in conductance by 8 %) [5], up-regulation of hyperpolarizing current (increase in conductance by 10 %) [9], up-regulation of the Na^+-Ca^{2+} exchanger current (an increase in scaling factor for I_{NaCa} by 5 %) [10]. Cell capacitance was increased by 30% [11], as was cell size [12]. Hypertrophy induced T-type calcium current was also considered in the hypertrophic model with a conductance of 2×10^{-7} µS. Steady state kinetics were obtained from [13]. The time constants and formulation for the current were taken from [14]. A 16.5 % decrease in SR activity was introduced [15].

Both normal and hypertrophic models were integrated using a simple forward Euler method with a time step of 0.1 µs which gave stable solutions.

Firstly, single cell models were integrated using a single stimulus. New and changed currents related to simulation of hypertrophy were measured. We then numerically integrated both normal and hypertrophic models to obtain S1S2 restitutions. This was done by applying 10 stimuli at 1 s intervals and consequently applying a premature stimulus. The final DI_n and APD_{n+1} were noted. We noted time profiles for potential and intracellular calcium, $[Ca^{2+}]_i$. We also obtained dynamic restitution curves. This was done by applying 10 stimuli at a given pacing interval and noting the final DI_n and APD_{n+1}. The pacing interval was progressively reduced. This was continued until the potentials in the normal and hypertrophic models failed to oscillate. The corresponding AP profiles and $[Ca^{2+}]_i$ profiles were noted. APs were initiated in the cell models by applying a stimulus of duration 5 ms and of strength 0.6 pA [16] for normal, and 1.8 pA for hypertrophic cases.

The virtual strand was taken to be 8 mm. Electrotonic interaction between cells was simulated through diffusive coupling and 1D models were developed by incorporating single cell models into a parabolic partial differential equation (PDE) of the form

$$C\frac{dV}{dt} = D\frac{\partial^2 V}{\partial x^2} - I_{ion}$$

where Iion is the total cellular ionic current, V is the cell membrane voltage, C is cell capacitance, and D is the diffusion constant. The space step was 0.1 mm. No flux boundary conditions were imposed at each end of the strand. Diffusion constant was set to a value of 0.1 cm2/s. Periodic waves were stimulated in the strand by applying a periodic SI to 7 nodes situated at one end of the strand. Strength and duration of the stimulus were the same as in the single cell models. A total of 10 stimuli were applied during each simulation. With the chosen value we obtained a CV of 14.2 cm/s in normal strand [17]. The corresponding CV for hypertrophic strand was 9.2 cm/s.

In case of single cell models, we measure APD90 and calcium transient duration and amplitude. We measure single cell restitution from the final APD and DI. In case of the virtual strand simulations, we measure the APD and CV as a function of time and space. We repeat strand simulations for increasingly smaller SI, i.e. for higher pacing rates.

3 Results

We incorporate all the modifications mentioned in the previous section and construct normal and hypertrophic models.

A single stimulus was applied to the single cell models to elicit a solitary AP. The resulting APD90 obtained is 41.5 ms for normal and 73.2 ms for hypertrophy. There is a 77 % increase in APD induced by hypertrophy. APD profiles are compared in Figure 1. Resting $[Ca^{2+}]_i$ increased from 78 nM for normal to 88 nM for hypertrophic case. Corresponding $[Ca^{2+}]_i$ peak are 347 nM and 282 nM. A significant reduction in $[Ca^{2+}]_i$ transient amplitude from 276 nM to 195 nM is induced by hypertrophy, a change of 30 %. Due to remodelling, the conductance I_{Na} of was upregulated. This increased the peak current by 16 %. I_{NaCa} increased by 19 %. The new I_{CaT} current that is not in the normal cell model gave current amplitude of 0.00142 pA during the hypertrophic AP. I_f increased due to up regulation by 13 %. Down regulation of I_{to} caused the peak current during AP to reduce by 20 %. Profiles of APD, $[Ca^{2+}]_i$ and all the changed currents are shown in Figure 1. Input resistance is increased by 223 %. The upstroke velocity for normal was 132 V/s and for hypertrophic case was 90 V/s, a reduction of 32 %. The peak overshoot increased marginally from 34.5 mV to 37 mV. Thus, in hypertrophy, APD increases significantly, $[Ca^{2+}]_i$ transient amplitude decreases, sodium current amplitude increases, and potassium current amplitudes decrease.

Single cell restitution was obtained for normal and hypertrophic models as described in the previous section. Restitution curves were constructed by plotting DI_n against APD_{n+1}. Protocol followed was as described in methods. We also noted the potential and $[Ca^{2+}]_i$. Profiles and the restitution curves are shown in Figure 2. In both cases, as the final DI decreases, APD is seen to increase. For normal case, at values of

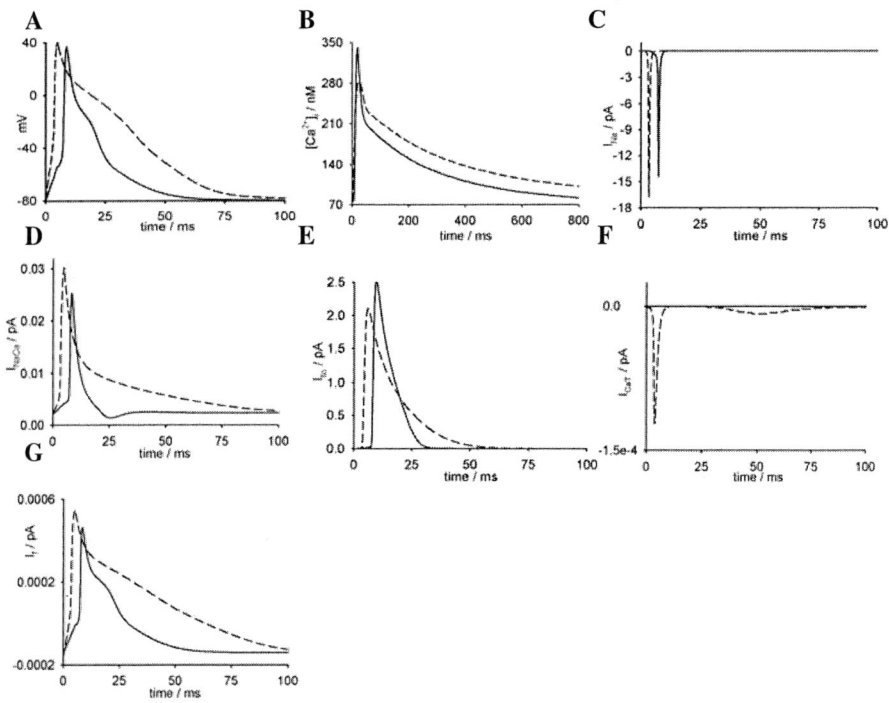

Fig. 1. Single cell simulation for a solitary AP and current profiles. Solid line represents quantities from normal cell model and long dashed line represents quantities from hypertrophic model. **A:** Solitary AP for normal and hypertrophic models. APD is considerably increased in hypertrophic case. **B:** $[Ca^{2+}]_i$ transients for normal and hypertrophic cases. The amplitude in hypertrophic case is reduced. **C:** I_{Na} current profiles. The absolute peak amplitude of the current is increased. **D:** I_{NaCa} current profiles. **E:** I_{to} current profiles. The peak value of I_{to} is reduced in the hypertrophic case. **F:** I_{CaT} current profile. This current is only present in the hypertrophic case. **G:** I_f current profiles. The amplitude of this current is increased in the hypertrophic case

DI lower than 210 ms, the resulting APD is very small. In the hypertrophic case, the smaller the value of DI, the larger is the resulting APD. The restitution curves show a region of negative slope.

Single cell dynamic APD restitution for normal and hypertrophic models was obtained as described in Methods section. SI was progressively decreased. At short SI we saw that the APD increased monotonically with time, with each successive stimulus. For values of SI lower than 230 ms in normal case and 450 ms in hypertrophy case we saw that APD increased monotonically. For values of SI 400 ms or lower in the hypertrophy case, we see that the AP fails to return to resting potential after a few stimuli. Final values of APD and DI were noted for constructing restitution curves.

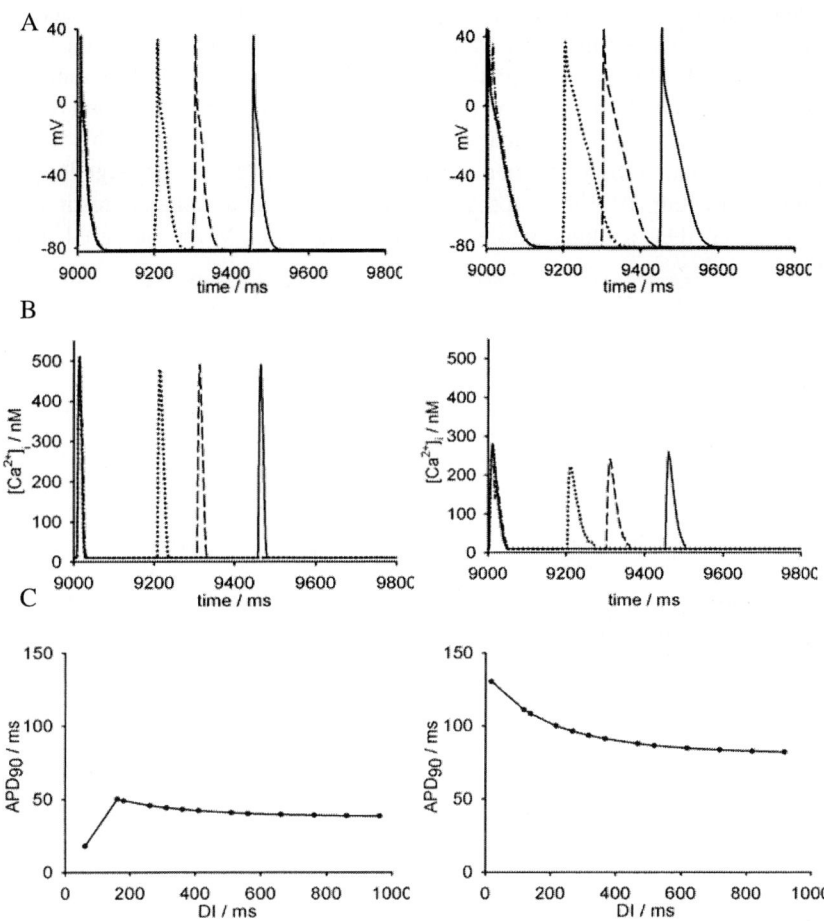

Fig. 2. S1S2 restitution (see text for detail) for normal and hypertrophic cell models. Left panels show normal and right panels show hypertrophic model. **A:** Final 2 APs were plotted for SI 550 ms, 350 ms, 220 ms and 25 ms. Profiles for SI = 550 ms is the solid line, for 350 is the long dashed line, for 220 s is the dotted line and for 250 ms is dashed dotted line. The increase in APD as DI decreases is more evident in the hypertrophic (right hand panel) case. **B:** Final 2 profiles for $[Ca^{2+}]_i$ transients were plotted. Both profiles are for the same values of stimuli used in **A**. **C:** Restitution plots for normal (left panel) and hypertrophic (right panel) cases. In the hypertrophic case, the APD increases monotonically as the DI is reduced

Figure 3 shows the restitution curves for normal and hypertrophic models. The dynamic APD restitution agrees qualitatively with experimental results showing a negatively sloped region at small DI [18].

Virtual strands of 8 mm length were paced at one end at successively decreasing pacing intervals. Space-time plots were obtained and APD and conduction velocity were measured at different times and at all spatial locations through the strands. This is shown in Figure 4. At high pacing rates (SI = 220 ms in normal case, and SI = 400 ms in hypertrophy case), we see that conduction block starts to occur. This is shown in Figure 5. Although the stimulated region becomes excited, the propagation of the APD does not occur. If SI is such that the DI corresponds to dynamic restitution of negative slope, then a monotonic change in APD and CV are observed in both cases.

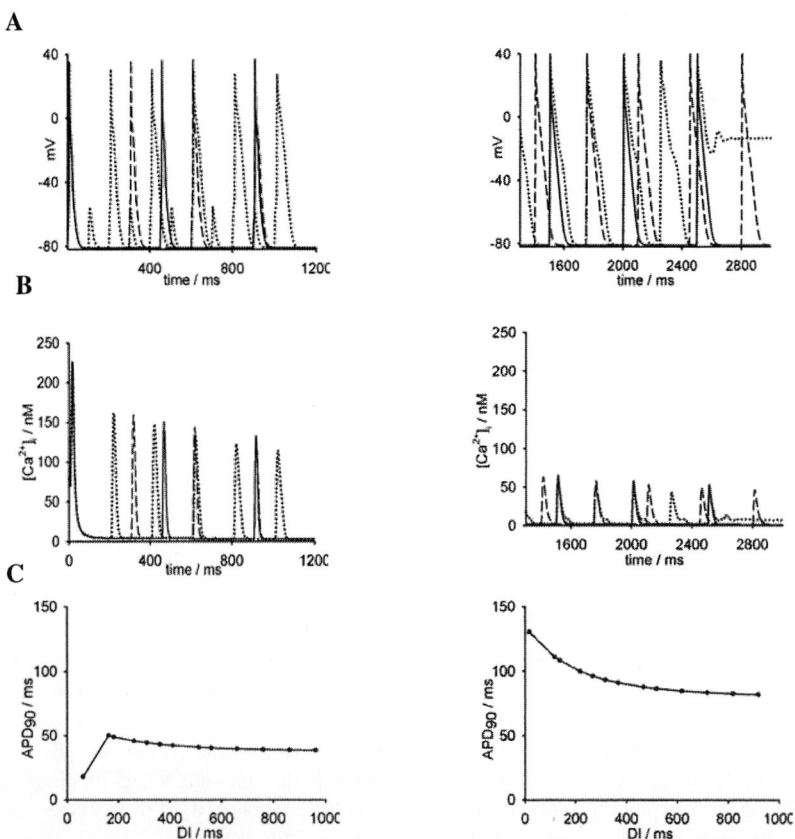

Fig. 3. Dynamic pacing of single cell models. Left hand panels represent profiles for normal model and right hand for hypertrophic. Profiles in **A** and **B** are represented by a solid line for SI 550 ms, long dashed for SI 350 ms, and dotted for SI 200 ms. **A:** AP profiles for dynamic pacing. The APDs increase monotonically at faster pacing rates. **B:** $[Ca^{2+}]_i$ transient profiles. **C:** Dynamic restitution curves for normal (left hand panel) and hypertrophic cases (right hand panel)

In normal case, APD increases monotonically and corresponding CV decreases. The difference in values at consecutive beats is larger closer to the stimulation site. Values come closer as the AP propagates along the strand. A similar phenomenon is seen to occur in the hypertrophy case, albeit at a much larger value of SI. At lower values of SI, conduction block is seen to occur.

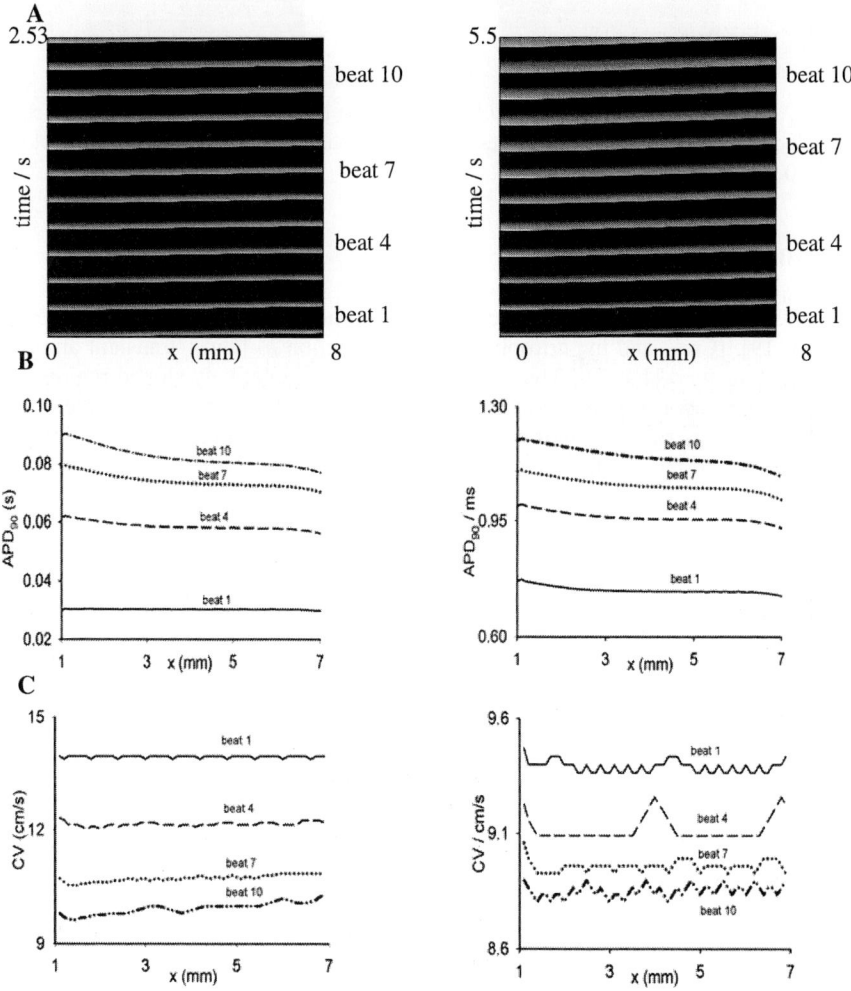

Fig. 4. Homogenous strands of normal (left panels) and hypertrophic models (right panels) paced at a constant SI. A total of 11 stimuli were applied to obtain 11 propagating waves. **A:** Homogenous strand of normal and hypertrophic tissue paced at SI 230 ms and 500 ms respectively. Beat numbers 1, 4, 7 and 10 are marked. **B:** APD was measured at beats 1, 4, 7 and 10. APD was measured at each point along the strand. **C:** Variation of CV along the strand. CV was measured for beat numbers 1, 4, 7 and 10 and at each point on the strand

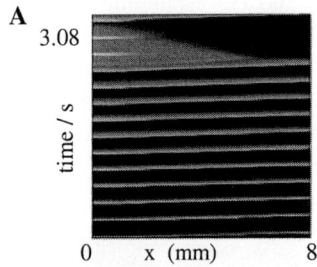 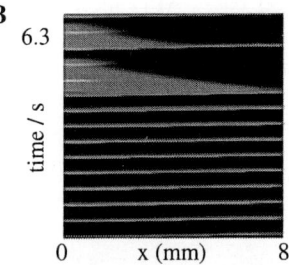

Fig. 5. At high rates of pacing (*i.e.* low SI), conduction block is observed. **A:** Normal strands show block at SI 220 ms or lower values. **B:** Hypertrophic strands shows conduction block at SI of 450 ms or lower

4 Conclusions

The cell models allow us to reproduce the AP and the hypertrophy induced increase in APD [4, 19]. $[Ca^{2+}]_i$ and hypertrophy induced reduction in $[Ca^{2+}]_i$ transient amplitude have been obtained [10, 20]. The peculiar property of negatively sloped region in the restitution curve was reproduced [1, 18]. The property of monotonic increase in APD at fast pacing was seen to favor conduction block in 1D simulations. Thus this property of single cell models favors arrythmogenesis in spatial models.

Acknowledgements

This research was supported by Medical Research Council (UK) and British Heart Foundation (UK) grants.

References

1. Panfilov A.V., Zemlin C.W.: Wave propagation in an excitable medium with a negatively sloped restitution curve. Chaos. **12(3)** (2002) 800-806.
2. Pandit S.V., Clark R.B., Giles, W.Y., Demir S.S.: A mathematical model of action potential heterogeneity in adult rat ventricular myocytes. Biophysical Journal. **81** (2001) 3029-3051.
3. Linz, K.W.: Meyer R.: Profile and kinetics of L-type calcium current during the cardiac ventricular action potential compared in guinea-pigs, rats and rabbits. Pflugers Arch - Eur J Physiol. **439** (2000) 588-599.
4. Li, Q., Keung, E.C.: Effects of myocardial hypertrophy on transient outward current. Am. J. Physiol. **266** (1994) H1738-H1745.
5. Alvarez, J.L., Aimond, F., Lorente, P., Vassort, G.: Late post-myocardial infraction induces a tetrodotoxin-resistant Na^+ current in rat myocytes. J. Mol. Cell. Cardiol. **32(7)** (2000) 1169-1179.
6. Zhang, H., Holden, A.V., Kodama, I., Honjo, H., Lei, M.: Mathematical models of action potentials in the periphery of and centre of the rabbit sinoatrial node. Am. J. Physiol. **279** (2000) H397-H421.

7. Winslow, R.L., Rice, J., Jafri, S., Marban, E., O'Rourke, B.: Mechanisms of altered excitation-contraction coupling in canine tachycardia-induced heart failure II. Model studies. Circ. Res. **84** (1999) 571-586.
8. Padmala, S., Demir, S.S.: Computational Model of the Ventricular Action Potential in Adult Spontaneously Hypertensive rats. Journal of Cardiovasc. Electrophysiol. **14(9)** (2003) 990.
9. Fernandez-Velasco, M., Goren, N., Benito, G., Blanco-Rivero, J., Bosca, L., Delgado, C.: Regional distribution of hyperpolarization-activated current (If) and hyperpolarization-activated cyclic nucleotide-gated channel mRNA expression in ventricular cells from control and hypertrophied rat hearts. J. Physiol. (Lond). **553** (2003) 395-405.
10. Gomez, A.M., Schwaller, B., Porzig, H., Vassort, G., Niggli, E., Eger, M.: Increased exchange current but normal Ca^{2+} transport via I_{NaCa} exchange during cardiac hypertrophy after myocardial infraction. Circ. Res. **91** (2002) 323-330.
11. Yokoshiki, H., Kohya, T., Tomita, F., Tohse, N., Nakaya, N., Kanno, M., Kitabatake, A.: Restoration of action potential duration and transient outward current by regression of left ventricular hypertrophy. J. Mol. Cell. Cardiol. **29** (1997) 1331-1339.
12. Qin, D., Zhang, Z., Caref, E.B., Boutjdir, M., Jain, P., El-Sherif, N.: Cellular and Ionic Basis of Arrhythmias in Postinfarction Remodeled Ventricular Myocardium. Circ. Res. **79(3)** (1996) 461-473.
13. Izumi, T., Kihara, Y., Sarai, N., Yoneda, T., Iwanaga, Y., Inagaki: Reinduction of T-Type Calcium Channels by Endothelin-1 in Failing Hearts In Vivo and in Adult Rat Ventricular Cells. Circulation. **108(20)** (2003) 2530-2535.
14. Dokos, S., Celler, B., Lovell, N.: Vagal Control of Sinoatrial Rhythm: a Mathematical Model. Journal of Theoretical Biology. **182(1)** (1996) 21-44.
15. Arta, Y., Geshi, E., Nomizo, A., Aoki, S., Katagiri, T.: Alterations in sarcoplasmic reticulum and angiotensin II receptor type I gene expression in spontaneously hypertensive rat hearts. Jpn. Circ. J. **63** (1999) 367-37.
16. Ward, C.A., Ma, Z., Lee, S.S., Giles, W.R.: Potassium currents in atrial and ventricular myocytes from a rat model of cirrhosis. Am. J. Physiol. **273** (1997) G537–G544.
17. Meiry, G., Reisner, Y., Feld, Y., Goldberg, S., Rosen, M., Ziv, N., Binah, O.: Evolution of action potential propagation and repolarisation in cultured neonatal rat ventricular myocytes. J. Cardiovasc. Electrophysiol. **12(11)** (2001) 1269-77.
18. Nanasi, P.P., Pankucsi, C., Banyasz, T., Szigligeti, P., Papp, J.G., Varro, A.: Electrical restitution in rat ventricular muscle. Acta Physiol. Scand. **158(2)** (1996) 143-53.
19. Shimoni Y., Firek, L., Severson, D., Giles, W.R.: Short-term diabetes alters K^+ currents in rat ventricular myocytes. Circ. Res. **74** (1994) 620-628.
20. Cerbai, E., Barbieri, M., Li, Q., Mugeli, A.: Occurence and properties of the hyperpolarization activated current I_f in ventricular myocytes from normotensive and hypertensive rats during aging. Circulation. **94** (1996) 1674-1681.

Virtual Ventricular Wall: Effects of Pathophysiology and Pharmacology on Transmural Propagation

Oleg V. Aslanidi[1], Jennifer L. Lambert[2], Neil T. Srinivasan[2] and Arun V. Holden[1]

[1] School of Biomedical Sciences, University of Leeds, LS2 9JT, UK
[2] School of Medicine, University of Leeds, LS2 9JT, UK
oleg@cbiol.leeds.ac.uk
http://cbiol.leeds.ac.uk

Abstract. Effects of pathophysiological conditions and pharmacological intervention on transmural propagation are computed for the virtual ventricular wall. ST depression during sub-endocardial ischaemia and unidirectional functional block in the vulnerable window during Class III drug action are explained by changes induced in the transmural dispersion of action potential duration.

1 Introduction

Electrophysiological recordings from cells and tissue isolated from endocardial, mid-myocardial (M-cell) and epicardial regions of the ventricular wall show marked transmural differences. These include differences in action potential (AP) shape and duration [1], as well as differential responses to changes in pacing cycle length, pharmacological intervention and pathophysiological conditions (e.g., ischaemia). The importance of dispersion of action potential duration (APD) for the initiation of arrhythmias is well recognized [2-4], and the respective transmural differences can be a potent arrhythmogenic source. Even if cell coupling reduces the transmural differences, they can be enhanced by pathophysiology, such as ischaemia [5], or by pharmacological blocking of repolarising K$^+$ currents (primarily I_{Kr} and I_{Ks}), which can increase the transmural APD dispersion and trigger re-entrant arrhythmias [6].

We use a computational model – virtual ventricular wall – to study electrophysiological changes of the transmural APD dispersion in response to pathophysiological conditions and pharmacological intervention – primarily, ST depression during sub-endocardial ischaemia and unidirectional propagation block in the vulnerable window (VW) during action of Class III drugs.

The link between the transmural heterogeneity of the ventricular tissue and clinically recorded electrocardiograms (ECGs) under normal and abnormal conditions has been explored recently: Antzelevitch et al. [7,8] related transmural APD differences to development of QT dispersion, and Gima and Rudy [9] explained how the APD dispersion accounts for ST elevation during global ischaemia [10]. We use a similar approach to dissect the electrophysiological mechanisms of ST depression during sub-endocardial ischaemia. Although depression of the ST-segment in ECGs is used clinically as an index of sub-endocardial ischaemia [11,12], the impact of the associated transmural electrophysiological changes is still not known.

Experiments suggest that ventricular tissues with large transmural differences in APD are more vulnerable to re-entry, and clinical studies show that pharmacological treatment that increases the APD dispersion can be proarrhythmic [6,13]. We compare electrophysiological properties of tissues treated with two different Class III drugs: amiodarone and d-sotalol. Proarrhythmic d-sotalol increases the transmural heterogeneity by preferentially increasing APD in M-cells [14,15], whereas amiodarone, the safest among Class III drugs, decreases the heterogeneity by increasing APD in endo- and epicardial cells [6]. However, electrophysiological mechanisms underlying the low arrhythmogenicity of amiodarone at the tissue level are poorly understood. We study the differential effects of the Class III drugs on transmural propagation, APD dispersion and vulnerable properties of the virtual ventricular wall.

2 Virtual Ventricular Wall

Virtual ventricular tissues are physiologically detailed reaction-diffusion models of ventricular tissues, that have proved to be an effective tool for simulating normal and abnormal ventricular propagation patterns, and for proposing hypotheses that can be tested experimentally [4,16-18]. In this paper we study transmural propagation in virtual ventricular wall consisting of three compact regions: endo-, M- and epicardial.

A one-dimensional (1D) model describing profiles of the membrane voltage, V (mV), through the virtual wall is based on the nonlinear cable equation [4, 16-18]:

$$\frac{\partial V}{\partial t} = \frac{\partial}{\partial x}\left(D\frac{\partial V}{\partial x}\right) - I_{ion}. \tag{1}$$

Here $0 \leq x \leq L$ is spatial coordinate through the virtual wall (mm), t is time (ms). $L = 15$ mm is the thickness of the virtual wall. D is the effective diffusion coefficient (mm^2 ms^{-1}), that characterizes electrotonic spread of voltage, primarily through the intercellular resistive gap junctions. I_{ion} is the total membrane ionic current (μA μF^{-1}). The latter can be described in biophysical detail by the Luo-Rudy dynamic (LRd) ventricular cell model [19], which includes equations for time and voltage-dependent current flow through ion channels, pumps and exchangers in the cell membrane, as well as for Ca^{2+} dynamics within the cell.

Note that "1D virtual wall" refers to an idealization of the real 3D ventricular wall. Similar to the transmural wedge experiments [7, 8] and previous modelling studies [9], our simulations correspond to the situation during a normal heart-beat, where the Purkinje system ensures near-simultaneous excitation of the endocardium, resulting in a planar transmural wave-front parallel to the endocardial surface. 1D model is sufficient for simulating such a planar wave.

Transmural differences in the density of two repolarizing ionic currents, the slow-delayed rectifier potassium current I_{Ks} (μA μF^{-1}) and the transient outward potassium current I_{to} (μA μF^{-1}), through the virtual ventricular wall are introduced to represent three cell types: endo-, M- and epicardial [9]. Primarily, the density ratio I_{Ks}:I_{Kr} (here I_{Kr} is the rapid-delayed rectifier potassium current) is 11:1, 4:1 and 35:1, and maximum conductance of the I_{to} current is 0.0, 0.2125 and 0.25 mS μF^{-1} for endo-, M- and epicardial cells, respectively.

For geometric simplicity we assume that each cell type composes a uniform region occupying a third of the wall. As the exact proportion of endo-, M- and epicardial cells within the wall is not known, this assumption constitutes a first approximation of the transmural cellular structure. It also allows direct comparison of our simulations with the results on ST elevation in globally ischaemic ventricular wall [9]. The diffusion coefficient is set to a uniform value of 0.06 mm^2 ms^{-1} through the whole wall, except for a 5-fold decrease at the boundary between M- and epicardial regions [9], giving a solitary action potential transmural propagation velocity of 0.45 m s^{-1} and a transmural propagation time of ~33 ms.

The equation (1) is solved numerically using the explicit Euler method with time step $\Delta t = 0.005$ ms and space step $\Delta x = 0.1$ mm. The endocardial end of the wall is stimulated 5 times by 0.5 ms, -100 μA μF^{-1} current pulse stimuli at a basic cycle length (BCL) of 500 ms, leading to the transmural action potential propagation. Ectopic (S2) stimuli are applied at different time intervals following the last AP, resulting in either propagation failure, unidirectional functional propagation block or bidirectional propagation. Vulnerable window is defined as the range of S2 intervals leading to the unidirectional block (see in Fig. 1).

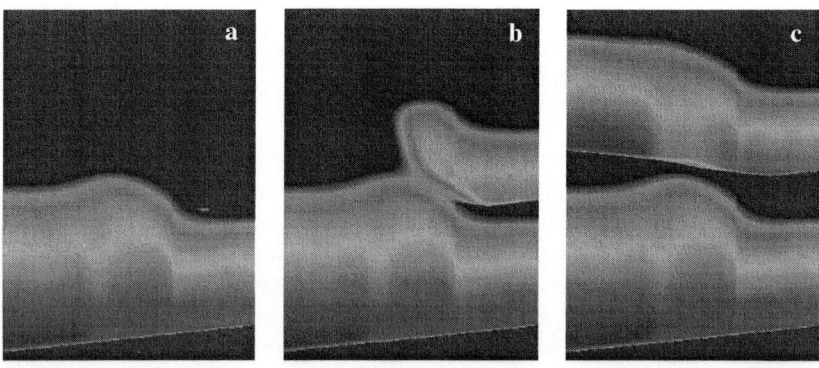

Fig. 1. Transmural AP propagation and VW definition in the virtual ventricular wall. The wall is stimulated from the endocardial end (at the left), leading to the AP propagation. Following S2 stimulation with different timing results in either (**a**) propagation failure for S2 = 160 ms, (**b**) unidirectional block for S2 = 180 ms or (**c**) bidirectional propagation for S2 = 200 ms. Space-time plots are shown, the membrane voltage is colour-coded using the standard rainbow palette. Each panel presents 400 ms of activity in 15 mm thick wall. In this illustration the S2 stimuli are applied to the epicardial region at $x = 12$ mm

Sub-endocardial ischaemia is an ischaemic region of spatial extent $l < L$ ($l = L$ corresponds to global ischaemia), in which the ATP concentration (mM), the extracellular potassium concentration $[K^+]_o$ (mM) and pH are changed [9,20]. In the normal conditions ATP = 10 mM, $[K^+]_o$ = 4.0 mM and pH = 7.5, in the ischaemic conditions ATP = 3 mM, $[K^+]_o$ = 10.0 mM and pH = 6.5. Transmural differences in the density of ATP-sensitive potassium current $I_{K(ATP)}$ ($\mu A\ \mu F^{-1}$) are also accounted for by varying the half-saturation coefficient for this current, k_{ATP} (mM): k_{ATP} equals 0.0625, 0.125 and 0.25 mM in the endo-, M- and epicardial regions, respectively [9].

The effects of amiodarone and d-sotalol are incorporated as changes in the density of the rapid-delayed rectifier potassium current I_{Kr} ($\mu A\ \mu F^{-1}$) – the primary target for the Class III drug action – and the L-type depolarizing calcium current $I_{Ca,L}$ ($\mu A\ \mu F^{-1}$), which reproduces relative alterations of APD in the three cell types [6,14,15]. For the d-sotalol model I_{Kr} is depressed by 40% in endo-, by 100% in M-cell and by 65% in the epicardial region; for the amiodarone model I_{Kr} is uniformly depressed by 50% throughout the wall and $I_{Ca,L}$ is depressed by 40% in the M-cell region.

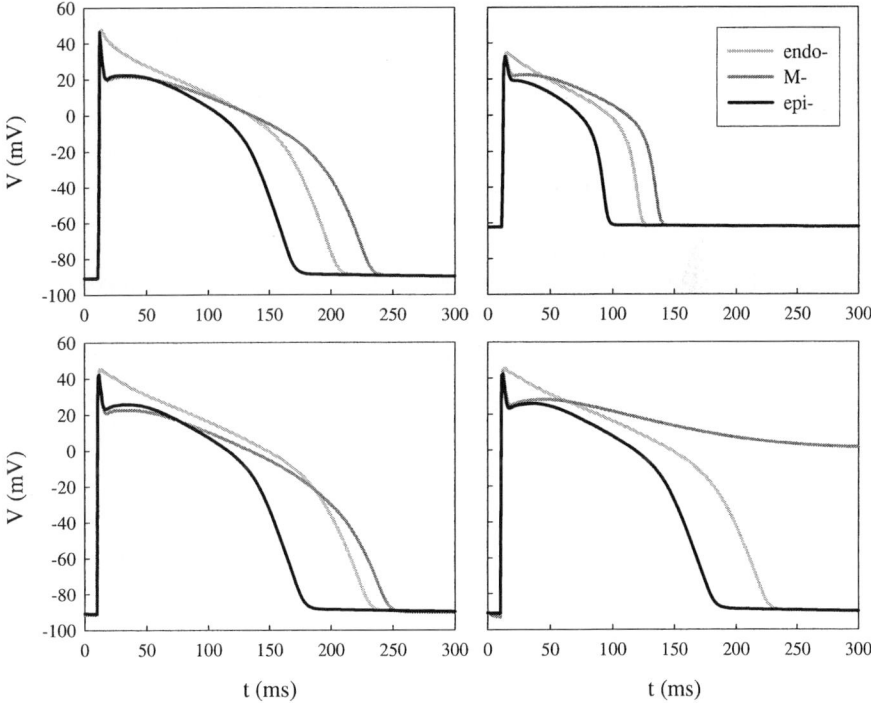

Fig. 2. Action potentials in endo-, M- and epicardial cells. APs for each of the three cell types are shown under the normal (top left) and ischaemic (top right) conditions, and under the effects of amiodarone (bottom left) and d-sotalol (bottom right)

Single-cell action potentials generated by endo-, M- and epicardial cells under various conditions studied in this paper are illustrated in Fig. 2.

The extracellular potential generated by the spatial membrane voltage distribution $V(x, t)$ within the 1D virtual ventricular wall is estimated using the expression [9]:

$$\Phi_e(x^*,t) = -K \int \frac{\partial V(x,t)}{\partial x} \cdot \frac{\partial}{\partial x}\left(\frac{1}{x^* - x}\right) dx. \qquad (2)$$

Here $K = 1.89$ mm^2 is a positive constant, $x^* = 20$ mm is the distance from epicardial end of the tissue to an in line "electrode" site. The time profile of Φ_e constitutes an approximation for the ventricular component of the ECG – pseudo-ECG.

3 Results

3.1 Pathophysiology

Our simulations have separated the spatial and cellular mechanisms of ST depression caused by transmural AP propagation through a heterogeneous virtual ventricular wall during sub-endocardial ischaemia. ST depression results from predominantly positive transmural spatial gradients in the membrane voltage, $\partial V/\partial x$, and hence, in negative values of the integral Φ_e (2), during ventricular repolarization (Fig. 3). The gradients are produced by an abnormal transmural repolarization sequence caused by a decrease of APD in the ischaemic region. ST depression is facilitated by elevation of the pseudo-ECG baseline (see in Fig. 3), which results from a negative spatial gradient of the resting membrane potential between the normal and the ischaemic regions.

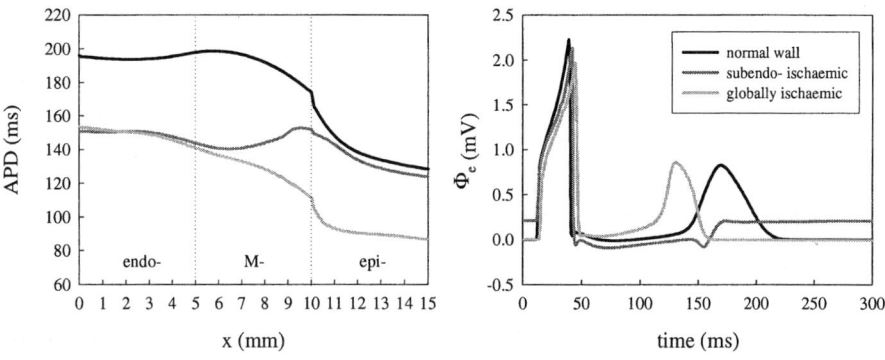

Fig. 3. Transmural distributions of APD in the normal, globally ischaemic and sub-endocardially ischaemic virtual ventricular walls (left), and respective pseudo-ECGs (right). Depth of the sub-endocardial ischaemia is $l = 10$ mm, such that the ischaemic region extends over the whole endocardial and M-cell regions, which results in decrease of APD

The cellular mechanisms of ST depression can be dissected by simulating separate components of the sub-endocardial ischaemia – acidosis, anoxia and hyperkalaemia. Our simulations show that sub-endocardial elevation of [K$^+$]$_o$ alone results in the transmural APD dispersion leading to ST-depression, whereas both sub-endocardially low pH and low ATP generate transmural APD distributions and ECG patterns resembling those of the normal virtual wall (Fig. 4). We conclude that the primary cellular mechanism underlying ST depression is sub-endocardial hyperkalaemia – elevation of the extracellular potassium concentration $[K^+]_o$. Its effects on the electrophysiological properties of the ventricular wall are mediated through the K^+-sensitive membrane currents regulating the cellular AP shape and duration.

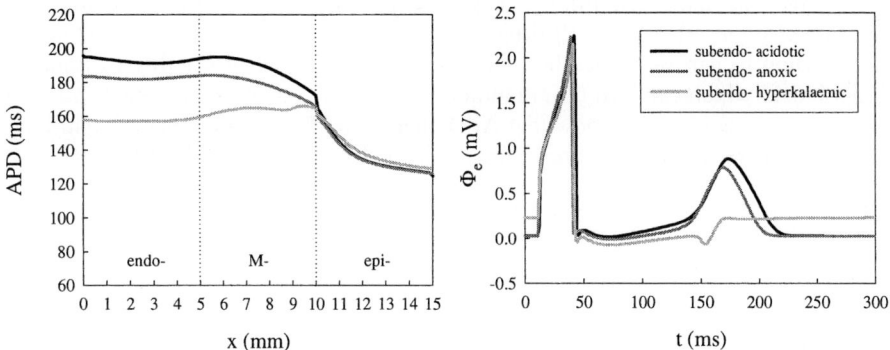

Fig. 4. Role of the ischaemic components – acidosis, anoxia and hyperkalaemia – in ST depression. Transmural distributions of APD (left) and pseudo-ECGs (right) are shown. $l = 10$ mm

3.2 Pharmacology

Fig. 1 illustrates AP propagation through the virtual ventricular wall and defines the vulnerable window. VWs computed for the walls treated with amiodarone and d-sotalol are shown in Fig. 5. The shape and extent of the VWs demonstrate clear correlation with the respective transmural APD dispersions within the wall. As d-sotalol increases the APD dispersion by predominant increase of APD in M-cells, the VW in

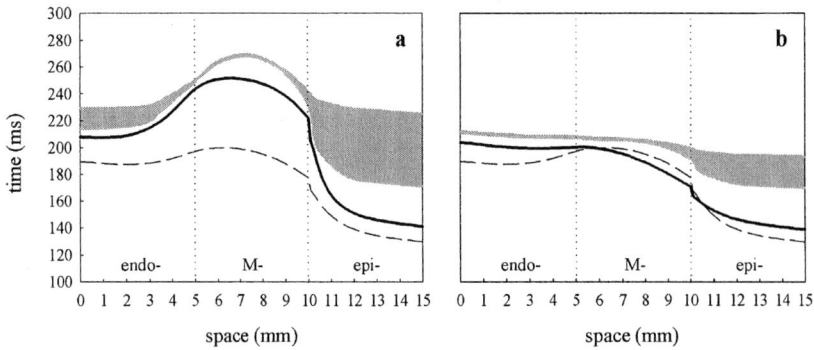

Fig. 5. Effects of (**a**) d-sotalol and (**b**) amiodarone on the virtual ventricular wall. Spatial distributions of APD through the normal wall (dashed line) and the wall treated by the drugs (solid lines) are shown along with the respective VWs (gray areas)

the endo- and epicardial regions, where unidirectional block persists until the M-cell region is fully repolarized, is wide. Amiodarone, however, decreases the dispersion by prolonging APD in endo- and epicardial cells and decreasing APD in M-cells, which results in a narrow VW similar to that of the normal tissue. Pacing the virtual wall at different BCL shows that the transmural APD dispersion is larger at low rates, especially with d-sotalol (Fig. 6).

We conclude that an electrophysiological explanation for the safety of amiodarone in comparison to other Class III drugs lies in relatively low transmural APD dispersion leading to narrow vulnerable window, and hence, low probability of unidirectional block (which can lead to initiation of re-entry in 2D and 3D) in the ventricular wall. Our simulations also show that APD change in M-cells is the major contributor to the transmural dispersion of repolarization, and hence, extent of the VW.

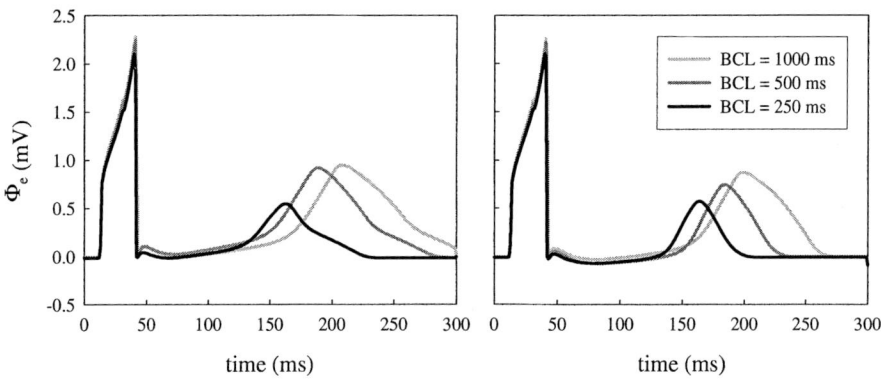

Fig. 6. Effects of BCL on the transmural APD dispersion. Pseudo-ECGs corresponding to the transmural AP propagation are computed for the virtual wall treated with d-sotalol (left) and amiodarone (right). T-wave prolongation is an index of increase in the APD dispersion [15]

4 Human Model

Although the LRd is the most experimentally validated of all cardiac cell models, it is adopted to description of AP properties in guinea-pig ventricular cells. However, clinical studies are performed on human patients, and hence, an adequate model of human cells, which accounts for the transmural variations, is required to compare computational results to these data. Such a model has been developed recently [21].

We use the equation (1) along with the description of the ionic current I_{ion} provided by the model [21] for endo-, M- and epicardial cells in order to simulate human virtual ventricular wall. The computational set-up is similar to that used in case of the LRd cellular models. As the human cell models have little difference in APD between the endo- and epicardial cells, the 1D virtual human wall does not generate a positive T-wave in the pseudo-ECG (2). Therefore, simulating ST-depression with this model is not feasible (as the ST-segment cannot be defined).

Results of simulating the effects of Class III drugs on the human virtual wall are illustrated in Fig. 7. The model reproduces the effects of d-sotalol and amiodarone on the APD dispersion, as seen in our LRd-based simulations, but the resultant VWs are very narrow (< 1 ms) throughout the virtual wall. We use dynamic restitution curves for single endo-, M- and epicardial cells to test rate-dependence of the APD dispersion. Fig. 8 shows that the transmural APD dispersion is large with d-sotalol and relatively small with amiodarone at all tested rates, which is in agreement with the experiments [15]. However, contrary to the experiments [15], the APD dispersion between the model endo-, M- and epicardial cells [21] does not substantially change with increasing BCL in the range from 1 to 10 seconds (see in Fig. 8).

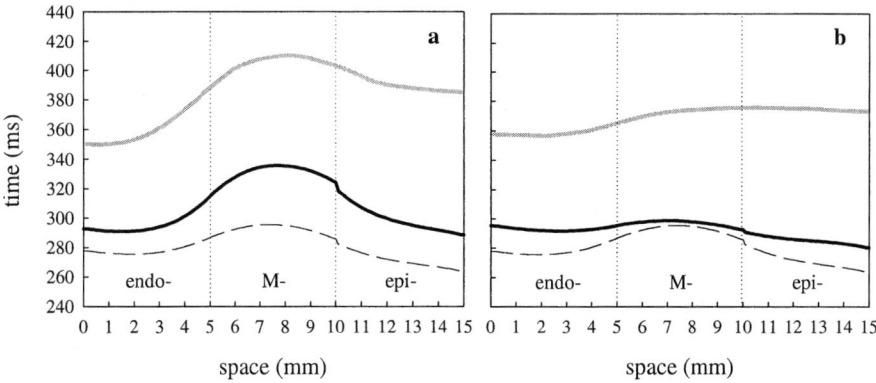

Fig. 7. Effects of (**a**) d-sotalol and (**b**) amiodarone on the human virtual ventricular wall. Spatial distributions of APD through the normal wall (dashed line) and the wall treated the drugs (solid lines) are shown along with the respective VWs (narrow gray areas)

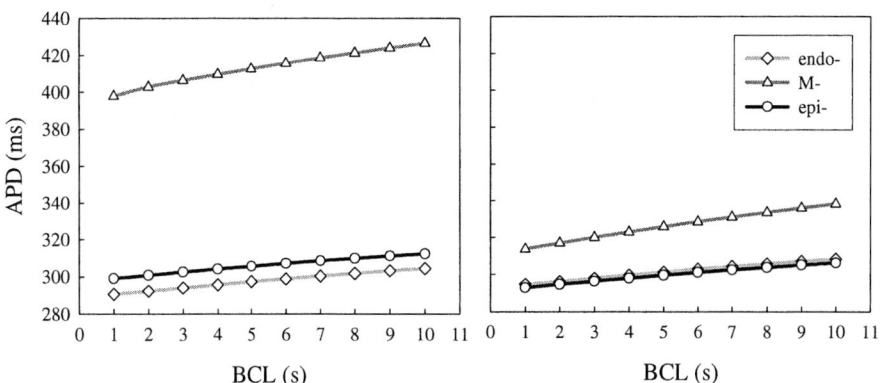

Fig. 8. Effects of BCL on APD in single human endo-, M- and epicardial cells treated with d-sotalol (left) and amiodarone (right). APD is measured after the cells are paced 5 times

5 Conclusion

We conclude that the effects of pathophysiology (sub-endocardial ischaemia) and pharmacology (Class III drugs) on the transmural propagation can be explained by changes in the transmural APD dispersion leading, primarily, to ST depression and increase of the vulnerable window in the virtual ventricular wall. ST depression results from relative changes of APD in endo-, M- and epicardial cells unequally effected by the ischaemia, and the increase/decrease of the tissue vulnerability in response to d-sotalol/amiodarone results from respective changes in the APD dispersion and probability of the unidirectional propagation block. Although our computational model has proved to be an effective tool for studying these effects, it cannot be used for simulating non-planar transmural or intramural waves, such as re-entrant vortexes [22,23]. Development of 2D and 3D models constitutes the next step of our research.

Acknowledgements

This research was funded by the MRC and EPSRC (UK).

References

1. Bryant, S.M., Wan, X., Shipsey, S.J. Hart, G. Regional differences in the delayed rectifier current (I_{Kr} and I_{Ks}) contribute to the differences in action potential duration in basal left ventricular myocytes in guinea-pig. Cardiovasc. Res. 40 (1998), 322-331.
2. Han, J., Moe, G.K. Nonuniform recovery of excitability in ventricular muscle. Circ. Res. 14 (1964), 44-60.
3. Burton, F.L., Cobbe, S.M. Dispersion of ventricular repolarization and refractory period. Cardiovasc. Res. 50 (2001), 10-23.
4. Clayton, R.H., Holden, A.V. Propagation of normal beats and re-entry in a computational model of ventricular cardiac tissue with regional differences in action potential shape and duration. Prog. Biophys. Mol. Biol. 85 (2004), 473-499.
5. Viswanathan, P.C., Rudy, Y. Cellular arrhythmogenic effects of congenital and acquired long-QT syndrome in the heterogeneous myocardium. Circulation 101 (2000), 1192-1198.
6. Akar, F.G., Yan, G.-X., Antzelevitch, C., Rosenbaum, D.S. Unique topographical distribution of M cells underlies re-entrant mechanism of Torsade de Pointes in the Long-QT syndrome. Circulation 105 (2002), 1247-1253.
7. Antzelevitch, C., Shimizu, W., Yan, G.X., Sicouri, S., Weissenburger, J., Nesterenko, V.V., Burashnikov, A., Di Diego, J., Saffitz, J., Thomas, G.P. The M cell: its contribution to the ECG and to normal and abnormal electrical function of the heart. J. Cardiovasc. Electrophysiol. 10 (1999), 1124-1152.
8. Antzelevitch, C., Yan, G.X., Shimizu, W. Transmural dispersion of repolarization and arrhyth-mogenicity: the Brugada syndrome versus the long QT syndrome. J. Electrocardiol. 32 (1999), 158-165.
9. Gima, K., Rudy Y. Ionic current basis of electrocardiographic waveforms – a model study. Circ. Res. 90 (2002), 889-896.
10. Kleber, A.G. ST-segment elevation in the electrocardiogram: a sign of myocardial ischaemia. Cardiovasc. Res. 45 (2000), 111-118.

11. Li, D., Li, C.Y., Yong, A.C., Kilpatrick, D. Source of electrocardiographic ST changes in subendocardial ischemia. Circ. Res. 82 (1998), 957-970.
12. Horacek, B.M., Wagner, G.S. Electrocardiographic ST-segment changes during acute myocardial ischemia. Card. Electrophysiol. Rev. 6 (2002), 196-203.
13. Huikuri, H.V., Castellanos, A., Myerburg, R.J. Sudden death due to cardiac arrhythmias. N. Engl. J. Med. 345 (2001), 1473-1482.
14. Sicouri, S., Moro, S., Litovsky, S., Elizari, M.V., Antzelevitch, C. Chronic amiodarone reduces transmural dispersion of repolarization in the canine heart. J. Cardiovasc. Electrophysiol 8 (1997), 1269-1279.
15. Drouin, E., Lande, G., Charpentier, F. Amiodarone reduces transmural heterogeneity of repolarization in the human heart. J. Am. Coll. Cardiol.32 (1998), 1063-1067.
16. Clayton, R.H., Holden, A.V. Computational framework for simulating the mechanisms and ECG of re-entrant ventricular fibrillation. Physiol. Meas. 23 (2002), 707-726.
17. Kohl, P., Noble, D., Winslow, R.L. Hunter, P.J. Computational modelling of biological systems: tools and visions. Philos. Trans. Roy. Soc. A 358 (2000), 579-610.
18. Aslanidi, O.V., Bailey, A., Biktashev, V.N., Clayton, R.H., Holden, A.V. Enhanced self-termination of re-entrant arrhythmias as a pharmacological strategy for antiarrhythmic action. Chaos 12 (2002), 843-851.
19. Luo, C.H., Rudy Y. A dynamic model of the cardiac ventricular action potential. I. Simula-tions of ionic currents and concentration changes. Circ. Res. 74 (1994), 1071-1096.
20. Shaw, R.M., Rudy, Y. Electrophysiologic effects of acute myocardial ischaemia: a theoreti-cal study of altered cell excitability and action potential duration. Cardiovasc. Res. 35 (1997), 256-272.
21. ten Tusscher, K.H., Noble, D., Noble, P.J., Panfilov, A.V. A model for human ventricular tissue. Am. J. Physiol. Heart Circ. Physiol. 286 (2004), H1573-1589.
22. Hyatt, C.J., Mironov, S.F., Wellner, M., Berenfeld, O., Popp, A.K., Weitz, D.A., Jalife, J., Pertsov, A.M. Synthesis of voltage-sensitive fluorescence signals from three-dimensional myocardial activation patterns. Biophys. J. 85 (2003), 2673-2683.
23. Clayton, R.H., Holden, A.V. Effect of regional differences in cardiac cellular electrophysiology on the stability of ventricular arrhythmias: a computational study. Phys. Med. Biol. 48 (2003), 95-111.

Electrophysiology and Tension Development in a Transmural Heterogeneous Model of the Visible Female Left Ventricle

Gunnar Seemann[1], Daniel L. Weiß[1], Frank B. Sachse[2], and Olaf Dössel[1]

[1] Institut für Biomedizinische Technik, Universität Karlsruhe (TH),
Kaiserstr. 12, 76128 Karlsruhe, Germany,
Gunnar.Seemann@ibt.uni-karlsruhe.de
www.ibt.uni-karlsruhe.de
[2] Nora Eccles Harrison Cardiovascular Research and Training Institute,
University of Utah, Salt Lake City, UT, USA

Abstract. Electrophysiological heterogeneity within human ventricles is mainly based on differences of ion channel characteristics inside the wall. This influences also properties of cellular tension development.

In this work, knowledge about transmural heterogeneity was transferred to an electro-mechanical heart model composed of a human model describing electrophysiology and of a model for the development of tensions. The heterogeneity was included in the cardiomyocyte model by varying ion channel kinetics and density on basis of measured data. The properties of the heterogeneous electro-mechanical model were demonstrated in a realistic model of left ventricular geometry and fiber orientation using a monodomain approach for describing electrical interaction.

This study indicated the necessity of incorporating regional heterogeneity to model human cardiac electro-mechanics with qualitative good agreement to measured data. The heterogeneity leads to a homogenization of the mechanical process due to increasing time to peak tension from epicardium towards endocardium.

1 Introduction

Electrical excitation causes mechanical contraction during a heart cycle. This electro-mechanical coupling is controlled by free cytoplasmic calcium. Electrical excitation and repolarization propagation depends on tissue type and distribution, geometry of the heart, and heart rate. Furthermore, fiber orientation, distribution of gap junctions and pathologies influence the activity of the heart. A heterogeneity in the ventricular myocardium is present caused by transmurally changing ion channel kinetics and distributions, influencing mainly plateau and repolarization phase of action potential (AP) and the development of tension.

Three principal ventricular cell types can be distinguished: subendocardial, midmyocardial (M), and subepicardial cells. They differ in electrophysiological properties, in their respond to pharmacological agents, and in the pathological expression [1]. A spike-and-dome morphology of AP is present for subepicardial cells vanishing throughout the wall towards the endocardium. The action potential duration (APD) is longest in M cells. Altogether, the transmural differences of transmembrane voltage are important factors for generating the positive monophasic T wave in transmural electrocardiograms (ECG) [2].

Some pathologies are based on defects mainly influencing the transmural heterogeneous balance by changing electrophysiolgical characteristics dramatically in parts of the tissue like M cells. E.g. Long QT Syndrome, Brugada Syndrome, and the genesis of torsades de pointes were attributed to the disarrangement of the transmural heterogeneity [1].

Mechanical function of a cardiomyocyte is initiated by the electrical excitation. The varying concentration of intracellular calcium triggers the development of tension in the force generating units. Thus, onset of tension follows nearly the same spatiotemporal sequence as the electrical excitation.

An electrophysiological model with heterogeneous parameters in combination with a tension development model in an anatomical model was applied. This anatomical model is based on the left ventricle (LV) of the Visible Female data set including realistic fiber orientation. The aim of this work was to investigate the effects of heterogeneity on the electromechanical properties of the tissue.

2 Materials and Methods

The mathematical description of cardiac electromechanical processes was based in this work on an anatomical model of the LV of the Visible Female data set including realistic fiber orientation, a cellular model for describing the physiological properties and a model to calculate the electrical interactions of the cells in the tissue. The cellular model consisted of two sub-models: One model described the electrophysiology of the myocyte and a second model quantified the function of the contractile units. Both sub-models describe the status of myocytes with a set of nonlinear-coupled partial differential equations. The electrical coupling of cells was achieved by incorporating a monodomain model.

2.1 Anatomical Model

Highly detailed three-dimensional (3D) anatomical models are not yet producible with standard medical tomographic systems. The photographic images of the Visible Human Project [3] were used to obtain a precise model. The images of a 59 year old female are basis for the heart model in this work. These images are transversal cryosections with a resolution of 0.33 mm. The distance between the data is also 0.33 mm. The images were pre-processed to obtain a 3D data set. Afterwards, this data set was segmented and classified using different techniques of digital image processing, e.g. thresholding, region-growing, morphological operators and interactively deformable contours [4]. Figure 1 shows the model consisting of cubic voxels in a surface based transparent visualization.

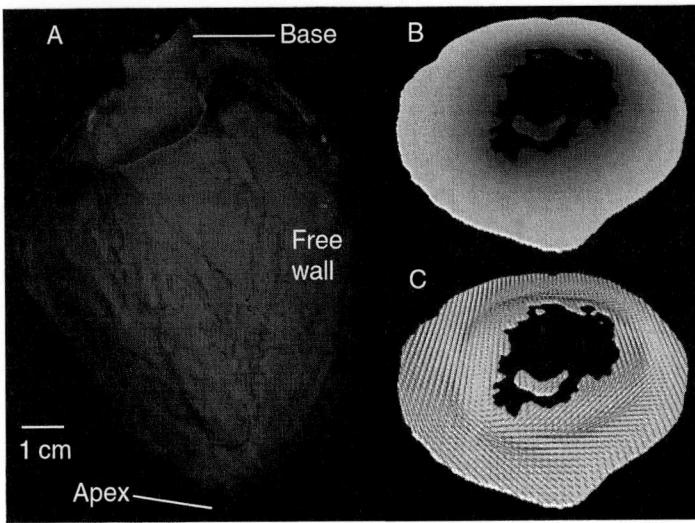

Fig. 1. A: Left ventricle of the Visible Female data set in a transparent frontal view. B: Central cross-sectional slice of the model describing the variation of tissue types from endocardium to epicardium with dark gray to white, respectively. C: Vectors showing the fiber orientation in the same exemplary slice

The orientation of muscle fibers was included into the model allowing the incorporation of anisotropic electrical and mechanical properties (fig. 1 C). The orientation was constructed with a rule-based method that was derived from anatomical studies [5]. The orientation of the fibers varies from subepicardial ($-75°$) via midmyocardial ($0°$) to subendocardial myocardium ($55°$) [6].

2.2 Heterogeneous Electrophysiology

The electrophysiological behavior of cells was simulated with a modified Priebe-Beuckelmann model of human ventricular cardiomyocytes [7]. This model is based on the fundamental work of Hodgkin and Huxley [8]. Both models describe the electrophysiological behavior of an excitable cell with a set of nonlinear-coupled differential equations. These equations were solved with the forward Euler method and reconstruct intra- and extracellular ion concentrations, ion-flows through the cell membrane, states and dynamic changes in ionic channels, and the transmembrane voltage. The model includes also detailed descriptions of the behavior of intracellular structures, i.e. the sarcoplasmic reticulum and calcium buffers. Especially, the accurate description of the intracellular calcium concentration $[Ca^{2+}]_i$ is of importance for this work, since it is the link between electrical and mechanical activity. The transport of ions is time dependent and influenced by gradients of the concentrations and the electrical field and is represented in the model by a set of membrane currents I_{mem}.

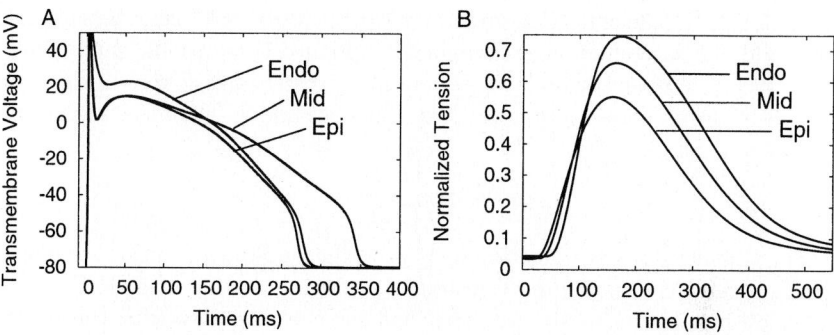

Fig. 2. Electromechanical properties of a single cell located in subendocardium, midmyocardium, and subepicardium. A: Action potential. B: Tension development

The temporal changes of the transmembrane voltage V_m is defined by

$$\frac{dV_m}{dt} = \frac{1}{C_m}(I_{mem} + I_{inter}) \tag{1}$$

with the membrane capacity C_m and the intercellular stimulus current I_{inter}.

The transmural heterogeneous characteristic is based mainly on differing ion channel expression. The detailed description of the measurement data and the obtained model is discussed in [9]. The transient outward potassium current I_{to} is largest in subepicardial areas and decreases towards endocardium. Differences in the expression of slow-delayed rectifier potassium current I_{Ks} were measured, describing a reduction of channel density in M cells. A varying density of the potassium inward rectifier current I_{K1} were measured. The largest density was found in subepicardial myocytes. A recent study reported a decreased expression of sodium-calcium exchangers (I_{NaCa}) towards endocardium.

The location of M cells varies within the ventricular wall [1]. M cells were found in more epicardial regions in the lateral wall, more endocardial areas in the anterior wall and throughout the wall near the valves. We placed the M cells in the end of the first quarter of the distance between endocardium and epicardium and defined a smooth transition of changing ion channel characteristics inside the whole ventricle (fig. 1 B).

Figure 2 A shows the transmembrane voltage for this model in subendocardial, M, and subepicardial cells after a stimulation with a frequency of $1\,Hz$.

2.3 Tension Development

Tension development in the contractile elements of myocytes is provoked and modulated by the concentration of intracellular calcium $[Ca^{2+}]_i$ commonly resulting from an electrical excitation. In this work the tension development was calculated with the model of Glänzel et al. [10, 11]. The model describes the binding of calcium to troponin C, the configuration change of tropomyosin, and the interaction of myosin and actin with a set of 14 state variables, which are coupled by rate coefficients. The sarcomere length, state variables, the sarcomere stretch, and the sarcomere stretch velocity influence these coefficients. The cooperativity mechanisms cross-bridge–troponin, cross-bridges–cross-bridges, and tropomyosin–tropomyosin are incorporated in the model. The resulting normalized tension T_n is given by:

$$T_n = \frac{\alpha T_{AM}}{T_{max}} \qquad (2)$$

with the sarcomere overlap function $\alpha = \alpha(\lambda)$, the sum of the tension developing states T_{AM}, and the maximum tension T_{max} during resting stretch. The original tension generating model was modified for the interplay with the utilized electrophysiological model [12].

Because sarcomere density and structure is assumed to be homogeneous in ventricles, it was proposed that inhomogeneous mechanical behavior is only due to electrical heterogeneity [13]. A study with guinea-pigs suggested that peak amplitude of cell shortening is largest in epicardial cells and smallest in midmyocardial ones [14]. On the other hand, another study with canine showed that peak amplitude of cell shortening is largest in endocardial cells and smallest in epicardial ones during unloaded cell shortening [13]. Also the time to peak and latency to onset of contraction were increasing from epicardium to endocardium.

Figure 2 B illustrates the simulated tension provoked by the change of $[Ca^{2+}]_i$ for isolated subendocardial, M, and subepicardial cells. The maximum tension is largest in subendocardial cells and decreases towards epicardium. This is consistent with the measurement data of Cordeiro et al. [13].

2.4 Anisotropic Conduction Model

The myocardium consists primarily of discrete myocytes, arranged in an oriented and laminated structure [6, 15]. Myocytes are enclosed by the sarcolemma, which delimits extra- from intracellular space and are of irregular shape, but a dominant principal axis can be assigned.

The intracellular space of myocytes is coupled by gap junctions, located at the intercalated disks. The distribution of gap junctions combined with the shape and orientation of the myocytes leads to a macroscopic anisotropic electrical conductivity.

The bidomain diffusion model treats the electrical behavior of the tissue in two domains, the intra- and extracellular space, separated by the cell membrane [16]. Poisson's equation for fields of stationary electrical current is fulfilled in each domain. The domains are coupled by the transmembrane voltage V_m.

The bidomain model can be reduced to a monodomain model for the special case of intra- and extracellular conductivity having the same anisotropy ratios. The monodomain model is described by the equation

$$\nabla (\sigma \nabla V_m) = \beta I_{mem} - I_{si} \qquad (3)$$

with the combined conductivity tensor σ, the surface to volume ratio of the membrane β, and an externally applied current source I_{si}.

The combined conductivity σ consists of conductivities for intra- and extracellular components and for gap junctions for longitudinal and transversal direction. σ was adapted to the resolution of $0.33\,mm$ using numerical experiments with higher resolution. The transversal conductivity was set to $0.15\,S/m$ and the longitudinal to $0.95\,S/m$.

2.5 Computational Environment

The presented models were combined to an electromechanical coupled heart model. The simulation was performed on 10 Apple XServe G5 dual 2 GHz processor cluster nodes using multiprocessing techniques [17]. The model consists of approximately 16 million cubic volume elements with a side length of $0.33\,mm$. Approximately 6.5 million of these elements describe excitable tissue, the remainder blood and surrounding tissue. The simulation required 2.6 GB of main memory and needed for a $600\,ms$ interval with a temporal increment of $20\,\mu s$ seven hours of calculation time. The monodomain model was discretized with a finite difference method and the integration of the cellular models was achieved with the Euler method. The software was implemented with C++.

Stimulation of the ventricular model was initiated at the subendocardium in voxels modeling myocytes with connections to Purkinje fiber ends. These points were placed semi-automatically, since the cardiac conduction system was not visible in the cryosection images of the Visible Female data set. The virtual Purkinje fiber ends are positioned by identifying the endocardial border of the tissue. The most apical point on the endocardium has to be selected manually. Starting from this point, a binary method defines the Purkinje fiber ends towards the base of the heart. In these points, intracellular currents were applied in a specific temporal sequence. The application of currents started at the most apical point and wandered basal with approximately $3\,m/s$.

3 Results

The transmembrane voltage distribution in the LV is shown in fig. 3 at different time steps after the initial activation during one heart cycle. The excitation started at subendocardial, apical points (fig. 3 A–B). Afterwards, the depolarization front wandered basal and from endocardial to epicardial regions (fig. 3 C–D). No significant transmural gradient of V_m was present during the plateau phase (fig. 3 E–F). The repolarization was mainly homogeneous (fig. 3 G), but the final repolarization was located in deep subendocardium near the M cells (fig. 3 H). These characteristics were also clearly evident in the cross-sectional slice in fig. 4 with more quantitative detail.

Figure 5 shows the developed tension in the model at different time steps. The onset of tension was more homogeneous than the electrical activation but still starts near the endocardium (fig. 5 A–D). The maximum tension was also larger in endocardial areas compared to epicardial during the peak of the tension (fig. 5 E–F). During relaxation, tension in endocardium was largest (fig. 5 G–H). The same information is shown in a cross-sectional slice in fig. 6.

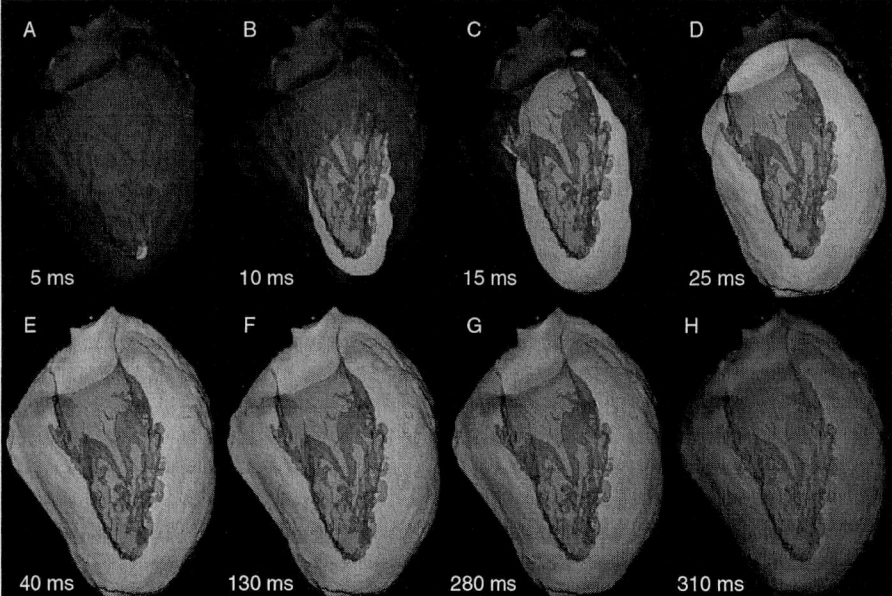

Fig. 3. Excitation and repolarization propagation in the Visible Female left ventricle at different time steps after initial activation. The transmembrane voltage distribution is illustrated gray coded. Dark gray is resting ($-80\,mV$) and white depolarized ($10\,mV$)

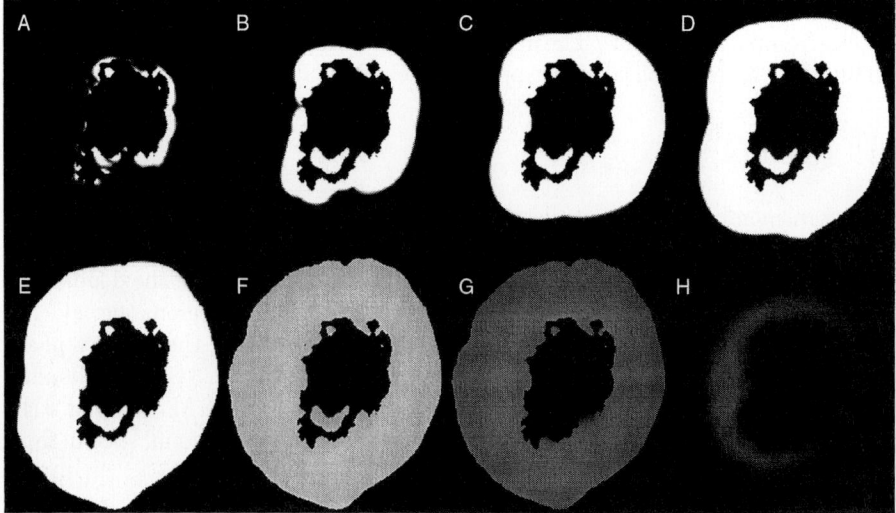

Fig. 4. Central cross-sectional slice of the left ventricle. Same information as in fig. 3. The action potential heterogeneity is mainly visible during repolarization phase

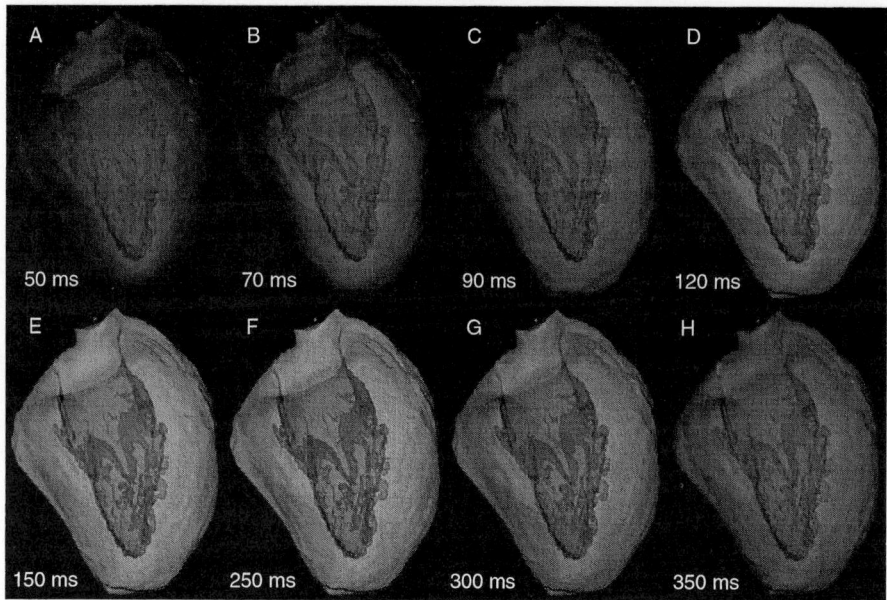

Fig. 5. Gray coded tension development in the Visible Female left ventricle at different time steps after initial electrical activation. Dark gray is resting (20 %) and white maximum normalized tension (70 %)

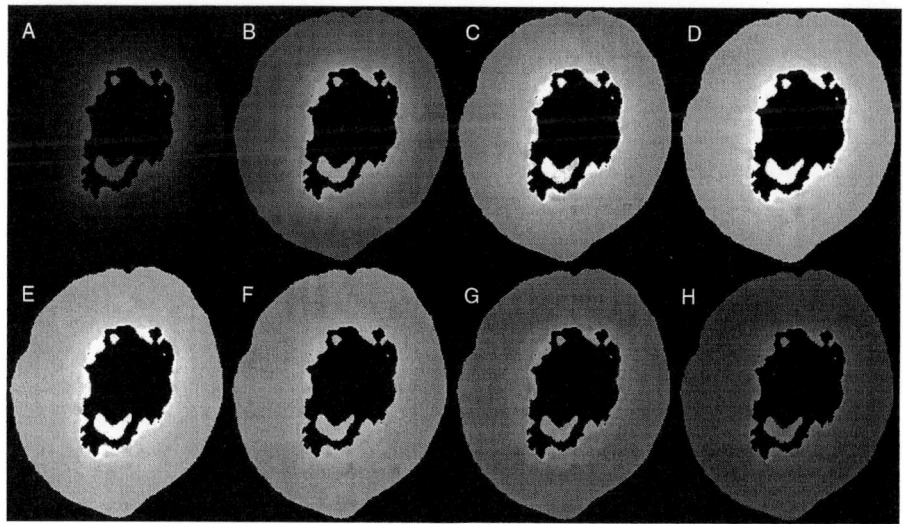

Fig. 6. Central cross-sectional slice of the left ventricle. Same information as in fig. 5

Figure 7 demonstrates the distinctions in the electromechanical properties in two different situations: The first results were obtained from simulations

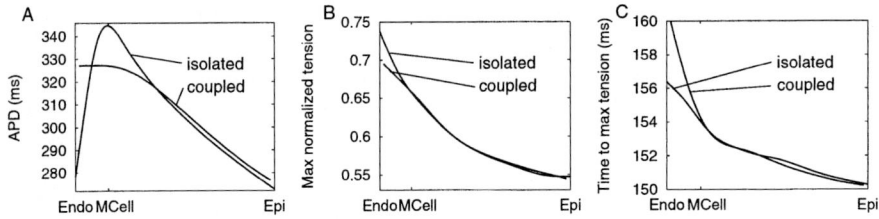

Fig. 7. Differences between heterogeneous single cell models and heterogeneous coupled model of the left ventricle. A: Action potential duration B: Normalized maximum peak tension C: Time to maximum peak tension

with isolated cells of different regions from endo- to epicardium and the others were derived from the heterogeneous and electrically coupled anatomical LV model.

Figure 7 A shows that the APD distribution occurring in the simulation with the isolated cells was smoothened in the LV model due to electrotonic interaction. This led to a prolongation of APD mainly in endocardial cells resulting in the vanishing of repolarization in the LV model close to the M cells.

The distribution of transmural maximum tension (fig. 7 B) was not as much influenced by the electrical coupling as the APD. The reason for this is that the calcium transient is a few milliseconds after the depolarization, but the electrotonic interaction acts mainly on the repolarization phase. The largest differences in transmural direction were visible in the subendocardial area.

The time to peak tension shown in fig. 7 C demonstrates the importance of electrophysiological heterogeneity. The time to peak tension is longer in subendocardial compared to subepicardial myocytes leading to a homogenization of the tension development. We suggest that the time to peak for subendocardial cells was more prolonged compared to the isolated cell simulations due to the coupling enhancing the effect of mechanical synchronization.

4 Conclusion

The presented model was suitable to simulate the general effects of the electrophysiological excitation and the tension. The simulated tension variation induced by electrophysiological heterogeneity shown in fig. 2 is in qualitative good agreement with recently published data for canine LV [13]. Both the simulation and the measurements showed that tension, respectively the unloaded cell shortening in the measurements, is largest in endocardial areas and decreases towards epicardium (fig. 7 B). Also, the simulated increasing time to maximum tension from subepicardial to subendocardial cells (fig. 7 C) is consistent with measurements [13]. This results in the synchronization of the tension across the ventricular wall due to the electrophysiological heterogeneity. Further experiments have to be carried out to validate the proposed characteristics.

The calcium handling approach of the electrophysiological model has to be adapted to recent descriptions with new measurement data to reproduce accurately the calcium transient in human ventricular myocytes. Furthermore, we neglected in this study the sarcomere length dependency of the tension. This might reduce the gradient in maximum tension (fig. 7 B), because tension is scaled by the fiber-strain due to overlap of actin and myosin.

The influence of heart rate and pathological behavior as well as deformation on the model will be examined in future. Such models have the potential to enable specific diagnosis and therapy advancements, to assist drug development and to improve understanding of the complex cardiovascular system.

References

1. Antzelevitch, C., Yan, G., Shimizu, W., Burashnikov, A.: Electrical heterogeneity, the ECG, and cardiac arrhythmias. In Zipes, D.P., Jalife, J., eds.: Cardiac Electrophysiology. From Cell to Bedside. 3 edn. W. B. Saunders Company, Philadelphia (1999) 222–238
2. Weiß, D.L., Seemann, G., Dössel, O.: Conditions for equal polarity of R and T wave in heterogeneous human ventricular tissue. In: Proc. BMT. Volume 49-2/1. (2004) 364–365
3. Ackerman, M.J.: Viewpoint: The Visible Human Project. J. Biocommunication **18** (1991) 14
4. Sachse, F.B., Werner, C.D., Stenroos, M.H., Schulte, R.F., Zerfass, P., Dössel, O.: Modeling the anatomy of the human heart using the cryosection images of the Visible Female dataset. In: Proc. Third Users Conference of the National Library of Medicine's Visible Human Project, Bethesda, USA (2000)
5. Sachse, F.B., Frech, R., Werner, C.D., Dössel, O.: A model based approach to assignment of myocardial fibre orientation. In: Proc. Computers in Cardiology. Volume 26., Hannover (1999) 145–148
6. Streeter, D.D.: Gross morphology and fiber geometry of the heart. In Bethesda, B., ed.: Handbook of Physiology: The Cardiovascular System. Volume I. American Physiology Society (1979) 61–112
7. Priebe, L., Beuckelmann, D.J.: Simulation study of cellular electric properties in heart failure. Circ. Res. **82** (1998) 1206–1223
8. Hodgkin, A.L., Huxley, A.F.: A quantitative description of membrane current and its application to conduction and excitation in nerve. J. Physiol. **177** (1952) 500–544
9. Seemann, G., Sachse, F.B., Weiß, D.L., Dössel, O.: Quantitative reconstruction of cardiac electromechanics in human myocardium: Regional heterogeneity. J. Cardiovasc. Electrophysiol. **14** (2003) S219–S228
10. Glänzel, K., Sachse, F.B., Seemann, G., Riedel, C., Dössel, O.: Modeling force development in the sarcomere in consideration of electromechanical coupling. In: Biomedizinische Technik. Volume 47-1/2. (2002) 774–777
11. Sachse, F.B., Glänzel, K., Seemann, G.: Modeling of protein interactions involved in cardiac tension development. Int. J. Bifurc. Chaos **13** (2003) 3561–3578
12. Sachse, F.B., Seemann, G., Chaisaowong, K., Weiß, D.: Quantitative reconstruction of cardiac electromechanics in human myocardium: Assembly of electrophysiological and tension generation models. J. Cardiovasc. Electrophysiol. **14** (2003) S210–S218

13. Cordeiro, J.M., Greene, L., Heilmann, C., Antzelevitch, D., Antzelevitch, C.: Transmural heterogeneity of calcium activity and mechanical function in the canine left ventricle. Am. J. Physiol. **286** (2003) H1471–H1479
14. Wallis, H.L., Sears, C., Bryant, S.: Regional differences in excitation-contraction coupling in the guinea-pig left ventricle. J. Physiol. **544.P** (2002) 53P–54P
15. LeGrice, I.J., Smaill, B.H., Chai, L.Z., Edgar, S.G., Gavin, J.B., Hunter, P.J.: Laminar structure of the heart: Ventricular myocyte arrangement and connective tissue architecture in the dog. Am. J. Physiol. **269** (1995) H571–H582
16. Henriquez, C.S., Muzikant, A.L., Smoak, C.K.: Anisotropy, fiber curvature and bath loading effects on activation in thin and thick cardiac tissue preparations: Simulations in a three-dimensional bidomain model. J. Cardiovasc. Electrophysiol. **7** (1996) 424–444
17. The Message Passing Interface (MPI) standard: (www-unix.mcs.anl.gov/mpi)

Reentry Anchoring at a Pair of Pulmonary Vein Ostia

L. Wieser[1,*], G. Fischer[1], F. Hintringer[2], S.Y. Ho[3], and B. Tilg[1]

[1] Institute for Biomedical Signal Processing and Imaging,
University for Health Sciences, Medical Informatics and Technology (UMIT),
Hall in Tirol, Austria
leonhard.wieser@umit.at
http://bm.umit.at
[2] Department for Cardiology, University Hospital Innsbruck,
Innsbruck, Austria
[3] National Heart & Lung Institute,
Imperial College and Royal Brompton & Harefield Hospitals, London, UK

Abstract. Recent findings in a sheep model of atrial fibrillation support the hypothesis that an organized micro-reentry could be the maintaining mechanism of the arrhythmia (mother wavelet). According to these studies we constructed a two dimensional computer model of tissue in the region around a pair of pulmonary vein ostia and investigated anchoring of a reentry wave at these ostia. We used the Luo Rudy phase I ionic current model to describe membrane kinetics and generated two different stages of electrical remodelling of the cells by varying the slow inward calcium current. Our attempt to initiate a stable reentry failed for cells with higher action potential duration and higher rate adaption. By simulating a higher stadium of electrical remodelling we finally were successful, and we were able to produce a periodic reentry. This led us to the conclusion that a low rate adaption (high electrical remodelling) facilitates organized activity in the atria.

1 Introduction

Current understanding of atrial fibrillation is mainly based on the hypothesis that the excitation pulse of cardiac activation is split into a number of irregular propagating wave fronts (multiple wavelet hypothesis) [1], [2], [3]. Though, it is not fully clarified how these multiple wavefronts are maintained. Experiments on sheep hearts show that a single micro-reentry circuit can act as driving mechanism (mother wavelet hypothesis) [4], [5]. In the tissue forming the leading reentry pathway signals of high rate and periodic activity were recorded, which supports the conceptual model of an organized activation pattern as the driving mother wavelet.

---

* The study was supported by the Austrian Science Fund (FWF) under the grant P16759-N04.

In three of the seven experiments investigated in [4] the anchor for the driving reentry circuit was close to the pulmonary vein (PV) ostium (in the sheep the PVs have one common ostium), but the possibility of an anatomical reentry around the ostium was excluded in this study. On the other hand the data in [6] indicate that the shortest fibrillation cycle length in humans is recorded around the right and left pair of PVs.

We hypothesized that an organized reentry pathway in human atria can be formed by an activation front travelling around a pair of PV ostia thus doubling the pathway length. A functional block in the isthmus between the ostia is maintained because the region is stimulated with double frequency since the activation front enters twice within one cycle.

It is well known that a shorter action potential duration (APD) favors initiation of a stable reentry around an anatomic obstacle because the wavelength of activation ($\lambda = \text{APD} \cdot c$, where c is the conduction velocity) becomes smaller than the necessary pathway length around the obstacle. It is also known that reducing the slow inward current which is mainly carried by Ca^{2+} ions has two effects: on one hand APD is decreased, on the other hand rate adaption is reduced. In this study we want to quantify this change of rate adaption and investigate its influence on the possibility to initiate a stable reentry around a pair of PV ostia.

2 Methods

2.1 Bioelectric Model

Intercellular coupling in the homogeneous tissue (domain Ω) was modelled by the monodomain equation in two dimensions x and y,

$$\sigma \left(\frac{\partial^2 V}{\partial x^2} + \frac{\partial^2 V}{\partial y^2} \right) = A_\text{m} \left(C_\text{m} \frac{\partial V}{\partial t} + I_\text{Ion} \right) \qquad \text{in } \Omega , \qquad (1)$$

with homogeneous Neumann boundary conditions,

$$\frac{\partial V}{\partial n} = 0 \qquad \text{on } \partial \Omega . \qquad (2)$$

Here, V is the membrane potential, σ is the intercellular conductivity, A_m is the surface to volume ratio of the cells, C_m is the membrane capacitance per area unit, and I_Ion is the ionic current per area, as obtained by the ionic current model (see next section). The finite difference method was applied for discretization in the spatial domain ($\Delta x = 160$ μm). The Crank-Nicolson method was used for time integration with a step size of 20 μs. Therefore, we had to solve a linear system with unchanged left hand side at each time step, which was done by a preceding LU decomposition, and forward- and backward substitutions during simulation.

The numerical model was implemented in C++, and each simulation was executed at a single processor machine with 2.8 GHz. For a typical model size used in this study (about 58000 nodes) and 4 seconds of simulated cardiac activity time we found execution times of around 165 minutes for the linear system part and 160 minutes for the part of the ionic current model, i.e. about 5 and a half hours for the whole simulation.

2.2 Single Cell

We used the Luo-Rudy phase I (LRI) model for computing the ion dynamics in the cell membrane [7]. Since this model was originally based on data from canine ventricular cardiomyocytes, we had to make a few modifications to approximate membrane kinetics of atrial cells in the stadium of early electrical remodelling. Thus, we reduced conductivity G_{si} of the mainly calcium related slow inward current I_{si}, similar as in [8] and [9]. In order to have two different values we set G_{si} to 30% and to 40% of its original value, i.e. $G_{si} = 0.27 \frac{S}{m^2}$ and $G_{si} = 0.36 \frac{S}{m^2}$. Throughout this paper, these settings will be referred to as $G_{si}^{30\%}$ and $G_{si}^{40\%}$, respectively.

To prove the correlation between slow inward calcium current and rate adaption we considered an isolated cell which was stimulated at constant cycle length. We measured APD as a function of cycle length for both values of G_{si}, similar as in [10]. For this purpose we used APD_{90} which is defined as the time from the first upstroke until transmembrane potential loses 90% of its amplitude.

2.3 Tissue Model

We wanted to create a model with comparable dimensions to human anatomy in the region of the PV ostia. Thus, we used a rectangular patch of size 51.2 mm × 32 mm with two holes representing the ostia (see Fig. 1). The distance between the two holes at their narrowest point was 2.56 mm. According to the data presented for formalin fixed preparations in [11] one can estimate that in about 50% of the adult human hearts this muscular separation is smaller than 3 mm. The ostial diameters were made 10 mm, which was again motivated by data from [11]. In this study a mean venous orifice diameter of 12.5 mm at the venoatrial junction is reported. Isotropic properties were assumed for the patch. Furthermore, we set $A_m = 100 \frac{1}{mm}$ and $C_m = 0.01 \frac{\mu F}{mm^2}$.

For $G_{si}^{40\%}$ we used a tissue conductivity $\sigma = 0.14 \frac{S}{m}$. Note that this conductivity leads to a conduction velocity of about 75 $\frac{cm}{s}$, if we measure a plane activation front. This value corresponds to data reported in [12] for the tissue around the PV ostia. In the case $G_{si}^{30\%}$ we turned conductivity to $\sigma = 0.20 \frac{S}{m}$ in order to keep the wavelength of activation constant. This choice is motivated as follows: Assuming a reduction of APD of between 20% and 25% due to the reduction of G_{si}, we should obtain a conduction velocity that is a factor 1.18 to 1.25 higher. From analytical considerations in a 1D cable model (in our model we have 1D wave propagation) [13] we get a quadratic dependence between conduction velocity and conductivity, i.e.

$$\frac{c^2}{\sigma} = \text{const.} \tag{3}$$

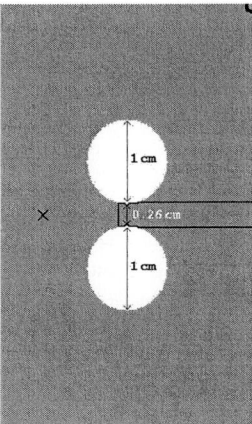

Fig. 1. Model geometry: stimulus sites are the upper right corner (quarter circle) and the framed area (for details see Sect. 2.4), transmembrane potential is recorded everywhere, and with a higher sample rate at the x-marked point

By this equation we obtain a modification factor for σ between 1.39 and 1.56, therefore a conductivity of $\sigma = 0.20\ \frac{S}{m}$ seems appropriate. We tested this choice *a posteriori* in a 1D cable model. For each of the cases $G_{si}^{30\%}$ and $G_{si}^{40\%}$ we applied 30 consecutive stimuli at one end of the cable, using a stimulus rate that was comparable to the period duration of the reentry in the 2D model. Thus, we could compare the distributions in space of the activations.

By keeping activation wavelength constant we could be sure that the influence of APD shortening on the anchoring process is excluded, thus we were able to investigate the effect of rate adaption only.

2.4 Stimulation Protocols

We had to use different time intervals for the stimulation protocols for $G_{si}^{30\%}$ and $G_{si}^{40\%}$ due to the different APDs. In both cases we applied two consecutive stimuli (S1, S2) at the upper right corner of the tissue to put the cells under fibrillation like conditions (high rate activation). After that we stimulated a rectangular area containing the inter-ostial isthmus (see black frame in Fig. 1) at the time when the upper border of this area was fully recovered from activation and the bottom border was not yet excitable (S3). By this means we obtained an activation front propagating in only one, namely in the upper direction. Note, that we were stimulating at the so called vulnerable window in order to achieve a unidirectional block [14]. Additionally, since the inter-ostial isthmus was included in the stimulated area, it was still active when the wave front had circled the upper PV ostium and reentered from the left hand side. Our aim was to establish a functional block in the isthmus region. Furthermore, we applied a lower stimulus current for S1 and S2 because cells were fully recovered in these cases, and a smaller current was sufficient to initiate wave propagation. Details of the stimulation protocol are presented in Tab. 1.

Table 1. Stimulation protocols

	distance to upper edge [mm]	distance to left edge [mm]	area [mm^2]	I_{stim} [$\frac{A}{m^2}$]	time ($G_{si}^{30\%}$)	time ($G_{si}^{40\%}$)
S1	31	0	1 × 1	−0.5	0	0
S2	31	0	1 × 1	−0.5	150	210
S3	15	24	17 × 3	−0.75	260	340

Position is given by distance to the upper and to the left edge of the tissue

3 Results

3.1 Slow Inward Calcium Current and Rate Adaption

If we stimulate an isolated cell with constant frequency, using the LRI-model, we find that APD$_{90}$ converges to a constant value after a few cycles. Figure 2 shows the action potentials in each case for a typical sinus rate and flutter rate, taken after 10 seconds of stimulation. A summary of all the results is presented in Tab. 2. Apparently, cells at $G_{si}^{30\%}$ are less affected by the higher stimulation rate than cells at $G_{si}^{40\%}$. Thus, we can conclude that a lower G_{si} decreases rate adaption, similar as for atrial cell model in [10].

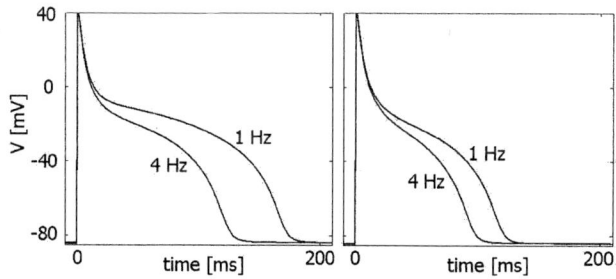

Fig. 2. Action potentials of a single cell for $G_{si}^{40\%}$ (left panel) and $G_{si}^{30\%}$ (right panel) are shown at sinus rate (1 Hz) and a typical flutter rate (4 Hz)

Table 2. APD$_{90}$ for different parameter settings

rate [Hz]	absolute values [ms]		relative to reference	
	$G_{si}^{30\%}$	$G_{si}^{40\%}$	$G_{si}^{30\%}$	$G_{si}^{40\%}$
0.5	118	169	100%	100%
1	117	168	99%	99%
2	110	155	93%	92%
4	94	121	80%	72%
6	81	98	69%	58%

Results at 0.5 Hz taken as the reference

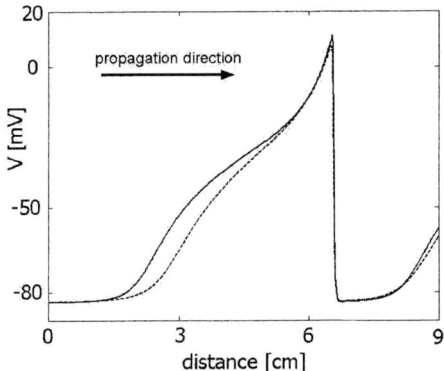

Fig. 3. Transmembrane potential along the 1D cable for $G_{si}^{30\%}$ (dotted line) and $G_{si}^{40\%}$ (solid line) after the 30th stimulus (applied at the left end of the cable). For both cases we obtain an activation wavelength of about 56 mm

3.2 Wavelength of Activation

We used a 1D cable model in order to compare the activation wavelengths for each setting. The stimulus was applied at one end of the cable, and its rate was 80 ms for $G_{si}^{30\%}$ and 93 ms for $G_{si}^{40\%}$, which was motivated by the period durations obtained in the 2D models (see Figs. 5 and 8). Figure 3 shows the transmembrane potential along the cable after the 30th stimulus. The time frame is chosen such that activation occurs at 6.5 cm in both cases. Note, that the curves intersect again at -83 mV (resting potential at -85.5 mV), which finally leads us to the conclusion that the wavelengths of activation are approximately the same.

3.3 Anchoring at Pulmonary Vein Ostia

To investigate the rate adaption dependent probability of anchoring at PV ostia we first consider the simulation with lower G_{si}.

Results for $G_{si}^{30\%}$: A sequence of snapshots of the activation pattern is presented in Fig. 4. The first picture (upper left) is taken immediately after the stimulus (S3) and thus highlights the stimulated area. The following pictures show the activation pattern each time when the wavefront reaches the top and the bottom of the two PV ostia. Note, that the inter-ostial isthmus never is fully recovered, so the wavefront is always blocked at this site and runs around both ostia. The activation pattern approaches more and more the regular shape of the bottom right panel (isopotential lines nearly orthogonal to propagation direction), since the pathway of the wavefront is the same each cycle. Finally we obtain a stable reentry and a self-maintained functional block in the isthmus.

Figure 5 shows the convergence of the action potential to periodic activity at a single cell, taken from the x-marked point in Fig. 1. In the left panel we plot the period duration against time, having the period duration defined as the

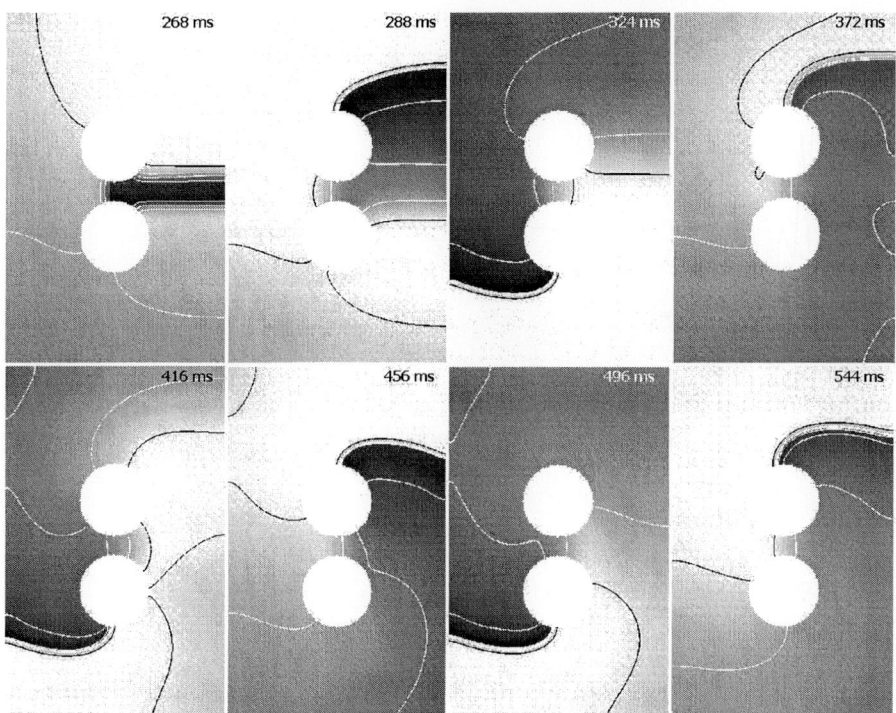

Fig. 4. Initiation of the reentry around both ostia. The pictures show distributions of transmembrane potential at times indicated in the upper right corner of each panel. A linear grey scale from white (-85 mV) to black (15 mV, fully activated) is used. There are 3 white isopotential lines indicating 0 mV, -25 mV and -50 mV, and one black isopotential line at -78.5 mV which roughly represents the border between partially and fully recovered cells

time between one action potential upstroke to the next. In the right panel we investigate the action potential shape, and calculate rms-error of each action potential with the last one serving us as reference. Convergence occurs in less than 2 seconds in both panels.

Figure 6 highlights the mechanism of the functional block between the ostia. At advanced simulation time (i.e. at already periodic activity), the action potentials of three different single cells are plotted. Locations of these cells are middle isthmus (solid line), edge of the isthmus (dashed line) and bulk medium (dotted line). Cells in the isthmus do not recover entirely because of double stimulation frequency and are therefore never fully activated, which is the cause of the functional block.

Results for $G_{si}^{40\%}$: The activation sequence for the stimulation protocol in this case is shown in Fig. 7. After the first cycle at 436 ms there is only a small excitable gap just in front of the propagating wave. This gap is small enough for

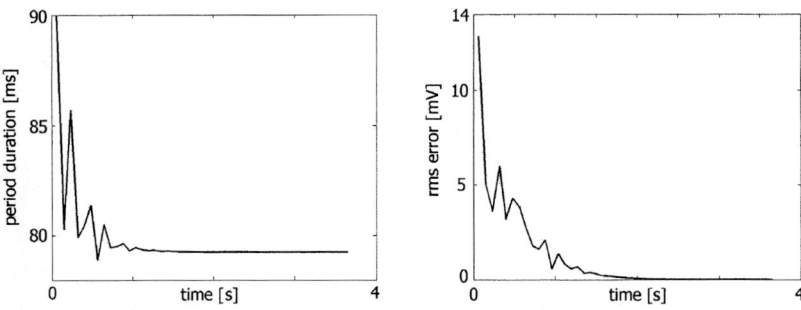

Fig. 5. Convergence to periodic activity at $G_{si}^{30\%}$: *Left Panel:* APDs at the x-marked point in Fig. 1 are calculated and plotted against time, *Right Panel:* Rms-error of each action potential with a reference (last action potential within simulation time) is plotted against time

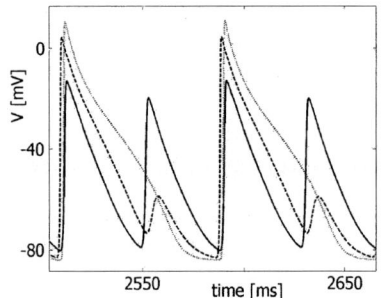

Fig. 6. Functional block in the inter-ostial space: The solid line shows transmembrane potential in the mid-isthmus region (high frequency, low amplitude, incomplete recovery), the dashed line indicates the potential at the border of the isthmus (medium amplitude and an after-depolarization like dome, full recovery), and the dotted line represents action potential in the bulk medium (high amplitude, full recovery)

the tip of the spiral wave to drift away from the upper ostium (474 ms) for a few milliseconds. During the next four half cycles (see panels for 508 ms, 558 ms and 607 ms) the functional block in the isthmus is maintained, but at 679 ms the spiral wave tip drifts away again. This time the half cycle around the upper ostium takes too long, and the tissue in the isthmus region is already excitable. Thus, the wavefront can pass through the ostia (721 ms) and propagates towards the border of the tissue because the excitable gap becomes too small (head meets tail). At this point the activation terminates.

The attempt to produce the same stable activation pattern as for $G_{si}^{30\%}$ fails with this stimulation protocol. Variations were made for the size of the stimulated area, for the stimulation time and for the intercellular conductivity σ but none of the simulations produced a stable reentry pattern. Therefore, we can conclude

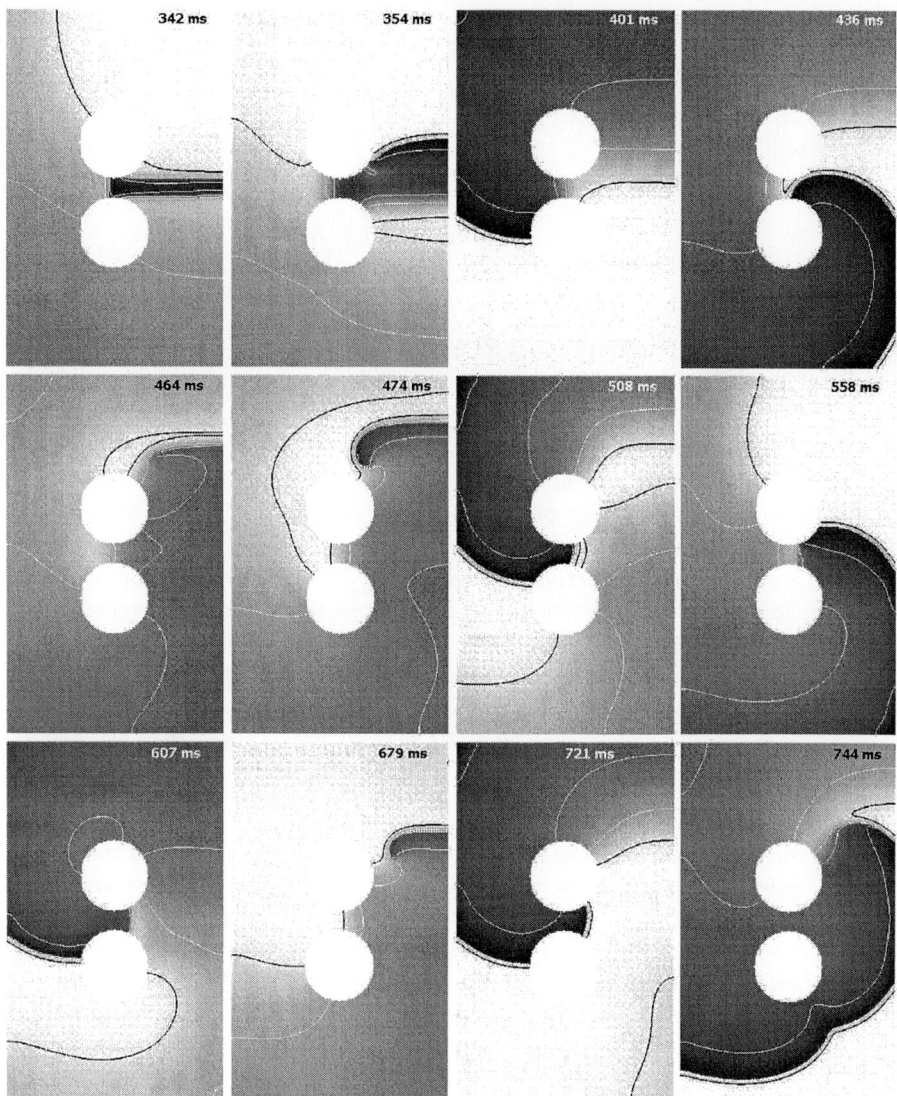

Fig. 7. Attempt to initiate reentry at $G_{si}^{40\%}$: The grey scale and the isopotential lines are the same as in Fig. 4. Simulation time is highlighted at the upper right corner of each panel. Activation terminates at about 750 ms because the spiral wave tip reaches the right border after 4 and a half cycles

that the initiation of such a regular pattern is less probable for higher rate adaption.

Since all our attempts failed to initiate stable reentry, we want to check whether a stable pattern for $G_{si}^{40\%}$ is generally possible or not (without having the effects of the initiation process). Therefore, we consider the simulation

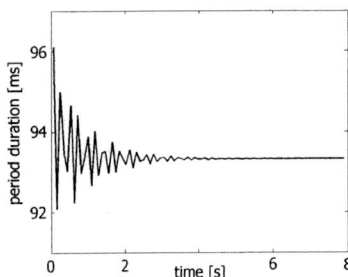 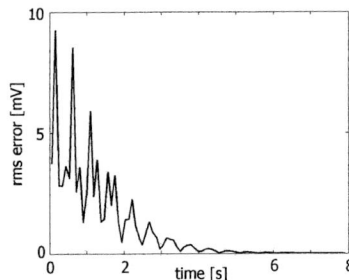

Fig. 8. Convergence to periodic activity at $G_{\text{si}}^{40\%}$: The plots are analogous to Fig. 4, but the convergence occurs more slowly than in Fig. 4

shown in Fig. 4 where a stable reentry is achieved, and store the state of the tissue at time 4000 ms. We then switch tissue conductivity from $\sigma = 0.20\ \frac{\text{S}}{\text{m}}$ to $\sigma = 0.14\ \frac{\text{S}}{\text{m}}$ and conductivity of the slow inward current G_{si} from 30% to 40% of the original value, and continue the simulation. By this means we are able to produce a stable periodic activation even for $G_{\text{si}}^{40\%}$. Time evaluation of period duration and rms-error is shown in Fig. 8. It can be seen that convergence occurs slower than in the simulation for lower G_{si}. This supports the assumption that a higher rate adaption reduces the probability of the formation of an organized micro-reentry.

4 Discussion

In this study we examined the influence of cellular rate adaption on the formation of a periodic reentry pattern around a pair of PV ostia. In order to exclude the effect of action potential shortening from the analysis two different conductivity values σ were used for $G_{\text{si}}^{30\%}$ and $G_{\text{si}}^{40\%}$ with a ratio of 1.4. This setting led to similar activation wavelengths of about 56 mm for both cases in a one dimensional cable (see Fig. 3), indicating the accurate compensation of action potential shortening in the 2D simulations. With this setting we were able to induce a periodic reentry for low rate adaption ($G_{\text{si}}^{30\%}$) by the stimulation protocol described in Tab. 1, while we could not induce it for higher rate adaption ($G_{\text{si}}^{40\%}$). For $G_{\text{si}}^{40\%}$ a periodic pattern was obtained only by a relatively artificial manipulation, i.e., the periodic reentry was induced at low adaptation and then the parameters in the simulation were switched to the setting used for higher rate adaption. Due to the abrupt change of parameters again a transition to a new periodic pattern was induced. Here, convergence took more than twice the time as needed for $G_{\text{si}}^{30\%}$ also supporting the hypothesis that rate adaption inhibits the formation of periodic micro-reentries.

As any computer model also our model provides only an approximation of the true biophysical phenomenon. The ionic current model used was originally based on data from canine ventricular cells. Similar as in [8], [9] the slow inward current was reduced for simulating electrical remodeling in the atria. The model dimension slightly underestimate the true dimensions as the data is based on formalin fixed preparations (10 to 20 % shrinking). No fibrous structures were included in the model.

However, the major goal of the study was to investigate the influence of electric remodeling (associated with a reduced rate adaption) on the formation of periodic reentries as experimentally observed in the animal model [4]. The study protocol enables the investigation of this effect at a significantly dumped contribution of action potential shortening. One can expect that similar results will be obtained for a modified tissue geometry including fibrous structures. Concluding, we can state that a high degree of electrical remodeling favors the formation of mother reentries as the driving mechanism of atrial fibrillation.

References

1. Nattel, S.: New ideas about atrial fibrillation 50 years on. *Nature* **415** (2002) 219–226.
2. Gray, R., Persov, A., Jalife, J.: Spatial and temporal organization during cardiac fibrillation. *Nature* **392** (1998) 75–78.
3. Schilling, R.: Which patient should be referred to an electrocardiologist: supraventricular tachycardia. *Heart* **36** (2002) 299–304.
4. Mandapati, R., Skanes, A., Chen, J., Berenfeld, O., Jalife, J.: Stable microreentrant sources as a mechansism of atrial fibrillation in the isolated sheep heart. *Circulation* **101** (2000) 194–199.
5. Chen, J., Mandapati, R., Berenfeld, O., Skanes, A., Gray, R., Jalife, J.: Dynamics of wavelets and their role in atrial fibrillation in the isolated sheep heart. *Cardiovasc Res* **48** (2000) 220–232.
6. Pappone, C., Rosiano, S., Oreto, G., Tocchi, M., Gugliotta, F., Vicedomini, G., Salvati, A., Dicandia, C., Mazzone, P., Santinelli, V., Gulletta, S., Chierchia, S.: Circumferential radiofrequency ablation of pulmonary vein ostia: A new anatomic approach for curing atrial fibrillation. *Circulation* **102** (2000) 2619–2628.
7. Luo, C., Rudy, Y.: A model of the ventricular cardiac action potential. Depolarization, repolarization, and their interaction. *Circ Res* **68** (1991) 1501–1526.
8. Ten Tusscher, K., Panfilov, A.: Reentry in heterogeneous cardiac tissue described by the Luo-Rudy ventricular action potential model. *Am J Physiol Heart Circ Physiol* **284** (2002) H542–H548.
9. Blanc, O., Virag, N., Vesin, J., Kappenberger, L.: A computer model of human atria with reasonable computation load and realistic anatomical properties. *IEEE Trans Biomed Eng* **48** (2001) 1229–1237.
10. Ramirez, R.J., Nattel, S., Courtemanche, M.: Mathematical analysis of canine atrial action potentials: rate, regional factors, and electrical remodeling. *Am J Physiol Heart Circ Physiol* **279** (2000) H1767–HH1785.
11. Ho, S.Y., Cabrera, J., Tran, V., Farre, J., Anderson, R., Sánchez-Quintana, D.: Architecture of the pulmonary veins: relevance to radiofrequency ablation. *Heart* **86** (2001) 265–270.
12. Harrild, D.M., Henriquez, C.S.: A Computer Model of Normal Conduction in the Human Atria. *Circ Res* **87** (2000) e25–e36.

13. Winfree, A.T.: Rotors, fibrilation and dimensionality. In *Computational Biology of the Heart*, pages 101–135. Edited by Panfilov, A.V. and Holden, A.V., John Wiley & Sons (1997).
14. Shaw, R.M, Rudy, Y.: The vulnerable window for unidirectioinal block in cardiac tissue: characterization and dependence on membrane excitability coupling. *J Cardiovasc Electrophysiol* **6** (1995) 115–131.

A Method to Reconstruct Activation Wavefronts Without Isotropy Assumptions Using a Level Sets Approach

Felipe Calderero[1], Alireza Ghodrati[2], Dana H. Brooks[2], Gilead Tadmor[2], and Rob MacLeod[3]

[1] Department of Signal Theory and Communications,
Technical University of Catalonia (UPC), Barcelona, Spain
[2] Department of Electrical and Computer Engineering,
Northeastern University, Boston, MA, USA
[3] Nora Eccles Harrison Cardiovascular Research and Training Institute (CVRTI),
University of Utah, Salt Lake City, Utah, USA

Abstract. We report on an investigation into using a Level Sets based method to reconstruct activation wavefronts at each time instant from measured potentials on the body surface. The potential map on the epicardium is approximated by a two level image and the inverse problem is solved by evolving a boundary, starting from an initial region, such that a filtered residual error is minimized. The advantage of this method over standard activation-based solutions is that no isotropy assumptions are required. We discuss modifications of the Level Sets method used to improve accuracy, and show the promise of this method via simulation results using recorded canine epicardial data.

1 Introduction

Inverse electrocardiography (ECG) estimates the electrical activity of the heart from potential measurements on the body surface. Because of smoothing and attenuation in the body, the measured potentials on the body surface can obscure significant detail about the heart's electrical activity. Thus conventional electrocardiography fails to detect heart problems in many situations [1]. A possible improvement is to model the electrical properties of the torso volume conductor and attempt to explicitly estimate features of cardiac electrical behavior; this is known as inverse electrocardiography. This problem is considered by many research groups [1, 2, 3, 4, 5]. However, the inverse problem of ECG is ill-posed and we need to add constraints to get a stable solution. The single most important feature of the heart's electrical activity is the activation wavefront, which passes through the heart muscle once per cardiac cycle and triggers, after some delay, the mechanical contraction of the muscle. The time that this wavefront passes through any given point in the heart is called the activation time. The problem of finding activation time has been studied using both activation-based models [3, 4, 5] and potential-based models [6].

The advantage of activation-based models is the reduction of the unknowns to the arrival time of the wavefront at each point on the epicardial and endocardial surfaces. Potential-based models instead treat the value of the potential at each point on the relevant surface at each time instant as a free variable. However, activation-based models depend on isotropy / homogeneity assumptions and a fixed shape of the temporal waveform in order to form a tractable forward model. Potential-based models are less restrictive but imply a high-order parameterization and thus require considerable smoothing (regularization). A method that is used frequently in inverse ECG is Tikhonov regularization, which indeed smooths the solution because of the type of 2-norm constraints employed. It is difficult to include the physical and geometric constraints imposed by the central physiological feature, namely wavefront behavior, except via indirect and somewhat coarse models [7, 8].

Our goal here is to investigate the possibility of estimating the activated region on the epicardium at each time instant using a Level Sets based inverse solution [9, 10]. The forward model we use is potential-based, with a very simple two-level model to characterize the potential distribution given the wavefront. The potential advantage is that we maintain some benefits of activation-based solutions without requiring isotropy assumptions.

In this work we use two constraints. The first assumes that the potentials on the heart can be effectively approximated by two values, representing the activated and inactivated regions respectively. This assumption is of course a rather crude approximation in both the activated and non-activated regions, and ignores the transition area between the two regions. But since we are looking for activation time this assumption may be useful, and we follow similar assumptions used in activation-based solutions [3, 4, 5]. The second constraint is a spatial constraint applied by the Level Sets method. Level Sets were first proposed in [9] to solve inverse problems when a constant-value inhomogeneity is enclosed in a constant-value background by evolving a boundary, starting from an initial region, such that the residual error is minimized. Modifications of the original Level Sets method were needed to improve reconstruction quality. New constraints were added to the Level Sets evolution to improve the shape of the recovered wavefront and to enhance sensitivity to regions of the epicardium whose effect on the residual error was otherwise too weak. In addition, we filtered the residual error to reduce the effect of the error introduced by the two-level quantization on the wavefront evolution.

Section 2 introduces the Level Sets Method applied to the inverse problem of electrocardiography. We first present the formulation of linear inverse problems in terms of Level Sets, as proposed in [9]. Second, we discuss practical problems implementing this method for inverse electrocardiography. In Section 3, we report on improvements obtained by adding new spatial constraints to the evolution and by filtering the residual error. Finally, Section 4 discusses our results, summarizes our conclusions, and gives some suggestions for future research.

2   Level Sets Method

2.1   Level Sets Formulation

The Level Sets Method, as described in [10], is a curve/surface evolution technique, based on a function whose dimension is one higher than the boundary of interest. The zero level set of this function is iteratively guided by a well designed speed function to evolve to an unknown desired contour. It naturally provides an opportunity for geometrical and spatial constraints. A particular set of inverse problems, known as obstacle reconstruction problems, can be formulated in terms of Level Sets [9]. In these problems, the solution consists of an unknown region, simply or multiply connected, with some characteristic that differs from the surrounding background. The solution only has two possible reconstruction values: one for the unknown region and another for the background. Applying Level Sets, the zero level set will evolve to the boundary of this region. Hence, Level Sets evolution adds geometrical and spatial constraints, without any *a priori* assumption about the connectedness of the region. Besides that, the Level Set boundary can split and merge naturally and provide multiple connectivity without any additional complexity. On the other hand, it turns a possibly linear problem into a decidedly nonlinear problem (although non-linearity is common to all activated-based inverse methods in ECG). In addition, there is no theoretical proof on convergence (only practical results, see [9]), and the solution depends on the algorithm initialization.

We use the approach described in [9]. Let ϕ, be the function whose level set $\phi = 0$ is taken as the contour of interest (here the activation wavefront location). The general Level Sets evolution equation is:

$$\phi_t + F|\nabla \phi| = 0 \qquad (1)$$

where F is the speed in the outward normal direction and ϕ_t is the time derivative of ϕ. The key issue in using Level Sets in most problems is determining the speed function F. In inverse problems F should be defined such that the solution moves toward minimizing the norm of the residual [9].

Our forward model for ECG is:

$$y = Ax + n \qquad (2)$$

where A is a forward matrix, obtained here by the boundary element method (BEM), x holds the heart potentials, y holds the body surface potentials and n is white Gaussian noise. It is shown in [9] that the residual error is monotonically descending if the speed is defined as follows:

$$F = -A^T(Ax - y). \qquad (3)$$

This evolution can be seen as a flow in the steepest descent direction of the residual error $||Ax - y||_2^2$.

Thus, the Level Sets evolution equation for inverse problems at iteration $n+1$ is:

$$\phi(n+1) = \phi(n) - \Delta t \cdot (A^T(A \cdot x(n) - y) \cdot |\nabla \phi(n)|) \qquad (4)$$

where x is initialized and then updated in each iteration as the zero level of ϕ.

To approximate the inverse problem in electrocardiography as an obstacle reconstruction problem formulated in terms of Level Sets, we divide the heart surface in two regions: activated and inactivated areas. The potential in each area is assumed to be constant with two different values, obtained independently for each time instant from a dataset of cardiac mapping ECG data. The zero level set is evolved to estimate the boundary between activated and inactivated regions.

2.2 Level Sets Practical Implementation and Initial Results

An inverse ECG Level Sets implementation has to overcome some practical problems. First, an accurate heart geometry model is needed. In this work we used the Utah Cardiovascular Research and Training Institute (CVRTI) Heart Geometry Model (a 3D, non-uniform triangulated grid). We further interpolation the surface to improve the model, and thus, the Level Sets solution, removing large triangles in the superior region and near the apex and some non-differentiable points in the original.

Another practical problem was the initialization of the Level Sets function, because of the solution's dependence on the starting value. Our solution ensured that the activated area was included inside the initial guess, and we centered it on the activated area recovered by Tikhonov regularization at each time instant. To obtain the activated area from the Tikhonov solution, the potential histogram was computed and the middle point of its two first maximums was used as a threshold (ensuring that the maxima were different enough that one belong to the activated potentials and the other to the inactivated set). Finally, we chose the value of evolution step size to ensure that the evolution didn't stop prematurely, but rather remained sensitive to the curve boundaries.

We first applied the Level Sets method in this straight-forward manner. We used an epicardial electrocardiogram dataset recorded during tank experiments by our collaborators at CVRTI in Utah [12]. From these epicardial potentials we computed the potentials on the torso surface using the linear model in Eq. 2, and added Gaussian white noise to achieve a 30dB signal to noise ratio (SNR). A realistic homogeneous torso volume conductor forward model matrix was computed by the BEM method with dimension 711 × 620: 620 nodes in the heart geometry model mapped to 711 electrodes on the torso.

The results obtained were not satisfactory. Although the algorithm provided some information about the location of the activated area, the shape of the wavefront was not geometrically reasonable. A lack of geometric constraints (causing, for instance, non-physiological aberrations such as inactivated nodes inside the depolarized region) and, a lack of sensitivity (few activated nodes, in general, on the side and back of the heart) were obvious.

3 Improvement of the Level Sets Method Performance

To improve on these results, we made several modifications to the standard Level Sets algorithm. The first two modifications were rather straight-forward attempts to reinforce the spatial constraints and improve the sensitivity of the Level Sets approach. First, we adopted a "restart" method, reinitializing the Level Sets function every 10 iterations to avoid excessive deformation. In addition, to improve the sensitivity, a new constraint was added, which at each restart pushed the zero level set inwards. In other words, assuming that the next activated region would be inside the current activated region, the zero level set was forced to evolve even when the error was small at a specific node. Specifically, the Level Sets function was rebuilt as equal to a signed distance function

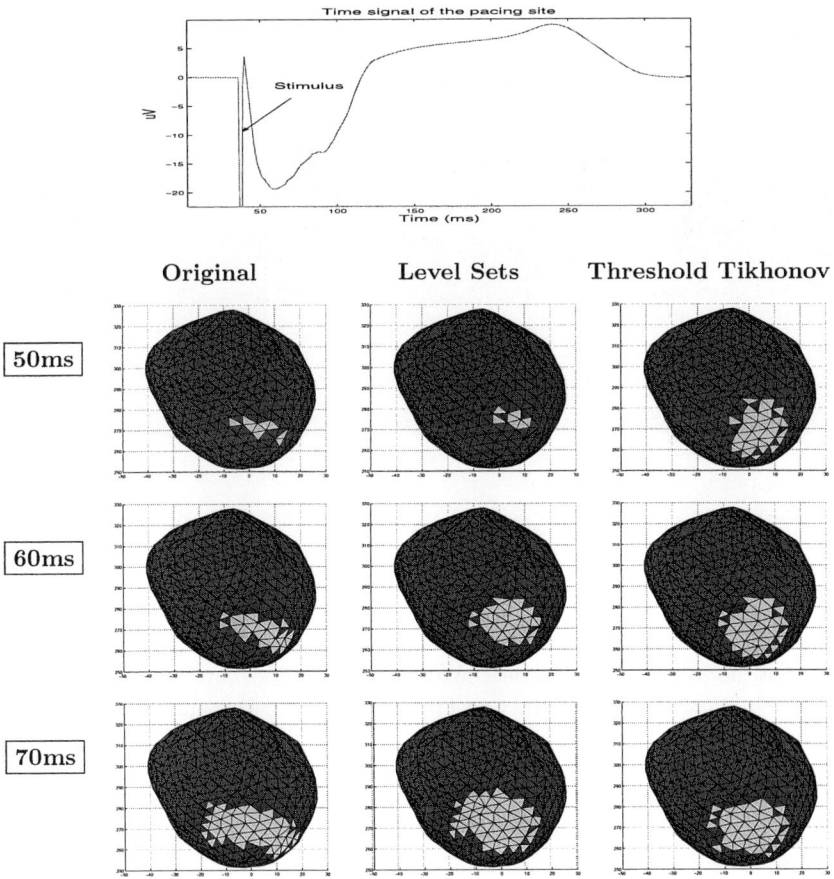

Fig. 1. Top panel: Time signal of the pacing site. Bottom: Activated and inactivated areas of the original data, and of the Level Sets and thresholded Tikhonov reconstructions. Time instants 50, 60 and 70ms, as seen on the top panel's waveform, are shown

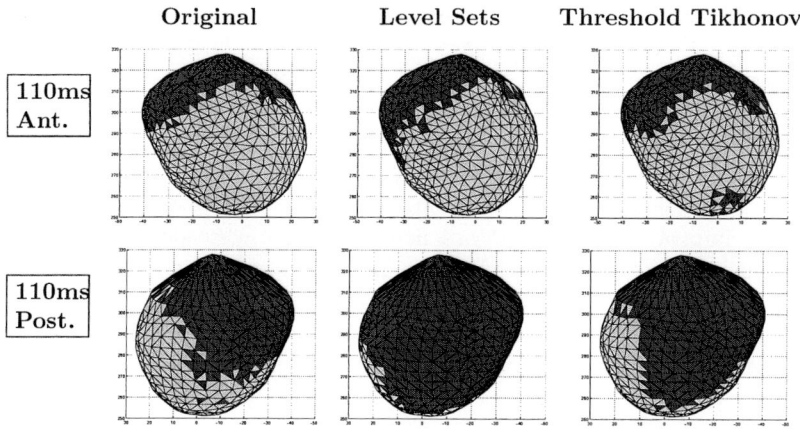

Fig. 2. Same format as bottom panel of previous figure, except that two views (Anterior, top, and Posterior, bottom) are shown for the same time instant, 110ms (using the time markings on the waveform shown in previous figure), later in QRS

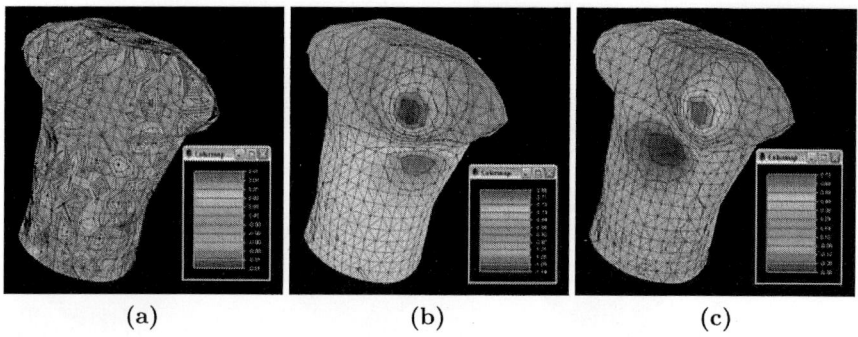

Fig. 3. (a) Front view of the residual error of the Tikhonov solution. (b) Front view of the residual error induced only by the two-level approximation of the original potentials. (c) Front view of Level Sets residual error. Time instant 70ms.

from each node to a zero level set. We shrank the zero level set first by simply taking the inward values at the border between positive and negatives points as the zero level set for the reinitialization.

In Fig. 1 and 2, we show the inverse solutions for different time instants of the electrocardiogram dataset described in Section 3. The time waveform at the pacing site is shown in the top panel of Fig. 1. The bottom panel of this figure contains maps at three time instants. Fig. 2 shows the same comparison for anterior and posterior views at a later time instant in QRS. From these results, we can conclude that Level Sets solution, after these modifications, is slightly better than Tikhonov in terms of shape information, capturing the anisotropy of the propagating front. This improvement is especially visible in the earlier time

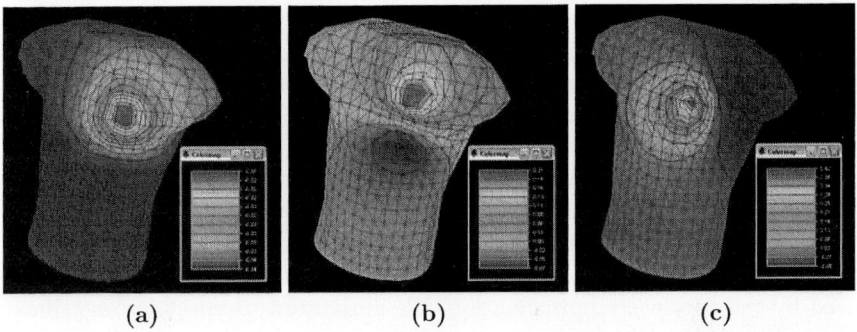

Fig. 4. (a) Front view of the first singular vector of the matrix U, from the Singular Value Decomposition: $A = U\Sigma V^T$. (b) Front view of the second singular vector of the matrix U. (c) Front view of the third singular vector of the matrix U. Time instant 70ms

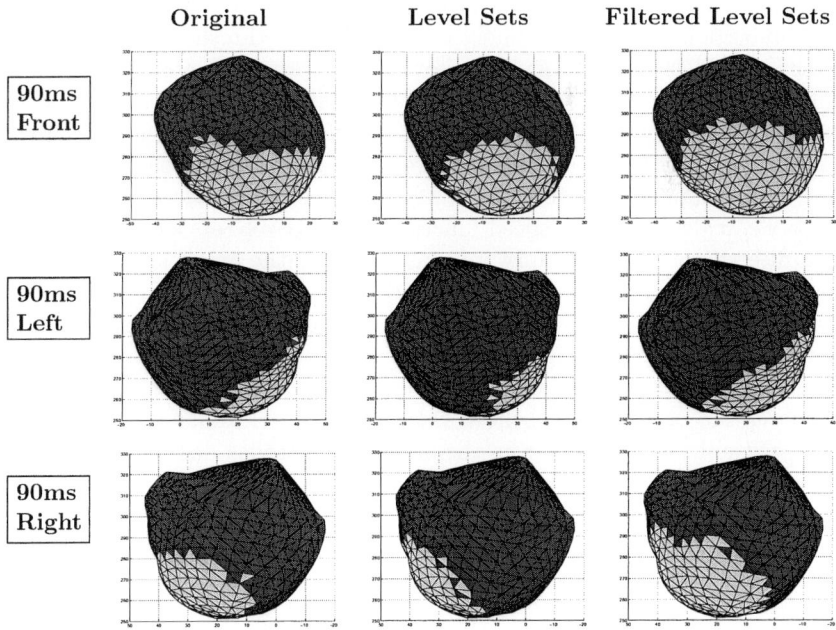

Fig. 5. Original activated and inactivated areas, Level Sets and filtered Level Sets ($k = 3$) activated and inactivated areas. Data from ECG dataset, time instant 90ms from front and side views, respectively

instants, when the propagating front has a characteristic elliptical shape. The solution also preserves some physiological behavior of the depolarization wavefront: closed activated area and no isolated activated or inactivated nodes inside inactivated or activated regions respectively. In this sense, Tikhonov fails for later time instants, as seen in Fig. 2. In addition, Tikhonov is more sensitive to threshold level variation than Level Sets, due to its smoothing. A small threshold variation can cause a large change in the area of the Tikhonov-estimated activated region, while the Level Sets solution would hardly change. On the other hand, the Level Sets method was rather insensitive to some regions, especially in the back and sides of the heart. The Tikhonov solution behaved slightly better in those areas, as it can be seen in the posterior view of Fig. 2.

Studying this effect, we noted that the residual error was highly spatially correlated, unlike the Tikhonov error. This is due to the systematic error introduced by the two-level approximation. We illustrate this effect for one time instant in Fig. 3 [13]. We can see that the Level Sets residual error varies smoothly and slowly in space, so that it is mainly a low spatial frequency phenomenon. We decreased the effect of this error on the inverse solution by minimizing a filtered version of the residual error, where the low spatial frequencies of the residual error were removed. The idea is to concentrate the Level Sets iterations on matching the components of the data that are in a subspace orthogonal to these low-frequency components, since this is the error due to mismatch of the activation region rather than simply the effect of the thresholding itself.

In Fig. 4 [13], the three singular vectors of A corresponding to the three largest singular values are shown as torso maps. We observe that the dominant residual error components are similar to these first singular vectors. Hence, we can apply the Level Sets method by projecting the residual error onto the subspace spanned only by the singular vectors corresponding to indices higher than some small value k, i.e., calculating the speed function of the Level Sets evolution equation with a filtered version of the forward matrix, where the first k singular values of A have been set to zero.

In Fig. 5 we show the results for one time instant when we filter the residual to remove components in the subspace spanned by the first 3 singular vectors. We note that the activated area, especially in the right view, more closely approximates the original. We believe this is because of enhanced sensitivity to regions of the heart farthest from the anterior electrodes due to removal of the low-frequency threshold-induced residual error.

4 Discussion, Conclusions, and Future Work

The purpose of this work was to develop and evaluate an initial attempt at a Level Sets method that can be applied to non-invasive electrocardiography to reconstruct activation wavefronts on the epicardium. The principle attractive features are that we avoid any isotropy assumptions and develop a framework within which we can use spatial and geometric information that the physiology of the problem might provide. Essentially, our method is close to standard activation-based methods in terms of how we model the source. However, those methods depend on isotropy assumptions that Level Sets skips because it calculates a coarse model of epicardial potentials and uses that in a potential-based forward model.

As we describe, we introduced some modifications to improve the initial performance of the Level Sets algorithm. Adding new geometrical constraints increased spatial consistency and sensitivity. We also introduced a high-pass filtering of the residual error to remove part of the two-level quantization error, which helped to improve the sensitivity in the side and back areas of the heart. The geometrical constraints are visible in the results: evolution of the boundary and a closed activation area. Moreover the anisotropy of the wavefront is generally captured better than with a thresholded Tikhonov solution. The shape of the activated area recovered is better at some time instants than others but generally reflects the anisotropy induced by fiber direction. Finally, the method here uses the actual data to obtain the threshold levels; an independent method needs to be developed to estimate these values without *a priori* knowledge or to allow them to remain constant in time.

We are currently looking at several remaining aspects of this study. The residual filtering approach we used was just a first attempt, and we believe that a more careful study can lead to a more effective implementation. In addition there is a tradeoff between enhanced sensitivity and loss of robustness when filtering more singular vectors; thus we need an algorithm to estimate an appropriate number of singular vectors to remove. An idea of primary interest is to introduce more geometric physiological information into the model by incorporating fiber direction information, even from a different heart. The Level Sets speed function provides a perfect vehicle to include this *a priori* information. Relaxing the quantization of the heart potentials by defining a transition area (dividing the epicardial surface in three regions instead of only two), and/or modeling this region as an analytical function such as an arc-tangent [3], and anchoring the evolution around breakthrough's calculated using the Critical Point Theorem [4], are some other approaches to better incorporate known physiological constraints.

Acknowledgment

The work was supported in part by the National Institutes of Health, National Center for Research Resources, grant number 1-P41-RR12553-3.

References

1. Dana H. Brooks, Robert S MacLeod, *Electrical Imaging of the Heart: Electrophysical Underpinnings and Signal Processing Opportunities*, IEEE Sig. Proc. Mag., 14(1):24-42, 1997.
2. Rudy Y, Messinger-Rapport B : *The inverse problem of electrocardiography. Solutions in terms of epicardial potentials*, CRC Crit. Rev. Biomed. Eng., vol. 16, pp. 215-68, 1988.
3. G. J. M. Huiskamp and A. van Oosterom, *The depolarization sequence of the human heart surface computed from measured body surface potentials*, IEEE Trans. Biomed. Eng., vol. BME35, pp. 1047-1058, 1988.
4. G.J.M. Huiskamp and F.S. Greensite, *A new method for myocardial activation imaging*, IEEE Trans. Biomed. Eng., vol. 44, pp. 433-446, 1997.

5. Pullan A.J., Cheng L.K., Nash M.P., Bradley C.P., Paterson D.J., *Noninvasive electrical imaging of the heart: theory and model development*, Annals of Biomedical Eng., 29(10):817-36, October 2001.
6. T. Oostendorp and R.S. Macleod and A. van Oosterom, *Non-invasive determination of the activation sequence of the heart: Validation with invasive data*, Proc. IEEE Int. Conf. Eng. in Med. and Biol. Soc. 1997.
7. D.H. Brooks and G.F. Ahmad and R.S. MacLeod and G.M. Maratos, *Inverse Electrocardiography by Simultaneous Imposition of Multiple Constraints*, IEEE Trans Biomed Eng., Vol. 46, Number 1, Pages: 3–18, 1999.
8. B. Messnarz, B. Tilg, R. Modre, G. Fischer, F. Hanser, *A new spatiotemporal regularization approach for reconstruction of cardiac transmembrane potential patterns*, IEEE Trans Biomed Eng. ,Volume: 51 , Issue: 2 , Pages: 273–281, Feb. 2004.
9. Santosa F., *A Level-Set Approach for Inverse Problems Involving Obstacles*, ESAIM: Control, Optimisation and Calculus of Variations, Vol.1, pp. 17-33, January 1996.
10. Osher S., and Sethian J.A., *Fronts Propagating with Curvature-Dependent Speed: Algorithms Based on Hamilton-Jacobi Formulations*, Journal of Computational Physics, 79, pp. 12-49, 1988.
11. Sethian J.A., *Level Set Methods and Fast Marching Methods*, Cambridge University Press (Second Edition), 1999.
12. MacLeod R.S., Ni Q., Punske B., Ershler P.R., Yilmaz B., Taccardi B., *Effects of Heart Position on the Body-Surface ECG*, J. Electrocardiol., 33 Suppl: 229-237, 2000.
13. R.S. MacLeod and C.R. Johnson, *Map3d: Interactive scientific visualization for bioengineering data*, Proc. Int. Conf. IEEE Eng. Med. Bio. Soc. 1993, Pages: 30–31.

Magnetocardiographic Imaging of Ventricular Repolarization in Rett Syndrome[1]

Donatella Brisinda*, Anna Maria Meloni*, Giuseppe Hayek^, Menotti Calvani°, and Riccardo Fenici*

* Clinical Physiology–Biomagnetism Center, Catholic University of Rome – Italy
^ Department of Child Neurology and Psychiatry, University of Siena -Italy
° Sigma Tau S.p.A., Rome –Italy
feniciri@rm.unicatt.it

Abstract. Rett syndrome (RS) is a severe neurological disorder, predominant in females, with higher risk of sudden death (SD). So far for risk-assessment, heart rate variability (HRV), QT duration and its dispersion (QTd) were measured with ECG. However SD has occurred in RS also in absence of ECG abnormality. We aimed to evaluate the feasibility of magnetocardiographic (MCG) mapping as an alternative to study ventricular repolarization (VR) alteration in RS patients. 9 female (age: 1-34 years) RS patients were studied with an unshielded 36-channels MCG system. To assess VR, heart rate (HR)-corrected JT_{peak}, JT_{end}, QT_{end}, $T_{peak-end}$ intervals and QTd, were measured from both MCG and ECG signals. Moreover the magnetic field (MF) gradient orientation (α-angle) during the ST segment and three MF dynamic parameters were automatically evaluated from MCG T-wave. HRV parameters were evaluated from 12-lead Holter ECG. 15 age-matched normal controls (NC) were studied for comparison. HR-corrected JT_{peak}, JT_{end}, QT_{end} and $T_{peak-end}$ intervals, and QTd were longer in RS than in NC. The differences were more evident with clinical impairment (stage IV). MF gradient orientation and MF dynamic parameters were abnormal in RS patients. As compared to NC, HRV parameters were altered in the time-domain, although still within normal range in the frequency-domain. In RS, ECG recordings are often noisy and BSPM is difficult. On the contrary MCG mapping is easily feasible and discovers VR alteration not evident at the ECG. The diagnostic value of MCG in RS remains to be defined.

1 Introduction

Rett syndrome (RS) is a severe progressive neurodevelopment disorder, occurring almost exclusively in females, characterized by cortical atrophy, psychomotor regression, mental retardation, irregular breathing, hyperventilation[1,2] caused by dominant mutation of the MeCP2 gene, encoding the transcriptional repressor methyl-CpG-binding protein 2, related to Xq28 locus[3]. *RS Diagnostic Criteria World Group*[4,5]

---

[1] The research work was partially supported by MIUR grants # 9906571299_001, 2001064829_001, and by Sigma-Tau S.p.A. grant DS/2001/CR/#39. No conflict of interest.

differentiates four stages of clinical evolution (*Stage I-IV*), which are characterized by progressive deterioration of neural, respiratory and cardiac functions. Life expectancy of RS patients is uncertain. The survival rate drops to 70% by age thirty-five, and for the profoundly mentally retarded, it drops to 27%, due to autonomic nervous system (ANS) abnormality with associated cardiac, gastrointestinal and breathing problems[6]. In RS incidence of sudden death (SD) is greater than that in the general population[7-8], likely due to cardiac electrical instability, associated with ANS activity abnormality[9] and with reduced level of nerve growth factor (NGF) leading to a decline in number of choline acetyltransferase (ChAT)-positive cells that are necessary for the production of acetylcholine[10-11]. So far, HR-corrected QT interval duration (QTc), its dispersion (QTd) measured from 12-lead ECG and heart rate variability (HRV) parameters have been used as markers of electrical instability in RS[12-15]. However SD has been reported in RS pts also in the absence of ventricular repolarization (VR) alterations at the ECG[16]. Furthermore, being the RS pts restless, movement artifacts often disturb the ECG recordings and impair precise measurements of ECG parameters. For the same reason body surface electric mapping has never been attempted so far in RS. Alternatively to ECG, contactless magnetocardiography, which provides accurate multisite mapping of cardiac electrical activity without movement artifacts[17], can be used. Previous studies suggest that magnetocardiographic (MCG) recordings might contain information additional to 12-lead ECG[18-19]. Moreover it has been shown that MCG mapping is useful for precise quantitative evaluation of VR abnormalities and to identify markers of arrhythmogenic risk[20-21]. The aim of this study was limited to evaluate the feasibility of multichannel MCG mapping in RS pts, and its reliability to detect VR abnormalities, associated or not to alteration of HRV parameters, in the absence of significant ECG alterations.

2 Methods

2.1 Patients

9 female RS pts, aged 1 to 34 years, clinically classified in stage II (2), in stage III (4), and in stage IV (3)[4] were investigated, after parental written informed. consent. 15 age-matched normal controls (NC) were studied for comparison.

2.2 Study Protocol

The cardiac magnetic field (MF) component perpendicular to the sensor array surface was mapped in the supine position, from a 6 x 6 grid covering an area of 20 x 20 cm (Figure 1), with a 36-channel system, featuring DC-SQUID sensors, coupled to second-order axial gradiometers, with pick-up coil diameter of 19 mm, baselines of 50-70 mm, and intrinsic sensitivity of 20 fT / √Hz, in the frequency range of interest for clinical MCG signals (DC to 100 Hz)[22] (*CardioMag Imaging Inc.* Schenectady, NY) (Figure 1 A).

MCG signals (low-pass filtered at 100 Hz) were digitally recorded at 1 kHz (with 24 bits resolution). The relative position of the patient in respect of the sensors was defined with three laser pointers. Each MCG mapping lasted typically 90 seconds and was repeated twice to test for reproducibility. 12-lead ECG was simultaneously recorded (bandwidth: 0.05-100 Hz), with amagnetic electrodes. HRV parameters were calculated, in the time (TD) and frequency (FD) domains, according to standard protocols[23], from 12-lead ECG Holter *(H-scribe Digital Holter, Mortara Instruments, Inc.)*.

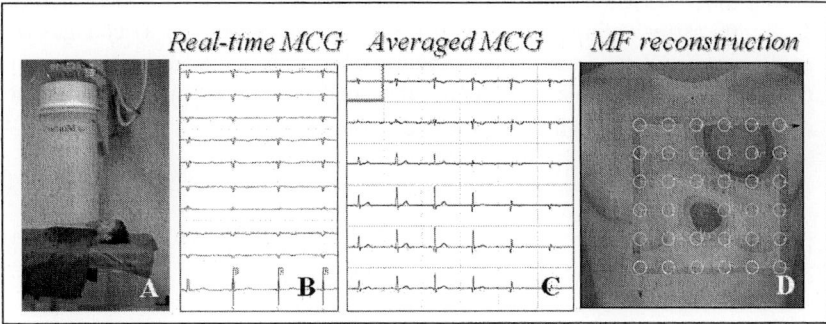

Fig. 1. Typical positioning of the patient under the MCG mapping system (A). Real–time MCG signals with one reference ECG (B). MCG averaged waveforms (C). Magnetic field reconstruction (D)

2.3 MCG Signal Processing and Analysis

MCG signals were automatically processed and analyzed with a **Windows-based** software *(CardioMag Image Inc)* and with the **UNIX-based** software developed by the Helsinki University of Technology *(NEUROMAG)*, as described elsewhere[20,24-26]. Briefly, MCG signals were adaptively filtered (Figure 1 B) and averaged (Figure 1 C) to eliminate the 50 Hz noise and improve the Signal/Noise ratio. After automatic (and/or interactive) baseline selection, MCG signals were analyzed as waveforms in the time domain and used to construct isofield contour maps by automatic interpolation, with a time resolution of 1 millisecond (msec). Contour maps were also constructed after time integration of specific intervals of interest (Figure 1 D). The software provides automatic measurements of ventricular time intervals. However, the Q wave onset, the J point, the T_{peak} and the T_{end} were also interactively edited, using a "butterfly" superposition of all MCG signals amplified at the resolution of 10 mm/pT (picoTesla) with a time scale of 200 mm/sec (Figure 2 A) and morphological analysis of the time evolution of the MF maps (Figure 2 B) to improve the timing accuracy.

2.4 Ventricular Repolarization Parameters

To assess VR, the following quantitative MCG parameters were evaluated:

1. The **JT_{peak}, JT_{end}, QT_{end}, and $T_{peak-end}$** intervals (Figure 2 A), all corrected for the heart rate (HR), and the **$QT_{dispersion}$**, measured automatically from MCG and manually from ECG signals. In order to correct to HR, the values were divided by the square root of the averaged R-R interval measured in seconds [corrected value = measured value (ms) / $\sqrt{RR}$ (sec)].

2. The **MF gradient (MFG) orientation,** measured at the integral of the second quarter from the J-point to the T_{peak} and at the T_{peak}, as the angles (α) between the direction of the largest MF gradient (vector between the maximum positive and negative magnetic poles) and the patient's right-left axis[26] (Figure 3).
3. The **dynamics of MF distribution**, in any floating time windows of 30 ms during the T-wave (starting when the MF strength is equal to 1/3 of that at the T_{peak}, arbitrarily defined T_{onset}, until the T_{peak}), quantified as: a) changes of the angle between + pole and - pole (abnormal if > 45 degrees); b) changes of the distance between + pole and - pole (abnormal if > 20 mm); c) changes of the ratio between the strength of + pole and - pole (abnormal if > 0.3)[25] (Figure 4).

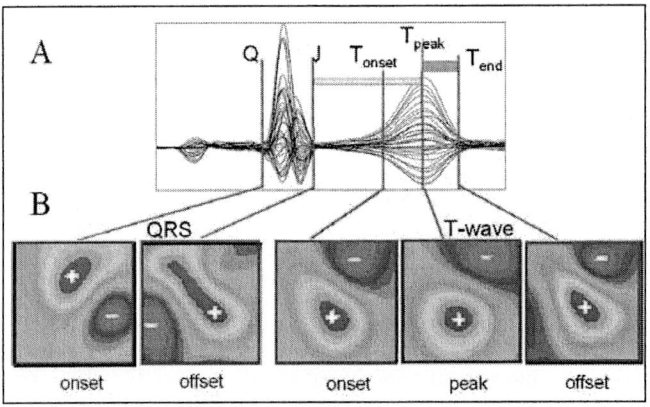

Fig. 2. "Butterfly" superposition of the 36 MCG averaged waveforms (A). Typical MF distribution at the onset and offset of the QRS, and at the onset, peak and offset of the T-wave (B)

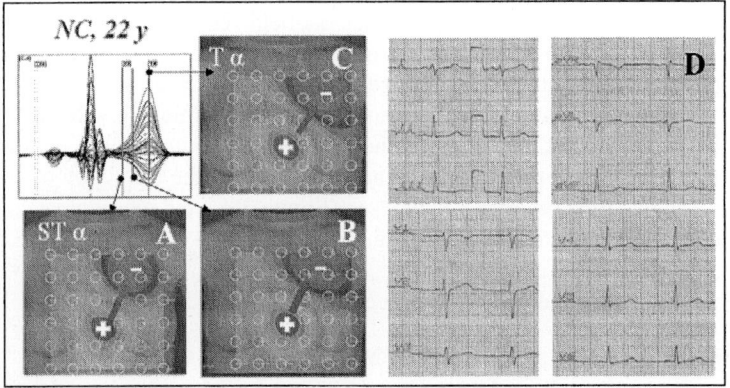

Fig. 3. Normal control (22 years old female). Examples of typical stability of the MF distribution and of the MF gradient orientation (angle α), measured at the second quarter of the ST (A), at the T-wave onset (B) and at the T-wave peak (C). In (D), 12-lead ECG

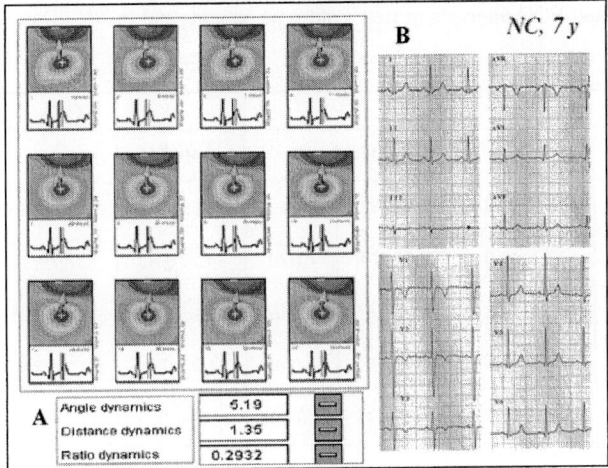

Fig. 4. Normal control (7 years old female). Example of automatic score analysis based on the MF dynamics during the T_{onset} -T_{peak} interval (A). In spite of the "juvenile" repolarization pattern at the 12-lead ECG (negative T-wave in V2 and V3) (B), all T-wave MF dynamic parameters are within normal range. The automatic classification is negative (green flags)

2.5 HRV Parameters

The following standardized parameters[23] were calculated:
pNN 50%: NN50 count divided by the total number of all NN intervals; **SDNN** (msec): Standard deviation of all NN intervals; **SDANN** (msec) Standard deviation of the averages of NN intervals in all 5 min segments of the entire recording; **r-MSSD** (msec): the square root of the mean of the sum of the squares of differences between adjacent NN intervals; **LF/HF ratio**: Low Frequency/High Frequency Ratio: LF [$msec^2$]/HF [$msec^2$].

2.6 Statistical Methods

Data are reported as mean ± S.D. Statistical analysis was performed with the unpaired two-tails Student t-test. A value of $p < .05$ was considered significant.

3 Results

3.1 Ventricular Repolarization

Average values of HR-corrected JT_{peak}, JT_{end}, QT_{end} and $T_{peak-end}$ intervals and of QTd are summarized in Table 1.

In general, all MCG intervals were shorter (p = n.s.) than corresponding ECG ones. In spite of the limited number of cases, significant differences were found between RS pts and NC for all parameters except JT_{peak}. However, only MCG evidenced significantly longer values of $T_{peak-end}$ and of QTd.

Table 1. MCG and ECG intervals in Rett syndrome patients and in NC. Data are presented as mean ± SD

HR-corrected	MCG			ECG		
	Rett	Normals	P	Rett	Normals	P
JT_{peak}	240.4±26.8	223.9±18.0	n.s.	248 ± 48.8	215.7 ± 26	0.05
JT_{end}	312.3±29.9	281.1±11.8	< 0.01	342.2 ± 43	307.9 ± 21	<0.02
QT_{end}	402.7±29.8	378.05 ± 15	< 0.02	428 ± 42.7	388.7 ± 21	<0.01
$T_{peak-end}$	71.8± 23.6	57.02 ± 10	< 0.05	94.2 ± 23.6	92.2 ± 20.9	n.s.
QT_d	18.6 ± 9.3	7.28 ± 1.46	< 0.001	33.8 ± 14.1	33.1 ± 17.3	n.s.

Table 2. MF orientation (α angle) and MF dynamics in RS patients and in NC. Data are presented as mean ± SD

	Rett	Normals	P value
ST α angle *(degrees)*	135.6±79	55.9±23.3	< 0.01
T α angle *(degrees)*	72.1 ± 2.9	60.8± 13.08	n.s.
MF +/- poles angle dynamics *(degrees)*	29.4 ± 38.3	4.8 ± 2.9	< 0.02
MF +/- distance dynamics *(mm)*	27.1 ± 28.1	7.6 ± 5.6	< 0.02
MF +/- ratio dynamics	0.68 ± 0.36	0.018 ± 0.09	< 0.01

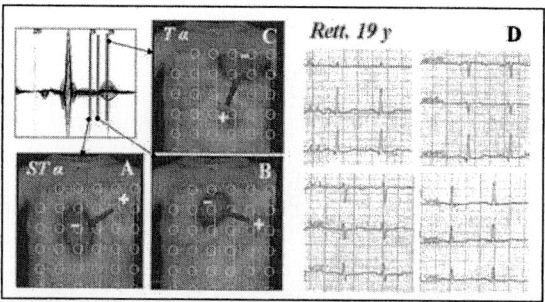

Fig. 5. Rett syndrome stage IV, (19 years old female). MCG mapping evidences clear-cut alteration of the MF gradient, during the ST interval (A and B), in spite of the absence of significant VR abnormalities at the 12-lead ECG at rest (D). The MF gradient at the T-wave peak is still within normal limits (C)

Taking into account the different degree of clinical impairment, a trend toward prolongation of JT_{end} and of QT_{end} values was found in stage IV patients, in respect of stage II and III patients. Among other MCG parameters (Table 2), the ST α-angle

(Figure 5) and the three T-wave MF dynamic parameters were significantly abnormal in RS patients (Figure 6), although only non-significant repolarization abnormalities were observed at the 12-lead ECG.

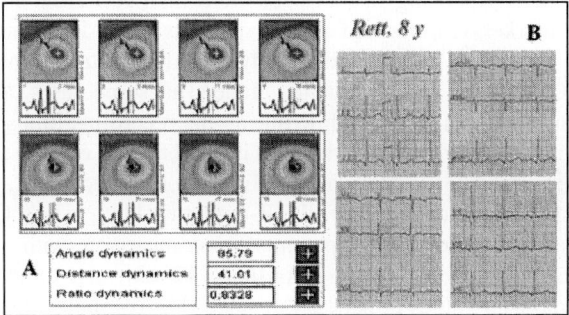

Fig. 6. Rett syndrome stage IV, (8 years old female). All MF dynamics parameters, during the T_{onset} -T_{peak} interval, are abnormal (red flags) (A), in spite of non-significant VR alterations (negative T wave in V2 and V3) at the 12-lead ECG (B), similar to those of the age-matched control (shown in Figure 4 B)

3.2 HRV Analysis

In all RS pts, independently of the clinical stage, HRV analysis in the time domain (pNN50%, SDNN, SDANN, r-MSSD) evidenced lower values as compared with normal age-matched subjects[27-28] (Table 3).

Table 3. HRV parameters in Rett Syndrome patients. Data are presented as mean ± SD

Stage	pNN50%	SDNN	SDANN	r-MSSD	LF/HF (day)	LF/HF (night)
II	3.13 ± 2.2	59.4±14.7	51.0± 12.7	22.7 ± 3.8	1.46 ± 0.4	2.5± 1.2
III	1.6 ± 0.8	62 ±17.9	51.5 ± 8.6	18.1 ± 4.6	4.1 ± 1.4	2.8± 1.2
IV	6.9 ± 0.6	109.8±7.4	93.1 ± 1.8	31.3 ± 3.1	2.0 ± 0.7	1.2± 0.1

In the FD, the total power was higher then 1900 $msec^2$ in all patients; however in stage II patients the LF/HF ratio was higher during the night than during daytime, whereas in stage III and IV, an inversion was observed, being the LF/HF ratio higher in daytime.

4 Discussion

RS is a severe neurodevelopmental disorder associated to higher risk of SD than in general population[7]. Cardiac electrical instability has been hypothesized as the potential mechanism for SD, in few studies, carried out with 12 lead-ECG or 24 hours ECG Holter monitoring[12-15], in which a reduced HRV and/or prolonged QTc interval were found. Although the ECG is the most widely used method to detect VR abnormalities, to assess the arrhythmogenic risk and predict cardiac death, especially in patients with ischemic heart disease (IHD) or long-QT syndrome[29-32], recent work has shown that MCG mapping is more sensitive than rest ECG in detecting VR alterations and markers for risk of SD in patients with IHD and with dilated cardiomyopathy (CMP)[20-21,33-34]. In this study, we demonstrated the feasibility of unshielded MCG mapping in RS patients. Indeed in some cases, although it was difficult to maintain the patient immobile during the recording, limbs movement did not affect significantly the quality of the MCG signals, good enough for quantitative analysis of VR, even when ECG was unreliable for artifacts. In agreement with previous studies[11-15], a prolongation of all HR-corrected JT_{peak}, JT_{end}, QT_{end} and $T_{peak-end}$ intervals was observed in RS patients in comparison with NC, measured independently from MCG and ECG recordings. Moreover a trend was observed toward more prolonged values in patients in stage IV. Absolute values of MCG VR intervals were shorter than those measured from ECG. This might be due to the partially different definition of the T-wave end by automatic MCG analysis as compared to manual ECG measurements. However MCG measurements evidenced significant differences for $T_{peak-end}$ and QTd, between RS patients and NC, which were not appreciable with ECG. This suggests that, as previously observed in patients with ischemic or dilated CMP[20-21,33], MCG mapping might be more sensitive than ECG in detecting early signs of VR dispersion in RS. Moreover abnormalities of ST α-angle and of MF dynamics, similar to those demonstrated in IHD and CMP, were found in RS patients in more advanced stages[4]. As concern HRV, we found that independently of the clinical stage of disease, TD parameters in RS were lower as compared with age-matched NC[27-28]. This might be a sign of parasympathetic impairment. However the FD parameters were still within normal range, although a non-physiological behavior of the LF/HF ratio was observed in patients in stage II. The evident limitation of this study is that the number of patients is too small to draw any conclusion about the statistical significance of the results. This is due to the fact that it was not easy to collect RS patients, in the condition to collaborate for the additional MCG procedure, unless with some sedation. On the other hand this was a feasibility study and we did not considered ethical to include patients needing sedation, without the "a priori" knowledge that MCG mapping could provide information useful for their risk stratification.

5 Conclusion

This is the first study reporting non-invasive MCG evaluation of RS patients. The MCG method provides easy and quick multi-site mapping of cardiac electromagnetic activity, without any contact or the need to undress the patient, thus avoiding some of the pitfalls, which impede sometime the recording of good quality ECG in non-collaborative patients. In conclusion, although the number of cases investigated in this feasibility study is too small to conclude that MCG mapping is more sensitive that ECG, it was observed that the MCG method evidence abnormality of VR dynamics, not detected by the 12-lead ECG. Thus MCG mapping might provide additional electrophysiological information, clinically useful for early non-invasive risk assessment, especially of uncooperative and restless RS pts, and to select more appropriate diagnostic and therapeutic approaches. The only limitation to a widespread use of MCG mapping is at the moment the cost of LT SQUID-based instrumentations. However low-cost MCG non-cryogenic systems, such as laser-pumped optic magnetometer[35], are under development and should be commercially available rather soon.

References

1. Jellinger KA. Rett syndrome- an update. *J Neural Transm.* 110: 681-701, 2003.
2. Dunn HG, MacLeod PM. Rett syndrome: review of biological abnormalities. *Can J Neurol Sci.* 28(1): 16-29, 2001.
3. Rosenberg C, Wouters CH, Szuhai K, Dorland R, Pearson P, Tien Poll-The B, Colombijn RM, Breuning M, Lindhout D. A Rett syndrome patient with a ring X chromosome: further evidence for skewing of X inactivation and heterogeneity in the aetiology of the disease. *Eur J Hum Genet.* 9(3): 171-177, 2001.
4. Hagberg BA, Witt-Engerstrom I. Rett Syndrome: A suggested staging system for describing impairment profile with increasing age towards adolescence. *American J Med genetics.* 24: 47-59, 1986.
5. Trevathan F. The Rett syndrome Diagnosis Criteria Working Group. Diagnostic criteria for Rett syndrome. *Ann Neurol.* 23: 425-428, 1988.
6. Naidu S. Rett syndrome. A disorder affecting early brain growth. *Ann Neurol*; 42 (1) :3-10, 1997.
7. Kerr AM, Armstrong DD, Prescott RJ, Doyle D, Kearney DL. Rett syndrome: analysis of deaths in the British survey. *Eur Child Adolesc Psychiatry.* 6 (suppl 1): 71-74, 1997.
8. Driscoll DJ, Edwards WD. Sudden unexpected death in children and adolescents. *J Am Coll Cardiol* 1985;5(6 Suppl):118B-121B.
9. Julu P, Kerr AM, Apartopoulos F, Alrawas S, Witt Engerstrom I, Jamal GA, Hansen S. Characterization of breathing and associated central autonomic dysfunction in the Rett disorder. *Arch Dis Child.* 85: 29-37, 2001.
10. Wenk GL, Hauss-Wegrzyniak B. Altered cholinergic function in the basal forebrain of girls with Rett syndrome. *Neuropediatrics.* 30 (3): 125-129, 1999.
11. Guideri F, Acampa M, Calamandrei G, Aloe L, Zappella M, Hayek Y. Nerve Growth Factor Plasma Levels and Ventricular Repolarization in Rett Syndrome. *Pediatr Cardiol.* 25(4): 394-396, 2004.
12. Sekul EA, Moak JP, Schultz RJ, Glaze D, Dunn JK, Percy AK. Electrocardiographic findings in Rett Syndrome: an explanation for sudden death? *The Journal of Pediatrics.* 125: 80-82, 1994.
13. Ellaway CJ, Sholler G, Leonard H, Christodoulou J. Prolonged QT interval in Rett syndrome. *Arch Dis Child.* 80: 470-472, 1999.
14. Guideri F, Acampa M, Hayek G, Zappella M, Di Perri T. Reduced heart rate variability in patients affected with Rett sindrome. A possible explanation of sudden death. *Neuropediatrics.* 30: 146-148, 1999.
15. Guideri F, Acampa M, Di Perri T, Zappella M, Hayek Y. Progressive cardiac disautonomia observed in patients affected by classic Rett syndrome and not preserved speech variant. *J Child Neurol.* 16: 370-373, 2001.

16. Dearlove OR, Walker RWM. Anesthesia for Rett syndrome. *Pediatr Anaesth.* 6:155-158, 1996.
17. Tavarozzi I, Comani S, Del Gatta C, Di Luzio S, Romani GL, Gallina S, Zimarino M, Brisinda D, Fenici R, De Caterina R. Magnetocardiography: current status and perspectives. Part II: Clinical applications. *Italian Heart J.* 3(2):151-165, 2002.
18. Wikswo JP, Barach J. Possible sources of new information in the magnetocardiogram. *Journal of Theoretical Biology.* 95:721-729, 1982.
19. Brockmeier K, Schmitz L, Bobadilla Chavez JD, Burghoff M, Koch H, Zimmermann R, Trahms L. Magnetocardiography and 32-lead potential mapping: repolarization in normal subjects during pharmacologically induced stress. *J Cardiovasc Electrophysiol.* 18: 615-626, 1997.
20. Korhonen P, Väänanen H, Mäkijärvi M, Katila T, Toivonen L. Repolarization abnormalities detected by magnetocardiography in patients with dilated cardiomyopathy and ventricular arrhythmias. *J Cardiovasc Electrophysio.*12: 772-777, 2001.
21. Korhonen P, Pesola K, Jarvinen A, Makijarvi M, Katila T, Toivonen L. Relation of magnetocardiographic arrhythmia risk parameters to delayed ventricular conduction in postinfarction ventricular tachycardia. *Pacing Clin Electrophysiol.* 25(9): 1339-1345, 2002.
22. Fenici R, Brisinda D, Meloni AM, Fenici P. First 36-channel System for Clinical Magnetocardiography in Unshielded Hospital Laboratory for Cardiac Electrophysiology. *International Journal of Bioelectromagnetism.* 5(1): 80-83, 2003.
23. Task force of the European Society of Cardiology and the North American Society of Pacing and Electrophysiology. Heart Rate Variability standards of measurement, physiological interpretation and clinical use. *Circulation.* 93: 1043-1065, 1996.
24. Brisinda D, Meloni AM, Fenici R. First 36-channel Magnetocardiographic Study of CAD Patients in an Unshielded Laboratory for Interventional and Intensive Cardiac Care. In Magnin I, et al, eds. *Lecture Notes in Computer Science.* 2674:122-131, 2003.
25. Brisinda D, Meloni A.M, Fenici P, Fenici R. Unshielded Multichannel Magnetocardiographic Study of Ventricular Repolarization in Healthy Subjects. *Biomed Tech.* 48(2): 165-167, 2004.
26. Hänninen H, Takala P, Mäkijärvi M, Montonen J, Korhronen P, Oikarinen L, Nenonen J, Katila T, Toivonen L. Detection of exercise induced myocardial ischemia by multichannel magnetocardiography in patients with single vessel coronary artery disease. *Ann. Noninv Electrocardiology.* 5: 147-157, 2000.
27. Goto M, Nagashima M, Baba R, Nagano Y, Yokota M, Nishibata K, Tsuji A. Analysis of heart rate variability demonstrates effects of development on vagal modulation of heart rate in healthy children. *The Journal of pediatrics.* 130(5): 725-729, 1997.
28. Umetani K, Singer D, McCraty R, Atkinson M. Twenty-four hour time domain heart rate variability and heart rate: relations to age and gender over nine decades. *J Am Coll Cardiol.* 31: 593-601, 1998.
29. Kardys I, Kors JA, van der Meer IM, Hofman A, van der Kuip DA, Witteman JC. Spatial QRS-T angle predicts cardiac death in a general population. *Eur Heart J.* 24: 1357-1364, 2003.
30. Kannel WB, Anderson K, McGee DL, Degatano LS, Stampfer MJ. Non-specific electrocardiographic abnormality as a predictor of coronary heart disease. The Framingham Study. *Am Heart J.* 113:370-376, 1987.
31. de Bruyne MC, Hoes AW, Kors JA, Hofman A, van Bemmel JH, Grobbee DE. Prolonged QT interval predicts cardiac and all-cause mortality in the elderly. The Rotterdam Study. *Eur Heart J.* 20:278-284, 1999.

32. Yan GX, Antzelevitch C. Cellular basis for the normal T wave and Electrocardiographic manifestation of the Long-QT syndrome. *Circulation.* 98:1928-1936, 1998.
33. Oikarinen L, Viitasalo M, Korhonen P, Vaananen H, Hanninen H, Montonen J, Makijarvi M, Katila T, Toivonen L. Postmyocardial infarction patients susceptible to ventricular tachycardia show increased T wave dispersion independent of delayed ventricular conduction. *J Cardiovasc Electrophysiol.* 12:1115-1120, 2001.
34. Steinberg BA, Roguin A, Allen E, Wahl DR, Smith CS, St John M. Reproducibility and interpretation of MCG maps in detecting ischemia. (Personal Communication, ACC March 2004].
35. Fenici R, Bison G, Wynands R, Brisinda D, Meloni AM, Weis A. Comparison of Magnetocardiographic Mapping with SQUID-based and Laser-pumped Magnetometers in Normal Subjects. Biomed Tech. 48(Suppl 2):192-194, 2004.

Insights into Electrophysiological Studies with Papillary Muscle by Computational Models

Frank B. Sachse[1], Gunnar Seemann[2], and Bruno Taccardi[1]

[1] Nora Eccles Harrison Cardiovascular Research and Training Institute,
University of Utah, UT, USA
fs@cvrti.utah.edu
http://www.cvrti.utah.edu
[2] Institut für Biomedizinische Technik,
Universität Karlsruhe (TH), Germany
http://www-ibt.etec.uni-karlsruhe.de

Abstract. Basic electrical properties and electrophysiological mechanisms of cardiac tissue have been frequently researched applying preparations of papillary muscle. Advantages of these preparations are the simplicity to satisfy their metabolic demands and the geometrical elementariness in comparison to wedge and whole heart preparations. In this computational study the spatio-temporal evolution of activation fronts in papillary muscle was reconstructed with a bidomain model of electrical current flow and a realistic electrophysiological model of cardiac myocytes. The effects of two different pacing sites were investigated concerning the distribution of extracellular potentials and transmembrane voltages. Results of simulations showed significant changes of the resulting wave fronts and the related potential distributions inside of the muscle and in the bath for the different pacing sites. Additionally, the results indicated that reliable measurements of activation times can be carried out only in regions adjacent to the wave front. These results can be applied for development of measurement setups and techniques for analysis of experimental studies of papillary muscle.

1 Introduction

Papillary muscles have been frequently applied in experimental studies to characterize electrical properties of myocardium and spread of electrical excitation under varying conditions [1–6]. In attempts to reconstruct measurement data and to support analysis of these data, the electrophysiological properties of papillary muscles were studied also with several computational models [7–11].

This computational study aims at guiding future development of measurement setups and techniques for analysis of electrophysiological experiments with papillary muscles. Particularly, this study parallels an experimental study of mechano-electrical feedback mechanisms in excised papillary muscle from rabbits [5]. A crucial component of the analysis in this and other experimental studies is the accurate detection of activation times from analysis of bath voltage measurements near to the surface of the muscle.

Extraction of local activation times was applied e.g. to create activation and isochrone maps as well as to determine conduction velocity. While detection of activation times in transmembrane voltage recordings is possible with simple numerical methods, detection in electrograms in the bath is challenging due to interferences from non-related electrical sources and a commonly significantly smaller signal-to-noise ratio. Further complexity is added by geometrical relationships between the positions of the measurement electrodes, the activation front and its spatial domain.

The computational study had two specific aims: (1) The spatio-temporal evolution of activation fronts and the related extracellular and bath voltage distributions were determined and analyzed for two different stimulus electrode arrangements. (2) The relationship between activation times detected in the bath and from the transmembrane voltage in the muscle, respectively, was investigated.

The computational study was carried out with an anisotropic bidomain model of papillary muscle in a bath. The model was created using simplified geometrical descriptions and applied in numerical experiments. The experiments were designed to reproduce experimental conditions, where a stimulus was given at one end of the preparation and electrograms are acquired at some points along its surface.

2 Methods

2.1 Experimental Conditions

In previous work we developed a software-controlled experimental setup for studying cardiac mechano-electrical feedback of excised papillary muscle in a physiological environment [5]. The setup allows measurement of intracellular and bath electrograms as well as tension of the muscle under various strain conditions. During the measurements the muscle was placed in a horizontal flow-through chamber filled with a modified Tyrode solution of temperature of 37^oC.

The measurement procedure was automated and necessitates only minor user interaction. Electrical signals were acquired from a set of positions, which were determined from a set of points given by the user through steering a pointer with the motorized manipulator. The coordinates of these points were read digitally from a motorized micro-manipulator. Commonly, the recording electrode served as pointer and points at the ends of the muscle were selected. The given coordinates were applied to describe line segments or quadrangles, which were discretized to define measurement positions using sampling of finite element shape functions [12].

2.2 Computational Model

A computational model was created to mirror relevant aspects of the previously described experimental setup. Central component of the model was an bidomain

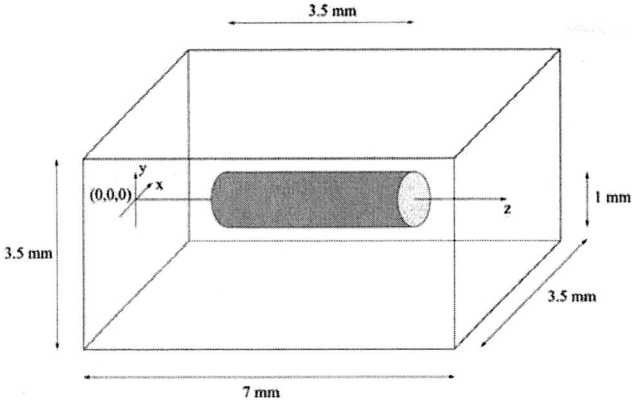

Fig. 1. Model of papillary muscle in bath. The muscle is represented by a cylinder, which is centered in the bath.

description used to reconstruct excitation propagation as well as the corresponding intra- and extracellular as well as bath potential distribution [13].

The bidomain model was based on a simplified geometrical description of the papillary muscle, the bath, and the reference electrode. The model included anisotropic intra- and extracellular conductivities as well as a biophysically detailed cellular electrophysiological model, i.e. the Noble-Kohl-Varghese-Noble model of a ventricular myocyte from guinea-pig [14]. For both domains the generalized Poisson's equation for electrical current fields was applied, which was discretized with the finite element method on hexahedral grids consisting of 70 x 70 x 140 elements in x-, y-, and z-direction (fig. 1). Each element was cubic with an edge length of 50 μm. A cylinder with a diameter of 1 mm and a length of 3.5 mm was rendered in the grids associated to the two domains to represent the papillary muscle. The long axis of the cylinder and the fiber orientation were chosen to be parallel to the z-axis. In summary, 24779 voxels were assigned to papillary muscle, 661221 voxels to bath.

Node variables were associated with the vertices of the hexahedrons [12]. In the finite element approach used in this work a trilinear polynomial was applied to interpolate intra-, extracellular and bath potentials as well as transmembrane voltages and conductivities inside of the hexahedrons. Integration of energy in the elements was carried out with 8-point Gaussian quadrature. A modification of this quadrature technique allowed to respect non-flow conditions at the boundary of the muscle's intracellular space.

The following conductivities were chosen (in S/m) [18]: myocardium extracellular longitudinal $\sigma_{e,l} = 0.375$ and transversal $\sigma_{e,t} = 0.214$, myocardium intracellular longitudinal $\sigma_{i,l} = 0.375$ and transversal $\sigma_{i,t} = 0.0375$, and bath $\sigma_b = 1.5$.

The Euler forward method with a temporal resolution of 10 μs was used to solve the ordinary differential equations associated with the electrophysiological model [15]. At each time step the Poisson equation attributed to the extracellular space and bath was solved with an over-relaxation method. An over-relaxation factor $\lambda = 1.85$ was chosen.

Two different pacing sites were chosen: central and superficial at the left end of the muscle. A stimulus current was applied extracellularly at $t = 0\ ms$ for a duration of $1\ ms$ at the center and a peripheral position, respectively, of the left circular face of the cylinder. The magnitude of the stimulus current was chosen in such a manner that a propagating front was initiated. Additionally, Dirichlet boundary conditions were defined on the right side of the bath, i.e. the extracellular potential at position (0,0,7) mm was set to zero.

Parallelization of computationally expensive tasks was achieved on basis of the OpenMP API [16]. The simulations were performed on a Silicon Graphics Origin 3000 compute server with 32 GB of main memory and 64 processors of type R14000/600 MHz. In both simulations 24 processors were employed, each for $\approx 40\ h$.

2.3 Detection of Activation Time

In this work the activation time in computed courses of transmembrane voltages and bath electrograms was detected by searching for the maximal and minimal temporal derivative, respectively [17, 18]. The search was restricted to a time window after the stimulus.

3 Results

The spread of excitation through the papillary muscle was simulated for central and superficial stimulation (figs. 2 and 4). The simulation was carried out for a time interval starting from stimulation to full excitation of the preparation.

Central Stimulation. In initial phases of the simulation a high curvature of the propagation front was found reflecting the anisotropy of conductivities (fig. 2, 2ms). This phase was followed by a decrease of curvature. In the last phase of spread, when the front reached the right end, a small curvature was found (fig. 2, 12ms).

Superficial Stimulation. The simulation showed activation fronts shaped significantly by the anisotropic properties of myocardium (fig. 4). The conduction velocity vector was composed of a large longitudinal and small transverse components. Full excitation of the preparation was found $2\ ms$ earlier than for central stimulation.

In both simulations, transmembrane voltages ranged between -92 and 48 mV as well as extracellular and bath potentials were between -25 and 10 mV. Both, the transmembrane voltages and extracellular potentials showed largest magnitudes of their spatial gradients at the propagation front (figs. 3 and 5). Isolines of extracellular potentials and transmembrane voltage were partly of similar shape inside of the muscle.

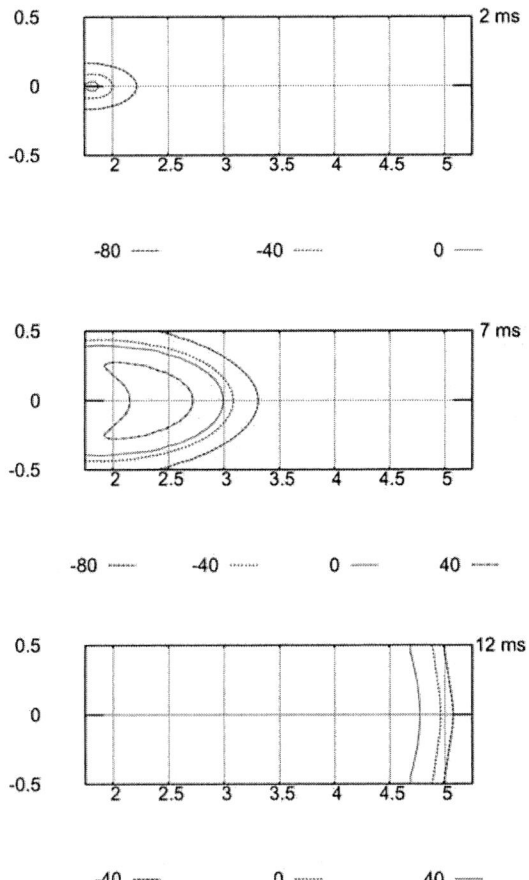

Fig. 2. Transmembrane voltages in XZ-slice through the long axis of muscle. The excitation was initiated with an electrical stimulus in the extracellular space at the center of the left border of the muscle, i.e. position (0, 0, 1.75). The three panels show isolines of voltages at 2, 7, and 12 ms. The style of the isolines codes voltages in mV. The scaling of the axes is in mm

Activation times at different positions were detected from the extracellular potentials and transmembrane voltages (fig. 6). The activation times detected in transmembrane voltages for central and superficial stimulation indicated differences of related activation fronts and boundary effects. A significant initial offset of ≈ 4.3 ms was found (fig. 6a), which was followed be a decrease due to the higher velocities for superficial stimulation. In the final phase of propagation, the activation time offset averaged to ≈ 2 ms. Reliable detection of activation times in extracellular electrograms was partly not possible due to their small slopes (fig. 6b,c). A non-constant offset between the activation time courses was observed. In general, the relationship between activation time and distance was found to be nonlinear particularly at the left and right end of the preparations.

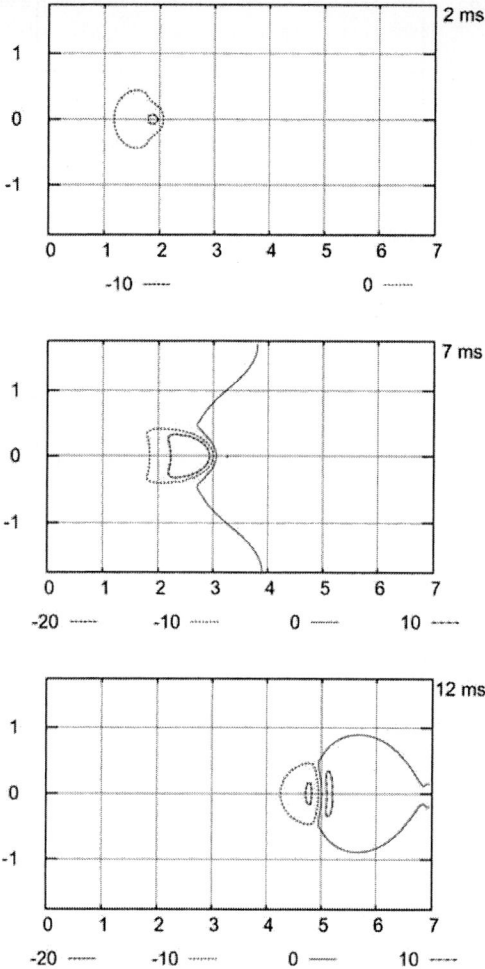

Fig. 3. Extracellular and bath potentials corresponding to fig. 2

4 Conclusions

The study revealed that excitation propagation in papillary muscle is a three-dimensional process, which can only poorly be described by uni-dimensional approximations. The simulations showed significant differences of the spatio-temporal evolution of activation fronts and related potential distributions resulting from central and superficial stimulation.

Already in the case of central stimulation, the curvature of the wave front, particularly near the stimulus site, led to radial differences of the transmembrane voltage and related detected activation times. These radial heterogeneities of transmembrane voltage led to radial heterogeneities of electrical fields in the bath and differences of therein detected activation times. Additional hetero-

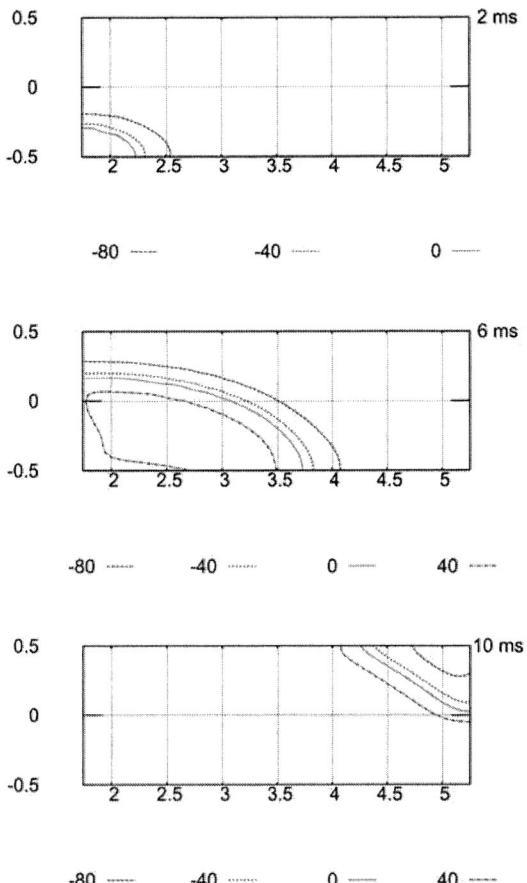

Fig. 4. Transmembrane voltages in XZ-slice through the long axis of muscle. The excitation was initiated with an electrical stimulus in the extracellular space at the lower left border of the muscle, i.e. position (-0.5, 0, 1.75)

geneities with a significant non-radial component were found for the superficial stimulation. Thus, particularly at the ends of the preparation the linearity of the relationship of activation time and distance degrades significantly as the distance of measurement position to the preparation increases. These findings confirms that conduction velocity can be detected with a given accuracy only in specific areas, which are defined by their neighborhood to the activation front.

For both cases of stimulation, the wave front was initially convex and afterwards its curvature decreased. In case of superficial stimulation the wave front was concave during the final phase of simulation (fig. 4 10ms). Other studies reported a steady state of conduction velocity associated with a concave curvature for a cylindrical strand of cardiac muscle [17, 18]. A steady state was not reached in our study, which can be explained by the relatively short length of our preparation, i.e. 3.5 versus 12.8 mm. Additionally, the type and location of

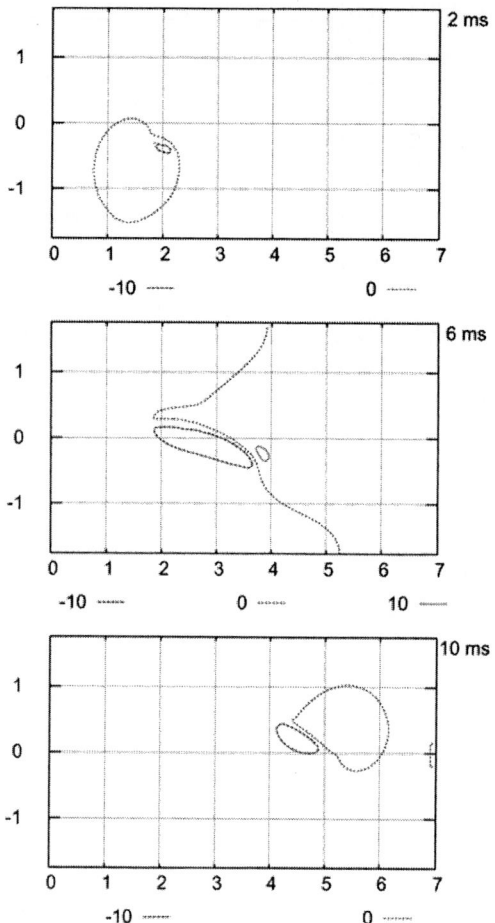

Fig. 5. Extracellular and bath potentials corresponding to fig. 4

stimulus electrodes applied in our study, i.e. central and superficial point versus ring electrodes, will increase the time until steady state is reached.

Future work will necessitate research concerning more efficient numerical methods to solve the bidomain model. Particularly, we are interested to reduce the high computational demand associated with solution of Poisson's equation associated to the extracellular space, e.g. by applying mesh-less techniques for solving of differential equations, and the high temporal resolution necessary to solve ordinary differential equations with the Euler method.

Acknowledgments

This work has been supported by the Richard A. and Nora Eccles Fund for Cardiovascular Research and awards from the Nora Eccles Treadwell Foundation.

(a)

(b)

(c)

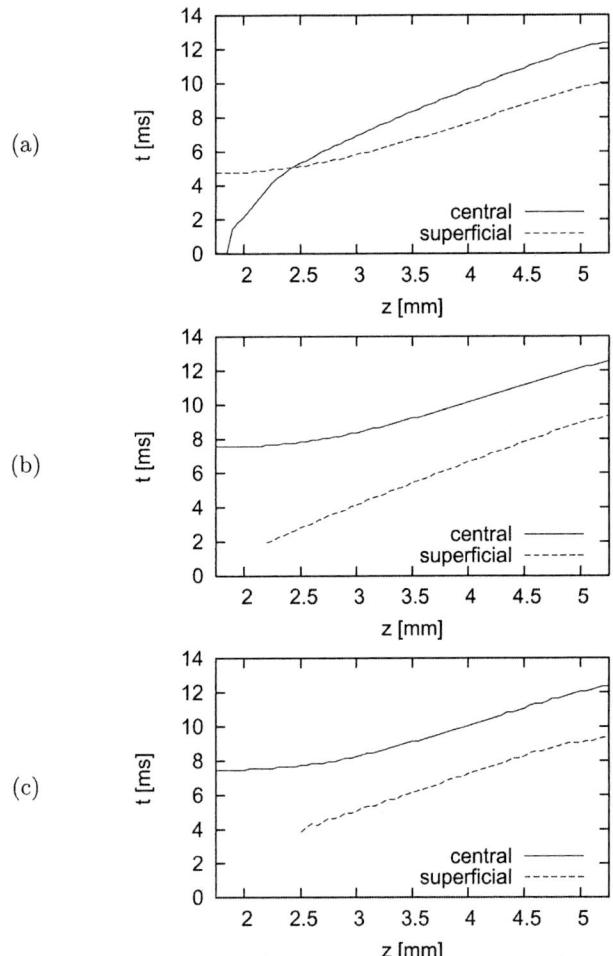

Fig. 6. Activation times detected for central and superficial stimulation. (a) The time of maximal derivative delivered activation from the transmembrane voltage at (0,0,z). The time of minimal derivative of the extracellular potentials in the bath at a distance of (b) 25 μm and (c) 275 μm to muscle surface determined activation

The authors gratefully acknowledge computing resources kindly provided by the Scientific Computing and Imaging Institute and the NIH NCRR Center for Bioelectric Field Modeling, Simulation, and Visualization.

References

1. Fleischhauer, J., Lehmann, L., Kléber, A.G.: Electrical resistances of interstitial and microvascular space a determinants of the extracellular electrical field and velocity of propagation in ventricular myocardium. Circ. **92** (1995) 587–594

2. Kléber, A.G., Riegger, C.B.: Electrical constants of arterially perfused rabbit papillary muscle. J. Physiol. **385** (1987) 307–324
3. Penefsky, Z.J., Hoffman, B.F.: Effects of stretch on mechanical and electrical properties of cardiac muscle. Am. J. Physiol. **204** (1963) 433–438
4. Saeki, Y., Kurihara, S., Komukai, K., Ishikawa, T., Takigiku, K.: Dynamic relations among length, tension, and intracellular Ca^{2+} in activated ferret papillary muscles. Am. J. Physiol. **275** (1998) H1957–H1962
5. Sachse, F.B., Steadman, B.W., Bridge, J.H.B., Punske, B.B., Taccardi, B.: Conduction velocity in myocardium modulated by strain: Measurement instrumentation and initial results. In: Proc. 26th Conf. IEEE EMBS. (2004)
6. Spear, J.F., Moore, E.N.: Stretch-induced excitation and conduction disturbances in the isolated rat myocardium. J. Electrocardiology **5** (1972) 15–24
7. Henriquez, C.S., Plonsey, R.: A bidomain model for simulating propagation in multicellular cardiac tissue. In: Proc. of the Annual International Conference of the IEEE Engineering in Medicine and Biology Society. Volume 4. (1989) 1266
8. Henriquez, C.S., Plonsey, R.: Simulation of propagation along a cylindrical bundle of cardiac tissue–II: Results of simulnts. IEEE Transactions on Biomedical Engineering **37** (1990) 861–875
9. Henriquez, C.S.: Simulating the electrical behaviour of cardiac tissue using the bidomain model. Critical Reviews in Biomedical Engineering **21** (1993) 1–77
10. Roth, B.J.: Action potential propagation in a thick strand of cardiac muscle. Circ. Res. **68** (1991) 162–173
11. Sachse, F.B., Seemann, G., Weiß, D.L., Punske, B., Taccardi, B.: Accuracy of activation times detected in simulated extracellular electrograms. In: Proc. Computers in Cardiology. (2004)
12. Schwarz, H.R.: Methode der finiten Elemente. 3 edn. Teubner, Stuttgart (1991)
13. Sachse, F.B.: Computational Cardiology: Modeling of Anatomy, Electrophysiology, and Mechanics. LNCS 2966. Springer, Berlin, Heidelberg, New York (2004)
14. Noble, D., Varghese, A., Kohl, P., Noble, P.: Improved guinea-pig ventricular cell model incorporating a diadic space, I_{Kr} and I_{Ks}, and length- and tension-dependend processes. Can. J. Cardiol. **14** (1998) 123–134
15. Press, W.H., Teukolsky, S.A., Vetterling, W.T., Flannery, B.P.: Numerical Recipes in C. 2 edn. Cambridge University Press, Cambridge, New York, Melbourne (1992)
16. Open Architecture Review Board: OpenMP: Simple, portable, scalable SMP programming (1997-2004)
17. Spach, M.S., Barr, R.C., Serwer, G.A., Kootsey, J.M., Johnson, E.A.: Extracellular potentials related to intracellular action potentials in the dog purkinje system. Circ. Res. **30** (1972) 505–519
18. Punske, B.B., Ni, Q., Lux, R.L., MacLeod, R.S., Ershler, P.R., Dustman, T.J., Allison, M.J., Taccardi, B.: Spatial methods of epicardial activation time determination in normal hearts. Annals Biomed. Eng. **31** (2003) 781–792

Induced Pacemaker Activity in Virtual Mammalian Ventricular Cells

Wing Chiu Tong and Arun V. Holden

Computational Biology Laboratory, School of Biomedical Sciences,
University of Leeds, Leeds LS2 9JT, UK
cbs7wct@leeds.ac.uk
http://www.cbiol.leeds.ac.uk

Abstract. The stability of induced pacemaker activity in a virtual human ventricular cell is analysed by numerical simulations and continuation algorithms, with the conductance of the time independent inward rectifying potassium current (I_{K1}) as the bifurcation parameter. Autorhythmicity is induced within a narrow range of this conductance, where periodic oscillations and bursting behaviour are observed. The frequency of the oscillations approaches zero as the parameter moves towards the bifurcation point, suggesting a homoclinic bifurcation. Intracellular sodium ($[Na^+]_i$) and calcium ($[Ca^{2+}]_i$) concentration dynamics can influence the location of the bifurcation point and the stability of the periodic states. These two concentrations function as slow variables, pushing the fast membrane voltage system into and out of the periodic region, producing bursting behaviour. Moreover, suppressing I_{K1} will prolong action potential duration and may introduce risks of developing stable periodic intermittency and arrhythmia. A genetically engineered pacemaker may appear an attractive idea, but simple analysis suggests inherent problems.

1 Introduction

Cardiac pacemaker cells are autorhythmic, while ventricular cells are excitable and normally require repetitive current flows from neighbouring excited cells to generate repetitive action potentials. These contrasting behaviours are the sum of the activities of different membrane channels. Channel properties such as availability, density and single channel conductance, can be modelled as sets of maximum conductance for each current. Holden and Yoda [1, 2] have argued that the ionic channel densities of excitable cells can act as bifurcation parameters and these cells can undergo a Hopf bifurcation into autorhythmicity by a reduction in potassium conductance.

Indeed, pacemaker activity has been induced in mammalian ventricular cells. Miake et al [3, 4] injected a negative dominant gene construct of the Kir2.1 channels into ventricular myocytes of adult guinea pigs. The expression of the construct produced non-functional channels; as a result, the maximum conductance

($\bar{G}_{K1}$) of the time independent inward rectifying potassium current (I_{K1}) was reduced. Cells with more than 80 % reduction in I_{K1} activity exhibited pacemaker activity; cells with moderate reduction had prolonged action potential duration (APD), depolarised resting membrane potential and reduced repolarisation rate. Over-expression of I_{K1} channels produced the opposite effects [4]. These studies suggest the possibility of genetically engineering a ventricular pacemaker by down-regulating I_{K1}.

Computational studies with ventricular cell models [5, 6] suggest that pacemaker activity induced by I_{K1} down regulation is carried by the sodium-calcium exchanger current (I_{NaCa}) and thus depends on intracellular sodium ($[Na^+]_i$) and calcium ($[Ca^{2+}]_i$) concentrations. As a consequence, the rate of the induced pacemaker activity may also respond to beta-adrenergic stimulation, as in natural pacemaker cells. Compared to the electronic pacemakers in clinical practice, a genetically engineered biological pacemaker that will not require surgical implantation and battery replacement every ten years, is regulated by the sympathetic nervous system, and can adapt to hormone changes, is an attractive idea.

However, for a functional pacemaker, stable periodic activity is crucial. Here we address this problem using numerical simulations and continuation algorithms to characterise the stability and bifurcations of the induced autorhythmic activities in a human ventricular cell model [7].

2 Human Ventricular Model

The human ventricular cell model by ten Tusscher et al [7] is used. For an isopotential single cell,

$$C_m \frac{dV}{dt} = -I_{\text{ion}} , \qquad (1)$$

where V is the transmembrane potential (mV); I_{ion} is the total current density ($\mu A\,cm^{-2}$), which is the sum of all currents of ion channels, pumps and exchangers; $C_m = 2\,\mu F\,cm^{-2}$, is the membrane specific capacitance. The equation was integrated using the forward Euler method with a variable time-step from 0.02 ms to 1 ms. The epicardial cell model, with parameters given in [7], was used so the results could be compared with other human epicardial cell models [8, 9]. In addition, an equivalent study was performed earlier using different cell types of the Luo-Rudy guinea pig ventricular cell model (LRd00) [10, 11]. Similar dynamics were observed among the cell types and their differences were marginal [12]. So, different cell types of the human model may also give similar results.

Down-regulation of I_{K1} was modelled by multiplying the standard value of $\bar{G}_{K1}$ (5.405 nS pF^{-1}) with a dimensionless fractional term x, where $0 \leq x < 1$ represents suppression. In what follows, x is referred to as g_{K1}. When required, stimuli were applied with a current density of -90 pA pF^{-1} for 0.5 ms, otherwise the cell was left unperturbed until steady state or periodic cycle was reached.

A solitary action potential evoked in the normal epicardial cell model and its restitution properties are illustrated in Fig. 1. Restitution is the relationship between action potential duration at 90 % repolarisation (APD$_{90}$) and diastolic

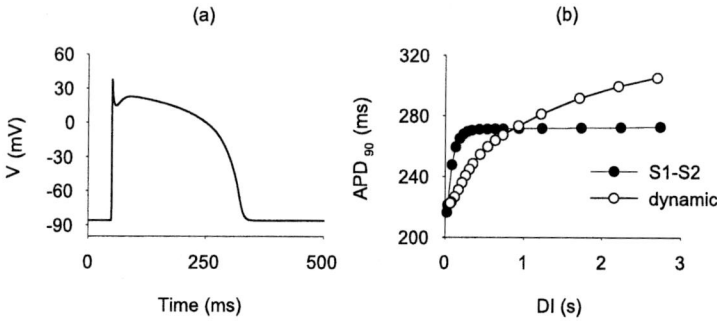

Fig. 1. (a) A solitary action potential excited by a stimulus in the epicardial cell model, $APD_{90} = 266$ ms. (b) The S1-S2 restitution and the dynamic restitution

interval (DI), defined as the interval between the time at 90% repolarisation of an action potential (AP) and the upstroke of the next AP. Two restitution protocols were used. S1-S2 restitution, a common method to define cellular properties, is found by applying a test stimulus (S2) at some DI after a train of 10 stimuli (S1) at 1 Hz. Dynamic restitution, which is more relevant to re-entry stability [13], is found by plotting the steady state APD_{90} against steady state DI during periodic pacing at different rates.

3 Down Regulation of I_{K1}

Pacemaker activity was induced in the human ventricular cell model. The membrane potential remained steady when I_{K1} was not suppressed ($g_{K1} = 1$) but autorhythmic action potentials appeared when I_{K1} was completely blocked ($g_{K1} = 0$) (Fig. 2a). The resting membrane potential depolarised as g_{K1} decreased; at $g_{K1} \approx 0.077$, i.e. $\sim 92\%$ reduction in I_{K1}, a bifurcation occurred separating the resting states and the oscillating states (Fig. 2b). Similar dynamics were also observed among different cell types of the LRd00 model: their bifurcation occurred at $g_{K1} \approx 0.3$, i.e. 70% reduction in I_{K1} [12]. Therefore, a greater reduction in I_{K1} is required for pacemaker activity to be induced in human ventricular cells.

However, ventricular cells with down regulated I_{K1} may become proarrhythmic. In fact, patients with Andersen syndrome (OMIM: #170390)[1], where mutations in I_{K1} channels are implicated, may experience frequent periodic paralysis and ventricular arrhythmias [4, 6]. Stimulated at 1 Hz, APD_{90} is prolonged from 266 ms to 315 ms as g_{K1} decreases, and peaks at 345 ms at the bifurcation

---

[1] OMIM (Online Mendelian Inheritance in Man) is a database of human genes and genetic disorders. It can be accessed through the National Center for Biotechnology Information. http://www.ncbi.nlm.nih.gov/

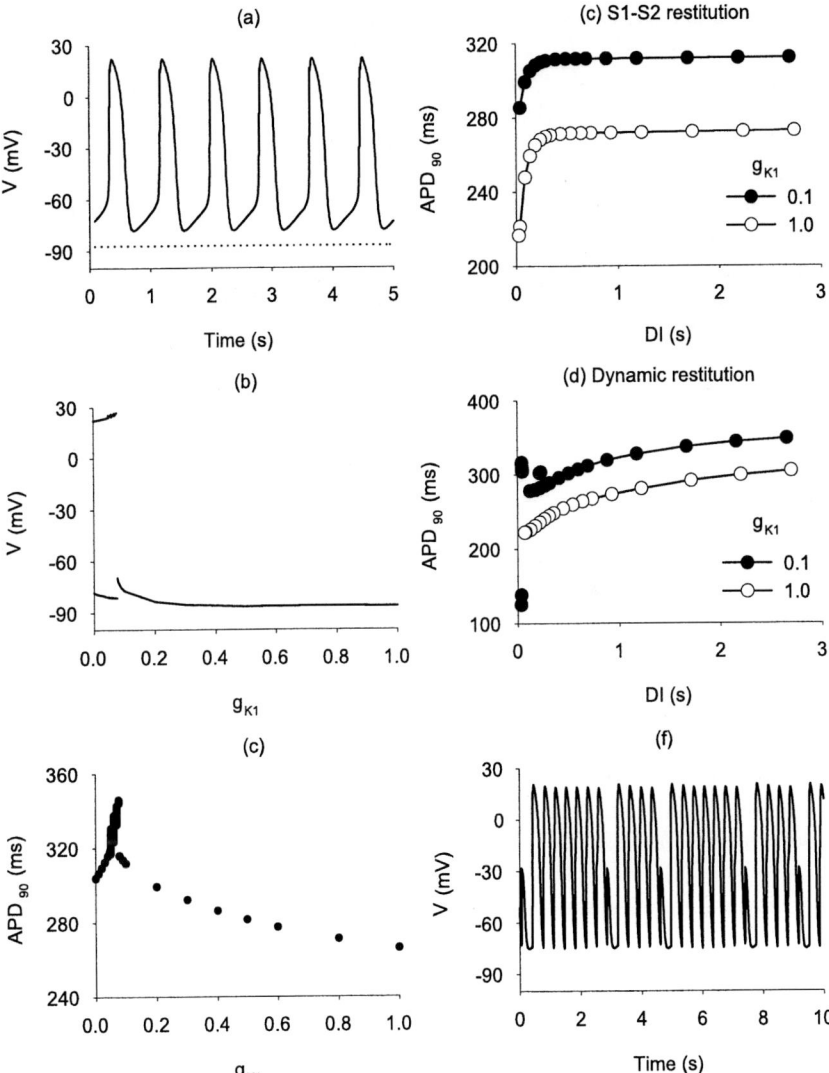

Fig. 2. (a) $V(t)$ with $g_{K1} = 0$ (*solid line*) and $g_{K1} = 1$ (*dotted line*) for the human epicardial cell model. Autorhythmicity is produced by complete block of I_{K1}. (b) The numerically computed bifurcation diagram with g_{K1} as the bifurcation parameter. The resting membrane potential depolarises as the parameter decreases and oscillations emerge at $g_{K1} \approx 0.077$. (c) Prolongation of APD_{90} with reducing g_{K1}. Where oscillations did not occur, steady state APD_{90} is determined by applying stimuli at 1 Hz. S1-S2 (d) and dynamic (e) restitutions of an epicardial cell model at $g_{K1} = 0.1$ and 1.0. Suppressing I_{K1} shifts both restitution curves upwards, and stable periodic intermittency is induced at small DI in dynamic restitution. (f) $V(t)$ of the intermittency after transients

point (Fig. 2c). Prolonged APD$_{90}$ increases the time interval of the vulnerable window where extrasystoles may occur.

Moreover, suppressing I_{K1} alters the APD$_{90}$ restitution properties, and may introduce a risk of developing stable periodic intermittency. At $g_{K1} = 0.1$, both the S1-S2 restitution (Fig. 2d) and the dynamic restitution (Fig. 2e) shifted upwards. Minimal change is found in the S1-S2 restitution. However, the dynamic restitution curve flattens when DI is less than 0.2 s and stable periodic intermittency develops for DI less than 50 ms. Fig. 2f shows an example of a stable periodic intermittency.

Down-regulation of I_{K1} in ventricular cell can push the cell from resting states into autorhythmic states, inducing pacemaker activity. However, potential risks for arrhythmogenesis increases as I_{K1} is suppressed, especially around the bifurcation point.

4 Bifurcation Analysis

For bifurcation analysis, continuation algorithms such as AUTO [14] – via XPP-AUT [15] a simulation tool with an AUTO interface – is used in addition to the numerical experiments, to further characterise the behaviour and its stability of the human ventricular epicardial cell model. Continuation algorithms are useful for exploring the behaviour of a system, as they can trace solutions within parameter space, identify different types of bifurcation points and determine their stability.

AUTO requires the system to start from a steady state solution or a periodic orbit. However, the human ventricular cell model is a complex system with stiff, high-order differential equations. As the cell model is electrically but not electrochemically neutral [16], the full system could not settle into a stable steady state in XPPAUT. Therefore, a reduced system without intracellular concentration dynamics was used for continuation analysis in XPPAUT, as the intracellular concentration dynamics are slow compared to the fast membrane voltage system. The intracellular concentrations were clamped at constant values: $[Na^+]_i = 11.6$ mM; $[Ca^{2+}]_i = 0.2\,\mu$M; intracellular potassium, $[K^+]_i = 138.9$ mM. A fourth order Runge-Kutta integrator (within XPPAUT) with a fixed time step of 0.02 ms was used to bring the reduced system into steady state. Again, g_{K1} was set as the bifurcation parameter.

Solutions from the numerical experiments on the full system (Fig. 3a) show the system bifurcates at $g_{K1} \approx 0.077$, at which point periodicity and bursting occur. Oscillations emerge with large periods and quickly decrease with further reduction in g_{K1}. Moreover, these oscillations appeared to be stable over long period of time (80 min). This suggests a homoclinic bifurcation rather than a Hopf bifurcation.

AUTO solutions of the reduced system (Fig. 3b) show that it bifurcates in the non-physiological range at a negative g_{K1} with a Hopf bifurcation and high frequency (low period); and ends at a homoclinic bifurcation close to $g_{K1} \approx 0.05$ as the frequency approaches zero (infinite period). Compared to the full system in

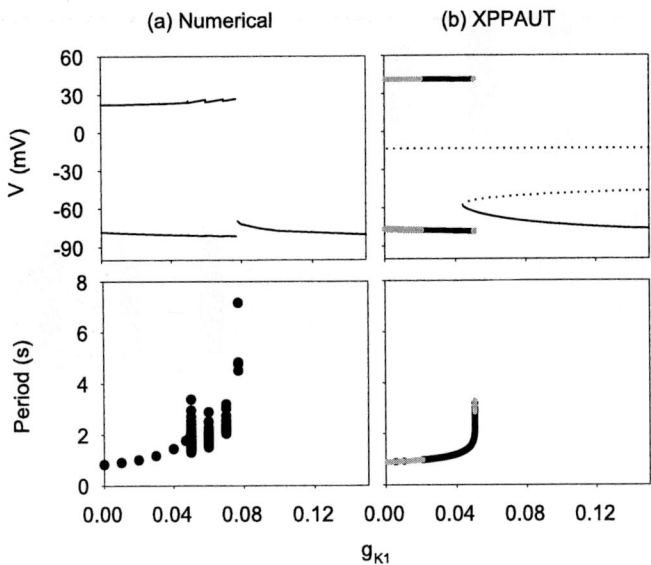

Fig. 3. Bifurcation diagrams with g_{K1} as the bifurcation parameter. (a) The full system with solutions computed numerically. Bifurcation point occurs at $g_{K1} \approx 0.077$. Periods are defined as the time interval between peaks (maximal V). In the range of g_{K1}: 0.05–0.077, multiple periods are observed and $V(t)$ shows bursting behaviour (see Fig. 4). (b) Stable (*black*) and unstable (*grey and dotted line*) solutions of the reduced system, which consists of only the fast membrane system and no intracellular ionic dynamics, are traced by the continuation algorithm, AUTO/XPPAUT. Only the physiological range of g_{K1} is shown. The bifurcation point occurs at $g_{K1} \approx 0.05$, less than the numerical solutions of the full system. Bursting is not seen in the reduced system

Fig. 3a, bursting is not observed in the reduced system. Also, stable oscillations only occur within a narrow range of g_{K1} and their stability is lost again when the system is close to the bifurcation point. Moreover, this bifurcation point coincides with the onset of bursting in the full system, suggesting that the dynamics of intracellular concentrations could influence the behaviour and stability of a virtual ventricular cell.

5 Bursting Behaviour

The bursting behaviour seen in the full system of the human ventricular cell is curious as it is common among neurons, smooth muscle, endocrine cells and embryonic cardiac cells [17, 18, 19, 20], but is rarely observed in adult cardiac myocytes. At $g_{K1} = 0.07$, $V(t)$ shows neuron-like bursting behaviour on a time scale of minutes with the action potential bursts repetitively interrupted by a long period of inactivity (Fig. 4a). The intervals between action potentials

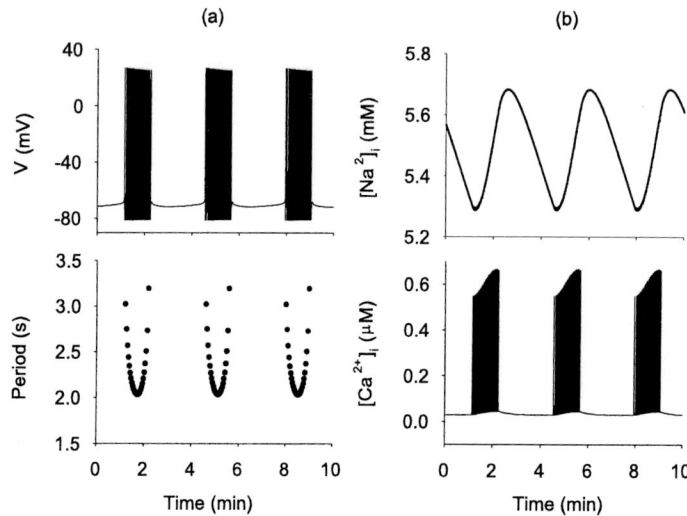

Fig. 4. Bursting behaviour of the full system seen at $g_{K1} = 0.07$ and after all the transients. (a) $V(t)$ shows periods of inactivity interrupting the oscillations (top). The periods of the oscillations follow a parabolic fashion (bottom). (b) Oscillations of $[Na^+]_i$ (top) and $[Ca^{2+}]_i$ (bottom) during bursting

decreases and increases during the bursting, and the action potentials undershoot below the voltage observed during the quiet period. These are characteristics of Type II parabolic bursting.

Type II bursting involves oscillations of at least two slow variables pushing the fast membrane system in and out of a homoclinic bifurcation [21, 22, 23]. As the intracellular concentration dynamics are slow compared to the fast membrane voltage system, these could be the variables that drive the bursting behaviour in the full system of the human ventricular cell model. We focused on $[Na^+]_i$ and $[Ca^{2+}]_i$ in particular as the pacemaker activity was suggested to be carried by I_{NaCa} [5]. Fig. 4b shows oscillations of $[Na^+]_i$ and $[Ca^{2+}]_i$ during bursting, and Fig. 5a shows that these concentrations form an orbit in the phase diagram, where bursting starts at low $[Na^+]_i$ and $[Ca^{2+}]_i$ and terminates as these concentrations accumulate.

In order to determine the dynamics of the fast membrane system with different combinations of intracellular concentrations, a two parameter bifurcation diagram with $[Na^+]_i$ and $[Ca^{2+}]_i$ as bifurcation parameters is computed at $g_{K1} = 0.07$ with XPPAUT and the reduced system of the human ventricular cell model (Fig. 5b). The left arm (between 5 mM of $[Na^+]_i$ and the cusp point) separates periodic and steady states solutions of the fast membrane system. The base of this arm swings from left to right as g_{K1} is reduced (not shown). In spite of the dynamics of this boundary, for stable steady solutions, the full system is represented by a point underneath the arm; for stable periodicity, the full system forms a single $[Na^+]_i$-$[Ca^{2+}]_i$ orbit above the arm.

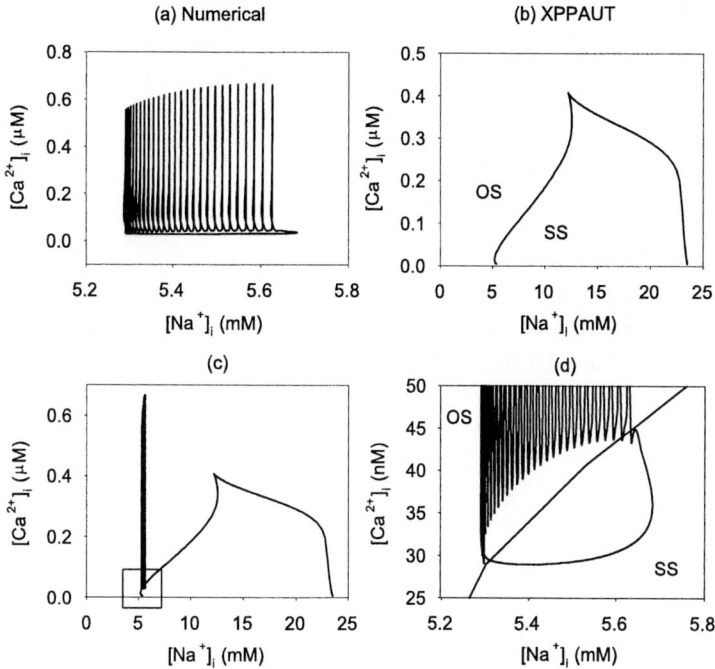

Fig. 5. (a) Numerical results of the dynamics between intracellular sodium ($[Na^+]_i$) and calcium ($[Ca^{2+}]_i$) concentrations during bursting in the full system at $g_{K1} = 0.07$. Bursting starts at low $[Na^+]_i$ and $[Ca^{2+}]_i$ and terminates as these concentrations accumulate. (b) Two parameter bifurcation diagram computed by XPPAUT using the reduced system at $g_{K1} = 0.07$. The arm between 5 mM of $[Na^+]_i$ and the cusp point separates steady states (*SS*) and the oscillatory states (*OS*) of the fast membrane system. (c) Superimposed the numerical results in (a) onto the two parameter bifurcation diagram in (b), the full system is right on the boundary between periodic and steady states. The details in the box are shown enlarged in (d). The full system is being pushed across and back from the boundary by $[Na^+]_i$ and $[Ca^{2+}]_i$ dynamics. Bursting starts clockwise as the system enters the *OS* region and terminates when it falls back into the *SS* region but unable to push across the boundary again

Superimposing the $[Na^+]_i$-$[Ca^{2+}]_i$ orbit during bursting in Fig. 5a onto the two parameter bifurcation diagram in Fig. 5b shows that at $g_{K1} = 0.07$, where bursting occurs, the system is on the boundary of periodic and steady states (Fig. 5c and Fig. 5d). Bursting initiates when both $[Na^+]_i$ and $[Ca^{2+}]_i$ are low. $[Ca^{2+}]_i$ accumulates much faster than $[Na^+]_i$ initially, pushing the system away from the boundary. However, as more $[Na^+]_i$ accumulates, the system is pushed back towards the boundary and eventually crosses the boundary into the steady states region.

Intracellular concentration dynamics are not only responsible for the bursting behaviour in the human ventricular cell model, but could also influence the location of the bifurcation point of g_{K1} and the stability of the pacemaker activity. For example, with higher $[Na^+]_i$, the bifurcation point of g_{K1} in the reduced system will shift to the right, but the oscillations become more unstable; with lower $[Na^+]_i$, the bifurcation point will shift to the left, but the oscillations with become more stable (results not shown).

6 Conclusion

Pacemaker activity induced in human virtual ventricular cells is addressed here using both numerical simulations and continuation algorithms. Autorhythmicity is induced within a very narrow range of g_{K1} (0–0.077), implying that more than 92 % block of I_{K1} may be required to induce pacemaker activity in human ventricular cells. Within this narrow range, less than two thirds of the g_{K1} (0–0.05) show apparent stable periodic oscillations, the remainder exhibiting bursting behaviour. Application of continuation algorithms to a reduced system without slow intracellular concentration dynamics shows that the range of g_{K1} for stable autorhythmicity is further restricted to $g_{K1} = 0.02$–0.05.

Intracellular concentration dynamics plays a critical role in the behaviour and stability of the induced autorhythmicity in ventricular cells. For example, the observed bursting behaviour is a product of the slow dynamics of intracellular concentrations driving the fast membrane system between periodic and steady states. Therefore, manipulating intracellular concentrations, maybe via the activities of the sodium-potassium pump and the sodium-calcium exchanger, could be tools to influence the stability and behaviour of the induced autorhythmicity in human ventricular cells.

In addition to the stability of induced pacemaker activity, suppressing I_{K1} in human ventricular cell will prolong APD_{90} and introduce risks of developing stable periodic intermittency and arrhythmia. The genetically engineered pacemaker suggested by Miake et al [3] may appear an attractive idea, but simple analysis suggests inherent problems.

WCT is supported by a British Heart Foundation research studentship (FS/03/075/15914).

References

1. Holden, A.V., Yoda, M.: Ionic channel density of excitable-membranes can act as a bifurcation parameter. Biol. Cyber. **42** (1981) 29-38
2. Holden, A.V., Yoda, M.: The effects of ionic channel density on neuronal function. J. Theor. Neurobiol. **1** (1981) 60-81
3. Miake, J., Marban, E., Nuss, H.B.: Biological pacemaker created by gene transfer. Nature **419** (2002) 132-133
4. Miake, J., Marban, E., Nuss, H.B.: Functional role of inward rectifier current in heart probed by Kir2.1 over expression and dominant-negative suppression. J. Clin. Invest. **111** (2003) 1529-1536
5. Silva, J., Rudy, Y.: Mechanism of Pacemaking in IK1-downregulated myocytes. Circ. Res. **92** (2003) 261-263

6. Tristani-Firouzi, M., Jensen, J.L., Donaldson, M.R., Sansone, V., Meola, G., Hahn, A., Bendahhou, S., Kwiecinski, H., Fidzianska, A., Plaster, N., Fu, Y.H., Ptacek, L.J., Tawil, R.: Functional and clinical characterization of KCNJ2 mutations associated with LQT7 (Andersen syndrome). J. Clin. Invest. **110** (2002) 381-388
7. ten Tusscher, K.H.W.J., Noble, D., Noble, P.J, Panfilov, A.V.: A model for human ventricular tissue. Am. J. Physiol. **286** (2004) H1573-H1589
8. Iyer, V., Mazhari, R., Winslow, R.L.: A Computational Model of the Human Left-Ventricular Epicardial Myocyte. Biophys. J. **87** (2004) 1507-1525
9. Priebe, L., Beuckelmann, D.J.: Simulation Study of Cellular Electric Properties in Heart Failure. Circ. Res. **82** (1998) 1206-1223
10. Faber, G.M., Rudy, Y.: Action potential and contractility changes in [Na+](i) overloaded cardiac myocytes: a simulation study. Biophys. J. **78** (2000) 2392-2404
11. Faber, G.M.: The Luo-Rudy dynamic model of mammalian ventricular action potential. (2000) http://www.cwru.edu/med/CBRTC/LRdOnline/
12. Benson, A.P., Tong, W.C., Holden, A.V., Clayton, R.H.: Induction of autorhythmicity in virtual ventricular myocytes and tissue. J. Physiol. (Proceedings). (2005) (to appear)
13. Koller, M.L., Riccio, M.L., Gilmour, R.F. Jr.: Dynamic restitution of action potential duration during electrical alternans and ventricular fibrillation. Am. J. Physiol. Heart. Circ. Physiol. **275** (1998) H1635-H1642
14. Doedel, E.J.: AUTO: A program for the automatic bifurcation and analysis of autonomous systems. Cong. Num. **30** (1981) 265-284
15. Ermentrout, G.B.: XPPAUT. http://www.math.pitt.edu/~bard/xpp/xpp.html
16. Hund, T.J., Kucera, J.P., Otani, N.F., Rudy, Y.: Ionic charge conservation and long-term steady state in the Luo-Rudy dynamic cell model. Biophys. J. **81** (2001) 3324-3331
17. Bub, G., Glass, L., Publicover, N.G., Shrier, A.: Bursting calcium rotors in cultured cardiac myocyte monolayers. Proc. Natl. Acad. Sci. U. S. A. **95** (1998) 10283-1.0287
18. Cohen, N., Soen, Y., Braun, E.: Spatio-temporal dynamics of networks of heart cells in culture. Physica. A. **249** (1998) 600-604
19. Cohen, N.: The development of spontaneous beating activity in cultured heart cells: From cells to networks. PhD thesis, Technion – Israel Institute of Technology. (2001)
20. Soen, Y., Cohen, N., Braun, E., Lipson, D.: Emergence of spontaneous rhythm disorders in self-assembled networks of heart cells. Phys. Rev. Lett. **82** (1999) 3556-3559
21. Bertram, R., Butte, M., Kiemel, T., Sherman, A.: Topological and Phenomenological Classification of Bursting Oscillations. Bull. Math. Biol. **57** (1995) 413-439
22. Rinzel, J., Ermentrout, B.: Analysis of neural excitability and oscillations. In: Koch, C., Segev, I. (eds.): Methods in Neuronal Modeling: From Synapses to Networks, 2nd eds. MIT Press, Cambridge, MA (1999) 251-292
23. Wang, X.-J., Rinzel, J.: Oscillatory and bursting properties of neurons. In: Arbib, M.A. (ed.): Handbook of brain theory and neural networks. MIT Press, Cambridge, MA (1995) 686-691

Transvenous Path Finding in Cardiac Resynchronization Therapy

Jean Louis Coatrieux[1], Alfredo I. Hernández[1], Philippe Mabo[2], Mireille Garreau[1], and Pascal Haigron[1]

[1] Laboratoire Traitement du Signal et de l'Image, INSERM, Université de Rennes 1,
Campus de Beaulieu, 35042 Rennes Cedex, France
http://www.ltsi.univ-rennes1.fr
[2] Centre Cardio-Pneumologique, CHU Pontchaillou, 35033 Rennes, France

Abstract. Cardiovascular diseases are a major health concern all over the world and, especially, heart failure has gained more importance in the recent years. Improving diagnosis and therapy is therefore critical and among the several resources at our disposal, implantable devices is expected to have a better rate of success. This paper is focused on two topics: (i) our views of the main challenges to face in order to reach these objectives and (ii) a specific target regarding the pose of leads for multisite pacemakers by means of virtual endoscopy pre-operative planning and path finding throughout the coronary venous tree.

1 Introduction

The cardiovascular disorders remain the most important cause of death in all countries. They cover a wide spectrum of causes among which coronary artery and cardiac valve diseases, abnormal excitation-contraction coupling or heart failure. Electrophysiological pathologies are among the most deeply investigated for a long time. They include cardiac arrhythmia and myocardial ischemia which both may originate from very distinct locations and have many underlying expressions (ectopic foci, spiral-like wavefronts, conduction blocks, ..). In all cases, early diagnosis must be based on a full exploration of the whole heart instead of focusing on the left ventricle. However, any improvement at the diagnosis stage will be of limited interest if subsequent sound therapies are not available. Advances have been made over years in the design of drugs but some of them have been associated with side effects and negative outcomes. Technological solutions rely on Cardiac Resynchronization Therapy and Radiofrequency Ablation. They both share, to be efficient, a lot of concerns regarding the definition of target sites where abnormal patterns are observed, the consequences of which are acute and overall dysfunctions in the heart pumping. They require advanced imaging techniques for their localization and for assisting the physician before, during and after the interventions.

This paper is focused on Cardiac Resynchronization Therapy (CRT) aimed at restoring the contractile coordination in hearts with severe heart failure (HF), sinus rhythm and ventricular conduction delay. It is performed by stimulating both the right and the left ventricles, pacing them simultaneously or with a small delay. Several clinical trials have shown that this technique is beneficial for acute as well as chronic disorders, improving the heart's performance, the capacity of the patient for exercise and reducing the mortality from heart failure, [Cazeau, 2001], [Leclercq, 2002], [Kass, 2003]. However, these studies point out that the main issue remains to identify and assess the most effective pacing sites in order to reduce the percentage of non-responding patients which may reach up to 25 to 30% of recipients. They have also shown that LV-lead positioning (either LV-only pacing or biventricular pacing) is without contest the most challenging task to carry out.

Our objective is therefore to better prepare the placement of CRT leads using the new Multislice Computed Tomography (MSCT) capabilities in imaging the heart. Section 2 provides a brief overview of the multiple challenges we must address at long range to understand how the major components (electrical, mechanical, etc.) work together and are regulated under normal and abnormal physiological conditions. Section 3 brings more clues on the CRT issues in pre-operative context. Section 4 reports some preliminary results achieved by using virtual endoscopy and a few perspectives are discussed in Section 5.

2 The Overall Heart Picture

The key dimensions [Coatrieux, 2004] for further advances in clinical diagnosis and therapy are reported figure 1. Only a few of them, that we consider as major issues to deal with, will be detailed here.

The identification of the disorder through non-invasive data recording is the first stage to go through. The most generic and relevant tool remains standardized ECG (or esophageal ECG) which can be coupled to clinical observations (symptoms, past history), impedance measurements (cardiac output), phonocardiogram (PCG)... Body Surface Potential Mapping (with or without the so-called direct-inverse problem solving) has been shown superior to the conventional 12-lead ECG for non-invasively identifying the sites of earliest endocardial activation and the further spread through the ventricles. However, all these resources only give first assumptions on the abnormal behaviours that are observed and their potential localizations.

The integration of multimodal imaging data is another critical issue [Roux, 1997]. It starts with the diagnosis tools providing the 2D, 3D and 4D elements to capture local, regional and global characteristics required to determine the morphological and functional patterns of the heart, either normal or abnormal. The progress in ultrasound techniques, and in Multislice CT allows now to acquire 3D time image sequences with high contrast and spatio-temporal resolutions. The major problems, beyond spatio-temporal registration methods aimed at deriving a common coordinate system, are to extract quantitative features that can be physically and physiologically interpreted with a proper anatomical reference. Accurate and reliable segmentation methods, fulfilling the time computation constraints in clinical practice, with robust motion estimation algorithms and perfusion parameters have to be combined in a sound information processing frame in order to get a full view of the status of the heart. The constraints imposed in intra-operative environment, the therapeutic nature of the intervention, are even more demanding due to the real-time responses required for registration, detection, guidance of instruments, etc.

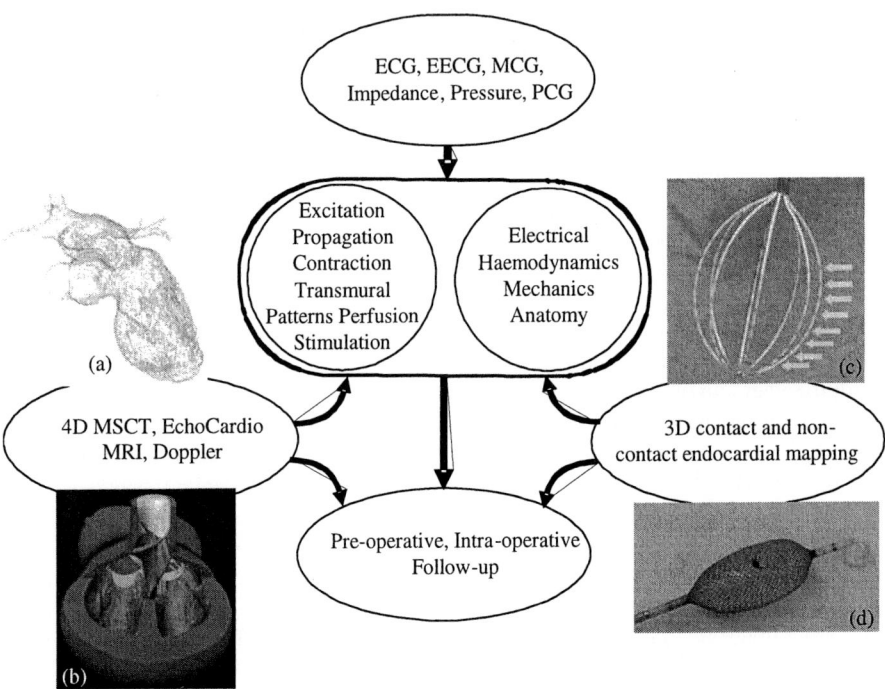

Fig. 1. Clinical diagnosis and therapy. Picture (a) is from [Garreau, 2004]. Picture (b) is from [Schleich, 2002]. Pictures (c) and (d) are from [de Boer, 2000]

Electrical mapping [Gepstein, 1997] [Schilling, 1998] [De Bakker, 2000] is a relevant complement of the imaging sources for electrophysiological tracking. The availability of catheter-based 3D non fluoroscopic contact (Carto, Biosense, or Constellation, Boston Scientific, Inc) and noncontact mapping (Ensite 3000, Endocardial Solutions, Inc) techniques allow in vivo assessment of the activation sequence with a relatively high spatial resolution. If the benefits of these techniques are clearly established in terms of electrophysiological insights, they have the inconvenient to be invasive, expensive and to increase the time duration of the exploration, and as such put more clinical demands.

The physiopathological in-silico modelling of the heart capable to fuse together the patient specific features (i.e electrical, mechanical and perhaps more importantly the electromechanical, mechanochemical, etc.) with the corresponding anatomical structures into generic models integrating the last data obtained through in-vitro, ex-vivo and in-vivo experiments is perhaps the grand challenge for tomorrow. A lot of efforts have been devoted to the restitution of the electro-physiological activity of the heart and two main model families can be distinguished (refer to [Bardou, 1996] [Virag, 2001] for full references): i) Simplified models, which are limited to the simulation of an action potential waveform, without taking into account any sub-cellular process, such as the Fitzugh-Nagumo's model (which was later improved by Aliev and Panfilov) or the model proposed by van Capelle and Durrer and ii) Electrophysiologically detailed models: which are based on the Hodgkin-Huxley approach for modeling ionic currents. A variety of models have been proposed for the later type, with increasing levels of detail and for specific myocardial tissues (i.e. ventricular, atrial, or Purkinje myocites). Two significant examples, for the ventricular myocite, are: the Beeler and Reuter's model, developed in 1977 which introduced the dynamics of internal calcium concentration and the models proposed by Luo and Rudy, with an increased number of ionic currents and a more detailed calcium dynamics. Bidimensional networks of cells can be built, where each node is described by one of these models and an extension of the cable equation is used to couple them. This approach has been readily applied since the 80s to model myocardial propagation.

More recently, large-scale electrical models have been developed [Noble, 1997] [Quan, 1998]. Some of these models have been mapped to 3D anatomical data [Hunter, 1996], [Rudy, 1995] [Sermesant, 2004] but the key issue remains the inverse problem, i.e. the identification of the system from the current observed data. However, even if it is not out of reach, we are still far to deal with the full complexity of cardiac mechanisms. To just take an example, the excitation-contraction coupling, which refers to the physiological processes linking myocite depolarisation and contraction, involves many structural and regulatory proteins whose nature and function are just emerging [Bers, 2002].

Merging the multifunctional models we need to face electrical, mechanical, hemodynamic facets, at different scales, distinct supports, time dynamics with the multimodal data that we have at our disposal, would directly impact our capability to diagnose and care.

Further advances should rely on the design of intelligent devices, implantable or not, able to handle the several variables required, with both real-time recording, processing, stimulation capabilities. Along this path, recent technological breakthroughs of implantable devices have been achieved: they concern biventricular pacing (or CRT), cardioverter-defibrillator (ICD, Implantable Cardioverter Defibrillator) or joint device (CRT-ICD).

The last fundamental component we wish to highlight concerns the pre-operative planning, intra-operative assistance and post-operative follow-up. With the advances in medical imaging, image processing and rendering, almost over three decades, the concept of computer-assisted surgery has emerged with the aim to reduce the duration of interventional procedures, make them more successful and secure [Taylor, 1996]. Many achievements have been reported from the early biopsy applications in the 80s based on simple instruments, straight line trajectories into rigid tissues. Image-guided therapies, with several marketed products, now go well beyond surgical procedures and address soft tissue applications as well as moving organs like the heart. They are of relevance for mitral valve replacement, ablation and pacing procedure. They share a number of methodological, technical and clinical features but also include several specific components, the main one for pacing being related to the venous coronary tree exploration that will be examined in the next paragraph.

3 Cardiac Resynchronization Therapy

In 1998, a way to overcome the problem of multisite pacing by using a transvenous technique has been proposed [Daubert, 1998], permitting to stimulate the left ventricle on a long-term basis. The overall procedure consists in positioning endocardial leads in the right atrium and the right ventricle, the left ventricle being paced via a lead passed through the coronary sinus to an epicardial vein on the free wall of the left ventricle. The implant success rate, in between 85 and 92%, although already high, is limited by the impossibility to access the target vein for lead placement, incorrect or suboptimal pacing site selection and possible electrode displacements [Alonzo, 2001], [Abraham, 2002].

The main objective is thus to assist cardiologists in improving and securing the implant techniques. From a clinical standpoint, it will rely on the study of the patient's coronary anatomy to define the target veins, to confirm their accessibility and to minimize the implant time. The pre-operative assistance, defined as Virtual Navigation [Haigron, 2004], will consist here to:

- Navigate through the patient's coronary venous tree so as to define the potential access paths for the pacing leads.
- Define which catheter and guide type should be used (with different diameters and curvatures, for example).
- Better define the optimal pacing site, based on the anatomo-functional information and on the electro-mechanical models of the cardiac activity.

Up to now, to our knowledge, there was no image-based planning of the implant procedure because no imaging source was capable to provide a full, 3D, time image sequence access to the venous tree of the heart. The pose of CRT is still directly performed, after the decision to implant a CRT, by using 2D venous coronary X-ray which leads to a partial and limited access to 3D anatomy.

The availability of MSCT is dramatically changing this situation: this 4D functional imaging CT Scanners can be used to obtain the basic structural and functional features required to achieve an optimal CRT planning. The LV can be paced transvenously through a subclavian vein, going successively via the cava vein, the right atrium, the coronary sinus and the great vein. The target location is a lateral or posterolateral vein. If lateral vein catheterization failed or in the case of poor pacing threshold, the LV lead is inserted into the great cardiac vein to pace the anterobasal wall or into the mid cardiac vein to pace the inferoapical region. Specifically designed coronary sinus leads are used. The injection of contrast medium in conventional X-ray angiography allows viewing the venous tree to be explored but it remains difficult to visually analyse due to the backward blood flow.

4 Looking Pre-operatively for a Left Ventricular Path: Preliminary Results

The procedure that has been worked out consists to select spatial sequences among 8 angio-scanners at our disposal (with minimal motion artefacts). The data were acquired on a Siemens Somatom volume zoom 4 detectors. Identical protocols were used with the following acquisition parameters: collimation of 0.6 mm, table displacement of 1.5 mm/rotation, reconstruction increment of 0.6 mm, size of the matrix 512x512 with about 250 slices and a pixel size from 0.33x0.33 to 0.4x0.4 mm. The resolution is 12 bits and the slice thickness of 1.25 mm.

In this section, we present one example of navigation which has been prepared following the anatomical pathway required for the implantation of the left lead of a biventricular pacemaker. As emphasized from the beginning, this path is the one that implies the highest difficulty along the pacemaker implantation procedure. The principles of the virtual navigation have been reported elsewhere [Haigron, 1996]. The interactive procedure has been retained here. It consists for the physician to position the virtual sensor inside the object of interest, to define the viewing direction and to set the detection threshold. These tasks are performed by using three orthogonal planes commonly used in marketed workstations. In short, the algorithm used for the image computation relies on a rough detection of the inner surface of the vessels, a linear interpolation of the subvolume around this point, a refined detection with subvoxel accuracy, the computation of the surface normal for shading. Experiments have been conducted to evaluate the influence of the threshold in different data volumes, for different objects and locations. It appears that its setting is not difficult to control by the user and provides good results within the window 70-120 Hounsfield units. In addition, the method performs a semi-automatic analysis of the computed image in such a way that the vascular branches can be identified, their locations stored and paths constructed with a possible backward exploration. Quantitative features, going from local lumen diameters, centerline positions, calcification volume if any, angles at bifurcations, are available during the navigation.

The virtual sequences that have been defined correspond to hundreds of computed images that would need to be displayed as video records. Only a few of them are depicted Figure 2 in order to show the high quality that can be reached. It must be emphasized that the size of the distal veins is very small (a diameter approximately of 3-4 voxels or even less) and that, even in such situations, the current tools at our disposal behave well. The resulting 3D trajectories of several paths, using a MIP display, are described Figure 3 for different viewpoints.

The feasibility of the search for candidate paths is the first point that is assessed. The image quality is high enough to show details as well as major information about the environment into which the navigation takes place.

However, the search for some structures is not always easy. For example, when the virtual sensor is located in the atrium, finding the coronary sinus can take some time. This is also the case when we are looking for a specific vein. From our experiments, several reasons can be invoked to explain that: (i) we must be fully familiar with the anatomy of the venous network; (ii) a learning curve is necessary before decoding the MSCT data which are providing a lot of insights in the heart but, conversely, lead to complex reading; (iii) the virtual images have also to be practiced and local views must be referred to global views simultaneously in order to facilitate the exploration.

The preliminary experiments conducted with our clinical partners point out that this pre-operative planning represents a relevant step for CRT. The time required to define the candidate paths may reach up to 10 minutes but with a learning phase it is foreseen to be significantly reduced. A quantitative validation is in progress to estimate the reduction gained in intra-oprative time and consequently in terms of irradiation benefits. Finding the candidate pathways, verifying if the stimulation site is accessible, estimating

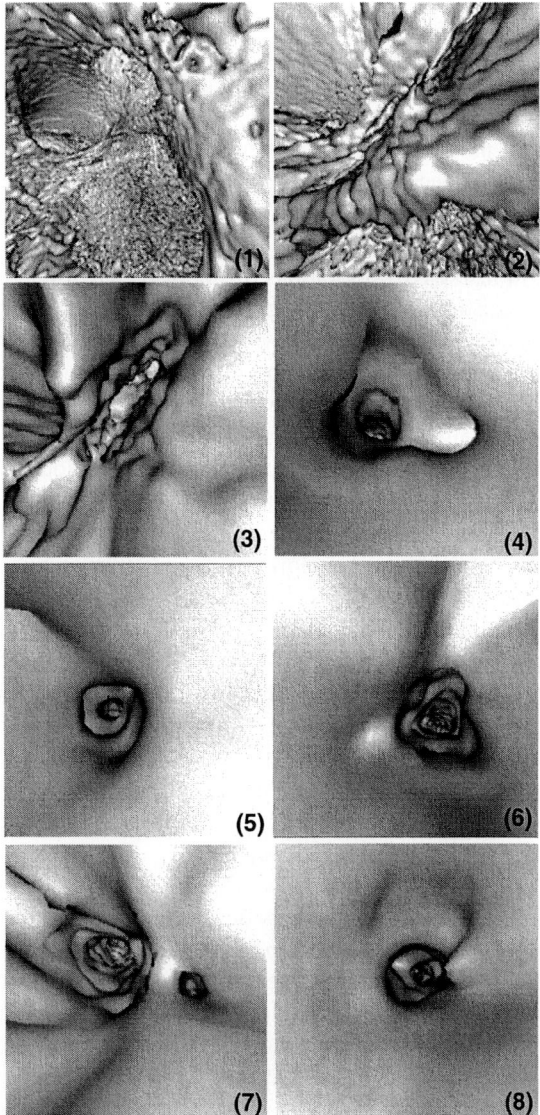

Fig. 2. Some virtual images computed during the navigation and showing the right atrium (1, 2), the postero-lateral and Marshall's veins (4, 5), the coronary sinus (3, 6), the great vein and the lateral veins (7, 8). The positions (1 – 8) are also reported in figure 3

feasibility of the implantation, learning the gestures to be carried out, selecting the proper instruments, etc. are all major components. Thus, they allow reducing risk for the patient, optimising and facilitating the implantation procedure and subsequently, reducing the time of medical personal and operating rooms required.

From the engineering standpoint point, the genericity of the solutions, early proposed for navigating into peripheral vessel networks and coronary arteries (where stenoses and calcifications may be present), has been demonstrated. The thresholding criteria involved in detection, interactively defined for a given data set, have been systematically studied and are not sensitive (i.e the interval of values is sufficiently large to provide enough robustness).

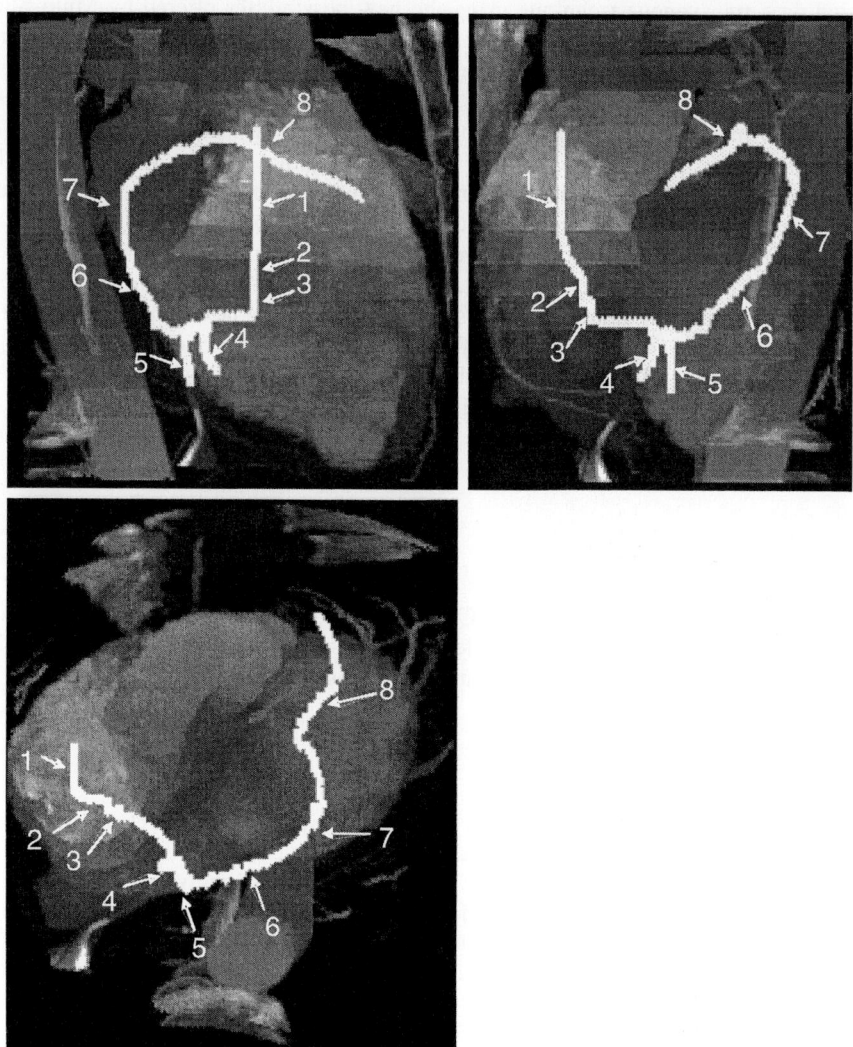

Fig. 3. Resulting 3D trajectories (in white) visualized using MIP from different points of view (lateral, apical and posterior). The ribs and other anatomical structures surrounding the heart have been suppressed using an appropriate ROI

5 Conclusion

The conjunction of electrical data, morphological and mechanical behaviour is very likely a source of additional progress. Insights into electromechanical coupling should improve the understanding of local, regional and global abnormalities and the localization of optimal stimulation or ablation sites. The present work was devoted to a subproblem in this overall frame: the pre-operative assistance of biventricular pacing based on a transvenous technique using the new possibilities offered by Multi-Slice Computed Tomography. These images are difficult to explore due to the many structures that are enhanced and to the complexity of the venous tree, close and strongly intermingled with the arterial network, and composed of thin, low contrast tube-like shapes. An interactive navigation has been proposed which leads to the definition of candidate venous paths toward the left ventricle. The joint measurements that can be carried out during the navigation (distance from reference entry points, diameters, angles) provide the means for a prior evaluation of path feasibility.

The 3D data sets examined so far were almost free from artefacts and blurring effects related to movement of the heart. However, these problems can become more critical when small structures like veins are concerned. The next generation of imaging device with an increase number of detectors and better motion-corrected reconstruction algorithms will minimize such problem in the near future. The next step is clearly the image guided intra-operative assistance of CRT. The poor contrast of fluoroscopic venous images makes this task very challenging for motion tracking and 2-D/3-D matching.

Acknowledgements

We would like to thank D Boulmier, C Leclercq, H Le Breton and all the physicians who have read and anatomically labelled the data sets. We are also indebted to B Le Bruno from Siemens France for her continuous assistance.

References

Abraham W.T, Fischer W.G, Smith A.L et al, Cardiac resynchronization in chronic heart failure, N.Engl.J.Med, 346, 1845-1853, 2002.
Alonzo C, Leclercq C, d'Allones F.R et al, Six years experience of transvenous left ventricular lead implantation for permanent biventricular pacing in patients with advanced heart failure: technical aspects, Heart, 86, 405-410, 2001.
Bardou.A, Auger P, Birkui P, Chasse J.L, Modeling of cardiac electrophysiological mechanisms: from action potential genesis to its propagation in myocardium, Crit.Rev.Biomed.Eng, 24, 141-221, 1996.
Bers.D, Cardiac excitation-contraction coupling, Nature, 415, 198-205, 2002.
Cazeau S, Leclercq C, Lavergne T et al, Effects of multisite biventricular pacing in patients with heart failure and intraventricular delay, N.Engl.J.Med, 344, 873-880, 2001.
Coatrieux J.L, Toward the living human: the challenge of multimodal and multiscale processing and modeling, Multimodal Bio-Medical Systems Workshop, IEEE/NLM/NSF, Bethesda, Oct 2004.

Coatrieux J.L, Roux C, Biomedical Imaging IV, IEEE EMBS Book Series, IEEE Press, NJ, 2002.
Daubert J.C, Pitter P, Le Breton H et al, Permanent left ventricular pacing with transvenous leads inserted into the coronary veins, Pace, 21, 239-245, 1998.
De Bakker J.M.T, Hauer R.N.W, Simmens T.A, Activation mapping: unipolar versus bipolar recording, In: Zipes D.P, Jalife J Eds, Cardiac Electrophysioloy: from cell to bedsite, 3rd Edition Philadelphia, Saunders, 1068-1078, 2000.
De Boer I H, Sachse F B, Mang S and Dössel O, Methods for Determination of Electrode Positions in Tomographic Images, International Journal of Bioelectromagnetism, 2, 2, 2000.
Garreau M, Simon A, Boulmier D, Guillaume H, Cardiac Motion Extraction in Multislice Computed Tomography by using a 3D Hierarchical Surface Matching Process, IEEE computers in Cardiology, Chicago, 2004.
Gepstein L, Hayan G, Ben-Haim S.A, A novel method for nonfluoroscopic catheter-based electroanatomical mapping of the heart: in vitro and in vivo accuracy results, Circulation, 95, 1611-1622, 1997.
Haigron P, Le Berre G, Coatrieux J.L, 3D Navigation in Medicine, Eng.Med.Biol.Mag, 15, 2, 70-78, 1996.
Haigron P, Bellemare ME, Acosta O, Goksu C, Kulik C, Rioual K, Lucas A. Depth-Map-Based Scene Analysis for Active Navigation in Virtual Angioscopy. IEEE Transactions on Medical Imaging , 23, 11, 1380-1390, 2004.
Hunter P.J, Nash M.P, Sands G.B, Computational electromechanics of the heart, in Computational Biology of the Heart, Wiley & Sons, 345-407, 1996.
Kass D.A, Predicting cardiac resynchronization response by GRS duration, J.Am.Coll.Cardiol, 42, 2125-2127, 2003.
Leclercq C, Kass D.A, Re-timing the failing heart: principles and current clinical status of cardiac resynchronization, J.Am.Coll.Cardiol, 39, 194-201, 2002.
Noble D, Winslow R.L, Reconstruction of the heart: network models of SA node-atrial interaction, in Computational Biology of the Heart, Wiley & Sons, 49-64, 1997.
Quan W, Evans S.J, Hastings H.M, Efficient integration of a realistic two-dimensional cardiac tissue model by domain decomposition, IEEE Trans.Biomed.Eng, 45, 3, 372-385, 1998.
Roux C, Coatrieux J.L, Contemporary perspectives on Three Dimensional Biomedical Imaging, IOS Press, Amsterdam, 1997.
Rudy Y, Insights from theoretical simulations in a fixed pathway, J.Cardiovasc.Electrophysiology, 6, 294-312, 1995.
Schilling R.J, Peters .S, Davies D.W, Simultaneous endocardial mapping in the left ventricle using a noncontact catheter: comparison of contact and reconstructed electrograms during sinus rhythm, Circulation, 98, 887-898, 1998.
Schleich JM, Dillenseger JL, Andru S, Coatrieux JL, Almange C, Understanding normal cardiac development using animated models, IEEE Computer Graphics and Applications, 22, 14-19, 2002.
Sermesant M, Rhode K, Anjorin A et al, Simulation of the Electromechanical Activity of the Heart Using XMR Interventional Imaging, Proceedings MICCAI 2004, LNCS 3217, Springer-Verlag, 786-794, 2004
Taylor R, Lavallée S, Burdea G, Mosges R, Computer-integrated surgery: technology and clinical applications, MIT Press, 1996.
Virag N, Blanc O, Kappenberger L, Computer simulation and experimental assessment of cardiac electrophysiology, Futura Publishing, 2001.

Methods for Identifying and Tracking Phase Singularities in Computational Models of Re-entrant Fibrillation

Ekaterina Zhuchkova[1] and Richard Clayton[2]

[1] Physics Faculty, Moscow State University, Leninskie Gory,
119992, Moscow, Russia
zhkatya@polly.phys.msu.ru
http://polly.phys.msu.ru/~zhkatya/
[2] Department of Computer Science, University of Sheffield, Regent Court,
211 Portobello Street, Sheffield, S1 4DP, UK
r.h.clayton@sheffield.ac.uk
http://www.dcs.shef.ac.uk/~richard

Abstract. The dangerous cardiac arrhythmias of tachycardia and fibrillation are most often sustained by re-entry. Re-entrant waves rotate around a phase singularity, and the identification and tracking of phase singularities allows the complex activity observed in both experimental and computational models of fibrillation to be quantified. In this paper we present preliminary results that compare two methods for identifying phase singularities in a computational model of fibrillation in 2 spatial dimensions. We find that number of phase singularities detected using each method depends on choosing appropriate parameters for each algorithm, but that if an appropriate choice is made there is little difference between the two methods.

1 Introduction

Cardiac cells are electrically excitable and a propagating sequence of electrical activation and recovery (the cardiac action potential) initiates the contraction and relaxation of cardiac tissue. During normal beats the cardiac pacemaker synchronizes the electrical and mechanical activity of the heart, but during an arrhythmia the electrical activity is self-sustaining. An excitation wave that propagates repeatedly along a closed path is termed re-entry, and re-entry is the mechanism that is believed to sustain many cases of the dangerous cardiac arrhythmias of ventricular tachycardia (VT), and ventricular fibrillation (VF). In a two-dimensional tissue sheet without any obstacles a single re-entrant wave adopts a spiral shape, and in three-dimensional tissue the wave adopts a scroll shape [1]. Mapping electrical activity on the surface of the ventricles during VF has revealed complex spatio-temporal activity [2], and direct evidence of re-entrant waves is seen only rarely [3]. This is attributed either to breakdown of re-entrant waves into multiple interacting wavelets or intermittent conduction of waves emanating from a mother rotor [4-6].

One of the ways to simplify the complex activity observed during experimental studies is to identify the tips of re-entrant waves on the heart surface, and this can be done by transforming the measured voltage distribution measured on the heart surface into phase [7, 8]. The tips of re-entrant spiral waves are surrounded by tissue in all phases of the activation-recovery cycle, and hence are phase singularities (PS). In an experimental study the phase singularities observed on the surface of ventricular tissue are intersections of filaments of PS around which re-entrant scroll waves rotate. Modeling studies show that filaments may be contained within the ventricular wall and intersect with the heart surface only briefly; this observation may explain the short lifetimes of PSs on the heart surface [9, 10].

Identification of PSs is a powerful technique for analyzing experimental data because the voltage signals recorded optically from the heart surface are often noisy, but transformation of voltage into phase removes the noise and enables accurate location of phase singularities [11]. PSs also provide a valuable link between experimental data and computational simulation [12]. However, different investigators have used different approaches to identify the location of PSs, and little is known about the differences between these methods [11-14]. In addition, although the tip trajectories of single spiral waves have been studied extensively in computational models with different parameter regimes [15], little is known about how PSs behave during the breakdown of a single re-entrant wave into fibrillation. The aim of this paper is to present preliminary results comparing two different approaches for locating PSs, and then to track phase singularities during the breakdown of a spiral wave into fibrillation.

2 Methods for Identifying Phase Singularities

In experimental studies the spatial distribution of phase has been used to identify PSs [7]. In computational studies more information is generally available, and PSs can be identified from the intersection of isolines of constant membrane voltage V_m and another variable associated with repolarisation such as a gating variable for the Ca^{2+} channel [16] or a line where $dV_m/dt = 0$ [13].

The spatial distribution of membrane voltage obtained from an experimental preparation or computational model can be mapped to a spatial distribution of phase by obtaining the current value of membrane voltage at each point $V_m(t)$ and a previous value of the membrane voltage $V_m(t\text{-}tau)$. If the delay tau is chosen to be a few ms, then a plot of $V_m(t\text{-}tau)$ against $V_m(t)$ for a spiral wave shows that the points follow a trajectory around a central reference point (Fig.1). If the coordinates $(V_{ref,x}, V_{ref,y})$ of the central reference point are obtained, then the phase at each point is the elevation above the V_m axis, and is given by [11]

$$phi(t) = \arctan\left[\frac{V_m(t-tau) - V_{ref,y}}{V_m(t) - V_{ref,x}}\right], \quad (1)$$

where **arctan** returns a value between –pi and +pi. Fig.2 shows the spiral wave simulation used to create the plot shown in Fig.1, together with the phase distribution obtained from Fig.1 using the mean value of V_m to give both $V_{ref,x}$ and $V_{ref,y}$.

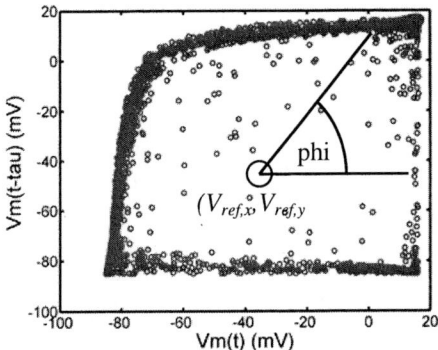

Fig. 1. Plot of $V_m(t-tau)$ vs $V_m(t)$ for a single spiral wave with a time delay *tau* of 5 ms. The phase angle *phi* is obtained for each point using the central reference point with co-ordinates $(V_{ref,x}, V_{ref,y})$ as shown

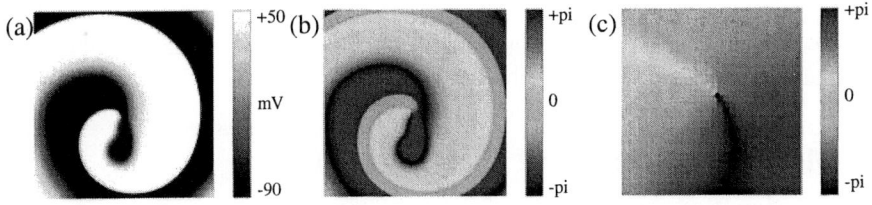

Fig. 2. Phase analysis of a single re-entrant spiral wave in a computational model (see text for details). (a) Distribution of membrane voltage. (b) Distribution of phase, obtained using a time delay of 5 ms. (c) Magnification of (b) showing the distribution of phase around the spiral wave tip and the phase singularity as a black point

As noted above, this approach is particularly useful for noisy experimental data, where the phase calculation acts as a filter to remove the noise. Different techniques can be used to identify phase singularities from the phase distribution, but all rely on identifying points that are surrounded by a complete cycle of phase from $-pi$ to $+pi$ [8, 11, 14]. The method used here is based on the concept of topological charge, n_t, and is described in detail elsewhere [8, 11, 14]. The mathematical definition of a PS is

$$n_t = \frac{1}{2\pi} \oint_C \nabla phi \cdot \vec{dl} \qquad (2)$$

where the line integral is taken over the path $\vec{l}$ on a closed curve C surrounding the singularity; n_t is an integer value with the sign depending on the chirality of phase surrounding PS. The integral (2) at location $[m,n]$ can be evaluated by the following convolution operation

$$\oint_C \nabla phi.\vec{dl} \propto \nabla_x \otimes k_y + \nabla_y \otimes k_x \qquad (3)$$

where $\otimes$ is a convolution operator, ∇_x and ∇_y are convolution kernels

$$\nabla_x = \begin{bmatrix} +1 & -1 \\ 0 & 0 \end{bmatrix} \quad \nabla_y = \begin{bmatrix} -1 & 0 \\ +1 & 0 \end{bmatrix}, \qquad (4)$$

and

$$k_x[m+1/2, n] = phi[m+1, n] - phi[m, n] \qquad (5)$$
$$k_y[m, n+1/2] = phi[m, n+1] - phi[m, n]$$

Equation (3) provides a way to obtain a two-dimensional array from the phase distribution, and phase singularities are defined as the nonzero elements in this array.

Using the second technique a PS can be determined as the intersection point of an isoline of constant membrane voltage $V_m = V_{iso}$ and a line where $dV_m/dt = 0$ [13]. So, at timesteps n and $n+1$ the membrane voltage at a point should satisfy

$$\begin{cases} V_m^n = V_{iso} \\ V_m^{n+1} = V_{iso} \end{cases}$$

The elements of the membrane voltage array that are in accord with these two simultaneous equations then designate phase singularities.

3 Cardiac Virtual Tissue

The cardiac virtual tissue used in this study was a simplified 2D model of a sheet of ventricular tissue, where action potential propagation was described using a monodomain formulation

$$\frac{\partial V_m}{\partial t} = D\left(\frac{\partial^2 V_m}{\partial x^2} + \frac{\partial^2 V_m}{\partial y^2}\right) - \frac{1}{C_m} I_{ion}, \qquad (6)$$

where D denotes a diffusion coefficient, C_m the specific membrane capacitance, and I_{ion} current flow through the cell membrane [17]. We used the three variable model described by Fenton and Karma to compute I_{ion}, with parameters set to reproduce the action potential duration (APD) restitution of the Beeler-Reuter model for canine ventricular cells [13]. In the model the APD restitution curve is steep, and so a single spiral wave is unstable, and breaks up into multiple wavelet re-entry [18].

4 Results

The performance of the first method with various reference points and time-delay (*tau*) is illustrated in Fig.3 and Fig.4 for a 2D tissue simulation where a single spiral wave has broken down into multiple wavelets 700 ms after initiation. The phase distribution in Fig.3 was obtained using (1). In Fig.4 white points label positive PSs (chirality = +1, clockwise rotation) and black negative PSs (chirality = -1, counter clockwise rotation).

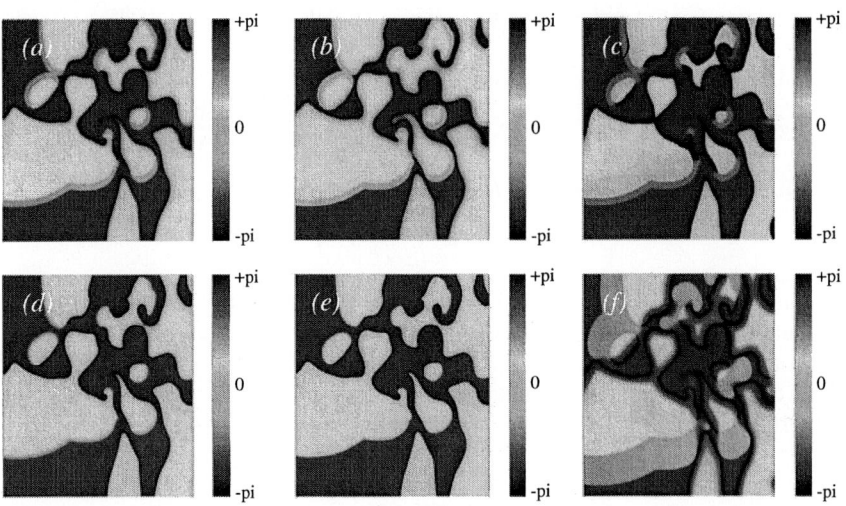

Fig. 3. Phase distribution at $t=700$ ms computed as (1). (a) $tau=5$ ms, $(V_{ref,x}, V_{ref,y})$ is equal to the mean value of $V_m(x,y,t)$ among all time records at the point (x,y). (b) $tau=5$ ms, $V_{ref} = -65$ mV. (c) $tau=5$ ms, $V_{ref} = 15$ mV. (d) $tau=2$ ms, V_{ref} =mean value of V_m. (e) $tau=1$ ms, V_{ref} =mean value of V_m. (f) $tau=25$ ms, V_{ref} =mean value of V_m

The phase distribution and PS location obtained using *tau* of 5 ms and three different $(V_{ref,x}, V_{ref,y})$ are shown in Fig.3(a)-(c) and in Fig.4(a)-(c) respectively. Although the overall phase distribution in each case is similar, there are two excess singularities in Fig.4(b) and three missing PSs in Fig.4(c) in comparison with Fig.4(a). Moreover, the coordinates of PSs obtained using 15mV as the reference point differ markedly from the coordinates of PSs in Fig.4(a) and 4(b).

Figures 3(d)-(f) and 4(d)-(f) demonstrate the change of phase distribution and PSs location with *tau*, which was equal to 2 (Fig.3(d), 4(d)), 1 (Fig.3(e), 4(e)), and 25 ms (Fig.3(f), 4(f)). Here the coordinates of the reference point were chosen as the mean value of $V_m(x,y,t)$. The PS location in Fig.4(a) and 4(d) are similar, but if *tau* is less then 2 ms (Fig.4(e)), there is an excess of singularities, similar to Fig.4(b). However, at $tau=25$ ms (Fig.4(f)) three PSs disappear similar to Fig.4(c).

The second technique for PS localization is illustrated in Fig.5, for the same 2D simulation. The figures represent contour plots of various V_{iso}: -10 mV (Fig.5(a)), -15 mV (Fig.5(b)), -20 mV (Fig.5(c)), -25 mV (Fig.5(d)), -30 mV (Fig.5(e)), -35 mV (Fig.5(f)). Blue snowflakes label PSs. PS location is strongly dependent on the choice of V_{iso}, and all parts of Fig.5 are different except Fig.5(e) and 5(f) which are similar and resemble Fig.4(b) and 4(e). Fig.5(c) is the exact copy of Fig.4(a) and 4(d). In some cases, very closely spaced PS pairs are identified and these are highlighted.

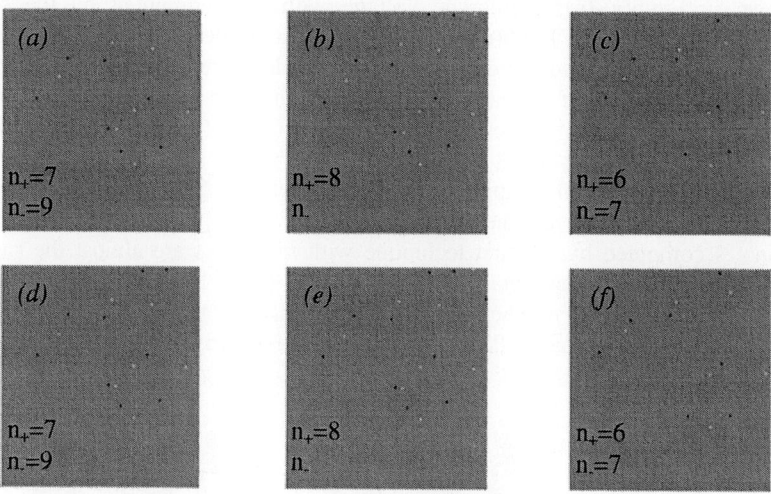

Fig. 4. Location of phase singularities corresponding to phase distribution in Fig.3. (a) *tau=5 ms*, $(V_{ref,x}, V_{ref,y})$ is equal to the mean value of $V_m(x,y,t)$ among all time records at the point *(x,y)*. (b) *tau=5 ms*, V_{ref} =-65 mV. (c) *tau=5 ms*, V_{ref} =15 mV. (d) *tau=2 ms*, V_{ref} =mean value of V_m. (e) *tau=1 ms*, V_{ref} =mean value of V_m. (f) *tau=25 ms*, V_{ref} =mean value of V_m

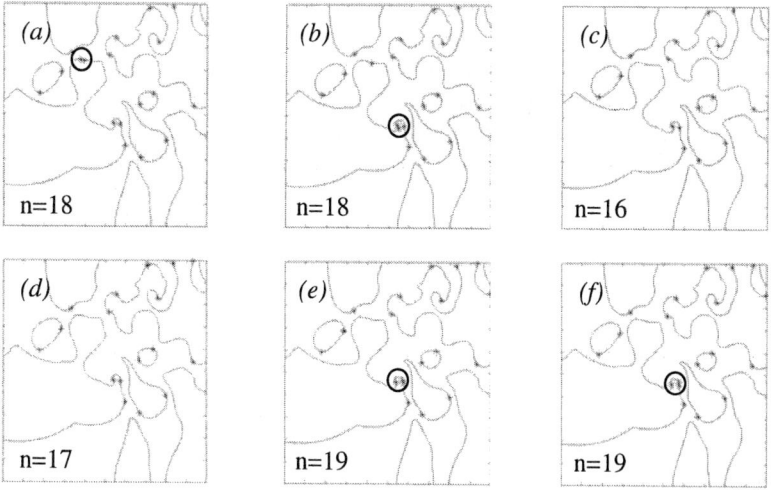

Fig. 5. Contour plots with PSs (blue snowflakes) computed identifying the intersection of two isolines at *t=700 ms*. (a) V_{iso}=-10 mV. (b) V_{iso}=-15 mV. (c) V_{iso}=-20 mV. (d) V_{iso}=-25 mV. (e) V_{iso}=-30 mV. (f) V_{iso}=-35 mV

A comparison of the two methods for PSs detection is presented in Fig.6. The distribution of membrane voltage with blue singularities obtained by the first method is shown in Fig.6(a)-(c). Membrane voltage with PSs computed by the second technique is displayed in Fig.6(d)-(f). At $t=50$ ms a single spiral wave and one PS exist, and the PS is located correctly by both methods. Here $tau=5$ ms (Fig.6(a)) and $V_{iso} =-20$ mV (Fig.6(d)). It is clear that these two figures are identical. The example of a more complicated case is presented in Fig.6(b), 6(c), 6(e), 6(f). Here $t=600$ ms, $tau=5$ ms (Fig.6(b)), $V_{iso} =-20$ mV (Fig.6(e)), $tau=2$ ms (Fig.6(c)), $V_{iso} =-30$ mV (Fig.6(f)). The PS location computed using a time-delay of 5 ms is similar to location of singularities obtained by the second method with $V_{iso} =-20$ mV, and the PS coordinates computed by the first technique with $tau=2$ ms are almost the same as coordinates of singularities obtained using $V_{iso} =-30$ mV. However, there are some differences between the number of PSs identified, and these differences can be attributed to the closely spaced PS pairs that are circled in Fig.6(b) and 6(e).

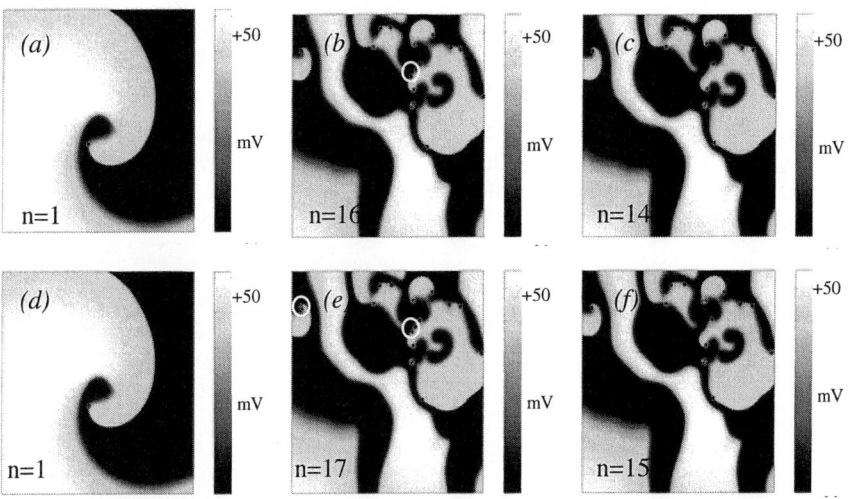

Fig. 6. Distribution of membrane voltage with singularities (blue snowflakes) obtained using both techniques for PSs detection. (a) $t=50$ ms, $tau=5$ ms, V_{ref} =mean value of V_m. (b) $t=600$ ms, $tau=5$ ms, V_{ref} =mean value of V_m. (c) $t=600$ ms, $tau=2$ ms, V_{ref} =mean value of V_m. (d) $t=50$ ms, $V_{iso}=-20$ mV. (e) $t=600$ ms, $V_{iso}=-20$ mV. (f) $t=600$ ms, $V_{iso}=-30$ mV

Trajectories of singularities computed by the first method with $tau=5$ ms and the mean value of membrane voltage as the reference point are displayed in Fig.7. Fig.7(a) shows PSs trajectories by 300 ms. By 111 ms a single positive (chirality=+1) PS (red trajectory) existed. At $t=111$ ms two additional PSs appeared: positive (yellow line) and negative (green) and a spiral wave began to break up. At 173 ms another two singularities were born: positive (magenta) and negative (black). At 175 ms the red and black PSs collided and disappeared. Continuing by a similar manner at 300 ms three singularities remained (a positive and two negative). However, at $t=1050$ ms the amount of positive PSs is a bit greater (Fig.7(b)). In Fig.7(b) red lines label trajectories of singularities of positive chirality and green – negative. In our simulations all PSs were created in pairs (87 pairs of positive-negative singularities) or due to interaction with a boundary (5 PSs). The PSs vanished, either due to positive-negative collision (57 pairs) or by collision with a boundary (54 PSs).

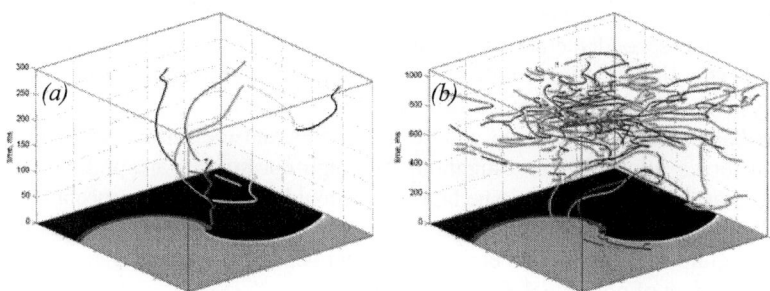

Fig. 7. PS trajectories computed using *tau=5 ms*, V_{ref} =mean value of V_m. (a) t ∈ [0; 300] ms. (b) t ∈ [0; 1050] ms. PSs trajectories of positive chirality are shown in red, trajectories of singularities of negative chirality are displayed in green

PS detection in 2D simulations of cardiac fibrillation. We have shown that both methods identify a broadly similar number of PSs, but the parameters used in each method can affect both the number of PSs that are detected as well as their location. PSs have already been identified in experimental data using the first method [7], and our study indicates that these results would be broadly comparable with the results of numerical studies where PSs are detected using the second technique.

The choice of *tau* and reference point is known to have a crucial influence on PS detection by the first method [8, 11, 14]. These values should be chosen such that the phase can be uniquely defined during the course of a spiral (scroll) wave rotation. The ideal reference point is one that is encircled by all trajectories independently of the originating spatial location. If one chooses a random point in the state space as the reference point, this point may lie within some trajectories and outside others. In this paper we compared the location of singularities obtained using the mean value of membrane voltage (that is supposed to be encircled by all trajectories) as the reference point (Fig. 4(a)), with PSs coordinates calculated using two example reference points.

As time-delay (*tau*) we chose four values, two of them (2ms and 25 ms) are often used in cardiac literature [8, 11, 14]. It was shown in [8, 11, 14] that for cardiac activation, if *tau* is on the order of the action-potential upstroke duration, the amount of trajectory folding (hence, nonunique calculation of phase) will reduce. Our results have demonstrated that assuming time-delay between 2 and 5 ms has a little effect (Fig.6(b), 6(c)) or no effect (Fig.4(a), 4(d)) on the PSs location. However, *tau* equal to 1 or 25 ms leads to additional or missing singularities relative to those obtained using *tau* of 5 ms and noticeable shift of PSs coordinates for time-delay of 25 ms.

Although an earlier study suggested that the choice of V_{iso} influences the location of the PS only slightly [19], our results have shown that this choice can be important (fig.5). The main effect is to identify additional closely spaced PS pairs.

The examples of $t=50$ ms, $t=600$ ms and $t=700$ ms (Fig.4-6) have demonstrated that using a time-delay of 5 ms in the first method for PSs detection produces a similar PS distribution to using $V_{iso} = -20$ mV in the second technique. In addition to the data shown in Fig.6, we also deployed each method at other times during the simulation. We found that if the number of PSs was insensitive to the choice of tau between 1 and 5 ms in the first method, then the choice of V_{iso} between -20 and -40 mV in the second method would also have little effect on the number of PSs detected. This finding suggests that the robustness of each method is variable, but a more systematic comparison of the two methods is needed.

The balance between created and destroyed PSs and the influence of boundaries are also topics for further investigation. Although our studies are preliminary, they indicate that the role of heart tissue boundaries could be important, especially for PSs death (54 singularities among 168 were destroyed due to collision with boundaries in spite of a quite large 12.5×12.5 cm domain), and hence these investigations could aid the development of possible defibrillation strategies.

6 Conclusions

In this paper we have compared two different methods for PSs detection, and using one of the techniques we have tracked singularities during the breakdown of a spiral wave into fibrillation (during approximately 1 s). Neither method is ideal since both require some parameter values to be chosen. However, there is clear advantage of using the first technique, which is based on topological charge. It immediately gives us knowledge about the sense of spiral wave rotation (chirality), and this is an additional feature that is extremely useful for PS tracking.

Acknowledgements

Ekaterina Zhuchkova is funded by INTAS fellowship 03-55-1920.

References

1. RA Gray, J Jalife: **Ventricular fibrillation and atrial fibrillation are two different beasts.** *Chaos* 1998, **8**:65-78.
2. J Jalife: **Ventricular fibrillation: Mechanisms of initiation and maintenance.** *Annual Review of Physiology* 2000, **62**:25-50.
3. JM Rogers, J Huang, WM Smith, RE Ideker: **Incidence, evolution, and spatial distribution of functional reentry during ventricular fibrillation in pigs.** *Circulation Research* 1999, **84**:945-954.
4. F Xie, ZL Qu, J Yang, A Baher, JN Weiss, A Garfinkel: **A simulation study of the effects of cardiac anatomy in ventricular fibrillation.** *Journal of Clinical Investigation* 2004, **113**:686-693.
5. PS Chen, TJ Wu, CT Ting, HS Karagueuzian, A Garfinkel, SF Lin, JN Weiss: **A tale of two fibrillations.** *Circulation* 2003, **108**:2298-2203.

6. AV Zaitsev, O Berenfeld, SF Mironov, J Jalife, AM Pertsov: **Distribution of excitation frequencies on the epicardial and endocardial surfaces of fibrillating ventricular wall of the sheep heart**. *Circulation Research* 2000, **86**:408-417.
7. RA Gray, AM Pertsov, J Jalife: **Spatial and temporal organization during cardiac fibrillation**. *Nature* 1998, **392**:75-78.
8. C Larson, L Dragnev, N Trayanova: **Analysis of electrically induced re-entrant circuits in a sheet of myocardium**. *Annals of Biomedical Engineering* 2003, **31**:768-780.
9. VN Biktashev, AV Holden, SF Mironov, AM Pertsov, AV Zaitsev: **Three-dimensional aspects of re-entry in experimental and numerical models of ventricular fibrillation**. *International Journal of Bifurcation and Chaos* 1999, **9**:695-704.
10. RH Clayton, AV Holden: **Filament behaviour in a computational model of ventricular fibrillation in the canine heart**. *IEEE Transactions on Biomedical Engineering* 2004, **51**:28-34.
11. AN Iyer, RA Gray: **An experimentalist's approach to accurate localization of phase singularities during re-entry**. *Annals of Biomedical Engineering* 2001, **29**:47-59.
12. M-A Bray, S-F Lin, RR Aliev, BJ Roth, JP Wikswo: **Experimental and theoretical analysis of phase singularity dynamics in cardiac tissue**. *Journal of Cardiovascular Electrophysiology* 2001, **12**:716-722.
13. F Fenton, A Karma: **Vortex dynamics in three-dimensional continuous myocardium with fibre rotation: Filament instability and fibrillation**. *Chaos* 1998, **8**:20-47.
14. M-A Bray, JP Wikswo: **Use of topological charge to determine filament location and dynamics in a numerical model of scroll wave activity**. *IEEE Transactions on Biomedical Engineering* 2002, **49**:1086-1093.
15. AT Winfree: **Varieties of spiral wave behaviour in excitable media**. *Chaos* 1991, **1**:303-334.
16. VN Biktashev, AV Holden: **Re-entrant waves and their elimination in a model of mammalian ventricular tissue**. *Chaos* 1998, **8**:48-56.
17. RH Clayton: **Computational models of normal and abnormal action potential propagation in cardiac tissue: Linking experimental and clinical cardiology**. *Physiological Measurement* 2001, **22**:R15-R34.
18. FH Fenton, EM Cherry, HM Hastings, SJ Evans: **Multiple mechanisms of spiral wave breakup in a model of cardiac electrical activity**. *Chaos* 2002, **12**:852-892.
19. RH Clayton, AV Holden: **A method to quantify the dynamics and complexity of re-entry in computational models of ventricular fibrillation**. *Physics in Medicine and Biology* 2002, **47**:225-238.

Estimating Local Apparent Conductivity with a 2-D Electrophysiological Model of the Heart

Valérie Moreau-Villéger[1], Hervé Delingette[1], Maxime Sermesant[2],
Hiroshi Ashikaga[3], Owen Faris[3], Elliot McVeigh[3], and Nicholas Ayache[1]

[1] EPIDAURE research project, INRIA Sophia Antipolis, France
[2] Computation Imaging Science Group, King's College, London, England
[3] Laboratory of Cardiac Energetics, National Heart Lung and Blood Institute,
National Institute of Health, Bethesda, Maryland, USA

Abstract. In this article we study the problem of estimating the parameters of a 2-D electrophysiological model of the heart from a set of temporal recordings of extracellular potentials. The chosen model is the reaction-diffusion model on the action potential proposed by Aliev and Panfilov. The strategy consists in building an error criterion based upon a comparison of depolarization times between the model and the measures. This error criterion is minimized in two steps : first a global and then a local adjustment of the model parameters. The feasibility of the approach is demonstrated on real measures on canine hearts, showing also the necessity to introduce anisotropy and probably a third spatial dimension in the model.

1 Introduction

Direct models of the electrical activity of the heart are numerous ([12, 3, 8]). Since *in vivo* measures are available([10, 4, 14]), a new challenge is to solve the inverse problem, that is to find the parameters of a model that best fit the measures obtained from a specific patient. Fitting a model on real measures is necessary for building a patient specific model suitable for diagnosis of electrical pathologies as well as for intervention planning.

When inspecting electrophysiological data, cardiologists often base their analysis on the depolarization and repolarization maps of the epicardium or endocardium ([14]). From those maps, expert eyes can detect different electrophysiological pathologies ranging from the presence of low conduction zones caused by infarcted tissue, to the occurrence of fibrillation caused by scrolling waves.

The aim of the research effort presented in this paper is to provide cardiologists with additional information for a better diagnosis and a better planning of therapies by finding the parameters of a cardiac electrophysiology model that can best explain electrophysiological observations (isochrones).

By inverting such a model, we can expect two important outcomes. First, we aim at estimating "hidden" physical parameters which help to better understand and quantify the heart physiology (conductivity for instance) from an original set of physical measurements (depolarization times). Second, with this set of parameters, we can use the direct model to study pathologies, to plan and even simulate some therapeutic protocols.

In vivo electric measures on the endocardium or epicardium ([10, 4]) consist in measuring of the extracellular potential, from which the depolarization times are computed. Very accurate models such as bidomain models ([6]) or Luo-Rudy models ([9]) provide excellent insight into the physiological phenomena provoking the electrical activity of the heart but are probably too sophisticated for our inverse problem. Indeed, these models are designed to capture very subtle modifications in the shape of the action potential whereas we only measure here the depolarization times. For this type of measures, a phenomenological model describing the action potential propagation is probably sufficient, such as the FitzHugh-Nagumo [5] model. Aliev and Panfilov developed a modified version suited to the cardiac action potential [1]:

$$\begin{aligned} \varepsilon^2 \partial_t u &= \varepsilon \mathrm{div}\,(D\nabla(u)) + ku(1-u)(u-a) - uz \quad (1.a) \\ \partial_t z &= -(ku(u-a-1)+z)) \quad (1.b) \end{aligned} \quad (1)$$

where u is a normalized action potential (between 0 and 1), z is a dynamic variable modeling the repolarization, k controls the repolarization, ε controls the coupling between the action potential and the repolarization variable z, and a controls the reaction phenomenon. The depolarization time of a point is computed as the first time such that $u(t) = 0.5$. A 3D anisotropic model based on the Aliev-Panfilov system was developed in the context of the ICEMA collaborative research action [2, 15].

The electrophysiological measures are usually available on the endocardium or the epicardium, so as a first methodical and essential stage before going on to the 3D problem, we treat a simplified and tractable problem by considering a surface model. In this manner, we simulate the Aliev and Panfilov model on a surface triangulation $\mathcal{S}$ with N vertices and L triangles. We name $\mathcal{V}$ the set of vertices and $\mathcal{T}$ the set of the triangles. Hence, the tridimensional propagation is simplified to a propagation on the 2D surface of the epicardium. Furthermore, the fiber directions are not relevant in the 2D model since they are not tangential to the epicardial surface, and we consider an isotropic propagation i.e. $D = d\,\mathrm{diag}(1,1,1)$ in system (1), where the diffusion coefficient d is proportional to a conductivity. System (1) is normalized, the model is only 2D and the 3 parameters a, k and d all influence the depolarization times. Hence it is not possible to estimate an electrical conductivity from the depolarization times and we will call d the *apparent conductivity* in the sequal. The temporal integration of the system (1) is done with an explicit Euler scheme. The spatial integration is performed with the finite elements method with linear triangular elements. The numerical issues and the implementation are described in [11].

In this article we present results on the inversion of the Aliev-Panfilov electrophysiological model leading to a regional estimation of apparent conductivities. In Section 2, we first achieve a coarse global estimation of the parameter k that properly scales the electrical propagation. In Section 3, we perform the regional estimation of the *apparent conductivity* by minimizing an error function between the measured and simulated depolarization times. In Section 4, a case study on dog hearts shows the efficiency of the presented approach for inverting the Aliev-Panfilov electrophysiological model. Finally in Section 5, we sum up this work and present its perspectives.

2 Global Estimation of the Parameter k

The parameter ε is chosen according to the grid size, and the parameters of the model a, k, or d can vary between different individuals or species. We choose to estimate the parameter k from the depolarization times while standard values are assigned to the other parameters.

As stated in [7], the velocity of the depolarization wave on a 1D domain can be expressed as follows

$$c = \sqrt{2kd}(0.5 - a) \qquad (2)$$

In 2D, this velocity is not constant in space. At each point in the mesh, it is equal to the velocity in 1D (Equation (2)) minus a term proportional to the curvature of the front [7]. Since we only need a global estimate of the propagation velocity on a surface, we neglect, as a first approximation, the front curvature and simply approximate the velocity c of the depolarization wave by its expression in Equation (2).

Luckily, the depolarization velocity can also be computed from the gradient of the measured depolarization times on the surface, $\boldsymbol{\nabla}_x t : 1/c = \|\boldsymbol{\nabla}_\mathbf{x} t\|$. Then, we can estimate a median value of the parameter k over the whole mesh: median $(\|\boldsymbol{\nabla}_\mathbf{x} t\|)^{-1} = \sqrt{2kd}(0.5 - a)$.

A direct inversion of this equation would be a comparison between a theoretical 1D velocity and an apparent velocity computed on a 2D surface. As a consequence, we use a velocity estimated from a first guess simulation, that we computed on the same mesh as the one used for the measures. As the velocity c is proportional to $1/\sqrt{k}$, a ratio between measured c^m and simulated c^s propagation velocity can be computed as follows.

$$\frac{\text{median} \|\boldsymbol{\nabla}_\mathbf{x} t^m\|}{\text{median} \|\boldsymbol{\nabla}_\mathbf{x} t^s\|} = \frac{c^s}{c^m} \approx \frac{\sqrt{k^s}}{\sqrt{k^m}}. \qquad (3)$$

The measured and the simulated depolarization times are denoted by t^m and t^s respectively. k^s is the value for the parameter k used to compute the first guess simulation and k^m is the value computed to adjust the measures. k^m can be computed as follows.

$$k^m = k^s \left(\frac{\text{median} \|\boldsymbol{\nabla}_\mathbf{x} t^s\|}{\text{median} \|\boldsymbol{\nabla}_\mathbf{x} t^m\|} \right)^2 \qquad (4)$$

3 Local Estimation of the Electrical *Apparent Conductivity*

With a simulation globally fitting the measures, a local adjustment of the model is possible. We choose the *apparent conductivity* d as the spatially varying parameter. Indeed, we can give a clinical interpretation of its variation: a region with a low *apparent conductivity* (AC) value is a region where the electrical wave does not propagate as fast as in the other regions and consequently may be pathological. The AC that we estimate cannot be compared to the electrical conductivity because we used normalized Aliev Panfilov equations. Moreover, we only estimate one parameter of the equation whereas the depolarization times also depend on a and k. Consequently, we detect variations of parameter d which are influenced by the other parameters.

Estimating the AC from patient specific data can be addressed as a data assimilation problem. None of the classical methods of data assimilation, like Kalman filtering and variational methods are truly suited for the model and the measures of our problem. Indeed, classical methods generally require an explicit functional relationship between the results of the model and the measures. Such a relationship is not available between action potentials and depolarization times since the depolarization time is an implicit function of the action potential.

In the discretized model ([11]), an AC value is assigned to each triangle. Consequently, we look for an AC map $(\mathbf{d}) = (d_j)_{0 \leq j \leq L-1}$, where L is the number of triangles in the triangulation. This AC map should minimize $C(\mathbf{d}) = \sum_{v \in \mathcal{V}} (t_v^m - t_v(d_0, \ldots, d_{L-1}))^2$ where $\mathcal{V}$ is the set of the vertices in the triangulation, t_v^m is the measured depolarization time at vertex v and $t_v(d_0, \ldots, d_{L-1})$ the depolarization time at vertex v resulting from a simulation with the conductivities $(d_0, \ldots, d_{L-1})$.

In order to have a robust estimation of the AC, we split the heart surface into different connected regions and estimate one AC value for each region. Let $(R_k)_{0 \leq k \leq K-1}$ be a partition of the surface in K regions. For each region R_k, $d_j = d_{R_k}$ for all j such that the jth triangle of the surface belongs to R_k. Then, the new minimization problem is to find $(\mathbf{d}) = (d_{R_k})_{0 \leq k \leq K-1}$ that minimizes $C(\mathbf{d}) = \sum_{v \in \mathcal{V}} (t_v^m - t_v(d_{R_0}, \ldots, d_{R_{K-1}}))^2$

We look for the minimum of $C(\mathbf{d})$ with respect to K variables: $d_{R_0}, \ldots d_{R_{K-1}}$. Instead of using a generic method to solve for this multidimensional minimization, we consider the causality of the electrical wave propagation: the depolarization times in one region mostly depend on the apparent conductivities of the regions that were depolarized before. Hence, we estimate the AC for one region after the other, following the order of depolarization. During the estimation of d_R, the conductivities of the other regions remain constant.

We transform a K-dimensional minimization problem to K successive one-dimensional minimization problems:

$$C(d_R) = \sum_{v \in \mathcal{V}} (t_v^m - t_v(d_R))^2 \qquad (5)$$

We simplify the criterion $C(\mathbf{d})$ by taking into account only the vertices of the region R because there are enough vertices in a region to provide a robust estimate. Equation (5) then yields $C(\mathbf{d}) = \sum_{v \in R}(t_v^m - t_v(d_R))^2$

The values of the function $\mathbf{t}(d_R)$ can only be computed after simulating the propagation. Therefore the derivative is computationally expensive to estimate. We favoured a minimization method that does not involve any derivative, an iterative inverse parabolic interpolation derived from the Brent method [13]. This very consistent method replaces the function to be minimized by a well-chosen parabola. The minimum of the function C is approximated by the easily and efficiently computed minimum of the parabola. Given three points on the curve $(d_a, C(d_a))$, $(d_b, C(d_b))$ and $(d_c, C(d_c))$, there is a unique parabola $f(x) = \alpha x^2 + \beta x + \gamma$ described by these points. It reaches its extremum at point x such that

$$x = d_b - \frac{1}{2}\frac{(d_b - d_a)^2(C(d_b) - C(d_c)) - (d_b - d_c)^2(C(d_b) - C(d_a))}{(d_b - d_a)(C(d_b) - C(d_c)) - (d_b - d_c)(C(d_b) - C(d_a))}. \quad (6)$$

From these remarks, we construct an iterative process which is a simplified version of Brent's method [13], to find the minimum from an initial bracketing of this minimum. We call a bracketing of the minimum of function C three points d_a, d_b and d_c such that $d_a < d_b < d_c$, $C(d_b) < C(d_a)$ and $C(d_b) < C(d_c)$. We repeat the parabolic estimation until we are satisfied with the computed value: if (d^k) is the sequence of successively estimated minima, we consider that convergence is reached when the difference between two successive estimations is smaller than a given precision value p i.e. $|d^{k+1} - d^k| < p$.

4 Results on *in vivo* Measures

The *in vivo* measures used in this section were acquired on adult male mongrel dogs using a multi-electrode epicardial sock during an artificial pacing on the

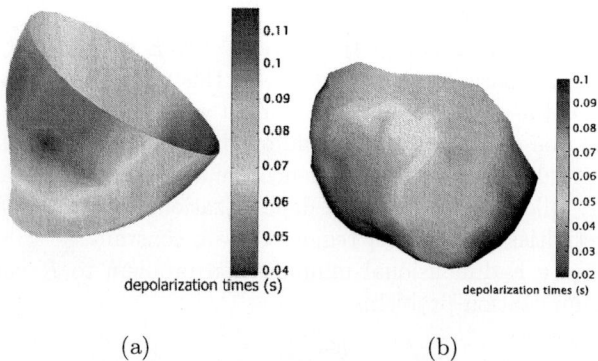

Fig. 1. Measured depolarization times. (a) Normal heart. (b) Case of an infarct on the anterior wall

right ventricle. The surgery, experimental layout and the data acquisition are described in [11, 15]. In this paper, we present two cases. The first case which is a normal heart, will be used to describe the procedure (Figure 1.a). The depolarization times were computed from a recording of electrical potentials on 128 electrodes and interpolated on a 192 vertices surface mesh. The second case is that of a heart with an anterior wall infarct (Figure 1.b). The depolarization times were computed from a recording of electrical potentials on 247 electrodes.

The first step toward a parameter estimation is a good initialization since the propagation is very sensitive to the localization of the pacing regions. We thus selected from the measures (Figure 1) the points with the smallest depolarization times to initialize the propagation.

4.1 Global Estimation of the Parameter k

Applying the method presented in Section 2 to the data of the normal heart, we obtained a global value of $k^m = 25.2$ starting from a crude initialization $k^s = 8$.

The absolute error between the simulated depolarization and the measured depolarization times before the automatic estimation of k is presented on Figure 2.a. After this estimation, the error is significantly lower as shown on Figure 2.b. Before the estimation, the mean error was 20.6 ms. After the automatic estimation, the mean error was 10 ms compared to the total duration of the depolarization wave which lasts around 120 ms.

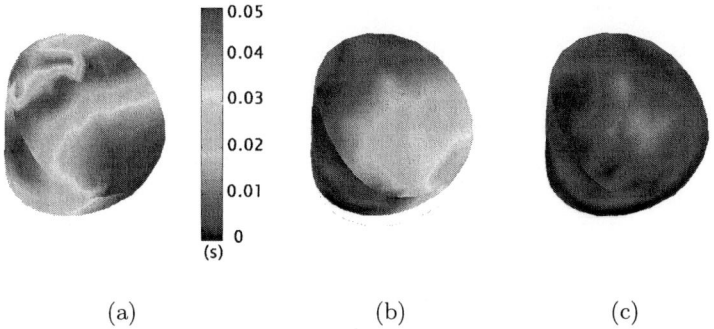

(a) (b) (c)

Fig. 2. Absolute error on the depolarization times between measures and simulations before (a) and after (b) the global automatic estimation and after the local estimation (c)

4.2 Local Estimation of the *Apparent Conductivity*

We now apply the presented method to perform the local estimation of the *apparent conductivity* (AC). We first need to partition the epicardium into different regions. We create a partition of the epicardium according to the electrical propagation. In this way, this partition is adapted to the particular artificial pacing of

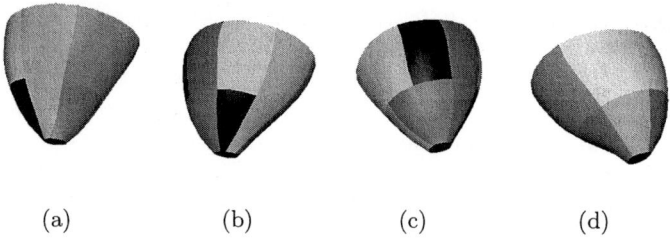

(a) (b) (c) (d)

Fig. 3. The regions chosen on the epicardium, according to the propagation of the depolarization wave. The large red region contains the pacing site

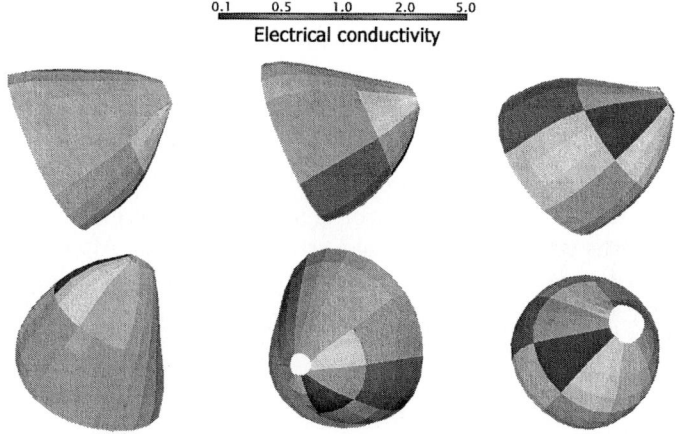

Fig. 4. *Apparent conductivity* map estimated from the first set of data

this experiment. In practice, we split the epicardium in successive regions following the isochrones of the depolarization times map as closely as allowed by the mesh resolution, and we then split these regions orthogonally to the isochrones. Figure 3 show a partition in 14 regions. We sort out the regions of Figure 3 in the order of their depolarization.

We then estimate one AC value for each region successively. The convergence on each region is quick and stable. Figure 4 presents the AC map that we obtain for the case of the normal heart.

4.3 Discussion

Although the variations of the computed AC for the normal heart do not have a physiological meaning, they closely reflect the asymmetry of the measures. These variations are probably due to the modeling of the epicardium as an homogeneous medium, without distinguishing the left and right ventricles nor taking into account the fibers direction.

Figure 5 displays the depolarization times simulated by the model before (5.a) and after (5.b) the local estimation of the AC, and compare them to the

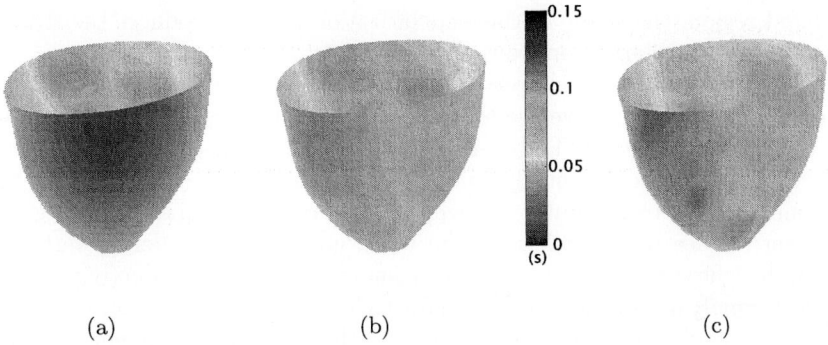

Fig. 5. Depolarization times before (a) and after (b) the local estimation compared with the measures (c). The absolute error on the depolarization times after the local estimation of the parameters is displayed Figure 2.c

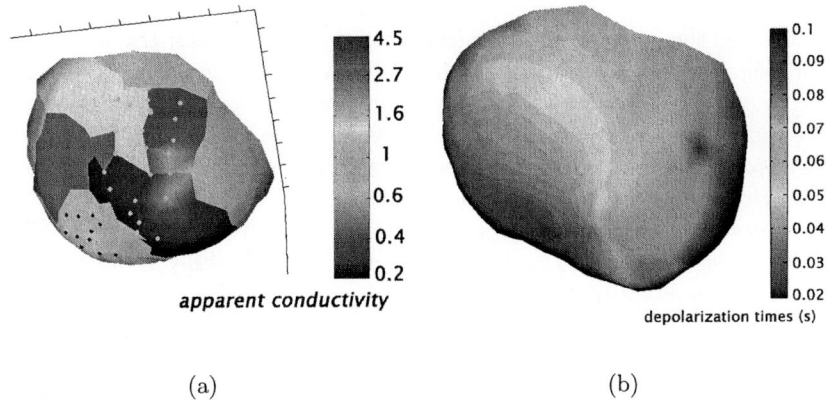

Fig. 6. AC estimated for the case of the anterior infarct (a). The points marked with a bright circle design the localization of the infarct. The points marked with a dark star design the pacing region. Depolarization times computed with these AC values (b)

measures (Figure 5.c). The depolarization times computed with a constant AC are in the proper range of values, but from Figure 5, when comparing these results with the measures (5.c), we notice that the shape of the depolarization front is much closer to the measures with the local adjustment.

The quality of this estimation is also assessed by the visualization of the absolute error (Figure 2.c) on the depolarization times in the epicardial surface. We can see on Figure 2 that the absolute error decreased significantly after both the global and the local estimation.

We also applied the AC estimation method on the case of an infarct on the anterior wall. The AC values are displayed on Figure 6.a, the purple circles correpond to the infarcted region. The depolarization times computed from a simulation taking into account these values are displayed on Figure 6.b. In the infarcted region, the shape of this depolarization front reproduces the shape of the measured depolarization front (Figure 1.b).

A large portion of the infarct is detected in the two regions with the lowest conductivity values, but we see that a part of the infarct is not detected as a low conductivity region. The heterogeneous infarct geometry in the heart wall can explain this observation: the infarct can be transmural (i.e. extending from the inner surface to the outer surface) or non-transmural (i.e. extending from the inner surface to somewhere in the wall), and when considering vertices in the mesh, where the infarct is non-transmural, electrical conductivity can be almost normal. In addition, a low conductivity is estimated in normal regions. As seen in the first case, this may be due to the modeling of the epicardium as an homogeneous medium. We are currently working on the inclusion of the fiber directions in this model.

5 Conclusions and Perspectives

We addressed the problem of estimating a set of parameters for the action potential propagation modeled by Aliev and Panfilov from measured depolarization times. In order to evaluate the quality of our results, we used a criterion based on the difference in depolarization times between the model and the measures. We first presented a procedure to globally estimate a set of parameters so that the electrical propagation in the model occurs in the same time and space scale as the measures. We then presented a method to locally estimate the electrical *apparent conductivity* (AC) region by region. We successfully estimated global and local parameters of the model from *in vivo* measures of a canine heart. The simulation of the model with these new values showed that the error on the depolarization times was significantly decreased. Moreover, the variations of the AC values that we computed are consistent with the measures. When this method was applied to an infarcted heart, a large part of the infarcted region was assigned a low AC value.

In order to have a fully automatic process, we still need to build automatically the epicardium's partition. The next step will be to estimate the parameters of a 3D model of the heart by establishing a correspondence between 2D measures and a 3D mesh. A proper physiological validation would require the application of our method to a benchmark of pathological and normal measures analyzed by experts. At that time, only the AC is estimated, but other measures, as the action potential duration, would allow us to estimate more parameters. An advantage of the proposed local estimation is that it is not dependent on the model since it only uses simulations of the direct model. Thus, it can easily be adapted to more complex models that can reproduce specific pathologies.

Acknowledgments

This work was partially funded by the scientific direction of INRIA through the Cooperative Research Action ICEMA2[1]. This action involved several partners including the INRIA research groups CAIMAN, EPIDAURE, MACS and SOSSO and was coordinated by Frédérique Clément (SOSSO). Finally, we would like to thank Tristan Picart for his proof-reading and his technical help.

References

1. R.R. Aliev and A.V. Panfilov. A simple two-variables model of cardiac exictation. *Chaos, Soliton and Fractals*, 7(3):293–301, 1996.
2. N. Ayache, D. Chapelle, F. Clément, Y. Coudière, H. Delingette, J.A. Désidéri, M. Sermesant, M. Sorine, and J. Urquiza. Towards model-based estimation of the cardiac electro-mechanical activity from ECG signals and ultrasound images. In *FIMH'01*, number 2230 in LNCS, pages 120–127. Springer, 2001.
3. M.E. Belik and A.D. McCulloch. Computational methods for cardiac electrophysiology. In N. Ayache, editor, *Computational Models for the Human Body*, Handbook of Numerical Analysis. Elsevier, 2004.
4. O. Faris, F. Evans, D. Ennis, P. Helm, J. Taylor, A. Chesnik, M.A. Guttman, C. Ozturk, and E. McVeigh. A novel technique for cardiac electromechanical mapping with MRI tagging and an epicardial electrode sock. *Ann. of Biomed. Engin.*, 31(4):430–440, 2003.
5. R.A. FitzHugh. Impulses and physiological states in theoretical models of nerve membrane. *Biophysical Journal*, 1:445–466, 1961.
6. D.B. Geselowitz and W.T. Miller. A bidomain model for anisotropic cardiac muscle. *Ann. biomed. eng.*, 3-4:191–206, 1983.
7. J.P. Keener. A geometrical theory for spiral waves in excitable media. *SIAM J. App. Math.*, 46(6):1039–1056, 1986.
8. V. Krinski, A. Pumir, and I. Efimov. Cardiac muscle models. In A. Scott, editor, *Encyclopedia of nonlinear science*. Routledge, 2004.
9. C.H. Luo and Y. Rudy. A dynamic model of the cardiac ventricular action potential - simulations of ionic currents and concentration changes. *Circ. Res.*, 74(6):1071–1097, 1994.
10. A. McCulloch, D. Sung, M.E. Thomas, and A. Michailova. Experimental and computational modeling of cardiac elctromechanical coupling. In *FIMH'01*, number 2230 in LNCS, pages 113–119. Springer, 2001.
11. V. Moreau-Villéger, H. Delingette, M. Sermesant, O. Faris, E. McVeigh, and N. Ayache. Global and local parameter estimation of a model of the electrical activity of the heart. Research Report 5269, INRIA, 2004.
12. D. Noble and Y. Rudy. Models of cardiac ventricular action potentials: iterative interaction between experiment and simulation. *Phil. Trans. R. Soc. Lond. A*, pages 1127–1142, 2001.
13. W.H. Press, B.P. Flannery, S.A. Teukolsky, and W.T. Vetterling. *Numerical Recipes in C*, chapter Minimization or Maximization of Functions, pages 290–352. Cambridge University Press, 1991.

---

[1] http://www-rocq.inria.fr/sosso/icema2/icema2.html

14. Y. Rudy. *Heart Physiology and Pathophysiology*, chapter Electrocardiogram and Cardiac Excitation, pages 133–148. Academic Press, 2001.
15. M. Sermesant, O. Faris, F. Evans, E. McVeigh, Y. Coudière, H. Delingette, and N. Ayache. Preliminary validation using in vivo measures of a macroscopic electrical model of the heart. In *International Symposium on Surgery Simulation and Soft Tissue Modeling (IS4TM'03)*, volume 2673, pages 230–243. INRIA Sophia Antipolis, Springer-Verlag, 2003.

Monodomain Simulations of Excitation and Recovery in Cardiac Blocks with Intramural Heterogeneity

Piero Colli Franzone[1], Luca F. Pavarino[2], and Bruno Taccardi[3]

[1] Dipartimento di Matematica, Universitá di Pavia and IMATI,
Istituto di Matematica Applicata e Tecnologie Informatiche,
Via Ferrata 1, 27100 Pavia, Italy
colli@imati.cnr.it
[2] Dipartimento di Matematica, Universitá di Milano,
Via Saldini 50, 20133 Milano, Italy
pavarino@mat.unimi.it
[3] Cardiovascular Research and Training Institute,
University of Utah, Salt Lake City, Utah
taccardi@cvrti.utah.edu

Abstract. Large scale simulations of an anisotropic and heterogeneous cardiac model in three dimensional myocardial blocks are presented. The Monodomain tissue representation used includes orthotropic anisotropy, intramural fiber rotation and homogeneous or heterogeneous intramural Luo-Rudy I membrane ionic models. Simulations of the entire QT interval for epicardial and endocardial pacing show that the effect of intramural heterogeneity on the dispersion of the action potential duration is mostly discernible along the epi- endocardial direction, while in the orthogonal directions the dispersion patterns have the same qualitative features of the homogeneous model.

1 Introduction

During a normal heartbeat, the ventricular transmembrane potential displays two main phases having different time and space scales: depolarization and repolarization. Repolarization exhibits a short rapid downstroke, a plateau and final, slower downstroke. While the excitation phase has been examined in considerable detail both experimentally and numerically much less is known concerning the recovery phase (see [8, 16, 7, 4]). Both phases are influenced by the fiber direction through which excitation is spreading and by the anisotropy of the intra and extracellular media. The study of these phases can be greatly enhanced by the use of computational models based on systems of differential equations. Previous studies considered simulations of the entire excitation and repolarization sequences mainly in 1D cables [10, 15, 17, 14] and 2D sheets, see e.g. [4]; only few simulation studies of a normal beat in 3D slabs are available in the literature, see e.g. [7, 9], even if reentry dynamics have been largely studied. This is mainly due to the high computational costs involved in large scale simulations of a full cardiac cycle in three dimensions, which require adaptive and parallel numerical techniques. In [5], we implemented an efficient parallel simulator and performed several numerical experiments in 3D on parallel architectures with both the Monodomain and the Bidomain models. In [6], a detailed comparison between the excitation and the repolarization sequences elicited by a local stimulus showed that the Monodomain model is adequate for a qualitative investigation of the repolarization sequences and of the patterns displayed by the action potential duration (APD) distributions. Recently, the electrophysiological consequences of the intramural heterogeneity of the APD have generated considerable interest and some controversy. A subpopulation of cells (M cells) has been discovered, displaying a longer APD than epicardial and endocardial ventricular cell types, mainly in "in vitro" experiments, see e.g. [18]. On the other hand, high degrees of intramural heterogeneity have not been detected in "in vivo" studies of normal hearts, see e.g. [1], where it is noted that the intercellular coupling in cardiac tissue is a factor affecting APD modulation. However, controversy still exists over the extent to which heterogeneity in repolarization is expressed across the normal ventricular wall. In this work, we use our parallel simulator to investigate the influence of intramural heterogeneity of the intrinsic properties of the cellular membrane on the repolarization sequences and on the APD dispersion.

2 Mathematical Models

From a macroscopic point of view, the cardiac tissue is conceived as the superposition of two averaged continuous media, the intra and the extracellular medium, whose anisotropy is characterized by the conductivity tensors $D_i(\mathbf{x})$ and $D_e(\mathbf{x})$. These tensors are anisotropic related to the direction of the cardiac fibers that rotates counterclockwise (CCW) from epicardium to endocardium and to the laminar organization of the heart muscle (see [11]). Therefore, at any point $\mathbf{x}$, it is possible to identify a triplet of orthonormal principal axes $\mathbf{a}_l(\mathbf{x})$, $\mathbf{a}_t(\mathbf{x})$, $\mathbf{a}_n(\mathbf{x})$, with $\mathbf{a}_l(\mathbf{x})$ parallel to the local fiber direction, $\mathbf{a}_t(\mathbf{x})$ and $\mathbf{a}_n(\mathbf{x})$ tangent and orthogonal to the radial laminae respectively and both being transversal to the fiber axis. Denoting by $\sigma_l^{i,e}$, $\sigma_t^{i,e}$, $\sigma_n^{i,e}$ the conductivity coefficients measured along the corresponding directions, then the conductivity tensors $D_i(\mathbf{x})$ and $D_e(\mathbf{x})$ related to *orthotropic anisotropy* of the media are given by: $D_{i,e} = \sigma_l^{i,e} \mathbf{a}_l \mathbf{a}_l^T + \sigma_t^{i,e} \mathbf{a}_t \mathbf{a}_t^T + \sigma_n^{i,e} \mathbf{a}_n \mathbf{a}_n^T$.

The intra and extracellular electric potentials u_i, u_e in the Bidomain model are described by a reaction-diffusion system, coupled with a system of ODEs for ionic gating variables $w \in R^Q$ and for the ions concentration $c \in R^p$. Denoting by $v = u_i - u_e$ the transmembrane potential and by $I_{tot} = -D_i \nabla u_i - D_e \nabla u_e$ the total current flowing in the two media, then, for an insulated cardiac domain H, (v, u_e, I_{tot}, w, c) satisfy the system:

$$\begin{cases} c_m \partial_t v - \mathrm{div}(D_e D^{-1} D_i \nabla v) + I_{ion}(v,w) - \mathrm{div}(D_e D^{-1} I_{tot}) = 0, \\ \partial_t w - R(v,w) = 0, \qquad w(\mathbf{x},0) = w_0(\mathbf{x}), \\ \partial_t c - S(v,w,c) = 0, \qquad c(\mathbf{x},0) = c_0(\mathbf{x}), \\ \mathbf{n}^T D_m \nabla v = 0, \qquad v(\mathbf{x},0) = v_0(\mathbf{x}), \\ I_{tot} = -D_i \nabla u_i - D_e \nabla u_e, \\ -\mathrm{div}(D \nabla u_e) = \mathrm{div}(D_i \nabla v), \qquad -\mathbf{n}^T D \nabla u_e = \mathbf{n}^T D_i \nabla v, \end{cases}$$

where $\partial_t = \partial/\partial t$, $c_m = \chi * C_m$, $I_{ion} = \chi * i_{ion}$, with χ the ratio of membrane area per tissue volume, C_m the surface capacitance and i_{ion} the ionic current of the membrane per unit area. Disregarding applied currents, from the current conservation law, we have $\mathrm{div} I_{tot} = 0$. It is well known that, assuming equal anisotropy ratio of the two media, the Bidomain system reduces to the Monodomain model. If we disregard the source term $\mathrm{div}(D_e D^{-1} I_{tot})$, then a Monodomain model is derived as a Relaxed Bidomain system without assuming that the two tensors are proportional. Therefore, we obtain the anisotropic Monodomain model by solving first a single parabolic reaction-diffusion equation for the transmembrane potential v with the conductivity tensor given by $D_m = D_e D^{-1} D_i$ and coupled with the same gating and concentration system

$$\begin{cases} c_m \partial_t v - \mathrm{div}(D_e D^{-1} D_i \nabla v) + I_{ion}(v,w) = I_{app}, \\ \partial_t w - R(v,w) = 0, \qquad w(\mathbf{x},0) = w_0(\mathbf{x}), \\ \partial_t c - S(v,w,c) = 0, \qquad c(\mathbf{x},0) = c_0(\mathbf{x}), \\ \mathbf{n}^T D_m \nabla v = 0, \qquad v(\mathbf{x},0) = v_0(\mathbf{x}), \end{cases}$$

and then solving an elliptic problem for the extracellular potential

$$-\mathrm{div}(D \nabla u_e) = \mathrm{div}(D_i \nabla v), \qquad -\mathbf{n}^T D \nabla u_e = \mathbf{n}^T D_i \nabla v.$$

We remark that the first equation is coupled with the system of ordinary differential equations in w, c and uncoupled from the elliptic equation in u_e; the system uniquely determines v, while the potential u_e is defined only up to an additive time-dependent constant related to the reference potential, chosen to be the average extracellular potential in the cardiac volume by imposing $\int_H u_e \, \mathrm{d}x = 0$.

3 Numerical Discretization

The cardiac volume H is discretized by a structured grid of hexahedral isoparametric Q_1 elements. A semidiscrete problem is obtained by applying a standard Galerkin procedure and choosing a finite element basis.

The time discretization is performed by a semi-implicit method using for the diffusion term the implicit Euler method, while the nonlinear reaction term I_{ion} is treated explicitly. The implicit treatment of the diffusion terms is essential in order to allow an adaptive change of the time step according to the stiffness of the various phases of the heartbeat. The ODE system for the gating variables is discretized by the semi-implicit Euler method and the explicit Euler method is applied for solving the ODE system for the ions concentration. We decouple the full system by solving the gating and ions concentration system first (given the potential $\mathbf{v}^n$ at the previous time-step)

$$\mathbf{w}^{n+1} - \Delta t\, R(\mathbf{v}^n, \mathbf{w}^{n+1}) = \mathbf{w}^n, \qquad \mathbf{c}^{n+1} = \mathbf{c}^n + \Delta t\, S(\mathbf{v}^n, \mathbf{w}^{n+1}, \mathbf{c}^n)$$

and then solving for $\mathbf{v}^{n+1}$

$$\mathcal{A}_{\Delta t}\mathbf{v}^{n+1} = \mathrm{M}\left[\frac{c_m}{\Delta t}\mathbf{v}^n - \mathrm{I}_{ion}^h(\mathbf{v}^n, \mathbf{w}^{n+1}, \mathbf{c}^{n+1}) + \mathrm{I}_{app}^{m,h}\right],$$

where $\mathcal{A}_{\Delta t} = \frac{c_m}{\Delta t}\mathrm{M} + \mathrm{A}$, with A the stiffness matrix, M the mass matrix and $\mathrm{I}_{ion}^h, \mathrm{I}_{app}^{(m,e),h}$ the finite element interpolants of I_{ion} and $I_{app}^{m,e}$, respectively. We employed an adaptive time-stepping strategy based on controlling the transmembrane potential variation $\Delta v = max(\mathbf{v}^{n+1} - \mathbf{v}^n)$ at each time-step, see [12]. The linear system at each time step in the discrete problems is solved iteratively using the PETSc parallel library [2] and a preconditioned conjugate gradient solver with block Jacobi preconditioner and ILU(0) on each block. The parallel machines employed are an IBM SP RS/6000 Power4 with 512 processors Power 4 - 1300 MHz, (www.cineca.it), and a Cluster Linux with 72 Xeon 2.4 GHz processors. More details on the parallel solver can be found in [5].

4 Results

The cardiac domain considered is a cartesian slab of dimensions $5 \times 5 \times 1\ cm^3$ modeling a portion of the left ventricle. A structured grid of hexahedral isoparametric Q_1 elements of size $h = 0.1\ mm$ was used in all computations. In the numerical tests, we have used the following parameters: $\chi = 10^3\ cm^{-1}, C_m = 10^{-3}\ mF/cm^2$, $\{\sigma_l^e, \sigma_l^i, \sigma_t^e, \sigma_t^i\} = \{2,\ 3,\ 1.35,\ 0.315\}\ m\Omega^{-1}cm^{-1}$ and $\sigma_n^e = \sigma_t^e/2$, $\sigma_n^i = \sigma_t^i/10$. These conductivity coefficients of the orthotropic anisotropy have been calibrated so that the associated propagation velocities $(\theta_l, \theta_t, \theta_n)$ of ideal plane wavefronts can be conservatively estimated as $(60,\ 25,\ 10)\ cm\ sec^{-1}$, respectively. These estimates are in accordance with the histological findings of [11]) supporting the idea that the cardiac tissue anisotropy could be orthotropic. The fibers rotate intramurally linearly, proceeding counterclockwise (CCW) from epicardium (-45°) to endocardium (75°), for a total amount of 120°. In this paper, we consider the phase I Luo-Rudy (LR1) model (see [12]), since it is one of the complex gating systems mostly used in recents 3D simulations. The initial conditions are at the rest and we apply an appropriate stimulus on a small area at the center of the slab (3 or 5 mesh points in each direction). Other than potentials and gating variables, at each time-step, we compute also the activation (ACTI) and the repolarization (REPO) times, defined as the times when the action potential (AP) crosses -60 mV during the upstroke and the downstroke, respectively; hence, the APD is defined as the difference APD = REPO - ACTI. . When homogeneous intrinsic properties of the cellular membrane are assumed, the slow inward current in the LR1 model is reduced by a factor 2/3, yielding an APD of about 266 msec. We also consider simulations with intramural heterogeneity of the cellular membrane. In order to reproduce qualitatively the APD transmural behaviour measured in wedge preparations, see [18], Fig. 4 and [13] Fig. 5, we performed 1D simulations using a suitable subdivision of the wall thickness with different membrane properties. More precisely, we subdivided the slab into four layers of thickness (0.1, 0.1,0.7 0.1) cm, respectively, proceeding from the endo- to epicardium and by multiplying the slow inward current I_{si} of the LR1 model by (7.66, 8.66,7.86.6.66), corresponding to intrinsic APDs of (295,324,301,266), respectively. Hence we assume that sub-endocardial and mid-myocardial layers display a longer APD than the epi- and endo-cardial cells. The piecewise constant line in Fig. 1 (dashed) displays the intrinsic intramural APD distribution of the cells. We first consider a one-dimensional model having uniform conductivity equal to the intramural cross-fiber conductivity σ_t with homogeneous or heterogeneous intrinsic properties of the cellular membrane. In the homogeneous case (left panel of Fig. 1), the excitation and recovery fronts reach a quasi-stationary propagation, apart from the acceleration during the starting phase of the propagation and also during the subsequent collision with the endocardium. The total times for activation and recovery are about 39 and 32 msec, respectively, and the APD dispersion amounting to about 7 msec, mostly concentrated around the stimulus and collision sites. Notice that repolarization moves slightly faster than activation in the homogeneous model.

In the heterogeneous case (central panel of Fig. 1), the activation time results practically unchanged, while we have a higher repolarization time of about 54 msec. The APD dispersion, amounting to 21 msec, is three times larger than the homogeneous case. Due to current conduction, the intrinsic APD differences between the four cell layers are strongly smoothed and reduced. We have also applied an endocardial stimulus to the heterogeneous 1D model (right panel of Fig. 1). The sequence of excitation results the same as the one elicited by the epicardial stimulus, while the repolarization process is completed in 19 msec and the APD dispersion amounts to 28 msec. Therefore, an endocardial stimulus in the heterogeneous case brings about a significant shortening of the recovery sequence and a higher APD dispersion than an epicardial stimulus. In other words, epicardial stimulation increases the dispersion of the recovery time. We remark that these simulations are limited to an action potential elicited by a single stimulus, a condition that emphasizes the APD dispersion, since it is well known that a periodic stimulation, at an increasing rate, results in shortening of APD with a reduced dispersion.

We consider now 3D simulations of the excitation and repolarization processes elicited by an epicardial central stimulation in an orthotropic slab, homogeneous in Fig. 2, heterogeneous in Fig. 3. In both cases, we show the spread of excitation (ACTI), the sequence of recovery (REPO) and the APD on the whole slab (bottom) and on 5 plane sections parallel to the epicardial face, located at z = 0 (endo), 0.25, 5, 0.75, 1 (epi) cm, respectively. We now briefly describe some common features of the homogeneous and heterogeneous models. The spread of excitation and recovery exhibit an acceleration in the direction across fibers and dimple-like inflections appear in the isochrone profiles, due to

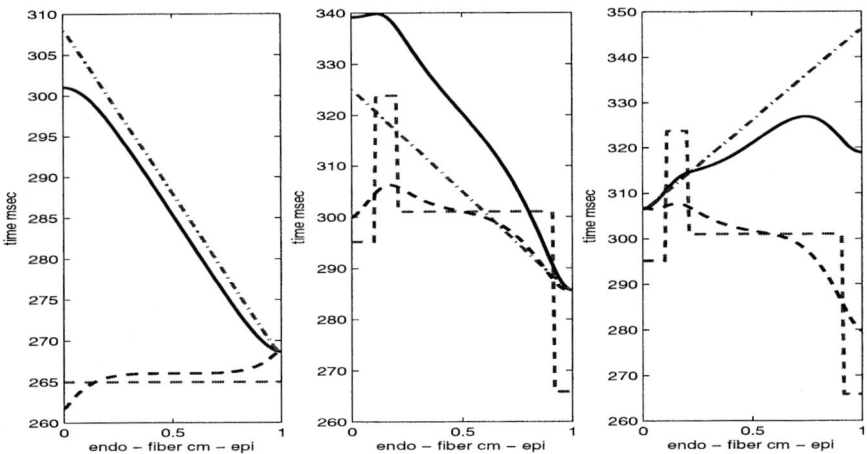

Fig. 1. Simulations along a fiber with conductivity coefficient σ_t of the intramural slab thickness. Activation, repolarization times and APD are displayed by dashed-dotted, continuous and dashed lines, respectively. Activation time has been shifted by the value of the repolarization time at the stimulus site. The piecewise constant line indicate the intrinsic APD of the cell layers. Left and Central Panels are related to an epicardial stimulation assuming homogeneous and heterogeneous intrinsic properties of the cellular membrane, respectively; the Rigth Panel refer to an endocardial stimulation for the heterogeneous case

the faster propagation of the fronts in deeper layers where the fiber direction rotates CCW relatively to the upper planes. The recovery isochrones on the epi, intramural and endocardial planes exhibit a somewhat smoother shape and slightly faster propagation compared with the excitation sequence. In particular, epicardial repolarization propagates across fibers faster than the excitation sequence, yielding a progressively APD shortening across fibers, as shown by Figs. 2,3. The APD patterns in both models are characterized by the following features: **i)** The APD distributions on the epicardial and intramural planes, exhibit a maximum located at the epicardial stimulation site or at the intramural points firstly reached by the excitation front, respectively; the level lines, surrounding these maxima, are elongated along the local fiber direction and display dog-bone shaped profiles. This indicates that APD decreases more rapidly when moving away from the center of the face in the cross-fiber direction than along fibers. **ii)** On the intramural sections (from subepicardial to midwall ones), two finger-shape valleys of decreasing APD values occur. These narrow valleys of relative APD shortening are located in the regions where excitation isochrones exhibit a dimple-like inflections. **iii)** On the endocardial plane the APD distribution displays a saddle point at the endocardial breakthrough; the APD increases reaching a maximum when moving away from the breakthrough point in a direction parallel to the endocardial fibers of 75° CW. On the other hand, on the transmural sections displayed in Fig. 4 we see considerable differences be-

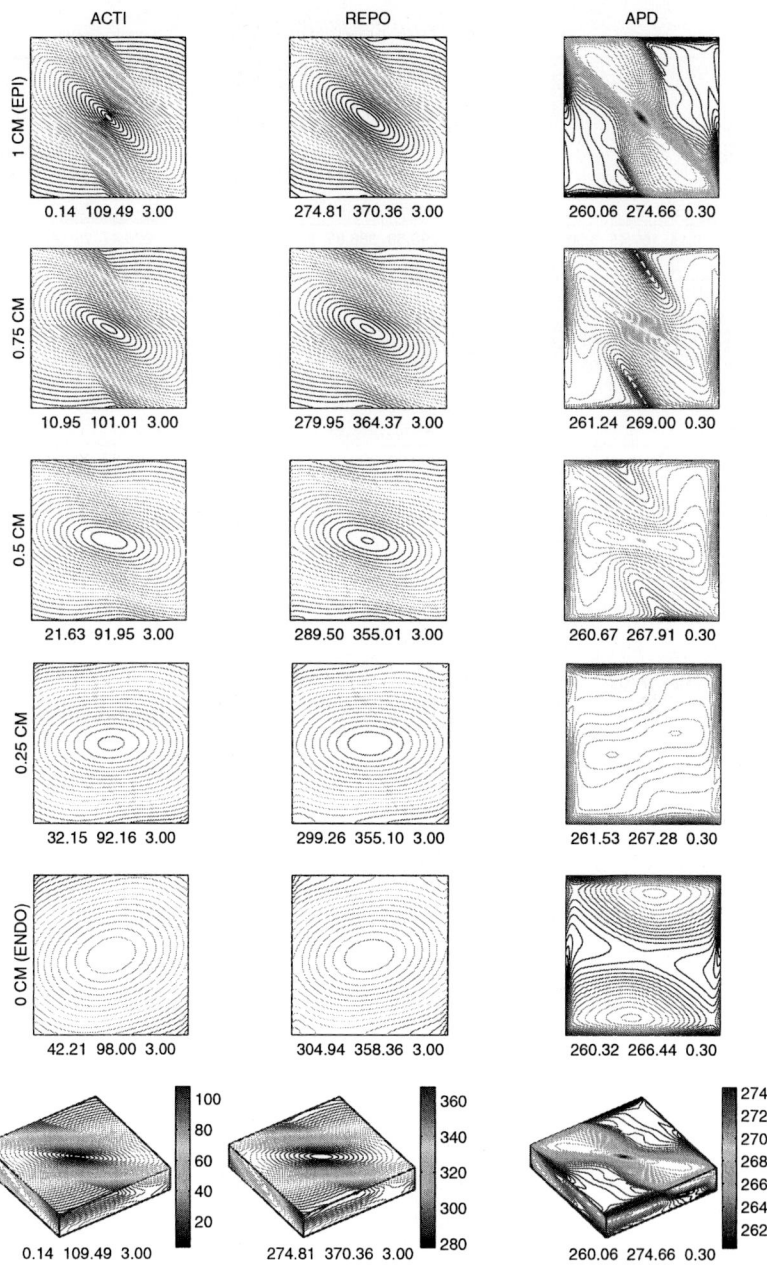

Fig. 2. Homogeneous tissue, orthotropic Monodomain slab $5 \times 5 \times 1 cm^3$. Isochrone lines of the depolarization time (first column ACTI), repolarization time (second column REPO) and action potential duration (third column APD) on 5 horizontal sections (z = 0, 0.25, 0.5, 0.75, 1 cm) and the whole slab; reported below each panel are the maximum, minimum and step in msec of the displayed map

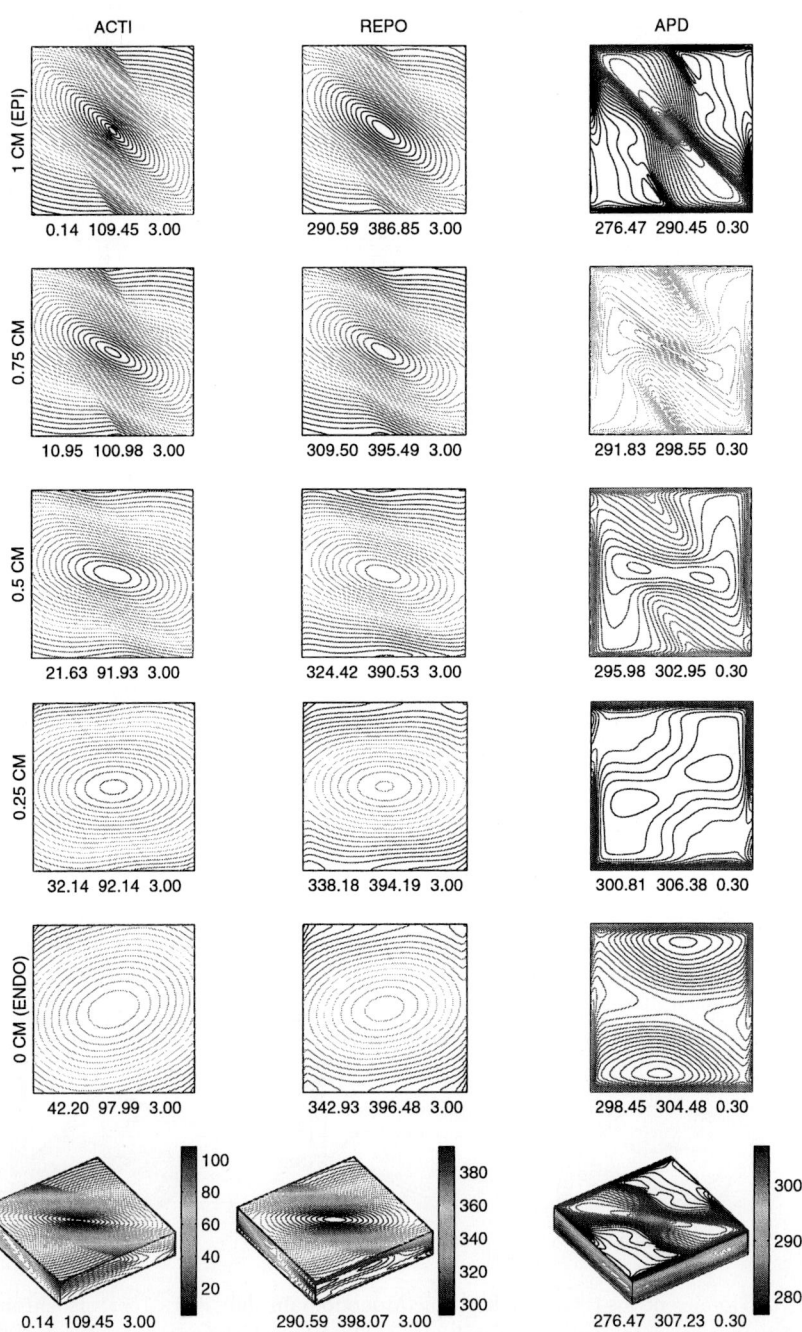

Fig. 3. Intramural heterogeneous tissue, orthotropic Monodomain slab $5 \times 5 \times 1 cm^3$. Same format as in Fig. 1

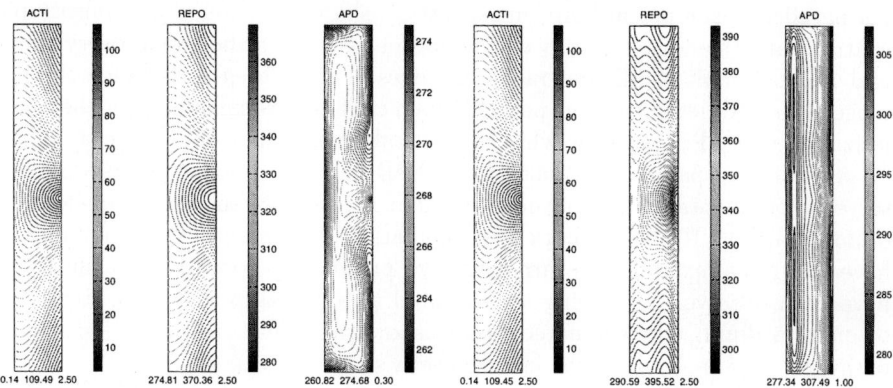

Fig. 4. Homogeneous tissue, orthotropic Monodomain slab $5 \times 5 \times 1 cm^3$. Isochrones of depolarization (first column ACTI), repolarization (second column REPO) times, action potential duration (third column APD) displayed on the transmural diagonal section perpendicular to the epicardial face and to the epicardial fiber direction. Left (right) three Panels refer to the homogeneous (heterogeneous) slab. Reported below each panel are the maximum, minimum and step in msec of the displayed map

tween the homogeneous and heterogeneous case. In both models the excitation isochrones on the transmural section show the presence of returning pathways, i.e. pathways that, starting from the epicardial stimulation site, proceed toward the endocardium but, about midway of the wall thickness, return toward the epicardial side. These pathways accelerate the propagation in epicardial areas where the excitation proceeds mainly across fibers. Return pathways of repolarization appear in the homogeneous slab whereas more complex recovery isochrone profiles are present in the heterogeneous model. In the heterogeneous model, the APD pattern shows parallel level lines stretched horizontally as opposed to the complex transmural APD pattern observed in the homogeneous slab. In planar sections parallel to the epicardial face, excitation and repolarization sequences and the spatial APD patterns elicited by endocardial pacing shared the same qualitative features as those described above for epicardial pacing in both the homogeneous and heterogeneous slab.

5 Conclusions

Our results show that, in spite of the homogeneous cellular membrane properties (i.e., all individual cells have the same intrinsic transmembrane action potential), the anisotropy of the media produces a spatial variation of the APD throughout the slab and the APD distribution exhibits anisotropic patterns strongly correlated with the excitation wave front motion and the front-boundary collisions.

The introduction of an intramural variation of the intrinsic cellular APD yields excitation and repolarization sequences and APD patterns which, on layers parallel to the epicardium, unexpectedly share the same main anisotropic spatial features encountered in the homogeneous slab, although recovery times and APDs exhibit a different range of values. The differences between the homogeneous and heterogeneous model remain confined transmurally for the repolarization and APD patterns while the excitation sequence does not change.

Anisotropic spatial variations of the APD along and across fibers were observed experimentally in 2D myocardial laminae in e.g. [8] and on the epicardium of dog hearts [3, 16]. We remark that simulation studies and experimental data have shown that excitation return pathways, proceedings toward the pacing level, have been observed for pacing sites located at any intramural level, from epi- to endo-cardium. Our simulated results show that clear repolarization return pathways are expected for the homogeneous slab.

In experimental studies in exposed and isolated dog hearts, the observed transmural dispersion of APD in the left ventricular wall is 30 msec at most, see e.g. [1]. Our unpublished experimental results confirm these findings, since during ventricular pacing with cycle length of 350 or 400 msec we observed 15-20 msec APD dispersion. In these preparations, the repolarization sequence was qualitatively similar to the activation sequence. When the pacing site was epicardial, both the excitation and the the repolarization "wave front" returned toward the epicardium in a transmural plane perpendicular to the epicardial fiber direction. However, further studies are needed to determine whether these findings occur consistently in varying experimental conditions. In this study, we have considered simulated beats by a single stimulus, a condition that emphasizes the APD difference and dispersion. Thus, our results show that transmural heterogeneities of APD cannot be detected from the epicardial pattern of the APD distribution.

References

1. Anyukhovsky E.P. et al.: The controversial M cell *J. Cardiovasc Electrophysiol*, 10: 244–260, 1999.
2. Balay S. and et al.: *PETSc Users Manual.* ANL TR anl-95/11 - rev. 2.1.5, 2002. http://www.mcs.anl.gov/petsc.
3. Burgess, M.J.B.and et al.: Nonuniform epicardial activation and repolarization properties of in vivo canine pulmonary conus. *Circ. Res.* 62 (2): 233–246, 1988.
4. Cates A.W. and Pollard A.E.: A model study of intramural dispersion of action potential duration in the canine pulmonary conus. *Ann. Biomed. Eng.*, 26: 567–576, 1998.
5. Colli Franzone P. and Pavarino L.F.: A parallel solver for reaction-diffusion systems in computational electrocardiology. *Math. Mod. Meth. Appl. Sci. (M3AS), 14 (6): 883–911*, 2004.
6. Colli Franzone P., Pavarino L.F. and Taccardi B.: Simulating patterns of excitation, repolarization and action potential duration with cardiac Bidomain and Monodomain models. *Submitted* , 2004.
7. Efimov I.R. et al.: Activation and repolarization patterns are governed by different structural characteristics of ventricular myocardium. *J. Cardiovasc. Electrophysiol.*, 7: 512–530, 1996.

8. Gotoh M. et al.: Anisotropic repolarization in ventricular tissue. *Am. J. Physiol.*, 41: 107–113, 1987.
9. Henriquez C. S and Penland R. C.: Impact of transmural structural and ionic heterogeneity on paced beats in the ventricle: a modeling study. *100 Years of Electrocardiology*, Schalit M.J. et al., Editors, Einthoven Foundation, 2002.
10. Joyner R. W.: Modulation of repolarization by electrotonic interactions. *Japan. Heart J.* 27: 167–183, 1986.
11. LeGrice I. J. et al.: Laminar structure of the heart: a mathematical model. *Am. J. Physiol. (Heart Circ. Physiol.)*, 272 (41): H2466-H2476, 1997.
12. Luo C. and Rudy Y.: A model of the ventricular cardiac action potential: depolarization, repolarization, and their interaction. *Circ. Res.*, 68: 1501–1526, 1991.
13. Poelzing S. et al.: Heterogeneous connexin43 expression produces electrophysiological heterogeneities across ventricular wall. *Am J. Physiol (Heart Circ. Physiol)*, 286: H2001-H2009, 2004.
14. Seeman G. et al.: Quantitative reconstruction of cardiac electromechanics in human myocardium. *J. Cardiovasc. Electrophysiol.*, 14: S219-S228, 2003.
15. Steinhaus B. M.: Estimating cardiac transmembrane activation and recovery times from unipolar and bipolar extracellular electrograms: a simulation study. *Circ. Res.* 64 (3): 449-462, 1989.
16. Taccardi B. et al.: Epicardial recovery sequences and excitation recovery intervals during paced beats. *PACE*, 22 (4) part II: 8–33, 1999.
17. Viswanathan,P. C. et al.: Effects of I_{Kr} and I_{Ks} heterogeneity on action potential duration and its rate dependence. *Circulation*, 99: 2466–2474, 1999.
18. Yan et al.: Characteristics and distribution of M cells in arterially perfused canine left ventricular wedge preparations. *Circulation*, 98: 1921–1927, 1998.

Spatial Inversion of Depolarization and Repolarization Waves in Body Surface Potential Mapping as Indicator of Old Myocardial Infarction

Paula Vesterinen[1,3], Helena Hänninen[1,3], Matti Stenroos[2], Petri Korhonen[1,3], Terhi Husa[1,3], Ilkka Tierala[1,3], Heikki Väänänen[2], and Lauri Toivonen[1,3]

[1] Helsinki University Central Hospital, Division of Cardiology, P.O. Box 340, FIN-00029 HUS
[2] Helsinki University of Technology, Laboratory of Biomedical Engineering, P.O. Box 2200, FIN-02015 TKK
[3] BioMag Laboratory, P.O. Box 340, FIN-00029 HUS

Abstract. To investigate the quantitative abnormalities induced by prior myocardial infarction (MI) on the whole electrical cardiac cycle, body surface potential mapping was recorded in 144 patients with prior MI and 75 healthy controls. QRS onset, offset and T-wave end were automatically determined from the averaged signal. Time integrals were calculated for the QRS wave and the STT wave. In MI patient group the average QRS and STT integrals showed strong negative correlation on the body surface ($r = -0.901, p < 0.001$) in contrast to the positive correlation in the control group ($r = 0.285, p < 0.001$). Sensitivity of an inverted QRS / STT integral relation to detect MI was 79%, as opposed to the sensitivity of the descriptive Minnesota code of 70%. Furthermore the degree of inversion correlated with left ventricular ejection fraction thus relating to the size of MI.

1 Introduction

Conventional electrocardiographic (ECG) criteria for diagnosing old myocardial infarction (MI) rely on descriptive features of the initial QRS wave. Yet, the sequential depolarization of the ventricle would imply that MI in different locations of the left ventricle would manifest at different time periods of the QRS. Furthermore, MI inevitably affects also the repolarization. To determine the abnormalities induced by MI on the whole electrical cardiac cycle, we quantitatively analyzed the electrocardiograms registered by body surface potential mapping (BSPM).

2 Methods

Altogether 144 patients with at least 1 prior myocardial infarction in the hospital records were recruited. All had presented with typical chest pain followed by

Table 1. Study population

	Age (years)	Left Ventricular Ejection Fraction (%)	Female / Male (N)
Patients	61 ± 10	41.0 ± 9.9	27 / 117
Controls	52 ± 12	61.8 ± 7.3	19 / 56

elevation of the cardiac CK-MB enzyme or troponin T. All had angiographically verified coronary artery disease and a local dysfunctional region in left ventricle, determined by cineangiography (108 patients), or by echocardiography (36 patients), along with left ventricular ejection fraction (EF) (Table 1). The patients were classified, on the basis of 12-lead ECG, as having a Q-wave myocardial infarction (QMI) or a non Q-wave myocardial infarction (NQMI), according to the Minnesota criteria 1-1 [1]. Of the patients 101 had QMI and 43 NQMI. As controls were included 75 subjects without any history of heart disease or Minnesota Q-waves in 12-lead ECG. The Minnesota code indicated MI with 70% sensitivity, and, by protocol definition, with 100% specificity.

BSPM with 120 unipolar leads covering the whole thorax, in addition to 3 limb leads, was recorded as reported previously [2]. Wilson's central terminal was used as reference potential.

The BSPM data were signal-averaged, and automatic identification of the QRS onset and offset was performed after bi-directional high-pass filtering of the depolarization wave [3]. T-wave end and apex were determined automatically, as described earlier [4]. Integrals over QRS and STT (from J-point to the end of T wave) were analyzed.

The average integral values in the MI patient and control groups were calculated for each lead. Scatter plot was constructed with the average QRS and STT integrals on x- and y-axes in each lead location. Spatial correlation between QRS and STT integrals was calculated, using Pearson correlation coefficient (r), for the group average values and for each study subject separately.

3 Results

The average QRS integral correlated negatively with corresponding average STT integral in the MI patient group ($r = -0.901, p < 0.001$) (Fig. 1). In the control group the average QRS integral correlated positively with the average STT integral ($r = 0.285, p < 0.001$) (Fig. 2). In subgroup analysis respective correlation was -0.904 for QMI patients ($p < 0.001$) and -0.888 for NQMI patients ($p < 0.001$).

In the analysis of each study subject separately, correlation between the QRS and STT integrals was negative in 114 MI patients (79%) and positive in 30 MI patients (21%). This correlation was negative in 24 controls (32%) and positive in 51 controls (68%). Thus, the sensitivity of the inverted correlation in detecting MI was 79% and the specificity was 68%.

The correlation of QRS and STT integrals was negative in 84 patients in QMI patient group (sensitivity 83%) and in 31 patients in NQMI patient group

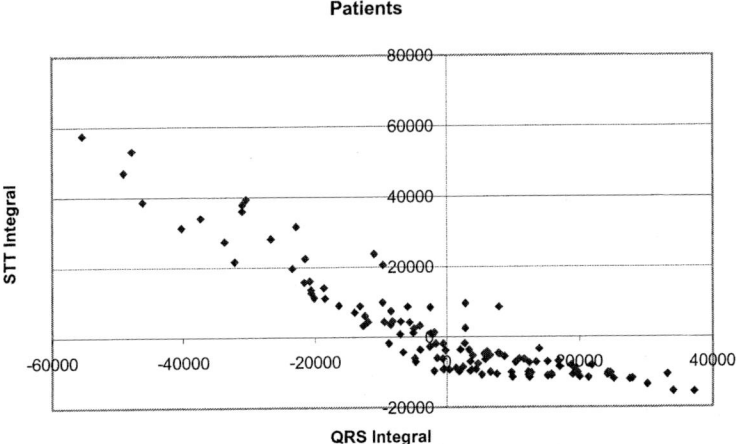

Fig. 1. Scatter plot of average QRS integral values against STT integral values in each BSPM lead in the MI patient group. Values are in μVms

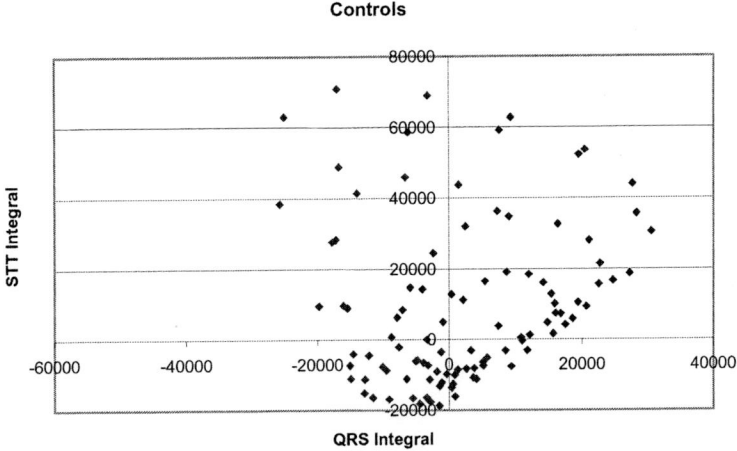

Fig. 2. Scatter plot of average QRS integral values against STT integral values in each BSPM lead in the control group. Values are in μVms

(72% sensitivity). There was a weak but significant positive correlation between the relationship of QRS to STT integrals and the left ventricular EF ($r = 0.219, p = 0.009$) in MI patients.

4 Discussion

The present study confirmed previous observations that the QRS and STT integrals are positively correlated in healthy controls [5]. Thus, the polarity and magnitude of the body surface potentials caused by depolarization and repolarization of the ventricles are mainly concordant in healthy subjects. This normally positive relationship of QRS and STT integrals is inverted in patients with prior MI. This indicates that the abnormalities in ventricular depolarization and repolarisation in old MI are spatially concordant though the values of the integral potentials are opposite. This close relationship of de- and repolarisation abnormalities in old MI is reflected as distinct line in the scatter plots whereas the values in controls are dispersed. Of importance is that the negative correlation of QRS and STT integrals is strong also in NQMI patients, suggesting that quantitative analysis of the QRS and STT deflections may detect MI in the absence of conventional qualitative criteria. In the light of our findings, cardiac mortality predicting ability of spatial QRS-T angle may be attributed to unrecognized MI [6]. Spatial inversion of QRS and STT integrals might even hold potential for risk stratification in post-MI patients, as does the angle between depolarization and repolarization wavefronts [7].

When the inversion of QRS and STT correlation was applied as an indicator of prior MI, the sensitivity of the method exceeded the sensitivity of the Minnesota code. The specificity fell behind the Minnesota code, which by definition was 100% due to the prerequisite of absent Q-waves for inclusion of controls.

The degree of the inversion of QRS and STT in each patient, expressed as correlation coefficient, showed a relationship with left ventricular EF. This finding indicates that the degree of inversion may be associated with the size of the MI.

5 Conclusions

In old myocardial infarction the ventricular depolarization and repolarization wave integrals are spatially inversely correlated as opposed to positive correlation in the healthy heart. The degree of inversion may relate to the size of the infarction. A quantitative analysis of the inversion of ventricular depolarization and repolarization waves may be combined with other electrocardiographic variables to improve detection of prior MI.

Acknowledgements

This work was supported by Finnish Cardiac Research Foundation, Aarne Koskelo Foundation, Paulo Foundation, and Helsinki University Hospital Research Funds.

References

1. Prineas RJ, Crow RS, Blackburn H: The Minnesota Code Manual of Electrocardiographic Findings, Standards and Procedures for Measurement and Classification. 1982, Wright, Bristol.

2. Simelius K, Tierala I, Jokiniemi T, Nenonen J, Toivonen L, Katila T: A body surface mapping system in clinical use. Med & Biol Eng. **34** (1996) (suppl) 107–108.
3. Korhonen P, Montonen J, Mäkijärvi M, Katila T, Nieminen MS, Toivonen L: Late fields of the magnetocardiographic QRS complex as indicators of propensity to sustained ventricular tachycardia after myocardial infarction. J Cardiovasc Electrophysiol **11** (2000) 413–420.
4. Oikarinen L, Paavola M, Montonen J, et al. Magnetocardiographic QT interval dispersion in postmyocardial infarction patients with sustained ventricular tachycardia: Validation of automated QT measurements. Pacing Clin Electrophysiol **21** (1998) 1934–42.
5. Montague T, Smith E, Cameron D, Rautaharju P, et al. Isointegral analysis of body surface maps: Surface distribution and temporal variability in normal subjects. Circulation **63** (1981) 1166–1172.
6. Kardys I, Kors J, van der Meer I, Hofman A, van der Kuip D, Witteman J. Spatial QRS–T angle predicts cardiac death in a general population. Eur Heart J **24** (2003) 1357–1364.
7. Zabel M, Acar B, Klingenheben T, Franz M, Hohnloser S, Malik M. Analysis of 12–lead T-wave morphology for risk stratification after myocardial infarction. Circulation **102** (2000) 1252–1257.

Dissipation of Excitation Fronts as a Mechanism of Conduction Block in Re-entrant Waves

Vadim N. Biktashev[1] and Irina V. Biktasheva[2]

[1] Department of Mathematical Sciences,
Liverpool University, Liverpool L69 7ZL, UK
vnb@liv.ac.uk
http://www.maths.liv.ac.uk/~vadim
[2] Department of Computer Science,
Liverpool University, Liverpool L69 3BX, UK

Abstract. Numerical simulations of re-entrant waves in detailed ionic models reveal a phenomenon that is impossible in traditional simplified mathematical models of FitzHugh-Nagumo type: dissipation of the excitation front (DEF). We have analysed the structure of three selected ionic models, identified the small parameters that appear in non-standard ways, and developed an asymptotic approach based on those. Contrary to a common belief, the fast Na current inactivation gate h is not necessarily much slower than the transmembrane voltage E during the upstroke of the action potential. Interplay between E and h is responsible for the DEF. A new simplified model emerges from the asymptotic analysis and considers E and h as equally fast variables. This model reproduces DEF and admits analytical study. In particular, it yields conditions for the DEF. Predictions of the model agree with the results of direct numerical simulations of spiral wave break-up in a detailed model.

1 Introduction

Contemporary detailed models of excitation propagation in heart tissue can reproduce many important conduction pathologies, including transient propagation blocks. Such blocks are involved in generation, transformation and termination of re-entrant circuits, the importance of which for cardiac pathologies has been recognized early[1]. In modern detailed models, the relevant phenomena include break-up of spiral waves[2], meandering patterns of spiral waves[3],[4], or spontaneous termination of re-entrant activity[5, 6]. Break-up of spiral waves is thought to be a key mechanism of transition from less dangerous arrhythmia to fibrillation[7, 8, 9]. Thus, it is important to understand, how such break-up, or, more generally, a spontaneous transient excitation conduction block may happen. The detailed mathematical models, in principle, answer this question, in the sense that they can, more or less accurately, reproduce the phenomenon. However, currently there is no other way to see how the possibility of conduction block changes with parameters but to repeat calculations, which may be rather extensive. Situation is even worse if we want to know what changes in

I_{Na} ionic gates tend to close well in some ranges of the transmembrane voltage, and the ranges of almost perfect closure of m and h overlap.

These considerations have lead us to a system of only two differential equations describing propagation of excitation, with transmembrane voltage E and the fast inactivation gate h as the key dynamic variables.

$$C_m \frac{\partial E}{\partial t} = I_{\text{Na,max}}(E) j h \theta(E - E_m) + D \frac{\partial^2 E}{\partial x^2}$$
$$\frac{\partial h}{\partial t} = \frac{1}{\tau_h(E)} \left(\theta(E_h - E) - h \right) \qquad (1)$$

where E is the membrane capacitance, $I_{\text{Na,max}}(E)$ is the maximal fast sodium current when all gates are open, j is the slow inactivation gate assumed almost unchanged during the front, D is the voltage diffusion coefficient, $\tau_h(E)$ is the characteristic time of the dynamics of the h-gate, E_h and E_m are the switch voltages of the h- and m-gates respectively ($E_m > E_h$), and $\theta()$ is Heaviside's perfect switch function. This is opposed to, say, 21 equations in Courtemanche et al. model. Some further simplification, in the form of replacing $I_{\text{Na,max}}(E)$ and $\tau_h(E)$ with constants, while retaining qualitatively correct behaviour of the solutions, has allowed exact analytical solutions. The details of the solutions have been described elsewhere[20, 21]. For our present purpose, the most interesting result is the excitability, measured say by the local instant value of gate j at the front,[1] that is necessary for propagation of a front with a given speed c:

$$j = \frac{C_m}{\tau_h I_{\text{Na,max}}} g\left(c\sqrt{\tau_h/D}, \frac{E_h - E_{\min}}{E_m - E_{\min}} \right). \qquad (2)$$

Here $E_{\min}$ is the pre-front value of the transmembrane voltage, and the dimensionless excitability g is defined as a nonlinear function of the dimensionless front speed

$$\sigma = c\sqrt{\tau_h/D}$$

and the dimensionless voltage load parameter

$$\beta = \frac{E_h - E_{\min}}{E_m - E_{\min}}$$

as

$$g(\sigma, \beta) = \frac{1 + \sigma^2}{(1 - \beta)\beta^{1/\sigma}}. \qquad (3)$$

Figure 2 illustrates these results, in comparison with the traditional simplified model. Panel (a) shows a typical behaviour of the front propagation speed in

---

[1] To avoid confusion, we stress here that the terminology we adopt may be different from other authors. Since in our approach gate j is considered as a slow variable, almost unchanged during the front, it is classified as an excitability condition. That is, it characterizes the ability for excitation, which is explicitly opposed to the variables E, m and h which change significantly during the front and thus represent excitation process proper.

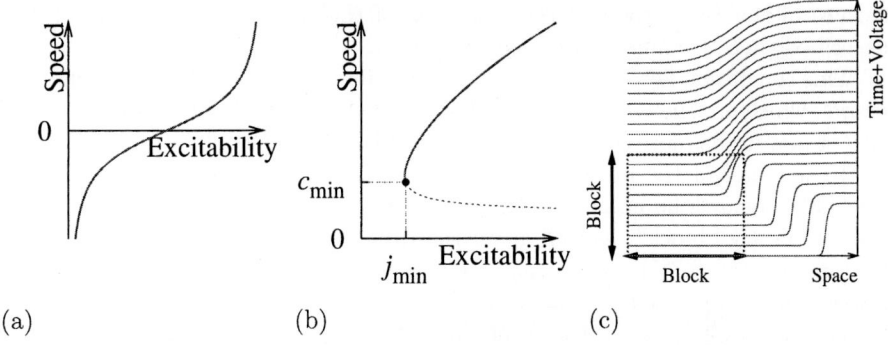

Fig. 2. Some analytical results on excitation propagation fronts, all graphs in arbitrary units. (a) Dependence of the front propagation speed on the instant value of an excitability parameter, in a FitzHugh-Nagumo-type model[15]. (b) Same, in our new simplified model based on detailed equations[20]; here the excitability parameter j is proportional to the product of the local value of I_{Na} conductivity, the slow inactivation gate, and the transmembrane voltage relative to I_{Na} reversal potential. (c) Dissipation of a front at a site with a temporary suppressed excitability and its failure to resume propagation after the excitability is recovered, in our new simplified model, in a setting similar to that on fig. 1

a traditional FitzHugh-Nagumo-type model, as a function of the instantaneous local value of a slow "excitability" parameter. In that class of simplified models, such parameters usually do not have a straightforward physiological connotation, as there only one slow variable is to represent all slow variables of detailed models at once. An essential feature of dependence shown on fig. 2(a) is that, as the excitability parameter varies, the propagation speed can be arbitrarily low, can be zero, and can even be negative, which corresponds to excitation front turning into a recovery front. Panel (b) illustrates what stands instead of this dependence in our new simplified model, where the excitability is varied via parameter j with other parameters fixed. The key feature of this dependence is that the excitability parameter given by equations (1,2) has a minimum as a function of speed, so for the front to propagate, excitability should not be less than a certain minimum j_{min}. For every value of excitability above that minimum, there are two solutions in the form of stationary propagating fronts. However, it appears that only the solution with the higher speed, shown with a solid line, is stable, while the solution with the lower speed, shown with a dashed line, is unstable [22]. Thus, a propagating front can have a speed no smaller than a certain c_{min}.

The New Model Describes Dissipation of Fronts. Thus we deduce that if the front, for any reason, is not allowed to propagate with a speed c_{min} or higher and/or if the local instant value of the excitability parameter is below j_{min}, then the stationary propagation would not be observed, and the only alternative is the front dissipation, as in the simplified model, this corresponds to a complete closure of the I_{Na} gates and the evolution of the transmembrane voltage E is described then by simply a diffusion equation.

This conclusion is confirmed by numerical simulations with the new simplified model, which are shown on fig. 2(c). Here the setting is similar to that of fig. 1(c), except now, to be more convincing, we did not use a complete block of excitability in the left half of the medium, but, rather, temporary decreased it to slightly (by 4.3%) below the critical value $j_{\min}$. The excitability in the left half after the temporary "block", as well as all the time in the right half of the medium, was slightly (by 8.7%) above $j_{\min}$. As a result, the excitation front reached the region with suppressed excitability, where it lost its sharp gradient, and after the excitability recovered, the front did not resume propagation but continued to spread diffusively. That is, it has shown exactly the same qualitative properties as observed in the full model (fig. 1c). So, our simplified model does take into account all the key factors involved in the front dissipation.

Application of the Propagation Condition to the Analysis of the Breakup of a Re-entrant Wave. So, our simplified model gives a necessary condition of propagation, in terms of the local excitability and the pre-front voltage. If the condition is not satisfied, the front cannot propagate and dissipates. Figure 3 shows a fragment of a simulation of a re-entrant wave in two-dimensional medium with the kinetics of Courtemanche et al. model[17], which is described in more detail in our recent work[5].

The top row shows distribution of the action potential, as it would be seen by an ideal optical mapping system (dark represents higher voltage). Propagation of a part of the re-entrant wave is blocked by the refractory tail of its previous turn. The wave then breaks up into two pieces, and the net result is there are now three free ends of excitation waves, i.e. three potential re-entry cores in place of one. The second row shows the profile of the transmembrane voltage along the dotted line on the upper panels. One can see first a reduction of the amplitude of the upstroke, and then the loss of the sharp upstroke altogether. The third row shows the profile of the factor of the I_{Na} due to the fast gates; the sharp peaks represent the excitation front. And the bottom row shows the profile of the slow gating variable j, which in our interpretation represents, together with the pre-front voltage, the conditions for the front propagation. The instant maximum of this profile is at the front, as the excitability restores before the fronts and falls after the front. The first shown moment $t = 4100\,\mathrm{ms}$ is when the excitability at the front drops down as low as the critical value, which is designated by a dotted horizontal line on the bottom row panels. If the front went slower at this moment, then excitability ahead of it would recover and it could propagate further. However, as predicted by our simplified model, the front speed cannot decrease below a certain minimum. So the front cannot slow down and "wait" until the excitability is recovered, but has to run further towards even a less excitable area. As a result, the conditions of propagation are no longer met, and the front dissipates, which is seen as the loss of the sharp gradient of $E(x)$, or, clearer, as disappearance of the peak of the $m^3(x)h(x)$ profile. After that, even though excitability j recovers above the critical level, the front does not resume.

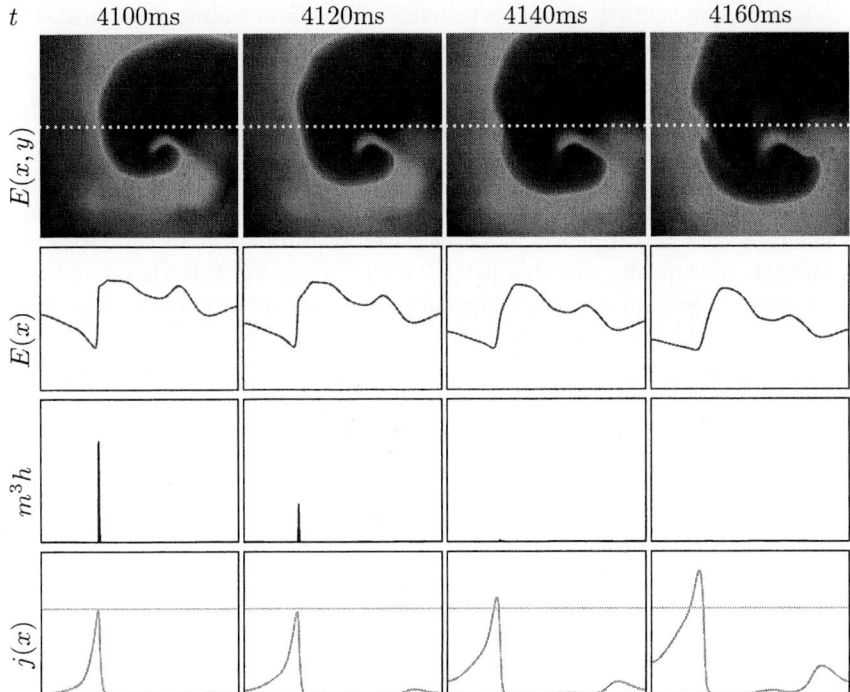

Fig. 3. Analysis of a break-up of a re-entrant wave in a two-dimensional (75 × 75 mm) simulation of a detailed model[17]. Top row: snapshots of the distribution of the transmembrane voltage, at the selected moments of time (designated above the panels). The other three rows: profiles of the key dynamic variables (designated on the left) along the dotted line shown on the top row panels, at the same moments of time. Dotted line here represents $j_{\min}$

Note that this analysis concerns only interaction of the front with the tail of the previous wave, and has nothing to do with the back of the new wave. Front dissipation occurs long before the wavelength reduces to zero. Of course, a break-up of a wave implies that its length vanishes eventually, but it will be long after the crucial events have already happened. Thus, the fate of the front here is determined already at the first snapshot, although it is not at all obvious in the voltage distribution.

3 Conclusions

Summary of Results

- For understanding the mechanisms of transient propagation blocks, such as occurring in re-entrant arrhythmia, it is important to bear in mind that the propagation speed in all circumstances has a *positive* lowest critical value, which is determined by the properties of the fast sodium channels. If the front is not propagated fast enough, say because excitability ahead of it recovers slower after the previous wave, then the front dissipates. After dissipation, the excitation front will not resume propagation even if excitability is restored.

- We have suggested a new simplified model that reproduces this behaviour of the excitation front, thus confirming the main physiological processes responsible for it. The simplified model is based on properties of the fast sodium current. Specifically, the dissipation of the excitation front is related to the simultaneous and mutually dependent dynamics of the transmembrane voltage E and the fast I_{Na} inactivation gate h.
- This particular mechanism of the propagation block is confined to the front, and has nothing to do with the wave back. That is, the propagation is blocked long before the wavelength reduces to zero.
- Apart from the transient propagation block, the new simplified model should be helpful in other cases concerning the margins of normal propagation. This includes initiation of excitation waves, which is the opposite of the propagation block, and the re-entrant waves around functional blocks, which imply juxtaposition of successful and unsuccessful propagation.
- FitzHugh-Nagumo type caricatures, although successfully describing successful propagation, fail to correctly describe propagation failure as it happens in reality or in detailed models. Thus using such models to describe any processes involving initiation of waves, block of propagation, or re-entrant waves, may misrepresent most important features. The new simplified model or its analogue should be used instead.

Limitations and Further Work. Model (1) has been obtained via a number of simplifications: freezing of slow processes, adiabatic elimination of fast processes, replacement of I_{Na} gates with perfect switches and replacement of $I_{Na,max}(E)$ and $\tau_h(E)$ with constants. Besides, the detailed models themselves are simplified, e.g. they are based on Hodgkin-Huxley description of Na channels rather than the more recent Markovian description. Validity of the results is therefore subject to one's ability to justify the simplifications and show that they do not alter the main properties. This is an ongoing work. We have shown recently that a formal asymptotic limit in Noble-1962 model naturally leads to (1) as a fast subsystem, and reproduces a single-cell action potential with a good accuracy [23]. We have also demonstrated that a similar asymptotic limit works in Courtemanche et al. model, and system (1) obtained in this way gives a reasonable estimate of the critical conditions of front dissipation, which can be further improved by taking into account the dynamics of m-gates instead of adiabatically eliminating them [24]. A Markovian, non-Hodgkin-Huxley description of I_{Na} involves a radically different description of the Na channels. Inasmuch as the old Hodgkin-Huxley description was reasonably accurate phenomenologically, one can expect that the main features should maintain; however, an ultimate answer to that can only be obtained via a further detailed study.

Acknowledgements. This work was supported in part by grants from EPSRC (GR/S43498/01, GR/S75314/01) and by an RDF grant from Liverpool University.

References

[1] V. I. Krinsky. Fibrillation in excitable media. *Cybernetics problems*, 20:59–80, 1968.
[2] A. V. Panfilov and A. V. Holden. Self-generation of turbulent vortices in a 2-dimensional model of cardiac tissue. *Phys. Lett. A*, 151:23–26, 1990.
[3] V. I. Krinsky and I. R. Efimov. Vortices with linear cores in mathematical-models of excitable media. *Physica A*, 188:55–60, 1992.
[4] V. N. Biktashev and A. V. Holden. Re-entrant waves and their elimination in a model of mammalian ventricular tissue. *Chaos*, 8:48–56, 1998.
[5] I. V. Biktasheva, V. N. Biktashev, W. N. Dawes, A. V. Holden, R. C. Saumarez, and A. M.Savill. Dissipation of the excitation front as a mechanism of self-terminating arrhythmias. *IJBC*, 13(12):3645–3656, 2003.
[6] I. V. Biktasheva, V. N. Biktashev, and A. V. Holden. Wavebreaks and self-termination of spiral waves in a model of human atrial tissue, 2005. In this issue.
[7] R. A. Gray and J. Jalife. Spiral waves and the heart. *Int. J. of Bifurcation and Chaos*, 6:415–435, 1996.
[8] J. N. Weiss, P. S. Chen, Z. Qu, H. S. Karagueuzian, and Garfinkel A. Ventricular fibrillation: How do we stop the waves from breaking? *Circ. Res.*, 87:1103–1107, 2000.
[9] A. Panfilov and A. Pertsov. Ventricular fibrillation: evolution of the multuple-wavelet hypothesis. *Phil. Trans. Roy. Soc. Lond. ser. A*, 359:1315–1325, 2001.
[10] J. B. Nolasco and R. W. Dahlen. A graphic method for the study of alternation in cardiac action potentials. *J. Appl. Physiol.*, 25:191–196, 1968.
[11] A. Karma, H. Levine, and X. Q. Zou. Theory of pulse instabilities in electrophysiological models of excitable tissues. *Physica D*, 73:113–127, 1994.
[12] E. M. Cherry and F. H. Fenton. Suppression of alternans and conduction blocks despite steep APD restitution: Electrotonic, memory, and conduction velocity restitution effects. *Am. J. Physiol. - Heart C*, 286:H2332–H2341, 2004.
[13] R. FitzHugh. Impulses and physiological states in theoretical models of nerve membrane. *Biophys. J.*, 1:445–456, 1961.
[14] J. Nagumo, S. Arimoto, and S. Yoshizawa. An active pulse transmission line simulating nerve axon. *Proc. IRE*, 50:2061–2070, 1962.
[15] H. P. McKean. Nagumo's equation. *Adv. Appl. Math.*, 4:209–223, 1970.
[16] J. J. Tyson and J. P. Keener. Singular perturbation theory of traveling waves in excitable media (a review). *Physica D*, 32:327–361, 1988.
[17] M. Courtemanche, R. J. Ramirez, and S. Nattel. Ionic mechanisms underlying human atrial action potential properties: insights from a mathematical model. *Am. J. Physiol.*, 275:H301–H321, 1998.
[18] A. L. Hodgkin and A. F. Huxley. A quantitative description of membrane current and its application to conduction and excitation in nerve. *J. Physiol.*, 117:500–544, 1952.
[19] D. Noble. A modification of the Hodgkin-Huxley equations applicable to Purkinje fibre action and pace-maker potentials. *J. Physiol.*, 160:317–352, 1962.

[20] V. N. Biktashev. Dissipation of the excitation wavefronts. *Phys. Rev. Lett.*, 89(16):168102, 2002.
[21] V. N. Biktashev. A simplified model of propagation and dissipation of excitation fronts. *Int. J. of Bifurcation and Chaos*, 13(12):3605–3620, 2003.
[22] R. Hinch. Stability of cardiac waves. *Bull. Math. Biol.*, 66(6):1887–1908, 2004.
[23] V .N. Biktashev and R. Suckley. Non-Tikhonov asymptotic properties of cardiac excitability. *Phys. Rev. Letters*, 93:168103, 2004.
[24] I. V. Biktasheva, R. S. Simitev, R. Suckley, and V. N. Biktashev. Asymptotic properties of mathematical models of excitability, 2005. Submitted to Phil. Trans. Roy. Soc. A.

Wavebreaks and Self-termination of Spiral Waves in a Model of Human Atrial Tissue

Irina V. Biktasheva[1], Vadim N. Biktashev[2], and Arun V. Holden[3]

[1] Department of Computer Science, Liverpool University, Liverpool L69 3BX, UK
ivb@csc.liv.ac.uk
http://www.csc.liv.ac.uk/~ivb
[2] Department of Mathematical Sciences, Liverpool University,
Liverpool L69 7ZL, UK
[3] School of Biomedical Sciences, Leeds University, Leeds LS2 9JT, UK

Abstract. We describe numerical simulations of spiral waves dynamics in the computational model of human atrial tissue with the Courtemanche-Ramirez-Nattel local kinetics. The spiral wave was initiated by cross-field stimulation protocol, with and without preliminary "fatigue" by rapid stimulation of the model tissue for a long time. In all cases the spiral wave has finite lifetime and self-terminates. However the mechanism of self-termination appears to depend on the initiation procedure. Spiral waves in the "fresh" tissue typically terminate after a few rotations via dissipation of the excitation front along the whole of its length. The dynamics of spiral waves in "tired" tissue is characterized by breakups and hypermeander, which also typically leads to self-termination but only after a much longer interval of time. Some features of the observed behaviour can not be explained using existing simplified theories of dynamic instabilities and alternanses.

1 Introduction

In this paper we continue to investigate the behaviour of re-entrant waves of excitation in a computational model of human atrial tissue, which we started in [1]. The model showed spontaneous break-ups and self-termination of spiral waves, which can have relationship to mechanisms underlying atrial fibrillation, a condition that adversely affects quality of life and bears potential life threat. So despite all the limitations coming from simplified geometry, homogeneity and isotropy of our model, these results were suggestive and promising, and warranted further investigation. The mechanisms of breakups and self-termination of spirals in our computational model are far from understanding. In this paper we extend the study to different types of initial conditions and to different methods of analysis. The purpose of considering different initial conditions is to assess their effect on the re-entry behaviour. In the analysis of results of simulation, we assess, in particular, the feasibility of the "slope-one" theory, started by Nolasco and Dahlen [2] and given much prominence recently [3–6]; we refer to it as Nolasco-Dahlen (ND) theory for brevity. This theory has been tested and confirmed on some simplified models and some experimental preparations, although its universal applicability so far remains questionable [6]. In this paper we deliberately avoid discussing the basis of the theory, but restrict ourselves to purely phenomenological analysis of the results of simulations. Thus the structure of the paper is as of an experimental paper, with Section 2 dedicated to the methods, Section 3 to the results, and Section 4 to their discussion.

2 Methods: 2D Model of Human Atrial Tissue

Excitation Kinetics of Cells. We used the human atrial action potential model by Courtemanche et al. [7] (CRN model) incorporated into a two-dimensional reaction-diffusion system of 21 partial differential equations.

Tissue Model. We modelled the tissue as a continuous, homogeneous, isotropic, monodomain syncytium, i.e. in terms of "reaction+diffusion" system of equations, with diffusion term only in the equation for the transmembrane voltage,

$$\frac{d\mathbf{u}}{dt} = \mathbf{f}(\mathbf{u}) + \mathbf{D}\nabla^2 \mathbf{u}$$

where $\mathbf{u} = (E, m, h, \dots)^T \in \mathbb{R}^{21}$ is the vector of the dynamic variables of the model, and $\mathbf{D} = \mathrm{diag}(D, 0, 0, \dots)$ is the matrix of diffusion coefficients, of which only the diffusion coefficient of the transmembrane voltage E is nonzero. This simplified description focuses on the excitation and propagation kinetics and allows interpretation in terms of numerous theories applicable to this kind of equations, while ignoring the additional complications due to geometry, anisotropy and heterogeneity of a real atrium. The diffusion coefficient of the transmembrane voltage $D = 0.03125\,\mathrm{mm}^2/\mathrm{ms}$ was set to give a plane wave velocity of $\approx 0.265\,\mathrm{mm}/\mathrm{ms}$. Different D produce identical behaviour, only on a different spatial scale. The problem was posed in a square $75\,\mathrm{mm} \times 75\,\mathrm{mm}$ with no-flux boundary conditions for E.

The Numerical Scheme. The partial differential equations were solved using explicit Euler scheme in time, with time step $\Delta t = 0.1\mathrm{ms}$, and simplest second-order approximation of the Laplacian space step $\Delta x = 0.2\mathrm{mm}$.

Initial Conditions. Spiral waves in this study were initiated by the **cross-field stimulation method**. This is a widely used method for initiation of spiral waves, as it is relatively easy to implement both in the experiments and in the simulations. Our numerics used one or more of plane waves (*conditioning waves*) initiated to propagate from right to left of the medium; then at a certain moment, when the recovery tail of a wave is somewhere in the middle of the medium, we excited the lower half of the medium by instantly raising the transmembrane voltage by 100 mV in the lower part of the medium. This creates a new excitation front in the right half of the medium where it has recovered but not in the left which is still refractory. This broken excitation front quickly develops into a spiral wave. This is different from the **phase distribution method** whereby one-dimensional calculations are used to record values of all dynamical variables in a plane periodic wave of a high frequency, $\mathbf{U}(\phi)$, where $\mathbf{U} \in \mathbb{R}^{21}$ is the vector of dynamic variables of the model and $\phi \in \mathbb{R} \mod 2\pi$ is the phase within the period, and then the initial conditions are set as $\mathbf{u}(x,y,0) = \mathbf{U}(\phi(x,t))$, where $\phi(x,y)$ is the distribution of the phase, chosen by will, e.g. corresponding to Archimedean spiral; thus the name of the method. The phase distribution method was used in our previous work [1] as it allowed quick generation of a spiral wave with the desired position of the core in the medium, and promised freedom from artificial inhomogeneities introduced by initial conditions. In present work, we used the cross-field stimulation as more realistic physiologically, and to see to what extent the behaviour of the spiral waves depends on the details of the initial conditions.

As in the case of phase-distribution method, the cross-field method can be implemented not only with 'fresh' medium but also with a 'tired' medium. The **fresh medium** was where the function $\mathbf{U}(\phi)$ and the conditioning waves are obtained by propagating a single wave through the medium, which prior to that was in the steady equilibrium 'resting state'. For the **tired medium**, this wave was the last wave in the series of a long (30 s long) series of rapid (period 300 ms) plane waves. There are 'superslow' variables in the model, which do not fully recover within a 300 ms excitation cycle; these changes accumulate over time which effectively amounts to change in the model parameters. Relevance of such changes in a particular model to any physiological condition is debatable, see [8] and references therein. Yet, these changes are relatively minor, and provide an example of physiologically feasible parameter variations, perhaps the best of what is achievable within the framework of this particular model and without involving further experimental data.

Thus, we had two sets of numerical experiments, with two different types of media, 'fresh' and 'tired'. This can be compared to two sets of simulations with similarly 'fresh' and 'tired' media but with phase-distribution initial conditions described in [1].

Processing of Results. We depict the front propagation patterns as snapshots of the excitation field and by isochrone maps. **Snapshots** show distribution of the fields of $E(x,y,t)$, the transmembrane voltage, and $o_i(x,y,t)$, the inactivation gate of the transient outward current, as functions of x,y at fixed selected values of time t. On each snapshot, the red component of the colour of a point corresponds to the value of E, with zero corresponding to $E = -100$ mV and maxium corresponding to $E = 50$ mV, and the green component of the colour corresponds to the value of o_i, with zero corresponding to $o_i = 0$ and the maximum corresponding to $o_i = 1$. By virtue of the excitation kinetics, the green and red colours are almost complement of each other. In black and white printed version, the regions with higher E (excited, systolic regions) are darker than the regions with higher o_i (unexcited, diastolic regions). **Isochrone maps** are collections of **isochrones**, i.e. instant positions of wavefronts or wavebacks, defined as fragments of isolines $E(x,y,t) = -40$ mV which satisfy condition of

$o_i(x,y,t) > 0.5$ (wavefronts) or $o_i(x,y,t) < 0.5$ (wavebacks) for a given value of t. Correspondingly, the **wavebreak points**, and spiral **wave tips** are defined as internal ends of these fragments, i.e. intersections of isolines $E(x,y,t) = -40\,\mathrm{mV}$ and $o_i(x,y,t) = 0.5$. The moment of **self-termination** of the arrhythmia was defined as the moment of ultimate disappearance of all tips; this inevitably lead to eventual return of the medium to the uniform resting state when last excitation waves reach boundaries. **Restitution curves** are usually defined as dependence of the **action potential duration (APD)** of a cell on the preceding **diastolic interval (DI)** of that cell. In our numerics, we defined APD as a continuous interval of time t when $E(x,y,t) > E_*$ at a given point (x,y) and for a certain fixed E_*; correspondingly, DI is the interval when $E(x,y,t) < E_*$. We have tried different values of E_*.

3 Results

3.1 Fresh Medium

We have made 7 simulations of spiral waves stimulated in the fresh medium, different in the initial position of the spiral wave with respect to the medium boundaries. This allowed us assess the effect of boundaries on the spiral wave dynamics. We describe in detail one simulation of this series. Figure 1 shows a collection of snapshots, from the moment of initiation by cross-field stimulation, $t = 0$, with an interval of 100 ms up to the frame $t = 1900$ ms, after which the spiral wave self-terminates, i.e. excitation fails to re-enter in the medium and eventually decays (not shown). As can be seen from the movie, but not necessary from the set of still pictures, the key event leading to the self-termination happens at around $t \approx 1400$ ms and is characterized by block of propagation, at which the wavefront "dissipates" along a long line, stretching almost up to the upper boundary of the medium.

To visualize this and other important events, we employed the method of isochrone maps. The relationship between the snapshot and isochrone representation is illustrated by fig. 2, where a front isochrone, a tip and a back isocrhone are shown for a selected snapshot from the previous series.

The isochrone maps, i.e. collection of such isochrones for a selected intervals of times, separately for fronts and backs, are shown on fig. 3. Whereas propagation of fronts is more or less smooth everywhere where the fronts propagate at all, the evolution of the back is highly irregular, and this irregularity has a tendency to increase with time. This is a display of the "dynamic inhomogeneity" discussed in [1]. At times the irregularity reaches a stage where visual propagation of the back is opposite to the propagation of the previous front at that point. In extreme cases, one can see islands of recovered medium before the overall back of the wave reaches that site. One such episode can be seen on the panel of back isochrones for $t = 600\ldots900$ ms in the left top quarter. This is the effect of "triggered recovery" [1,9]. The inhomogeneity of the propagation pattern of wavebacks, with the relative homogeneity of the wavefronts, is an evidence of high variability of the action potential durations. This inhomogeneity affects propagation of next

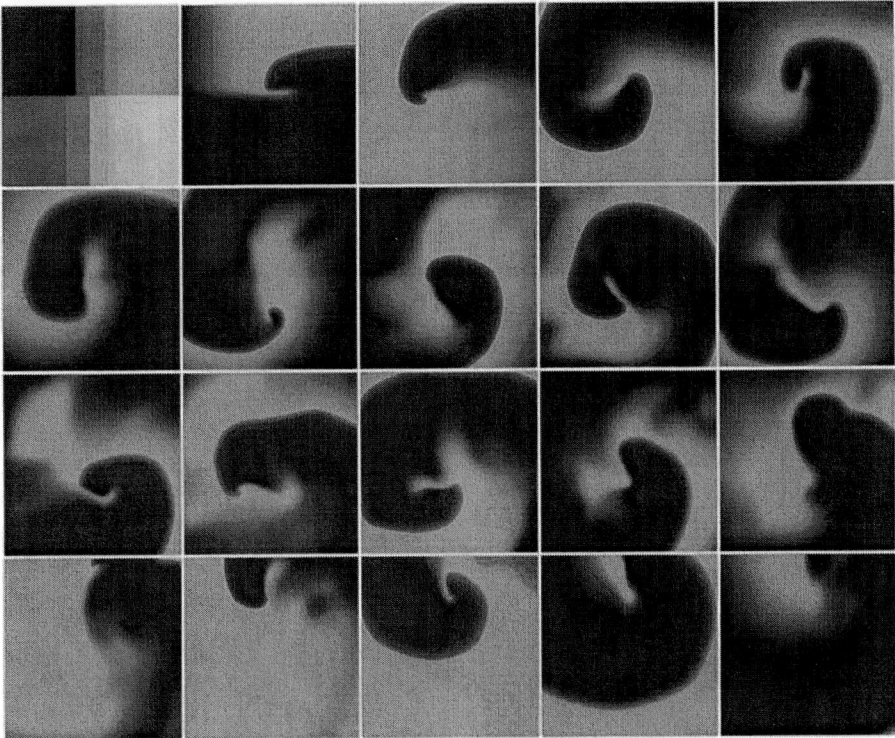

Fig. 1. (color online) Development of a spiral wave, initiated by cross-field stimulation in the fresh medium. Shown are snapshots with interval 100 ms, starting from 0 ms, in the "reading order" (left to right, then top to bottom)

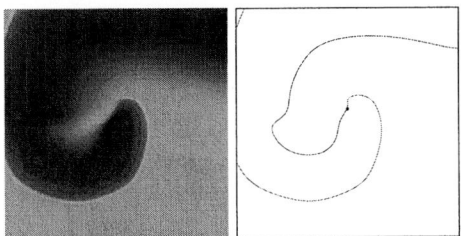

Fig. 2. (color online) Alternative representations of an excitation pattern. Left: snapshot $t = 300$ ms from fig. 1. Right: the same, represented by isoline $E = -40$ mV, the 'isochrone'. The dot on the line is the tip of the spiral, defined by the additional condition $o_i = 0.5$. This point splits the isoline to two parts, front (red) and back (blue)

wavefronts either by slightly delaying or speeding them up, and from time to time by blocking next fronts, which leads to sudden displacement of the functional block of the spiral wave, which results in the highly irregular, "hyper-meander"

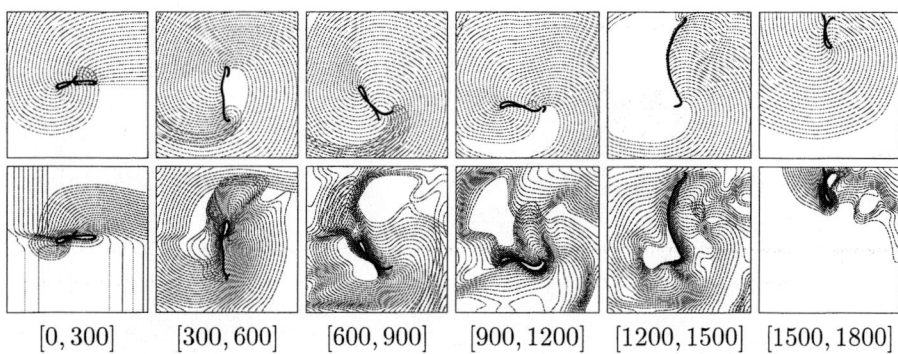

[0, 300]   [300, 600]   [600, 900]   [900, 1200]   [1200, 1500]   [1500, 1800]

Fig. 3. (color online) Isochronal maps corresponding to fig. 1. Time interval between isochrones is 10 ms. Fronts are on upper panels, backs are on lower panels. Bold points, blending together into thick black lines, are the free ends of the isochrones, i.e. tips of the spirals

trajectory of the spiral tip. This is likely to happen where the speed of recovery wave is slow, which shows as dense location of back isochrones. The above mentioned key event of front propagation block is clearly seen on front isochrone map $t = 1200\ldots1500$ ms as a long almost straight trajectory of the wavetip at which the wavefronts retract rather than protrude; this is a visual effect of wavefronts actually dissipating when reaching that line. This event is preceded by a density of wavebacks, which is seen on back isochrone maps of $t = 900\ldots1200$ ms (from approx. 1000 ms on) and $t = 1200\ldots1500$ ms (up until approx. 1300 ms), which occurs right at the site where subsequent fronts are blocked.

A distinct feature of the above discussed simulation is a relatively short lifetime of the spiral wave, 1776 ms. Other 6 simulations with different initial positions of the spiral showed similarly short lifetimes, from 1732 ms to 2732 ms, with average and standard deviation of 1900 ± 367 ms for all 7 simulations.

3.2 Tired Medium

Figures 4 and 5 show evolution of a spiral wave in a selected simulation with a tired medium. The lifetime of this spiral wave was 6992 ms, i.e. much longer than those in a fresh medium. Figure 4 shows snapshots in the first 3400 ms of that interval and fig. 5 shows isochrones in the first 1500 ms of it.

As in the fresh medium, there is prominent dynamic inhomogeneity caused by variability of APD and reflected by irregular patterns of the waveback isochrones. However, this time the irregularity seems less, although we didn't measure it by any formal measure. As a result, the dynamic imhomogeneity does not develop a complete block of the propagation and instead shows localized block of propagation, i.e. a breakup of the wave. The analysis of the record of the wavebreak points in the numerics of figs. 4,5 shows that the breakups lead to generation of new pairs of spiral waves which occur both nearly in the same place and each

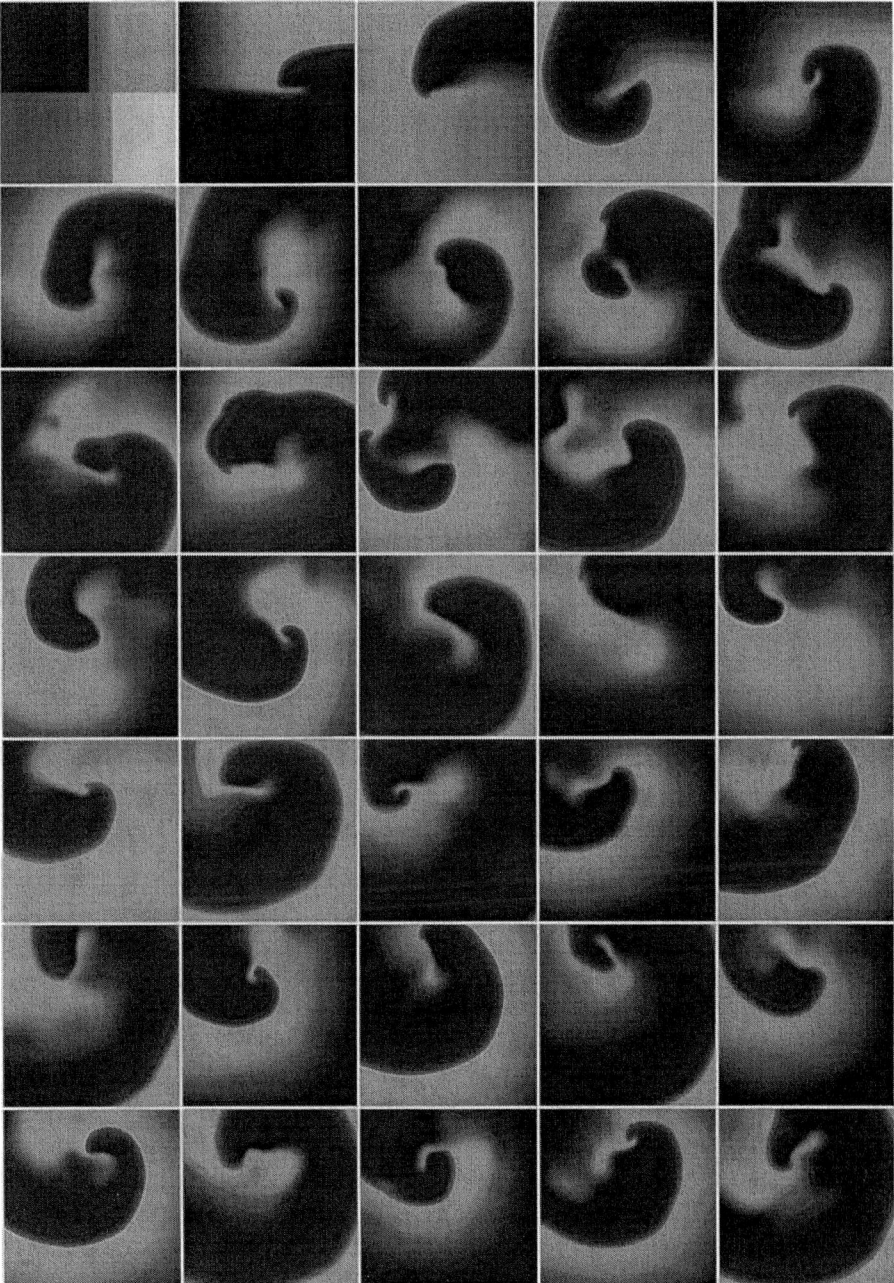

Fig. 4. (color online) Spiral wave in a tired medium. Shown are snapshots with interval 100 ms, starting from 0 ms

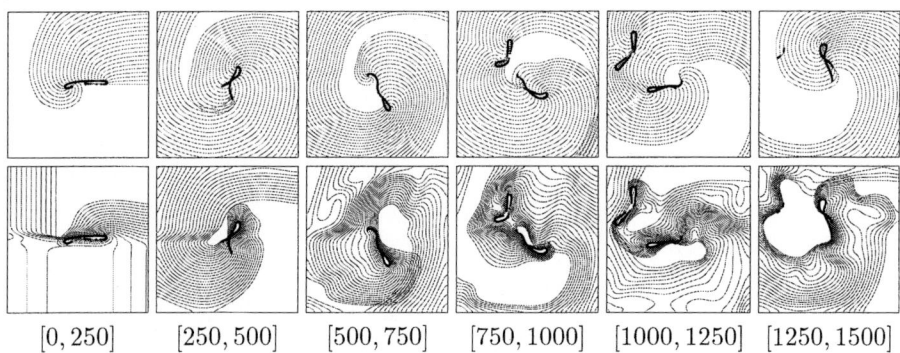

| [0, 250] | [250, 500] | [500, 750] | [750, 1000] | [1000, 1250] | [1250, 1500] |

Fig. 5. (color online) Isochronal maps of initial 1800 ms of evolution of a spiral wave initiated by cross-field method in a tired medium. Notations are the same as on fig. 3

exist for about one period before annihilating, in the intervals $t = 770\ldots 874$ ms and $t = 1132\ldots 1258$ ms (although at different definition of wavebreaks this could perhaps be considered as one pair of spirals which existed for two rotations).

Evolution in other two simulations with the tired medium showed similar types of evolution and similarly large lifetimes of the spirals, from 4034 ms to 9924 ms with average and standard deviation of 3 simulations 6983 ±2945 ms.

3.3 De-facto Restitution Properties

The ND theory and its variations aim to describe precisely the kind of process, when the action potential duration (APD) variability in response to the history of excitation causes instability of regular propagation of waves. This theory is based on the relationship between the diastolic interval (DI) and subsequent APD. To test how much this theory can be applied to our model, we have recorded values of transmembrane voltage, $E_j(t) = E(x_j, y_j, t)$, for a regular grid of 13×13 points (x_j, y_j), $j = 1\ldots 269$ regularly spread through the medium. We have tried different values of E_* for defining APD and DI. Figure 6 shows a typical electrogram in relation to voltages $E_* = -33$ mV, corresponding to $\bar{m}(E_*) = 1/2$, $E_* = -67$ mV, which corresponds to $\bar{h}(E_*) = 1/2$, and $E_* = -50$ mV which is the average of the two, i.e. in the middle of the fast Na current excitation window. We see that the high variability of action potential profiles makes any of these voltages not ideal for determining APD and DI, although, arguably, $E_* = -67$ mV looks more sensible than the other two.

The graphs of APD vs DI for the three selected values of E_* are shown on fig. 7. We believe that even bearing in mind all possible sources of errors, these graphs are an evidence that DI is not a good predictor of APD at all.

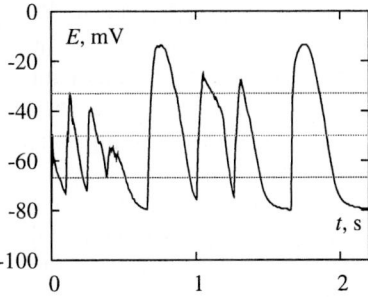

Fig. 6. An electrogram recorded at point with coordinates $(35.2, 35.2)$ mm from the left top corner in the numerics described in figs. 1 and 3. Horizontal lines show the levels at which detection of APD and DI was attempted

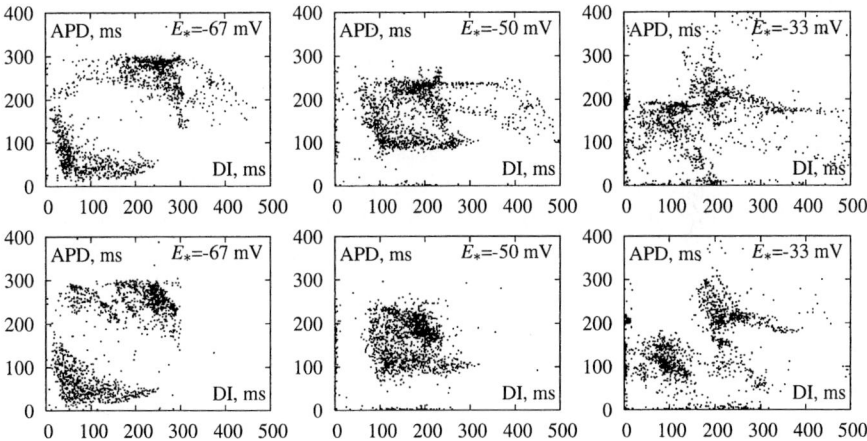

Fig. 7. Graphs of APD vs DI for electrograms at 289 points in the numerics shown on figs. 1 and 3, for different values of E_*. Top row: fresh medium. Bottom row: tired medium

4 Discussion

We have found that the difference in the initial condition does not qualitatively change the behaviour of spiral waves in this model of atrial tissue. As in [1], we see that the evolution of spiral waves is dominated by such factors as developing dynamic inhomogeneity, including triggered recovery, which causes localized or massive dissipation events of wavefronts, which in turn lead to hypermeander of the tip of the spiral, spontaneous generation of new spirals, and eventually to self-termination of all re-entrant activity. The new study confirms that the time to self-termination in a tired medium is much longer than in the fresh medium.

This finding seems to suggest a possible mechanism of proarrhythmic action of tissue fatigue. However, it is not certain how our "numerical fatigue" relates to physiology. So the theoretical aspect, the mechanism of the development of the dynamic inhomogeneity, may be even more important. The ND theory explains the dynamic inhomogeneity based on the assumption that an APD is determined by the immediately preceding DI. This works for some models [5]. However, this assumption does not seem to bear any resemblance to the processes happening in this model. Perhaps, correct predictors of APD can be found based on asymptotic analysis of the realistic excitability models, rather than phenomenology of simplified models. Also, there is no obvious explanation in ND theory to the difference between the behaviour in fresh and recovered medium, perhaps because that theory does not even aim to explain the process of propagation block, only development of a dynamic instability.

An interesting theoretical mechanism of finite lifetime of re-entrant waves was suggested long ago by Krinsky [10], within an axiomatic "tau-model", which has some properties that are relevant to cardiac tissue but not captured by FitzHugh-Nagumo type simplified models. Although the mechanism of finite lifetime re-entry of [10] is based on static inhomogeneity of tissue properties and thus not directly applicable to our case, it still can offer some food for thought. A new hope on further progress in understanding the phenomena described in this work comes from recent results on asymptotic properties of realistic models which make them different from traditional simplified models [11, 12].

Acknowledgements. This work was supported in part by grants from EPSRC (GR/S03027/01, GR/S43498/01, GR/S75314/01) and by an RDF grant from Liverpool University.

References

[1] I. V. Biktasheva, V. N. Biktashev, W. N. Dawes, A. V. Holden, R. C. Saumarez, and A. M.Savill. Dissipation of the excitation front as a mechanism of self-terminating arrhythmias. *IJBC*, 13(12):3645–3656, 2003.

[2] J. B. Nolasco and R. W. Dahlen. A graphic method for the study of alternation in cardiac action potentials. *J. Appl. Physiol.*, 25:191–196, 1968.

[3] M. Courtemanche, L. Glass, and J. P. Keener. Instabilities of a propagating pulse in a ring of excitable media. *Phys. Rev. Lett.*, 70:2182–2185, 1993.

[4] A. Karma, H. Levine, and X. Q. Zou. Theory of pulse instabilities in electrophysiological models of excitable tissues. *Physica D*, 73:113–127, 1994.

[5] M. Courtemanche. Complex spiral wave dynamics in a spatially distributed ionic model of cardiac electrical activity. *Chaos*, 6(4):579–600, 1996.

[6] E. M. Cherry and F. H. Fenton. Suppression of alternans and conduction blocks despite steep APD restitution: Electrotonic, memory, and conduction velocity restitution effects. *Am. J. Physiol. - Heart C*, 286:H2332–H2341, 2004.

[7] M. Courtemanche, R. J. Ramirez, and S. Nattel. Ionic mechanisms underlying human atrial action potential properties: insights from a mathematical model. *Am. J. Physiol.*, 275:H301–H321, 1998.

[8] J. Kneller, R. J. Ramirez, D. Chartier, M. Courtemanche, and S. Nattel. Time-dependent transients in an ionically based mathematical model of the canine atrial action potential. *Am. J. Physiol. Heart Circ. Physiol.*, 282:H1437–H1461, 2002.

[9] L. J. Leon, F. A. Roberge, and A. Vinet. Simulation of two-dimensional anisotropic cardiac reentry: effects of the wavelength on the reentry characteristics. *Ann. Biomed. Eng.*, 22:592–609, 1994.

[10] V. I. Krinsky. Fibrillation in excitable media. *Cybernetics problems*, 20:59–80, 1968.

[11] R. Suckley and V. N. Biktashev. Comparison of asymptotics of heart and nerve excitability. *Phys. Rev. E*, 68:011902, 2003.

[12] V. N. Biktashev and R. Suckley. Non-Tikhonov asymptotic properties of cardiac excitability. *Phys. Rev. Letters*, 93:168103, 2004.

Calcium Oscillations and Ectopic Beats in Virtual Ventricular Myocytes and Tissues: Bifurcations, Autorhythmicity and Propagation

Alan P. Benson and Arun V. Holden

Computational Biology Laboratory, School of Biomedical Sciences,
University of Leeds, Leeds LS2 9JT, UK
alan@cbiol.leeds.ac.uk
http://www.cbiol.leeds.ac.uk

Abstract. One mechanism for the onset of arrhythmias is abnormal impulse initiation such as ventricular ectopic beats. These may be caused by abnormal calcium (Ca^{2+}) cycling. The Luo-Rudy model was used to simulate the dynamics of intracellular Ca^{2+} ($[Ca^{2+}]_i$) handling and the initiation of ectopic beats in virtual ventricular myocytes and tissues. $[Ca^{2+}]_i$ in the reduced Ca^{2+} handling equations settles to a steady state at low levels of intracellular sodium ($[Na^+]_i$), but oscillates when $[Na^+]_i$ is increased. These oscillations emerge through a homoclinic bifurcation. In the whole cell, Ca^{2+} overload, brought about by inhibition of the sodium-potassium pump and elevated $[Na^+]_i$, can cause autorhythmic depolarisations. These oscillations interact with membrane currents to cause action potentials that propagate through one dimensional virtual tissue strands and two dimensional anisotropic virtual tissue sheets.

1 Introduction

Cardiac arrhythmias such as ventricular tachycardia and fibrillation are a major cause of morbidity and mortality in developed countries. Ca^{2+} overload in cardiac myocytes is well known to be a cause of delayed afterdepolarisations, where an action potential is followed by triggered activity [1, 2, 3, 4, 5]. This is a distinct arrhythmogenic mechanism from autorhythmicity, where spontaneous depolarisations occur from the resting membrane potential and do not necessarily require a preceding action potential [1]. Here we use virtual ventricular myocytes and tissues to investigate the role of Ca^{2+} in causing such autorhythmicity.

Normally, regular membrane potential (V) oscillations (the action potentials) in ventricular myocytes drive regular intracellular calcium ($[Ca^{2+}]_i$) oscillations, but under conditions of sarcoplasmic reticulum (SR) Ca^{2+} overload, Ca^{2+} is spontaneously released from the SR [3], causing $[Ca^{2+}]_i$ oscillations independently of V oscillations. These $[Ca^{2+}]_i$ oscillations can be arrhythmogenic if they interact with the cell membrane and produce transient inward currents that depolarise V past threshold. The transient inward currents that respond to spontaneous SR Ca^{2+} release could be the Na^+-Ca^{2+} exchange current I_{NaCa} and/or the Ca^{2+}-activated non-specific current $I_{ns(Ca)}$ [6,7]. Propagation of these autorhythmic depolarisations into surrounding tissue is dependent on the size of the autorhythmic focus and on the cell-cell coupling within the tissue.

One mechanism of inducing Ca^{2+} overload is block of the Na^+-K^+ pump current I_{NaK}, as seen during ischemia [8] and digitalis intoxication [9,10] for example. The consequent build up of $[Na^+]_i$ reduces the effectiveness of the Na^+-Ca^{2+} exchanger at removing Ca^{2+} from the cell and intracellular Ca^{2+} concentrations become elevated [9].

Mathematically, the conditions that lead to this type of autorhythmicity can be identified using bifurcation analysis: at a bifurcation the qualitative behaviour of a system changes, from a single stationary solution to oscillatory activity, for example. Several previous studies have used mathematical models of atrial and Purkinje fibre cells and tissues to investigate autorhythmicity brought about by abnormal Ca^{2+} handling. Varghese & Winslow [11] examined the stability of the equations describing the Ca^{2+} subsystem in the DiFrancesco-Noble model [12] of the cardiac Purkinje fibre. Using a clamped voltage between -40 and -100 mV and constant $[Na^+]_i$ (a step justified by the slow dynamics of $[Na^+]_i$ compared to $[Ca^{2+}]_i$), they found a single stationary solution for $[Ca^{2+}]_i$ at low values of $[Na^+]_i$. As $[Na^+]_i$ was increased, a supercritical Hopf bifurcation led to an unstable fixed point and stable periodic $[Ca^{2+}]_i$ oscillations. These oscillations were shown to alter regular V oscillations in the complete model [13]. Winslow et al. [14] showed that inhibition of the Na^+-K^+ pump in an atrial cell model [15] resulted in $[Na^+]_i$ overload and $[Ca^{2+}]_i$ oscillations driving V oscillations. When a compact subset of around 1000 $[Na^+]_i$ overloaded cells were placed in the centre of a two-dimensional (2-D) tissue composed of 512 × 512 cells, the depolarisations could propagate out into the surrounding quiescent tissue. Recently, Joyner et al. [16] examined how a spontaneously depolarising focus leads to excitation of sheets of atrial and ventricular cell models, while Wilders et al. [17] examined the effects of tissue anisotropy on propagation from an autorhythmic focus in a virtual ventricular sheet.

We used the Luo-Rudy dynamic (LRd00) model of the ventricular myocyte [18] to: (i) identify bifurcations that lead to autorhythmicity; (ii) induce $[Ca^{2+}]_i$ overload and ectopic beats in single myocytes and (iii) characterise the conditions required for propagation of these beats in one dimensional (1-D) and 2-D virtual ventricular tissues.

2   Numerical Methods

LRd00 models the ventricular action potential using an ordinary differential equation (ODE) that describes the rate of change of V:

$$\frac{dV}{dt} = \frac{-1}{C_m} I_{ion}, \tag{1}$$

where $C_m = 1\,\mu F\,cm^{-2}$ is membrane capacitance and I_{ion} is the sum of ionic currents through the cell membrane. I_{ion} is composed of voltage gated channel currents modelled using Hodgkin-Huxley formalism [19], as well as currents carried by other channels, pumps and exchangers. Ionic concentrations are modelled using the ODE:

$$\frac{d[B]}{dt} = \frac{-I_B \cdot A_{cap}}{Vol_C \cdot z_B \cdot F}, \quad (2)$$

where [B] is the concentration of ion B, I_B the sum of the currents carrying ion B, A_{cap} is capacitive membrane area, Vol_C the volume of the compartment whose concentration is being updated, z_B the valency of ion B and F is the Faraday constant. Parameter values and equations describing the ionic currents can be found in [18, 20, 21, 22, 23]. Equations describing gating variables were solved using the scheme of Rush & Larsen [24], and equations of the form (1) and (2) using an explicit Euler method. The source code, written in C/C++, can be found at http://www.cwru.edu/med/CBRTC/LRdOnline/LRdModel.htm. Two variants of the model were used: one describing the Ca^{2+} handling system, the other describing a single myocyte. These were incorporated into 1-D and 2-D tissues.

2.1 Calcium Handling Equations

We reduced the LRd00 equations to those describing the Ca^{2+} handling system by applying a V clamp and using an adiabatic approximation where $[Na^+]$ and $[K^+]$ remain constant [11, 25]. V-dependent gates take on their steady-state values and, as V, $[Na^+]$ and $[K^+]$ are constant, any ionic currents that contribute only to the rate of change of these variables were removed from the system. The LRd00 equations were therefore reduced to four ODEs describing the rate of change of total $[Ca^{2+}]$ in the network SR, junctional SR, intracellular and extracellular spaces. $[Ca^{2+}]_i$ is dependent on $[Na^+]$ via I_{NaCa} and on V via the driving force of the membrane currents and the V-dependent gates of the L-type Ca^{2+} channel (gates d and f) and the T-type Ca^{2+} channel (gates b and g). $[Na^+]_i$ and V were therefore treated as control parameters.

2.2 Modified Single Myocyte

The complete LRd00 virtual endocardial cell was modified to induce $[Ca^{2+}]$ overload: 100% I_{NaK} inhibition, $I_{ns(Ca)}$ increased twofold, $[CSQN]_{th}$ decreased to 7.0 mM, and the time constants of activation and inactivation of $I_{rel,jsrol}$ increased to 5 ms. In all Ca^{2+} handling and single myocyte integrations, a time step of $dt = 0.01$ ms was used.

2.3 One- and Two-Dimensional Tissues

A 15 mm 1-D tissue strand composed of equal fractions of endocardial, midmyocardial and epicardial tissue [22], and a 60 × 60 mm 2-D anisotropic endocardial tissue sheet were used to investigate the propagation of autorhythmic depolarisations. In the 1-D tissue, the rate of change of V is given by a reaction-diffusion equation:

$$\frac{\partial V}{\partial t} = D_x \frac{\partial^2 V}{\partial x^2} - \frac{1}{C_m} I_{ion}, \qquad (3)$$

and in the 2-D tissue:

$$\frac{\partial V}{\partial t} = D_x \frac{\partial^2 V}{\partial x^2} + D_y \frac{\partial^2 V}{\partial y^2} - \frac{1}{C_m} I_{ion}, \qquad (4)$$

where D is a diffusion coefficient, and x and y denote directions perpendicular to and parallel to fibre axis, respectively. No-flux boundary conditions were imposed at the edges of each geometry. Ca^{2+} overloaded tissue, composed of modified LRd00 myocytes as described in Sect. 2.2, was located on the endocardial border of the 1-D strand or the centre of the 2-D sheet. Diffusion coefficients of $D_x = 0.06\,\text{mm}^2\,\text{ms}^{-1}$ and $D_y = 0.1\,\text{mm}^2\,\text{ms}^{-1}$ were used, giving conduction velocities for solitary plane waves of 0.4 and 0.54 m s^{-1} in the x and y directions, respectively. A time step of $dt = 0.02$ ms and a space step of $dx = dy = 0.1$ mm were used in both 1-D and 2-D simulations. Computation time was decreased when running 2-D simulations by tabulating the values of V-dependent exponential functions for values of V between -100 and 100 mV with a resolution of 0.1 mV. Linear approximation was used for functions where V fell between the tabulated values.

3 Bifurcations in the Calcium Handling Equations

We determined behaviour of the Ca^{2+} handling equations by numerical integration over a period of 120 s, where $[Ca^{2+}]_i$ either settled to a stationary stable state or oscillated. Fig. 1A is a bifurcation diagram showing stationary stable states and amplitudes of oscillations, where V is clamped at -90 mV and $[Na^+]_i$ is the bifurcation parameter. The periods of the oscillations are shown in Fig. 1B. As $[Na^+]_i$ is increased to ~ 16.1 mM, large period oscillations emerge. The period of the oscillations decreases rapidly as the bifurcation parameter is further increased, indicative of a homoclinic bifurcation rather than the Hopf bifurcation identified in the DiFrancesco-Noble Purkinje fibre model by Varghese & Winslow [11]. Fig. 1C shows $[Ca^{2+}]_i$ oscillations, with $V = -90$ mV, $[Na^+]_i = 20$ mM and initial Ca^{2+} concentrations as in the normal LRd00 model ($[Ca^{2+}]_i = 79$ nM at $t = 0$ ms). In this case, minimum and maximum $[Ca^{2+}]_i$ is 171 and 928 nM, respectively, giving an amplitude of 757 nM and a period of 1194 ms. Fig. 1D shows the dynamics of the system in $[Na^+]_i$-V parameter space for values of $[Na^+]_i$ between 0 and 20 mM and for values of V between -110 and -40 mV.

4 Whole Cell Calcium and Voltage Oscillations

In Sect. 3 we showed that, under certain conditions, $[Ca^{2+}]_i$ in LRd00 can oscillate independently of membrane potential oscillations. By unclamping the voltage, it is possible to observe whether these $[Ca^{2+}]_i$ oscillations can drive transient inward currents strong enough to take the membrane potential past threshold

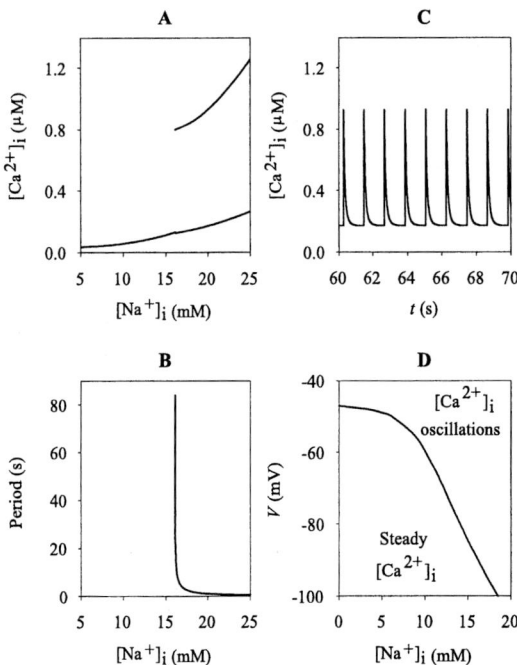

Fig. 1. Bifurcation analysis of the calcium handling equations in LRd00. **A:** Bifurcation diagram with V clamped at $-90\,\text{mV}$, showing steady state values and oscillation amplitudes of $[Ca^{2+}]_i$. For low values of $[Na^+]_i$, $[Ca^{2+}]_i$ settles to a steady state. Oscillations emerge when $[Na^+]_i$ is increased to $\sim 16.1\,\text{mM}$. **B:** The periods of the oscillations shown in **A**. As oscillations emerge the period is large, then decreases rapidly as $[Na^+]_i$ is further increased. **C:** $[Ca^{2+}]_i$ oscillations with V clamped at $-90\,\text{mV}$ and $[Na^+]_i$ at $20\,\text{mM}$. **D:** $[Ca^{2+}]_i$ dynamics in $[Na^+]_i$-V parameter space. Dynamics are classified according to behaviour during the first $120\,\text{s}$ of integration, and can either oscillate or settle to a steady state

to induce autorhythmic depolarisations. That is, we can observe whether the $[Ca^{2+}]_i$ oscillations can drive V oscillations.

$[Na^+]_i$ was clamped at $20\,\text{mM}$, a value that results in $[Ca^{2+}]_i$ oscillations in the Ca^{2+} handling system at LRd00 resting V of approximately $-88\,\text{mV}$ (see Fig. 1D). $[K^+]_i$ was clamped at $125\,\text{mM}$ [11]. Integration of this modified LRd00 model revealed that $[Ca^{2+}]_i$ oscillations drive V oscillations through the action of transient inward currents. Fig. 2A shows these autorhythmic depolarisations occurring during the first $10\,\text{s}$ of integration. The cell initially depolarises at around $t = 3.34\,\text{s}$, with the action potential having an upstroke velocity of $184\,\text{mV\,ms}^{-1}$, a peak V of $28\,\text{mV}$ and an action potential duration (APD_{90}) of $192\,\text{ms}$. Figs. 2B and 2C show the transient inward currents that cause V to increase past threshold. When I_{NaCa} is operating in forward mode, Ca^{2+} is extruded from the cell and Na^+ is brought in at a ratio of 1:3, and consequently the net current is inward (i.e. depolarising). $I_{ns(Ca)}$ carries both Na^+ and K^+ and so net $I_{ns(Ca)}$

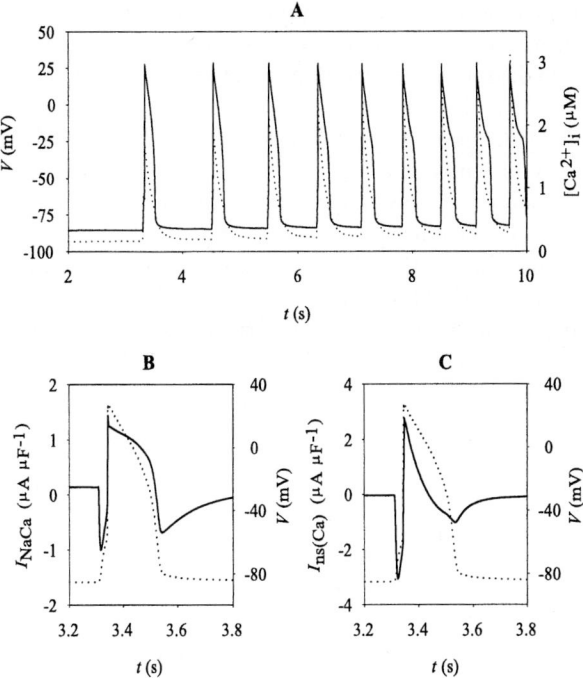

Fig. 2. Autorhythmic activity in a single modified LRd00 myocyte. **A:** $[Ca^{2+}]_i$ oscillations (dotted line) interact with the cell membrane, resulting in repetitive autorhythmic depolarisations (solid line). Both I_{NaCa} (**B**) and $I_{ns(Ca)}$ (**C**) are transiently inward, depolarising currents immediately before the depolarisation. In **B** and **C**, currents are shown as solid lines, V as dotted lines. Note that currents have been calculated for 1 cm² of membrane, and that negative current amplitudes indicate inward currents

is dependent on the sum of both these component currents. Under physiological conditions, the Na⁺ current carried by $I_{ns(Ca)}$ is a depolarising inward current, while the K⁺ current is a hyperpolarising outward current. Immediately before the depolarisation, both I_{NaCa} (Fig. 2B) and $I_{ns(Ca)}$ (Fig. 2C) become relatively large depolarising inward currents in response to Ca^{2+} release from the SR. This is in contrast to the other inward currents – the fast Na⁺ current and the L- and T-type Ca^{2+} currents – that increase in amplitude only in response to the depolarisation (not shown). Block of either I_{NaCa} or $I_{ns(Ca)}$ causes a reduction of the membrane response to spontaneous SR Ca^{2+} release, and V remains sub-threshold: the transient inward current is only large enough to take V to threshold when both I_{NaCa} and $I_{ns(Ca)}$ are included.

5 Propagation

Propagation of an action potential from a localised area into the surrounding tissue is dependent on several factors. In 1-D and 2-D tissues, both the size of the focus from which the action potential propagates and the cell-cell coupling within and between the autorhythmic and quiescent tissues are important. Additionally in 2-D tissues, the degree of anisotropy present and the curvature of the wavefront both affect propagation.

A minimum of 3.4 mm of Ca^{2+} overloaded tissue located on the endocardial border was required to produce repetitive propagation of action potentials through the 1-D heterogeneous tissue strand. Fig. 3A is a space-time plot showing these depolarisations emerging from the autorhythmic focus and propagating along the strand during the first 5 s of integration. The initial depolarisation propagates with a wavefront velocity of $0.4 \, m\,s^{-1}$, taking 28.8 ms to reach the epicardial border from the edge of the autorhythmic focus. The period of the depolarisations in the autorhythmic focus is $\sim$1072 ms during the first 5 s, decreasing with time presumably as ionic concentrations accumulate or deplete (due to Na^+-K^+ pump inhibition) and affect membrane conductance. As no-flux boundary conditions are used, the liminal length (the minimum amount of excited tissue located in the middle of a strand required to produce bi-directional propagation) is 6.8 mm. Here we used a spatially homogeneous diffusion coefficient of $D_x = 0.06 \, mm^2 \, ms^{-1}$. However, alterations in cell-cell coupling (modelled as the diffusion coefficient, D), either within the autorhythmic focus, within the quiescent tissue, or at the interface of the two, will affect the behaviour of these tissues and, therefore, the liminal length required for propagation [26].

In the 2-D anisotropic endocardial tissue, repetitive propagation of autorhythmic depolarisations from a central circular focus occurred with a minimum focus radius of 5 mm, giving a liminal area of 79 mm^2. This radius is larger than the 3.4 mm in the 1-D strand as the excitatory current provided by the autorhythmic focus must distribute over a larger area due to the curved border of the 2-D focus [27], even though increasing anisotropy has been shown to decrease the liminal area [17]. With LRd00 cell dimensions of 100 × 22 μm [18], the liminal area is composed of nearly 36,000 cells. This is in comparison to the liminal area of $\sim$1,000 cells found by Winslow et al. [14] using an atrial cell network model. Figures 3B-E show snapshots of the propagation of a single action potential through the tissue at times 910, 935, 960 and 1120 ms, respectively. As well as the direct effects of cell-cell coupling (i.e., the passive electrical properties of the tissue), the geometry of the wavefront also affects propagation [27]: the wavefront here is not a simple plane wave but is curved due to the circular autorhythmic focus, and this convex wavefront curvature acts to reduce conduction velocity. Thus the shape of the autorhythmic focus (and therefore the shape of the propagating wavefront) affects propagation. Additionally, the degree of anisotropy will affect conduction velocity through this mechanism, as the wavefront is less convex in the direction of lower cell-to-cell coupling, with this discrepancy increasing as the wave propagates (compare, for example, Figs. 3B and 3C). Both these mechanisms cause a decrease of conduction velocity; as the wavefront geometry changes with distance from the focus, wavefront conduction velocity will also change.

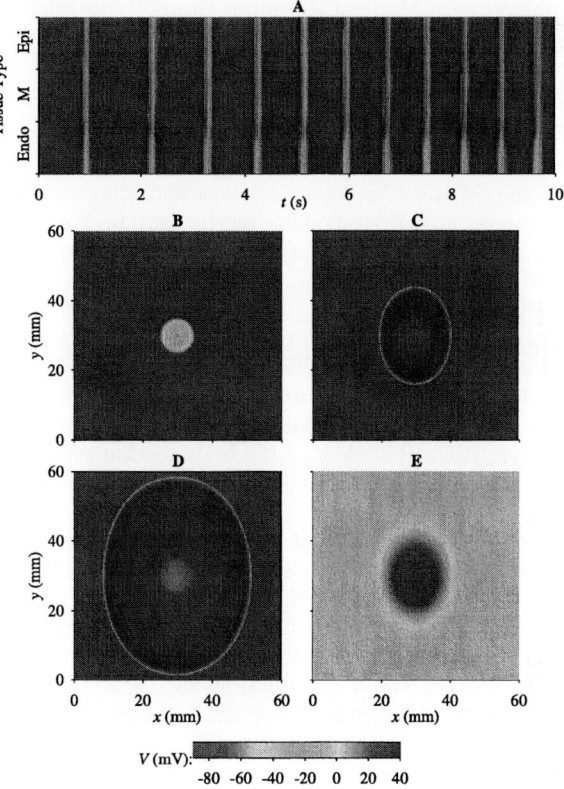

Fig. 3. Propagation of autorhythmic depolarisations in virtual tissues. **A**: Space-time plot showing propagation from a 3.4 mm autorhythmic focus located on the endocardial border of a 15 mm heterogeneous LRd00 virtual tissue strand. **B-E**: Snapshots showing propagation from an autorhythmic focus 5 mm in radius, located in the centre of a 60 × 60 mm 2-D anisotropic endocardial LRd00 virtual tissue sheet, at $t = 910$, 935, 960 and 1120 ms, respectively

6 Conclusions

We have shown that Ca^{2+} overload, brought about via inhibition of the Na^+-K^+ pump, can cause propagating autorhythmic activity in virtual ventricular myocytes and tissues of the LRd00 family. $[Ca^{2+}]_i$ in the Ca^{2+} handling equations settles to a steady state at low levels of $[Na^+]_i$, but oscillates when $[Na^+]_i$ is increased; these oscillations emerge via a homoclinic bifurcation. In the whole cell, the $[Ca^{2+}]_i$ oscillations interact with the membrane currents I_{NaCa} and $I_{ns(Ca)}$ to cause autorhythmic depolarisations that, if spatially localised, can propagate through 1-D heterogeneous virtual tissue strands and 2-D homogeneous anisotropic virtual tissue sheets. The shape of the 2-D focus and the degree of tissue anisotropy affect both the liminal area of the focus and the conduction of the wavefront through the 2-D tissue.

APB is supported by a Medical Research Council bio-informatics (computational biology) priority area research studentship (G74/63).

References

1. Wit, A.L., Rosen, M.R.: Afterdepolarizations and triggered activity: distinction from automaticity as an arrhythmogenic mechanism. In: Fozzard, H.A., Haber, E., Jennings, R.B., Katz, A.M., Morgan, H.E. (eds.): The heart and cardiovascular system: scientific foundations, 2nd ed. Raven Press, New York (1991) 2113-2163
2. Luo C.H., Rudy, Y.: A dynamic model of the cardiac action potential. II. Afterdepolarizations, triggered activity, and potentiation. Circ. Res. **74** (1994) 1097-1113
3. Bers, D.M.: Excitation-Contraction Coupling and Cardiac Contractile Force, 2nd ed. Klewer Academic Publishers, Dordrecht, The Netherlands (2001)
4. Verkerk, A.O., Veldkamp, M.W., Baartscheer, A., Schumacher, C.A., Klopping, C., van Ginneken, A.C., Ravesloot, J.H.: Ionic mechanism of delayed afterdepolarizations in ventricular cells isolated from human end-stage failing hearts. Circ. **104** (2001) 2728-2733
5. Spencer, C.I., Sham, J.S.: Effects of Na^+/Ca^{2+} exchange induced by SR Ca^{2+} release on action potentials and afterdepolarizations in guinea pig ventricular myocytes. Am. J. Physiol. **285** (2003) H2552-H2562
6. Noble, D., DiFrancesco, D., Denyer, J.: Ionic mechanisms in normal and abnormal cardiac pacemaker activity. In: Jacklet, J.W. (ed.): Neuronal and cellular oscillators. Marcel Dekker Inc, New York (1989) 59-85
7. Omichi, C., Lamp, S.T., Lin, S.-F., Yang, J., Baher, A., Zhou, S., Attin, M., Lee, M.-H., Karagueuzian, H.S., Kogan, B., Qu, Z., Garfinkel, A., Chen, P.-S., Weiss, J.N.: Intracellular Ca dynamics in ventricular fibrillation. Am. J. Physiol. **286** (2004) H1836-H1844
8. Carmeliet, E.: Cardiac ionic currents and acute ischemia: from channels to arrhythmias. Physiol. Rev. **79** (1999) 917-1017
9. Kass, R.S., Tsien, R.W., Weingart, R.: Ionic basis of transient inward current induced by strophanthidin in cardiac Purkinje fibres. J. Physiol. **281** (1978) 209-226
10. Gheorghiade, M., Adams, K.F., Colucci, W.F.: Digoxin in the management of cardiovascular disorders. Circ. **109** (2004) 2959-2964
11. Varghese, A., Winslow, R.L.: Dynamics of the calcium subsystem in cardiac Purkinje fibers. Physica D **68** (1993) 364-386
12. DiFrancesco, D., Noble, D.: A model of cardiac electrical activity incorporating ionic pumps and concentration changes. Phil. Trans. Roy. Soc. (Lond.) B **307** (1985) 353-398
13. Varghese, A., Winslow, R.L.: Dynamics of abnormal pacemaking activity in cardiac Purkinje fibers. J. Theor. Biol. **168** (1994) 407-420
14. Winslow, R.L., Varghese, A., Noble, D., Adlakha, C., Hoythya, A.: Generation and propagation of ectopic beats induced by spatially localized Na-K pump inhibition in atrial network models. Proc. Roy. Soc. (Lond.) B **254** (1993) 55-61
15. Earm, Y.E., Noble, D.: A model of the single atrial cell: relation between calcium current and calcium release. Proc. Roy. Soc. (Lond.) B **240** (1990) 83-96
16. Joyner, R.W., Wang, Y.-G., Wilders, R., Golod, D.A., Wagner, M.B., Kumar, R., Goolsby, W.N.: A spontaneously active focus drives a model atrial sheet more easily than a model ventricular sheet. Am. J. Physiol. **279** (2000) H752-H763

17. Wilders, R., Wagner, M.B., Golod, D.A., Kumar, R., Wang, Y.-G., Goolsby, W.N., Joyner, R.W., Jongsma, H.J.: Effects of anisotropy on the development of cardiac arrhythmias associated with focal activity. Pflug. Arch. **441** (2000) 301-312
18. Luo C.H., Rudy, Y.: A dynamic model of the cardiac action potential. I. Simulations of ionic currents and concentration changes. Circ. Res. **74** (1994) 1071-1096
19. Hodgkin, A.L., Huxley, A.F.: A quantitative description of membrane current and its application to conduction and excitation in nerve. J. Physiol. **117** (1952) 500-544
20. Luo C.H., Rudy, Y.: A model of the ventricular cardiac action potential: depolarization, repolarization, and their interaction. Circ. Res. **68** (1991) 1501-1526
21. Zeng, J., Laurita, K.R., Rosenbaum, D.S., Rudy, Y.: Two components of the delayed rectifier K^+ current in ventricular myocytes of the guinea pig type: theoretical formulation and their role in repolarization. Circ. Res. **77** (1995) 140-152
22. Viswanathan, P.C., Shaw, R.M., Rudy, Y.: Effects of I_{Kr} and I_{Ks} heterogeneity on action potential duration and its rate-dependence: a simulation study. Circ. **99** (1999) 2466-2474
23. Faber, G.M., Rudy, Y.: Action potential and contractility changes in $[Na^+]_i$ overloaded cardiac myocytes: a simulation study. Biophys. J. **78** (2000) 2392-2404
24. Rush, S., Larsen, H.: A practical algorithm for solving dynamic membrane equations. IEEE Trans. Biomed. Eng. **25** (1978) 389-392
25. Guckenheimer, J., Labouriau, I.S.: Bifurcation of the Hodgkin and Huxley equations: a new twist. Bull. Math. Biol. **55** (1993) 937-952
26. Benson, A.P., Holden, A.V., Kharche, S., Tong, W.C.: Endogenous driving and synchronization in cardiac and uterine virtual tissues: bifurcations and local coupling. Phil. Trans. Roy. Soc. (Lond.) A (to appear)
27. Fast, V.G., Kléber, A.G.: Role of wavefront curvature in propagation of cardiac impulse. Cardiovasc. Res. **33** (1997) 258-271

Left Ventricular Shear Strain in Model and Experiment: The Role of Myofiber Orientation

Sander Ubbink[1], Peter Bovendeerd[1], Tammo Delhaas[2,3], Theo Arts[1,2], and Frans van de Vosse[1]

[1] Eindhoven University of Technology, PO Box 513,
5600 MB Eindhoven, The Netherlands
[2] Maastricht University, PO Box 616,
6200 MD Maastricht, The Netherlands
[3] University Hospital Maastricht, PO Box 5800,
6202 AZ Maastricht, The Netherlands

Abstract. Mathematical modeling of cardiac mechanics could be a useful clinical tool, both in translating measured abnormalities in cardiac deformation into the underlying pathology, and in selecting a proper treatment. We investigated to what extent a previously published model of cardiac mechanics [6] could predict deformation in the healthy left ventricle, as measured using MR tagging. The model adequately predicts circumferential strain, but fails to accurately predict shear strain. However, the time course of shear strain proves to be that sensitive to myofiber orientation, that agreement between model predictions and experiment may be expected if fiber orientation is changed by only a few degrees.

1 Introduction

Adequate pump function of the heart relies on the function of many underlying systems, such as initiation of cardiac contraction through electrical depolarisation, exchange of oxygen, nutrients and waste products through the coronary circulation, and directioning of the blood flow through the heart valves. Malfunction of each of these systems leads to deterioration of pump function.

Many cardiac pathologies (e.g. ischemia, disturbed conduction) are reflected in abnormal deformations of the cardiac wall [11,13,14]. Clinically, deformation patterns of the heart can be assessed with MR tissue tagging, creating opportunities in determining the underlying pathology. However, the relation between the change in deformation pattern and the underlying pathology is not straight forward. A mathematical model, capable of predicting the forward relation between pathology and deformation, could, if used in an inverse analysis, be a useful clinical tool in determining causes of the pathology and a proper treatment.

Several models describing deformation in the heart have been proposed [2, 5, 7, 8, 12]. None of these models has already been used as a diagnostic tool. Even the first step towards this application, the complete prediction of deformation in the normal healthy heart, has not been made. In this study we investigate the capability of a previously published three-dimensional finite element (FE) model of cardiac mechanics [6], to predict the deformations of the myocardial wall of the healthy left ventricle.

2 Methods

2.1 Assessment of LV Wall Strain Using MR Tagging

In four healthy subjects (age 28 to 33 years) deformation patterns of the heart were assessed noninvasively using MR tagging. The experiments were performed at the University Hospital Maastricht, in a 1.5 T scanner (Gyroscan NT, Philips Medical Systems, Best, The Netherlands), with imaging parameters set as follows: echo time 10 ms, inter-tag distance 6 mm, slice thickness 8 mm, tag-width 2.5 mm, field of view 250 mm and image size 256 × 256 pixels. Images were acquired using ECG-triggering on the R wave during breath hold for a period of about 20 s. Five parallel short-axis slices of the heart were imaged, evenly distributed from apex to base (figure 1). Series of line-tagged images from the same slices were obtained, with time intervals of about 20 ms, using spatial modulation of the magnetization [1].

From the MR-images, displacement maps were determined [13]. Next, the Green-Lagrange strain tensor was determined with respect to begin ejection, and written in components with respect to a local cylindrical coordinate system.

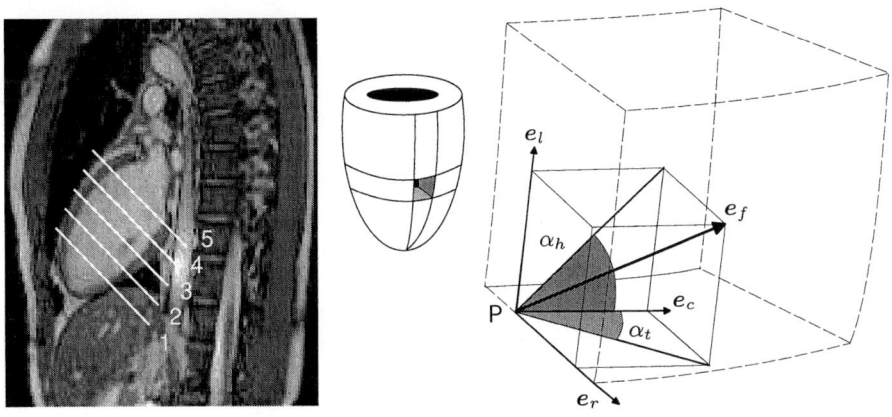

Fig. 1. Left: long-axis MR slice demonstrating the positions of the five short-axis slices, imaged in the LV of the heart, numbered from 1 (apex) to 5 (base). Right: illustration of the ellipsoidally shaped model of left ventricular mechanics showing helix (α_h) and transverse (α_t) fiber angles. The fiber angles at a point P are defined from projection of the fiber direction e_f on planes spanned by the local transmural e_r, longitudinal e_l and circumferential e_c direction

Strain components were averaged in transmural direction, with weight factors distributed according to a gaussian curve with the top at the midwall position. Finally, these midwall values were averaged in circumferential direction, yielding the average circumferential (E_{cc}) and radial (E_{rr}) normal strain, and circumferential-radial shear strain (E_{cr}) for each of the five slices.

To relate timing of the MR images to phases in the cardiac cycle, for each moment in each slice the midwall radius was determined. Using the interslice distance, the volume enclosed by the ventricular midwall was estimated. Differentiation of this volume with respect to time yielded an estimation of mitral inflow and aortic outflow. Zero crossings of this signal defined moments of transition between phases in the cycle.

2.2 Computation of LV Wall Strain Using the FE Mechanics Model

The finite element (FE) model of LV mechanics has been described before [6]. In short, a thickwalled geometry is assumed with endocardial and epicardial surfaces consisting of confocal ellipsoids. The helix angle α_h and transverse angle α_t are used to describe the base-to-apex component and the transmural component of the myofiber direction, respectively (figure 1). In the model, conservation of momentum is solved:

$$\nabla \cdot \boldsymbol{\sigma} = \mathbf{0} \qquad (1)$$

with $\boldsymbol{\sigma}$ representing the Cauchy stress, composed of a passive component ($\boldsymbol{\sigma}_p$) and an active component (σ_a) along the myofiber direction $\boldsymbol{e}_f$:

$$\boldsymbol{\sigma} = \boldsymbol{\sigma}_p + \sigma_a \boldsymbol{e}_f \boldsymbol{e}_f \qquad (2)$$

Passive material behaviour is modeled nonlinearly elastic, transversely isotropic and virtually incompressible. The active stress σ_a is assumed to depend on time elapsed since depolarization, sarcomere length and sarcomere shortening velocity. Depolarization is assumed to be simultaneous.

Several simulations were performed. In the simulation, indicated with ref, settings of α_h and α_t were adopted from Rijcken et al. [9, 10], who determined myofiber orientation by optimization for a homogenous distribution of fiber strain during ejection (figure 2). In view of reported sensitivity of wall mechanics to fiber orientation [2, 4], sensitivity of computed strains to settings of α_t and α_h was investigated. Sensitivity to the choice of α_t was assessed in simulations trans-2 and trans+2, where α_t was shifted by $-2°$ and $+2°$, and simulations trans*0.75 and trans*1.25, where α_t was multiplied by 0.75 and 1.25. Similar variations were applied to the helix angle α_h: in simulations helix-5 and helix+5 α_h was shifted by $-5°$ and $+5°$, whereas in simulations helix*0.9 and helix*1.1 α_t was multiplied by 0.9 and 1.1 (figure 2). The variations were chosen such, that the resulting distributions of the fiber angles were within the range of experimental data.

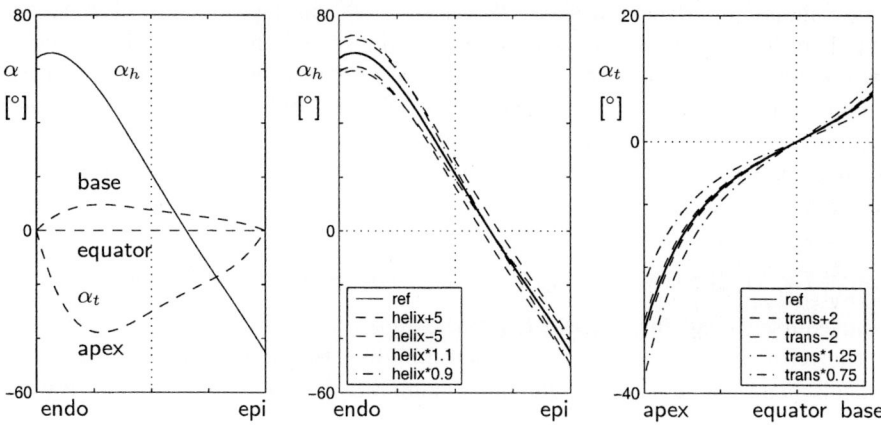

Fig. 2. Left: variation of the helix angle α_h (—) at the equator, and transverse angle α_t (- -) at base, equator and apex, from endocardial to epicardial surface, as in the reference simulation; middle: transmural variation of α_h at the equator for various simulations; right: base-to-apex variation of α_t at the midwall for various simulations

3 Results

3.1 LV Wall Strain Assessed with MR Tagging

Images were acquired over a time span of 600 ms. Since the cardiac cycle time was about 750 ms, no strains were determined for the last part of the filling phase and the initial part of the isovolumic contraction phase. Measured circumferential strain E_{cc} was similar in all four hearts. Typically, E_{cc} was similar in all five

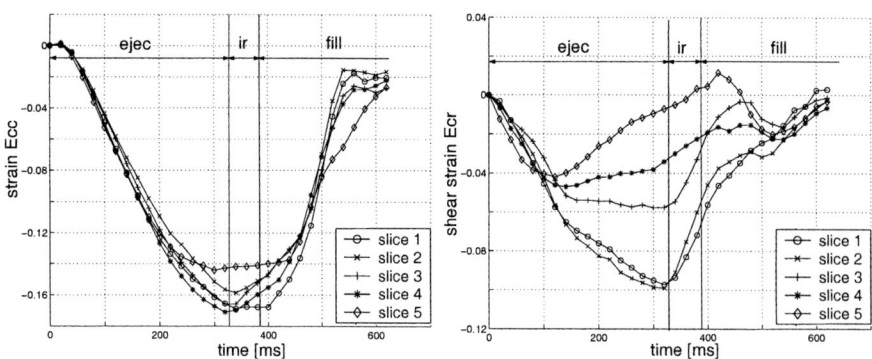

Fig. 3. Representative measured time courses of left ventricular midwall circumferential (left) and circumferential-radial shear (right) strain. Slices are numbered from apex (slice 1) to base (slice 5). Phases are indicated by ejec (ejection) ir (isovolumic relaxation) and fill (filling)

slices (figure 3). Measured shear strain E_{cr} was similar in 3 out of 4 hearts. Typically, E_{cr} decreased equally in all five slices until one third of the ejection period (figure 3). Thereafter, E_{cr} continued decreasing near the apex (slice 1), remained about constant near the equator (slice 3), and increased near the base (slice 5). Throughout the isovolumic relaxation phase, E_{cr} increased in all slices. During the first part of filling, E_{cr} converged towards the state at begin ejection in all slices.

3.2 LV Wall Strain Computed with the Reference Model

Time courses of strains E_{cc} and E_{cr} with respect to begin ejection, as computed in the ref simulation at lattitudes corresponding to slices 2 to 4 are shown in the left panel of figure 4. To facilitate comparison, the experimental data are

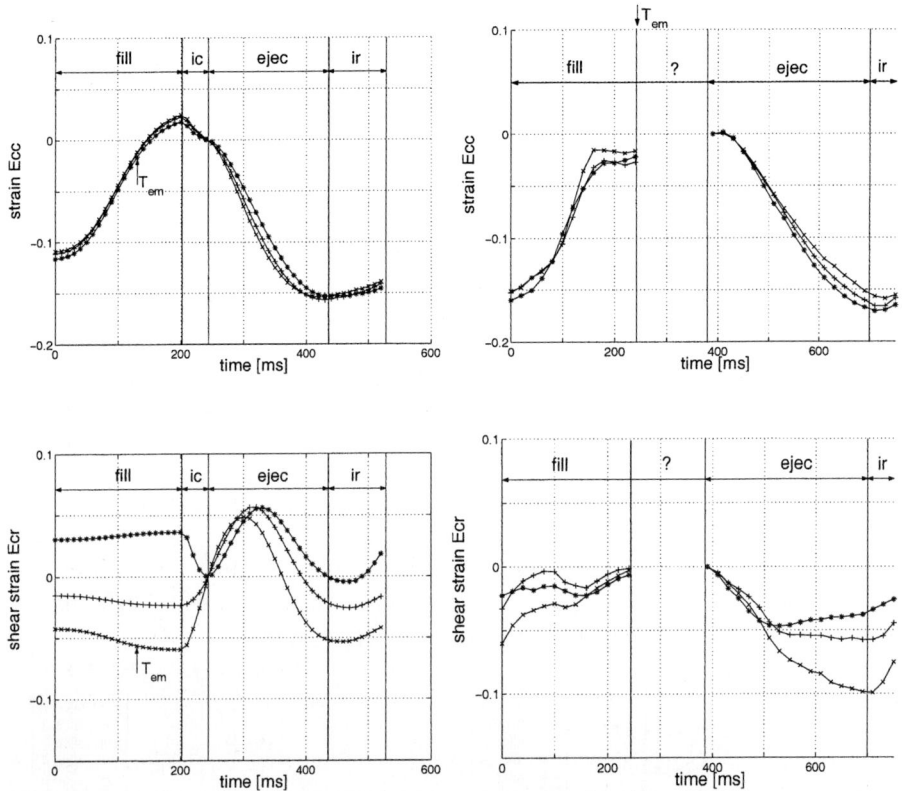

Fig. 4. Time courses of circumferential strain E_{cc} (top) and circumferential-radial strain E_{cr} (bottom) as predicted by the LV mechanics model according to the ref simulation (left) and as measured (right) for slices 2 (×), 3 (+) and 4 (⋆). Phases are indicated by fill (filling), ic (isovolumic contraction), ejec (ejection), and ir (isovolumic relaxation). No experimental data are available in the period indicated by ?

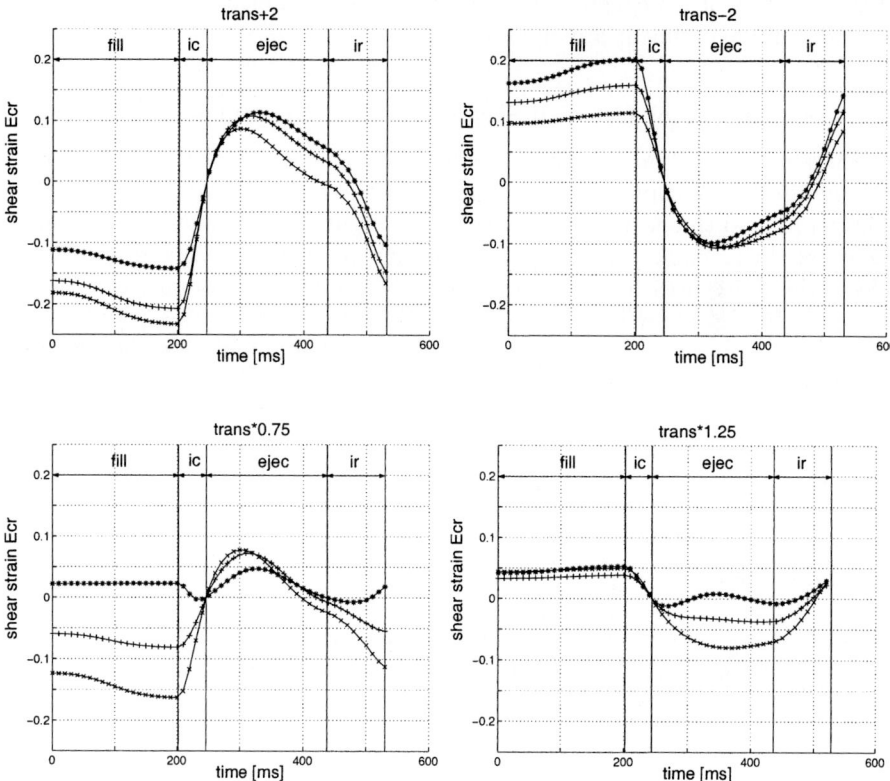

Fig. 5. Time courses of midwall E_{cr} for slices 2 (×), 3 (+) and 4 (⋆), as predicted by the LV mechanics model, according to simulations trans+2, trans-2, trans*0.75 and trans*1.25, representing for various settings of the transverse fiber angle. Note that range of the strain axis is twice the range of the strain axis in figure 4

redrawn in the right panel. The model predicts a virtually identical time course of E_{cc} for all three slices. E_{cc} increases during filling, decreases slightly during isovolumic contraction, decreases further during ejection, and remains nearly constant during isovolumic relaxation. The pattern is similar to the measured pattern, including the overall increase in strain from T_{em}, the moment the last MR image was acquired, to begin ejection.

In the model, E_{cr} is about constant during filling, with a small decrease for slices 2 and 3, and a small increase for slice 4 towards the end of filling. During isovolumic contraction, E_{cr} increases in slices 2 and 3 and decreases in slice 4. During the first part of ejection, E_{cr} increases similarly for the three slices. Thereafter, E_{cr} values decrease and diverge. During isovolumic relaxation, E_{cr} values increase towards the values at the beginning of filling.

Computed E_{cr} is quite different from measured E_{cr}. In particular, the model predicts a strong change of E_{cr} from T_{em} until begin of ejection, while variation in the experiment is negligible. During the first part of ejection, changes in E_{cr} in model and experiment are opposite.

3.3 Sensitivity of LV Wall Strain to Fiber Orientation

Simulations trans+2 en trans-2 illustrate that a change in offset in α_t predominantly affects the common change in E_{cr} in all three slices during isovolumic contraction and the beginning of the ejection (figure 5; top panels). A change in the slope of the longitudinal course of α_t (simulation trans*0.75 and trans*1.25) affects the differences in E_{cr} between the slices during isovolumic contraction (figure 5; bottom panels).

A change in offset of the helix angle distribution (simulations helix+5 and helix-5) changes the range in E_{cr} during ejection (figure 6; top panels) moderately. A change in slope of the transmural course of α_h (simulations helix*0.9 and helix*1.1) has a small effect on the distribution of changes in E_{cr} during isovolumic contraction (figure 6; bottom panels).

4 Discussion

Left ventricular wall strains, as predicted by an existing model of left ventricular mechanics, were compared with strains, derived from MR tagging measurements in healthy humans. Differences between measured and computed time course of E_{cc} in the equatorial region of the LV were small, but the differences in E_{cr} appeared significant. This is not surprising, since the change of E_{cc} is closely related to the change in cavity volume, while E_{cr} is related to the internal equilibrium of forces in the LV wall. Discrepancies in E_{cr} occurred predominantly between the moment of end of measurement T_{em} and about mid ejection. In the model, E_{cr} changed strongly during isovolumic contraction, when the LV transforms from the passive diastolic to the active systolic state. Apparently, mechanical equilibrium in the anisotropic activated tissue involves large forces in the passive matrix, and consequently a large deformation. In contrast, in the experiment, E_{cr} at T_{em} was about equal to that at the beginning of ejection. During the period from T_{em} until the end of isovolumic contraction, E_{cr} was not measured. However, E_{cr} at T_{em} may be considered representative for strain at end diastole: during the last part of the filling phase no shear deformation is to be expected since the tissue is passive and hence virtually isotropic. This expectation is supported by the time derivative of measured E_{cr} near T_{em}, which is about zero. Although no strains were measured during isovolumic contraction, the data suggest that no significant deformation of the passive matrix occurs during this phase either.

The parameter variations show that the extent to which the passive matrix is involved in force transmission is very sensitive to the orientation of the muscle fibers. The influence of the helix angle α_h and the transverse α_t on shear E_{cr} is illustrated in figure 7. As a consequence of the base-to-apex component of fiber orientation, expressed by α_h, shortening of subendocardial myofibers would cause

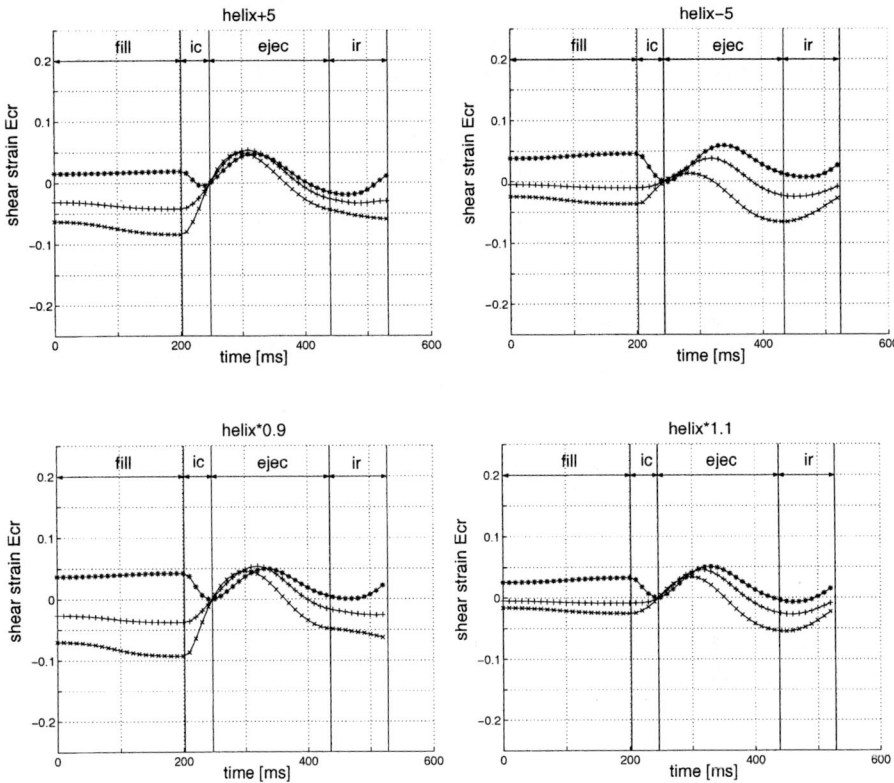

Fig. 6. Time course of the midwall E_{cr} for slices 2 (×), 3 (+) and 4 (⋆) as predicted by LV mechanics model according to simulations helix+5, helix-5, helix*0.9 and helix*1.1, representing various settings of the helix fiber angle

a clockwise apical rotation, when viewing the apex in apex-to-base direction. Shortening of subepicardial myofibers would cause a counterclockwise apical rotation. In the absence of an endo-to-epi component of fiber orientation, this shear load can only be counteracted by the passive myocardial tissue. Because the passive tissue has a low stiffness, mechanical equilibrium is reached at large shear strains E_{cr}. If, however, myofibers cross over between inner and outer layers of the wall, as expressed by a nonzero α_t, then these active, relatively stiff myofibers participate in the transmission of the shear load as well, and shear deformation is reduced. The role of the transverse angle has been identified before [2], but no realistic prediction of LV wall shear was obtained in that study.

Our study has its limitations. Strains are averaged in circumferential and radial direction, which reduces noise but also removes information on strain gradients in these directions. Circumferential averaging is in line with the rotational symmetry of our model of LV mechanics. Probably computed shear strains are less sensitive to geometry [4] than to timing of depolarization of the LV wall [6]

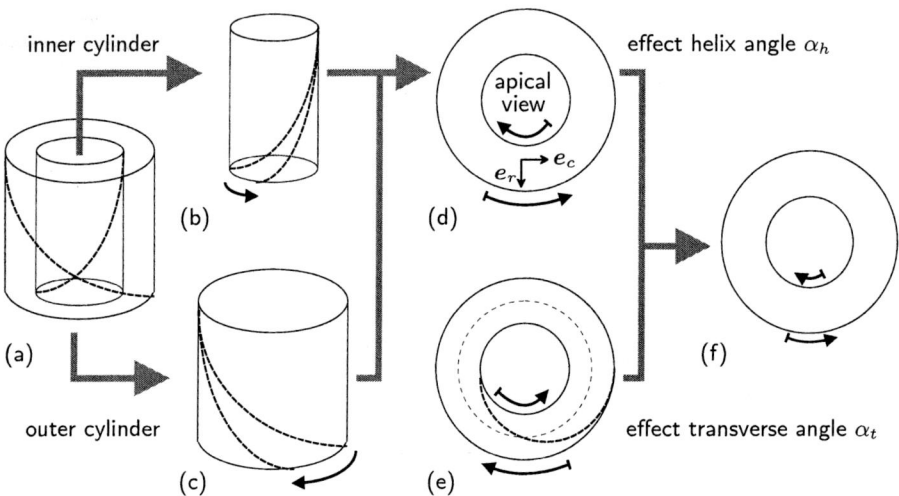

Fig. 7. Schematic illustration of the influence of helix angle α_h and transverse angle α_t on transmural shear in a cylindrical model of the LV. In absence of a transverse angle (a), shortening of the myofibers (- - -) would cause a clockwise apical rotation of the endocardium (b) and a counterclockwise rotation of the epicardium (c). In the commonly used apex-to-base view of short-axis images, this results in a positive circumferential-radial shear (d), i.e. the angle between line elements originally oriented in circumferential (e_c) and radial (e_r) direction will become less than 90°. This shear is counteracted by a negative circumferential-radial shear, as a result of the negative transverse angle in the apical part of the ventricle (e). The final shear (f) depends on the balance between the two effects

or passive shear stiffness. While the latter aspects remain to be investigated, we expect the sensitivity of shear strain to myofiber orientation to persist.

In literature, other models of the cardiac mechanics have been proposed [5,7,12], but none of them yielded an accurate representation of measured shear strains. Discrepancies were attributed to simplifying orthotropic passive behaviour to transversely isotropic behaviour [5], but sensitivity of the computed deformations to the choice of the passive material model was small [12]. Also, it was suggested that active cross-fiber stress development should be incorporated into the models. Indeed, introduction of active cross-fiber stress affected end-systolic shear strains significantly, but agreement with experimental values was not obtained [12].

The present study suggests that the main shortcoming in the above models lies in an unrealistic setting of myofiber orientation, in particular in neglecting the transmural component of myofiber orientation, expressed by a non-zero transverse angle. With histological sectioning techniques, fiber angles are obtained with an accuracy of ±10°. When using magnetic resonance diffusion tensor imaging, accuracy is about ±6° [3]. Our sensitivity study shows that the reliability of experimental data is too low to be able to yield reliable predictions of LV wall strain.

5 Conclusion

It is concluded that (1) the previously published three-dimensional mathematical model of cardiac mechanics [6] adequately predicts circumferential strain, but fails to accurately predict shear strain, (2) the time course of shear strain is very sensitive to the choice of the myofiber orientation, in particular to the choice of the transverse angle, and (3) agreement between model predictions and experiment may be expected by a change of the fiber orientation by only a few degrees.

References

1. Axel L. MR imaging of motion with spatial modulation of magnetization. *Radiology*, 171:841–845, 1989.
2. Bovendeerd P.H.M., Huyghe J.M., Arts T., van Campen D.H., Reneman R.S. Influence of endocardial-epicardial crossover of muscle fibers on left ventricular wall mechanics. *J. Biomech.*, 27:941–951, 1994.
3. Geerts L., Bovendeerd P.H.M., Nicolay K., Arts T. Characterization of the normal cardiac myofiber field in goat measured with MR diffusion tensor imaging. *Am. J. Physiol.*, 283, H126–H138, 2002.
4. Geerts-Ossevoort L., Kerckhoffs R.C.P., Bovendeerd P.H.M., and Arts T. Towards patient specific models of cardiac mechanics: a sensitivity study, in *Functional Imaging and Modeling of the Heart* Ed. Magnin I.E., Montagnat J., Clarysse P., Nenonen J., Katila T., Lyon, France, 81-90, 2003.
5. Guccione J.M., Costa K.D., McCulloch A.D. Finite element stress analysis of left ventricular mechanics in the beating dog heart. *J. Biomech.*, 10:1167–1177, 1995.
6. Kerckhofs R.C.P., Bovendeerd P.H.M., Kotte J.C.S., Prinzen F.W., Smits K., Arts T. Homogeneity of cardiac contraction despite physiological asynchrony of depolarization: a model study. *Ann. Biomed. Eng.*, 31:536–547, 2003.
7. Nash M. *Mechanics and material properties of the heart using an anatomically accurate mathematical model.* PhD thesis, University of Auckland, 1998.
8. Nash M.P., and Hunter, P.J. Computational mechanics of the heart. *J. Elast.*, 61:113–141, 2000.
9. Rijcken J. *Optimization of left ventricular muscle fiber orientation.* PhD thesis, University of Maastricht 1997.
10. J. Rijcken, P.H.M. Bovendeerd, A.J.G. Schoofs, D.H. van Campen, Arts T. Optimization of cardiac fiber orientation for homogeneous fiber strain during ejection. *Ann Biomed Eng.*, 27:289–297, 1999.
11. Stuber M., Scheidegger M.B., Fischer S.E., Nagel E., Steinemann F., Hess O.M., Boesiger P. Alterations in the local myocardial motion pattern in patients suffering from pressure overload due to aortic stenosis. *Circ.*, 100:361–368, 1999.
12. Usyk T.P., Mazhari R., McCulloch A.D. Effect of laminar orthotropic myofiber architecture on regional stress and strain in the canine left ventricle. *J. Elast.*, 61:143–164, 2000.

13. van der Toorn A., Barenbrug P., Snoep G., van der Veen F.H., Delhaas T., Prinzen F.W., Maessen J., Arts T. Transmural gradients of cardiac myofiber shortening in aortic valve stenosis patients using mri tagging. *Am. J. Physiol.*, 283:1609–1615, 2002.
14. Villarreal F.J., Lew W.Y.W., Waldman L.K., Covell J.W. Transmural myocardial deformation in the ischemic canine left ventricle. *Circ. Res.*, 689:368–381, 1991.

Cardiac Function Estimation from MRI Using a Heart Model and Data Assimilation: Advances and Difficulties*

M. Sermesant[1], P. Moireau[2], O. Camara[1], J. Sainte-Marie[2],
R. Andriantsimiavona[1], R. Cimrman[3], D.L.G. Hill[1,*],
D. Chapelle[2], and R. Razavi[1]

[1] Cardiac MR Research Group, King's College London,
5th Floor Thomas Guy House, Guy's Hospital, London, UK
[2] MACS project, INRIA Rocquencourt, France
[3] New Technologies Research Centre, Západočeská univerzita v Plzni,
Univerzitní 22, 306 14 Plzeň, Czech Republic
derek.hill@ieee.org

Abstract. In this article, we present a framework to estimate local myocardium contractility using clinical MRI, a heart model and data assimilation. First, we build a generic anatomical model of the ventricles including muscle fibre orientations and anatomical subdivisions. Then, this model is deformed to fit a clinical MRI, using a semi-automatic fuzzy segmentation, an affine registration method and a local deformable biomechanical model. An electromechanical model of the heart is then presented and simulated. Data assimilation makes it possible to estimate local contractility from given displacements. Presented results on adjustment to clinical data and on assimilation with simulated data are very promising. Current work on model calibration and estimation of patient parameters open up possibilities to apply this framework in a clinical environment.

1 Introduction

The integration of knowledge from biology, physics and computer science makes it possible to combine *in vivo* observations, *in vitro* experiments and *in silico* simulations. From these points of view, knowledge of the heart function has greatly improved at the nanoscopic, microscopic and mesoscopic scales [15, 18].

Due to the limitations of medical imaging, modelling capabilities and computational power, the validation of heart models with human *in vivo* data and furthermore their use in clinical applications are very challenging. We present in this paper a framework aiming at overcoming these difficulties by directly combining modelling of the heart, cardiac function estimation and parameter

---

* Corresponding author.

adjustment. The detailed application is the estimation of local contractility in the myocardium from displacements measured in medical images.

The 5 components presented in this paper are: medical imaging techniques to observe the heart *in vivo*, building a generic anatomical heart model, equations used to simulate the cardiac electromechanical behaviour, adjustment of a generic heart model to patient anatomy, and data assimilation method to estimate local contractility. We emphasise in each of these sections the advances made and the difficulties encountered.

2 Observations: Magnetic Resonance Imaging

Magnetic Resonance Imaging (MRI) is a successful and promising modality but it remains difficult to use, and is seldom used outside major research centres. A particular challenge in Cardiac MRI is that the heart is a moving organ, in a moving environment (breathing). This restricts the resolution that can be obtained, and often leads to inconsistent data, making subsequent analysis challenging.

2.1 Anatomical Imaging

Black-Blood Imaging is characterised by the suppression of the signal from flowing blood. This gives a good visualisation of the myocardium, which is of great interest for our modelling purpose. Unfortunately, due to the pre-pulse and the inversion time, black-blood imaging is essentially a single slice sequence for each breath-hold. However, new methods are emerging to allow multi-slice imaging.

Bright-Blood Imaging on the contrary generates high signal intensity for blood and can be used for dynamic (cine) images of a small number of slices in each breath-hold. Therefore, bright-blood acquisitions allow both morphological and functional assessment. A major drawback however is that delineating the epicardium remains difficult due to the poor contrast between the heart and the lungs. 4D images with bright blood are becoming available to provide approximately isotropic resolution. These sequences usually have inferior temporal resolution and contrast to 2D dynamic sequences.

2.2 Functional Imaging

Global Cardiac Function Analysis. The quantification of ventricular volumes, myocardial mass and ejection fraction using MRI are both accurate and reproducible in the hand of experienced users. However, the time required for acquisition and analysis of MR images hampers the introduction of cardiac MR into routine clinical use. Cardiac MR examinations are frequently more than one hour, and involve numerous breath-holds, which add to patient discomfort.

Fractional k-space filling methods or view-sharing strategies enable the imaging time to be considerably reduced without substantial loss of image quality and resolution. However, these techniques are mainly used for 2D imaging. The major issue with multiple 2D imaging is that the coverage of the ventricles requires multiple breath-holds. Inconsistencies in the different breath-hold positions can lead to errors in image interpretation.

3D multiphase single breath-hold imaging methods appear to be a promising alternative for functional imaging [21]. Nevertheless, compromises had to be made in terms of image quality, spatial and temporal resolutions.

Regional Cardiac Function Analysis. Ejection Fraction is a global parameter that assesses the status of the cardiac function with great efficiency. However, it is not specific enough for myocardial efficiency and contractility. The study of wall deformations provides more insights on the mechanical contraction of the heart. Tagging is a well-known method to track local deformations of a "printed" grid as it follows the heart contracting. It enables parameters such as twist, strain and strain rate to be derived. The extraction of motion is based on models or registration techniques [5]. Although accurate, it is time consuming and suffer from the low spatial resolution of tags. More recently, the development of harmonic phase (HARP) and displacement encoding stimulated echo (DENSE) methods makes it possible to quantify the displacement of each moving pixel inside the myocardium. These techniques though are currently limited to 2D displacements.

The direct relationship between myocardial motion and contractility is difficult to estimate directly from the images. A model-based approach could thus help to extract this hidden information.

3 Generic Anatomical Heart Model

The aim of this work is to provide a method for model-based analysis of the previously described medical images. The idea is to adjust a biomechanical model of the myocardium using these images in order to extract hidden parameters useful for diagnosis, like local contractility. To achieve the simulation of cardiac electromechanical activity, we need the myocardium geometry and the muscle fibre orientations as anatomical inputs. The geometry gives the domain on which to carry out computations. Fibre orientations are important for both the active and passive behaviour of the myocardium.

The difficulty for this step is to obtain both types of information for a particular myocardium. On the one hand, geometry can be extracted from anatomical medical images. But fibre orientations cannot be measured *in vivo* and current diffusion tensor images of fixed hearts are still noisy compared to the smoothness required by the electromechanical computations.

On the other hand, when fibre orientations are measured from dissection, the geometry is often not available, or in so deformed shape that adjustment of the model to the *in vivo* images becomes very challenging.

Due to these problems, our approach is to synthesise a generic anatomical model of the myocardium, composed of a simple geometry, close enough to *in vivo* observations, and of synthetic fibre orientations, generated according to the measurements available in the literature.

3.1 Heart Geometry

Left ventricle shape is close to a truncated ellipsoid, as shown by the use of this shape for left ventricle volume estimation from 2D images [19]. The right ventricle can also be approximated by a truncated ellipsoid. The generic heart model geometry is defined using different parameters for the radii of the left and right ventricles ellipsoids, their thickness and the height of the truncating basal plane (see Fig. 1).

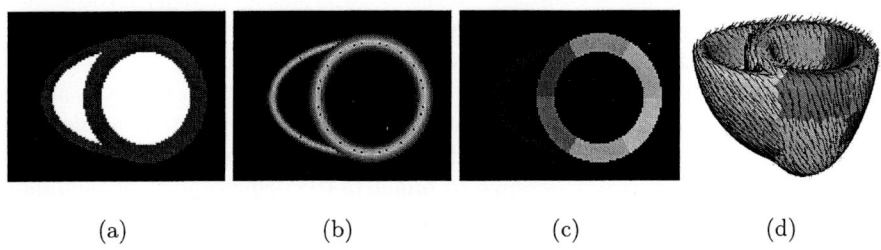

(a) (b) (c) (d)

Fig. 1. Generic anatomical heart model: equatorial short axis slice of (a) geometry, (b) fibres orientation (elevation angle), (c) AHA division. (d) Resulting mesh with fibre orientations (black segments) and AHA divisions (colours)

3.2 Heart Fibres Orientations

It is well known that muscle fibre orientations vary across the myocardial wall. Most diffusion tensor imaging and dissection analysis observed an elevation angle (angle between the fibre and the short axis plane) varying from $+90°$ on the endocardium to $-90°$ on the epicardium [11]. We analytically defined the model fibre orientation to follow a linear variation between these two values, this orientation being in the short axis plane at mid-wall.

3.3 Cardiac Anatomical Divisions

Accurate calibration, estimation and analysis of the model is made easier by subdividing it into different anatomical regions. Generating the model makes it possible to analytically divide it into the 17 regions of interest proposed by the American Heart Association (AHA) [4].

3.4 Myocardium Mesh

From the anatomical image, a triangulated surface is extracted using the marching cubes algorithm. This surface is used to create a tetrahedral mesh with the

INRIA GHS3D software[1]. Finally, fibre orientations and subdivisions are assigned to the mesh using rasterization, see details in [24]. The resulting mesh is presented Fig. 1d.

4 Cardiac Muscle Biomechanics

Modelling the myocardium behaviour is difficult because of its active, non-linear, anisotropic nature. Several constitutive laws were proposed for the active and passive properties of the myocardium [15, 18].

4.1 Myofibre Active Constitutive Law

A constitutive law of the electrically-activated myofibres was proposed by Bestel-Clément-Sorine [3]. Whereas most modelling endeavours rely on heuristic considerations [2, 8, 12], this law is based on a multi-scale approach taking into account the behaviour of myosin molecular motors, and the resulting sarcomere dynamics is in agreement with the sliding filament hypothesis introduced in [13]. Denoting by σ_c the active stress and by ε_c the strain along the sarcomere, this law relates σ_c and ε_c as follows:

$$\begin{cases} \dot{\tau}_c = k_c \dot{\varepsilon}_c - (\alpha|\dot{\varepsilon}_c| + |u|)\tau_c + \sigma_0 |u|_+ & \tau_c(0) = 0 \\ \dot{k}_c = -(\alpha|\dot{\varepsilon}_c| + |u|)k_c + k_0 |u|_+ & k_c(0) = 0 \\ \sigma_c = \tau_c + \mu \dot{\varepsilon}_c + k_c \xi_0 \end{cases} \quad (1)$$

where u represents the electrical input ($u > 0$: contraction, $u \leq 0$: relaxation). Parameters k_0 and σ_0 characterise muscular contractility and respectively correspond to the maximum value for the active stiffness k_c and for the stress τ_c in the sarcomere, while μ is a viscosity parameter.

The propagation of the action potential activating the muscle contraction can be modelled by non-linear reaction-diffusion equations, see [10] and references therein. However, the corresponding numerical computations are costly, and in particular make a combined electromechanical data assimilation procedure well out of reach. Hence so far we have mostly considered simplified activation patterns such as uniform activation (in space) or a planar wave travelling from apex to base.

4.2 3D Model of the Myocardium

The above active constitutive law was used within a rheological model of Hill-Maxwell type [6], as depicted in Fig. 2a. The element E_c accounts for the contractile electrically-activated behaviour governed by (1). In addition, an elastic material law is considered for the series element E_s, while E_p is taken viscoelastic. Based on experimental results, the corresponding stress-strain laws are assumed to be of exponential type for E_p [26], and linear for E_s [20].

---

[1] http://www-rocq.inria.fr/gamma/ghs3d/ghs.html

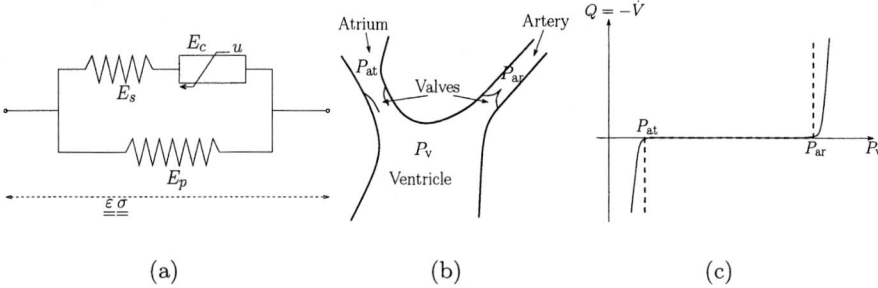

Fig. 2. (a) Hill-Maxwell rheological model. (b) Aortic valve model mechanism. (c) Formulation as a double contact problem, dashed: relation (2), solid: regularised

This rheological model is compatible with large displacements and strains and led to a continuum mechanics description of the cardiac tissue [7]. A study and simulations of a simplified 1D model derived from this continuum mechanics model are detailed in [6].

4.3 Modelling the Blood

The blood inside each ventricle is modelled as a pressure / volume system. The phases of the cardiac cycle (isovolumetric contraction, ejection, isovolumetric relaxation, filling) are distinguished through coupling conditions between the internal fluid and other parts of the cardiovascular system, namely the atrial cavities and the external circulation. With P_v, P_{ar} and P_{at} denoting the blood pressures in the ventricle, the artery, and the atrium, respectively, the ejection occurs when $P_v \geq P_{ar}$ whereas the mitral valve opens when $P_v \leq P_{at}$, see Fig. 2b. Denoting by Q the outgoing flow, the coupling conditions can be formulated as a (double) contact problem:

$$\begin{cases} Q \geq 0 \text{ when } P_v = P_{ar} & ejection \\ Q = 0 \text{ when } P_{at} < P_v < P_{ar} & isovolumetric\ phases \\ Q \leq 0 \text{ when } P_v = P_{at} & filling \end{cases} \quad (2)$$

To avoid numerical difficulties, we used a regularised form of this function as depicted by the solid line in Fig. 2c. External circulation is modelled by a Windkessel model [17, 25], and blood flow coming from the atria by a pressure P_{at}.

5 Data Assimilation

The aim of data assimilation is to incorporate measurements into a dynamic system model in order to produce accurate estimates of the current (and possibly future) state variables, parameters, initial conditions and input of the model.

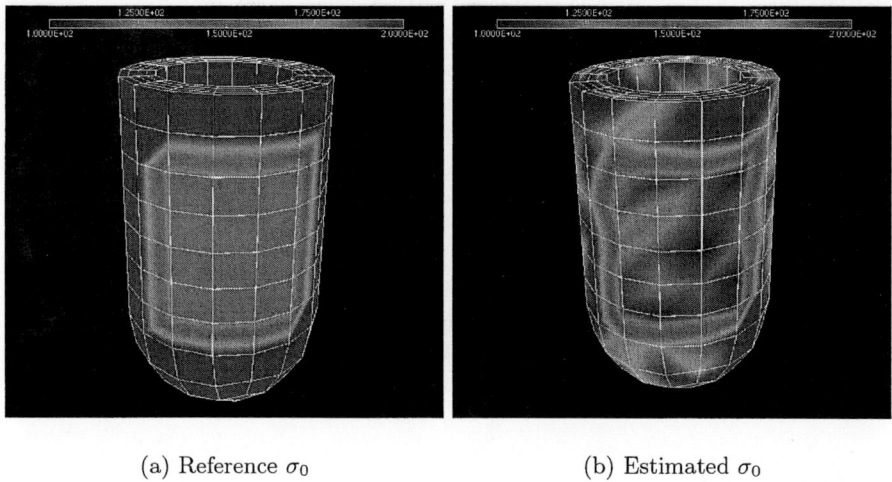

(a) Reference σ_0 (b) Estimated σ_0

Fig. 3. Contractility (σ_0) estimation from simulated data. The observations are the displacements on the epicardial and endocardial surfaces and the variational technique is used. The lower contractility region is well recovered by the data assimilation. The scale is going from 1.10^2 (central zone with reduced contractility) to 2.10^2 (normal contractility)

The symbol H denoting the observation operator, $Y(t)$ the available measurements and $X(t)$ the model response, the general objective of data assimilation is the minimisation of a cost function J performed over the set of parameters to be estimated

$$J = \int_I \|Y(t) - HX(t)\|_\Omega^2 dt + penalty \qquad (3)$$

$\|.\|_\Omega$ being a suitable norm associated with the problem formulation.

If I denotes the complete simulation time interval $[t_0, T]$, the assimilation technique is said to be variational and corresponds to an optimal control problem [9, 16]. If at each time step t_k, $I = [t_0, t_k]$, then the filtering technique is said to be sequential [14].

Due to the complexity of the model and to observability considerations, estimating all the quantities appearing in the complete electromechanical problem is out of reach. Hence we focus on parameters that are crucial for medical purposes in order to detect contraction troubles, in particular the parameters σ_0 and k_0.

Preliminary results in data assimilation have been presented in [23]. Results presented here have been obtained using numerically simulated observations assimilated with the complete 3D problem on a left ventricle model (Fig. 3).

Both sequential and variational techniques have been tested. Without considering the computational costs, the results given by the two methods are very similar. The data assimilation process validation is the following:

- The direct 3D problem is simulated with a given parameter σ_0.
- Observations $\{Y(M_k, t_p)\}_{k,p}$ are obtained using $Y(M_k, t_p) = HX(M_k, t_p)$, H being the chosen observation operator. Here the observations Y chosen are the displacement vectors for the points located on the epicardium and endocardium surfaces only.
- Starting from a given parameter $\hat{\sigma}_0$ different from the one used for the direct simulation, the data assimilation is carried out.

For this simulation we chose $\sigma_0(M)$ constant across the wall, with a region of different σ_0, visible on Fig. 3a. The result of the estimation of σ_0 is shown in Fig. 3b. The data assimilation process, initialised with a homogeneous distribution, recovered the spatial variations of σ_0 rather accurately. We used the variational technique (four iterations), briefly described in Appendix.

6 Toward Patient-Specific Cardiac Function Estimation

We demonstrated in the previous section that we could estimate local contractility in the myocardium using sparse known displacements and a biomechanical model, through data assimilation. Before being able to run this method on clinical data, we first have to adjust the model anatomy to the 3D patient anatomy.

6.1 Patient-Specific Heart Anatomy

Automatic segmentation of the myocardium from MRI is still very difficult, as the epicardium is not easily distinguished and the right ventricle is quite thin. Our approach is to deform the generic anatomical model designed in Section 3 into the first 3D image of the sequence. This is done in three steps: segmentation of the image blood pools, intensity-based registration for the affine adjustment and deformable model-based segmentation for local deformations.

As the clinically used MR sequences produce relatively homogeneous blood pool intensity, one can semi-automatically segment the ventricular blood pools using a combined boundary-based and regional-based fuzzy classification method [1], in order to ease the registration step.

Then, an automatic affine (15 parameters) registration algorithm is applied from the segmented image to the blood pools from the model geometry, using the cross correlation as similarity measure.

Finally, local adjustment is done with a deformable biomechanical model [24], using the affine transformation computed previously for initialisation. We use a Houbolt semi-implicit time integration which gives better results for large deformations, as the semi-implicit part regularises the way the mesh deforms. This allows precise adjustment of the generic anatomical model to the patient image (Fig. 4), but still preserves surface smoothness and element numerical quality: the aspect ratio mean (resp. standard deviation) on all the tetrahedra of the mesh is 0.674 (0.109) for the original mesh, 0.650 (0.113) after affine transformation, and 0.652 (0.114) after local deformation.

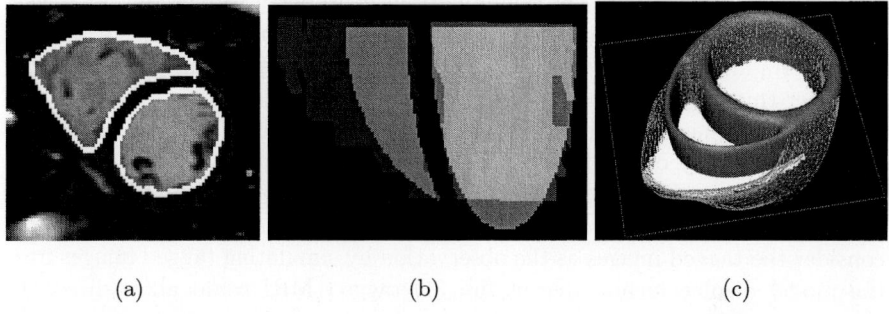

Fig. 4. Patient-specific model three steps: (a) semi-automatic segmentation of the blood pools (white contours), (b) affine registration between the model blood pools (grey) and the segmentation (green), (c) local adjustment (orange wire-frame) from the affine transform of the model (blue surface) using a deformable biomechanical model

To introduce prior information (e.g. fibre orientation) and make it easier to compare the results between normal and pathological cases, it is very advantageous to have correspondence between the reference mesh and patient-specific meshes. Moreover, it is difficult to obtain good quality meshes directly from automatically segmented medical due to the lack of smoothed boundaries. Therefore we chose to build a generic model and then deform it into the patient-specific data.

The whole process could be rather automated, as segmentation and registration parameters are robust with respect to the input images. The local deformation needs some visual control as no stop criterion has yet been implemented. Nevertheless, from our experience, the parameters in these three steps are quite constant across different images so it should be possible to minimise user input, and the overall time is considerably inferior to manual segmentation duration.

This patient-specific model can then be used for the data assimilation procedure, but there are requirements on the observations used.

6.2 Data Assimilation Difficulties

Measurements used to apply the data assimilation in section 5 are displacements in some points of the myocardium. The idea is to use the same data assimilation procedure, with displacements from tagged MRI.

In data assimilation, difficulties arise from various areas, from the available measurements to the complexity of the operator (type of variables, dimension, rank) and the natures of the spaces and norms used. Current work on these difficulties should help design the best possible operators to achieve this parameter estimation.

A difficulty is the invertibility of this observation operator because we want to obtain the state X and the parameters from the observations using a generalised inverse of H. The analysis of this invertibility property (observability) is very difficult in general both as regards surjectivity (whether there exists a set of parameters and variables which leads to the given observation) and injectivity (whether this set is unique).

Another difficulty in the choice of the observation operator lies in the fact that the efficiency of the filtering technique is highly dependent on the noise. To avoid adding noise in the different image processing steps, an idea could be to formulate an operator as close to the measurements as possible. We could consider the tagged images as the observation by simulating tagged images from the model displacements. Recent full 3D tagged MRI could allow direct 3D comparison [22].

7 Conclusion

This article presents a framework for automatic estimation of local contractility from MRI and a model of the myocardium, along with initial results. We detailed the medical images used and the information we can extract from it, the construction of a generic anatomical model of the ventricular myocardium integrating muscle fibre orientations and its subdivision into segments, the biomechanical modelling of the myocardium, and a data assimilation method to automatically adjust the parameters of the model from known displacements.

We demonstrated the capability of such a framework, with also pointing out the different difficulties at the theoretical and practical levels. The results so far obtained by combining modelling and data are very promising. Precise calibration of the model before data assimilation is difficult and additional measurements can help for this task. Progress in MRI, especially in flow measurements, open up possibilities to obtain patient-specific boundary conditions.

For data assimilation, the current work is on observation operators, for instance whether it can be written in a Lagrangian framework, as with tagged MRI. Lagrangian approach is more easily dealt with, but it is not suitable for all types of measurements, in particular for those directly related to the deformed geometry. Many clinical observations (e.g. cine MRI and ultrasound) are indeed more Eulerian in essence.

Future developments are planned to integrate different modalities. For instance, with patients undergoing electrophysiology studies, electrophysiology clinical data can be acquired, using XMR interventional imaging for instance. Such datasets open up possibilities to also adjust electrophysiology models. We point out that an interesting open problem concerns whether or not the electrical activity may also be estimated from displacements measurements of the myocardium. The proposed framework could give insights on this problem. Finally, coupling models and parameter estimation is valuable for intervention planning and therapy testing, owing to the predictive capacity of modelling.

Acknowledgements

The authors would like to thank for their collaboration the Cardiac MR Research Group in Guy's Hospital, London and the co-workers of the ICEMA collaborative research actions[2,3] funded by INRIA. Part of this work was done during the Summer Mathematical Research Center on Scientific Computing and its Applications (CEMRACS)[4]. The authors acknowledge grant support from EPSRC (M.S., O.C. and R.A.) and the use of software developed by the Epidaure project[5], INRIA.

References

1. R. Andriantsimiavona, L. Griffin, D. Hill, and R. Razavi. Simple cardiac MRI segmentation. In *International Society for Magnetic Resonance in Medicine Scientific Meeting*, volume 6, page 951, 2003.
2. T. Arts, P. Bovendeerd, A. van der Toorn, L. Geerts, R. Kerckhoffs, and F. Prinzen. Modules in cardiac modeling: Mechanics, circulation, and depolarization wave. In *Functional Imaging and Modeling of the Heart (FIMH'01)*, number 2230 in Lecture Notes in Computer Science (LNCS), pages 83–90. Springer, 2001.
3. J. Bestel, F. Clément, and M. Sorine. A biomechanical model of muscle contraction. In *Medical Image Computing and Computer-Assisted intervention (MICCAI'01)*, volume 2208 of *Lecture Notes in Computer Science (LNCS)*, pages 1159–1161. Springer, 2001.
4. M. Cerqueira, N. Weissman, V. Dilsizian, A. Jacobs, S. Kaul, W. Laskey, D. Pennell, J. Rumberger, T. Ryan, and M. Verani. Standardized myocardial segmentation and nomenclature for tomographic imaging of the heart. *Circulation*, 105:539–542, 2002.
5. R. Chandrashekara, R. Mohiaddin, and D. Rueckert. Analysis of 3-D myocardial motion in tagged MR images using nonrigid image registration. *IEEE Transactions on Medical Imaging*, 23(10):1245–1250, 2004.
6. D. Chapelle, F. Clément, F. Génot, P. Le Tallec, M. Sorine, and J. Urquiza. A physiologically-based model for the active cardiac muscle contraction. In *Functional Imaging and Modeling of the Heart (FIMH'01)*, number 2230 in Lecture Notes in Computer Science (LNCS), pages 128–133. Springer, 2001.
7. D. Chapelle, J. Sainte-Marie, and R. Cimrman. Modeling and estimation of the cardiac electromechanical activity. In *Proceedings of the ECCOMAS 2004 Conference*, 2004.
8. K. Costa, J. Holmes, and A. McCulloch. Modelling cardiac mechanical properties in three dimensions. *Philosophical Transactions of the Royal Society of London*, 359(1783):1233–1250, 2001.
9. P. Courtier and O. Talagrand. Variational assimilation of meteorological observations with the adjoint vorticity equation. *Quart. J. Roy. Meteorol. Soc.*, 113:1311–1347, 1987.

---

[2] http://www-rocq.inria.fr/who/Frederique.Clement/icema.html
[3] http://www-rocq.inria.fr/sosso/icema2/icema2.html
[4] http://smai.emath.fr/cemracs/cemracs04/index.php
[5] http://www-sop.inria.fr/epidaure/index.php

10. R. FitzHugh. Impulses and physiological states in theoretical models of nerve membrane. *Biophysical Journal*, 1:445–466, 1961.
11. E. Hsu and C. Henriquez. Myocardial fiber orientation mapping using reduced encoding diffusion tensor imaging. *Journal of Cardiovascular Magnetic Resonance*, 3:325–333, 2001.
12. P. Hunter, A. Pullan, and B. Smaill. Modeling total heart function. *Annual Review of Biomedical Engineering*, 5:147–177, 2003.
13. A.F. Huxley. Muscle structure and theories of contraction. *Progress in Biophysics & Biological Chemistry*, 7:255–318, 1957.
14. R.E. Kalman. A new approach to linear filtering and prediction problems. *ASME Trans.–Journal of Basic Engineering*, 82(Series D):35–45, 1960.
15. T. Katila, I. Magnin, P. Clarysse, J. Montagnat, and J. Nenonen, editors. *Functional Imaging and Modeling of the Heart (FIMH'01)*, number 2230 in Lecture Notes in Computer Science (LNCS). Springer, 2001.
16. J.L. Lions. *Contrôle optimal des systèmes gouvernés par des équations aux dérivées partielles*. Dunod, 1968.
17. D.A. MacDonald. *Blood flow in arteries*. Edward Harold Press, 1974.
18. I. Magnin, J. Montagnat, P. Clarysse, J. Nenonen, and T. Katila, editors. *Functional Imaging and Modeling of the Heart (FIMH'03)*, number 2674 in Lecture Notes in Computer Science (LNCS). Springer, 2003.
19. J. Mercier, T. DiSessa, J. Jarmakani, T. Nakanishi, S. Hiraishi, J. Isabel-Jones, and W. Friedman. Two-dimensional echocardiographic assessment of left ventricular volumes and ejection fraction in children. *Circulation*, 65:962–969, 1982.
20. I. Mirsky and W.W. Parmley. Assessment of passive elastic stiffness for isolated heart muscle and the intact heart. *Circul. Research*, 33:233–243, 1973.
21. D. Peters, D. Ennis, P. Rohatgi, M. Syed, E. McVeigh, and A. Arai. 3D breath-held cardiac function with projection reconstruction in steady state free precession validated using 2D cine MRI. *Journal of Magnetic Resonance Imaging*, 20(3):411–416, 2004.
22. S. Ryf, M. Spiegel, M. Gerber, and P. Boesiger. Myocardial tagging with 3D-CSPAMM. *Journal of Magnetic Resonance Imaging*, 16(3):320–325, 2002.
23. J. Sainte-Marie, D. Chapelle, and M. Sorine. Data assimilation for an electromechanical model of the myocardium. In K.J. Bathe, editor, *Second M.I.T. Conference on Computational Fluid and Solid Mechanics*, pages 1801–1804, 2003.
24. M. Sermesant, C. Forest, X. Pennec, H. Delingette, and N. Ayache. Deformable biomechanical models: Application to 4D cardiac image analysis. *Medical Image Analysis*, 7(4):475–488, 2003.
25. N. Stergiopulos, B.E. Westerhof, and N. Westerhof. Total arterial inertance as the fourth element of the windkessel model. *Am. J. Physiol.*, 276:H81–H88, 1999.
26. D.R. Veronda and R.A. Westmann. Mechanical characterization of skin - finite deformation. *Journal of Biomechanics*, 3:114–124, 1970.

Appendix

Variational data assimilation techniques are based on an iterative approximation of the optimality condition $\nabla_S J(S^*) = 0$, where S denotes the parameter set to estimate, leading to an adjoint problem. If the problem to solve is (A), in the absence of a penalty term in J, the adjoint state P is governed by (B):

(A) $\begin{cases} \dot{X} = F(X,S,t) \\ X(t_0) = X_0 \\ S \text{ unknown parameters} \end{cases}$ (B) $\begin{cases} \dot{P} + [\frac{\partial F}{\partial X}]^t P = H^t(HX - Y) \\ P(T) = 0 \end{cases}$

The numerical algorithm used is the following:

- Start from a first guess S_0 of the parameter set
- Start iteration n
- Integrate the direct model from 0 to T
- Integrate the adjoint model from T to 0
- Calculate the gradient $\nabla J(S_n) = \int_0^T \left[\frac{\partial F}{\partial S_n}\right] P \, dt$
- Compute $S_{n+1} = S_n + \rho_n \nabla J(S_n)$
- $n \to n+1$ until a stopping criterion on J is reached

This algorithm was used to make the local contractility evolve (with $S = \sigma_0(M)$) in the data assimilation results presented in section 5.

Assessment of Separation of Functional Components with ICA from Dynamic Cardiac Perfusion PET Phantom Images for Volume Extraction with Deformable Surface Models*

Anu Juslin[1], Anthonin Reilhac[2], Margarita Magadán-Méndez[1], Edisson Albán[1], Jussi Tohka[1,3], and Ulla Ruotsalainen[1]

[1] Tampere University of Technology, Institute of Signal Processing, Tampere, Finland
anu.juslin@tut.fi
[2] McConnell Brain Imaging Centre, Montreal Neurological Institute, Montreal, Canada
[3] Laboratory of Neuro Imaging, Division of Brain Mapping, Department of Neurology, UCLA, School of Medicine, Los Angeles, CA, USA

Abstract. We evaluated applicability of ICA (Independent Component Analysis) for the separation of functional components from $H_2^{15}O$ PET (Positron Emission Tomography) cardiac images. The effects of varying myocardial perfusion to the separation results were investigated using a dynamic 2D numerical phantom. The effects of motion in cardiac region were studied using a dynamic 3D phantom. In this 3D phantom, the anatomy and the motion of the heart were simulated based on the MCAT (Mathematical Cardiac Torso) phantom and the image acquisition process was simulated with the PET SORTEO Monte Carlo simulator. With ICA, it was possible to separate the right and left ventricles in the all tests, even with large motion of the heart. In addition, we extracted the ventricle volumes from the ICA component images using the Deformable Surface Model based on Dual Surface Minimization (DM-DSM). In the future our aim is to use the extracted volumes for movement correction.

1 Introduction

The Positron Emission Tomography (PET) using Oxygen-15-labeled water allows for noninvasive quantification of myocardial blood flow [1]. The analysis

---

* This work was supported by TEKES Drug 2000 technology program, Tampere Graduate School of Information Sciences and Engineering (TISE), Graduate School of Tampere University of Technology. Anu Juslin obtained a grant from the Cultural Foundation of Pirkanmaa for this work. Jussi Tohka's work was funded by the Academy of Finland under the grants no. 204782 and 104824 and by the NIH/NCRR grant P41 RR013642, additional support was provided by the NIH Roadmap Initiative for Bioinformatics and Computational Biology U54 RR021813 funded by the NCRR, NCBC, and NIGMS.

is based on estimating the time-activity curves (TACs) of the blood pool and the myocardial tissue from the dynamic PET images. The TACs describe the time-dependent uptake of the radiopharmaceutical in tissues. However, unlike in F-18 labeled 2-fluoro-2-deoxy-D-glucose (FDG) cardiac studies it is difficult to identify the anatomical structures in $H_2^{15}O$ images, because the ^{15}O labeled water is rapidly distributed over the entire thorax region producing images with low contrast between anatomical structures. The motion of the patient and the motion of the inner structures of the thorax region are also significant problems in the analysis of functional cardiac images [2]. Because the nature of the motion in the thorax region is non-rigid and it is composed of the motion of the heart and other tissues, the detection and correction of the movement artifacts is a very complicated task. Furthermore, the varying tracer uptake and noise in dynamic images form additional challenges for the extraction of the cardiac structures.

The evaluation of the image processing procedures in cardiac PET imaging is a difficult task. However, using realistic simulations for the image acquisition process and phantom images describing the human anatomy, it is possible to evaluate the performance of image analysis algorithms. In the thorax region the motion of the involved structures has to be taken into account as well as the dynamic behaviour of the tracer in the different regions, which make the generation of the cardiac perfusion PET phantoms very demanding.

Our long term goal is to develop a procedure to correct the motion artefacts between two $H_2^{15}O$ cardiac studies of the same patient. Our novel idea is to first enhance the contrast of the cardiac perfusion PET images, so that different functional components such as ventricles and myocardium could be separated from surrounding tissues and noise. We chose the Independent Component Analysis (ICA) method [3] for this separation task. The ICA method has been previously applied to robust extraction of the input function from myocardial dog PET images [4], but has never been applied to human cardiac studies. In the second phase we will automatically extract the volumes of the ventricles and the myocardium with deformable models. The extracted volumes can then be used for the realignment of two studies.

To reach this goal we need first to assess the performance of the methods with phantom data. In this study the first goal was to analyse how well the ICA method could separate different tissues when the myocardial flow varies. The myocardial flow can differ from 40 ml/min*100g in infarcted regions up to 500 ml/min*100g during physiological stress [5]. Further on, we studied the effect of the motion of the heart on the separation of the functional components and on the automatic volume extraction with deformable surface models. In the future, we will use the extracted volumes to correct image artifacts caused by the movement of the patient.

2 Materials

Two different phantoms were generated for this study. Simple numerical 2D dynamic cardiac phantom was created for assessing the sensitivity of the ICA

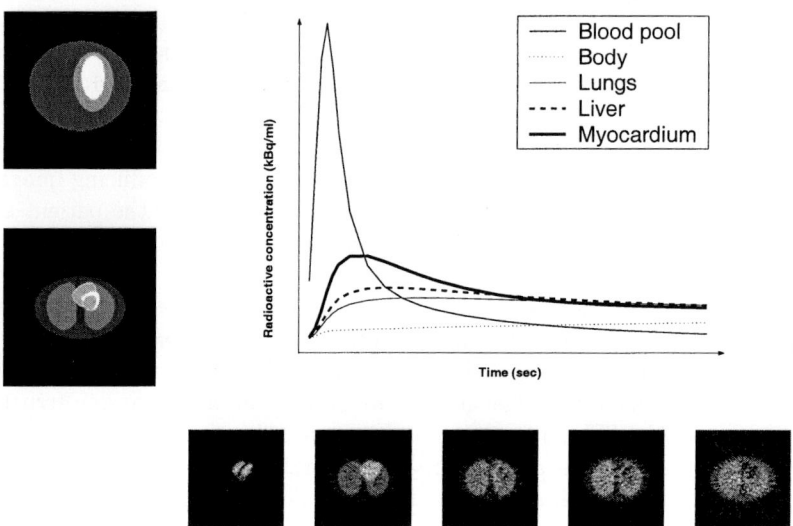

Fig. 1. In the left the anatomical geometry of the 2D phantom (Top) and the 3D phantom (Bottom). In the top right the TACs for different tissues used in the generation of the dynamic 3D cardiac perfusion PET images. In the bottom example of the simulated 3D dynamic phantom at 5 different time frames

separation of functional components with varying myocardial perfusion. In addition, with PET SORTEO, a Monte Carlo-based PET simulator [9], we generated realistic dynamic cardiac studies, using the MCAT phantom [7][8]. These simulated studies were used to investigate the impact of the heart and respiratory motions on the outcome of the separation and volume extraction processes.

The simple 2D phantom was composed of three different anatomical components: left ventricle, myocardium and body background (Fig. 1). The phantom consisted of 20 time frames (6 frames of 5 s, 6 frames of 15 s and 8 frames of 30 s) with virtual time dependent tracer distribution. We applied a measured blood TAC to calculate the tissue TACs using one block model [10]. The virtual perfusion values for the myocardium of the phantom were set from 10 ml/min*100g up to 500 ml/min*100g. For the body background the value of 5 ml/min*100g was applied in all cases. Over-dispersed Poisson noise with large variance [9] was added to the sinogram bin intensity values. The sinograms were reconstructed with filtered back projection (FBP) using Hann-filter with cut-off value 0.5.

For the evaluation of the motion effect, we generated a realistic dynamic 3D PET cardiac data set based on the MCAT phantom [7][8] using PET SORTEO software [9]. The MCAT phantom was used as an anatomical base for the phantom and the PET SORTEO software for simulating the PET dynamics with a realistic signal degradation. In our phantom we took into account 6 different anatomical structures: ventricles, atriums, myocardium, lungs, body and liver (Fig. 1). The MCAT phantom provided a possibility to simulate the motion of the heart and lungs over the time. We set the heart rate to be 60 beats per minute and the breathing to 12 cycles per minute. The resulting 4D phantom characterizes both the physiological behaviour of the heart and lungs and the dynamic behaviour of the tracer in the involved tissues. In addition, one dynamic study without the cardiac and respiratory motions was generated corresponding to the end-diastole phase.

The tissue TACs for 3D phantom were generated similarly as in the 2D phantom. The perfusion values were set to be 75 ml/min*100g for the myocardium, 25 ml/min*100g for the lungs, 35 ml/min*100g for the liver and for the body 5 ml/min*100g (Fig. 1). The simulation of dynamic PET acquisition was carried out using Monte Carlo-based 3D PET simulator PET SORTEO [9]. This simulation tool has been dedicated to full ring PET tomography. The simulation was performed for the Ecat Exact HR+ scanner operating in 3D mode. The ^{15}O imaging protocol lasted 6 minutes with the same frame times than in the 2D phantom. The raw data was reconstructed with FBP (3DRP with Hann apodizing window and the Nyquist frequency cutoff, scatter correction, online subtraction of randoms, arc correction, normalization, and attenuation correction). This resulted in 20 time frames of 128x128x63 voxels each, whose sizes were 3.52mm x 3.52mm x 2.43mm.

3 Methods

In this study the proposed approach to extract structures of interest from $H_2^{15}O$ cardiac PET images, split into two major steps. First, the tissues of interest in the dynamic images were separated using the ICA method. Secondly, the volumes of the left and right ventricles were extracted using the DM-DSM method [12][13] from the ICA component images. The results were evaluated both visually and quantitatively. The results of automatic segmentation were compared to the ground truth by computing the Jaccard similarity coefficient [15] (also known as the Tanimoto coefficient) between the automatically segmented structures and the ground truth structures. The Jaccard value ranges from 0 for volumes that have no common voxels to 1 for volumes that are identical.

3.1 Independent Component Analysis

Our aim was to separate different tissues from dynamic cardiac perfusion data for the volume extraction. This problem was considered as a Blind Source Separation Problem. In order to solve it, we applied the ICA method on the reconstructed dynamic cardiac images. ICA is a statistical method whose goal is to represent a set of random variables as linear combinations of statistically independent component variables [3]. The ICA can be formulated to be the estimation of the following linear model for the data:

$$x = \mathbf{A}s, \qquad (1)$$

where x is a random vector modelling the observations, s is a vector of the latent variables called the independent components, and $\mathbf{A}$ is an unknown constant matrix, called the mixing matrix. In this study, x was the vectorized form of the voxel intensity values from dynamic images and s was the vectorized form of the functional components, which we tried to separate from the dynamic images. The problem of ICA is then to estimate both the mixing matrix and the independent components using only observed mixtures.

To solve the ICA separation problem we used FastICA algorithm [14]. The FastICA algorithm is a computationally highly efficient method for performing the estimation of independent components. The resulting ICA component images were identifying those voxels to same component whose dynamic behaviour were similar. We considered that our mixture was composed of 4 different independent components. In the simple 2D phantom we knew that the amount of the source components was four (blood pool, myocardium, body and noise). In the more realistic 3D dynamic phantom we assumed that there were 4 different functional components in the field of view (blood pool, myocardium, lungs and body background including noise component).

The result of FastICA depends on the initialization of the mixing matrix $\mathbf{A}$. Conventionally the initialization of the FastICA has been done using a random matrix. The problem of using random initialization is that every run gives different result. For this reason we used fixed initialization to solve ICA problem with FastICA algorithm, because with fixed initialization we always end up to the same result. The initialization matrix $\mathbf{A} = (a_{ij})_{nxm}$ was defined in the following way: $a_{ij} = 1$ if $i = j$ and otherwise $a_{ij} = 0$, n was the number of the mixtures and m was the number of the source components. PCA (Principal Component Analysis) and whitening were used as pre-processing for ICA in order to de-correlate the input data and reduce the dimensionality of data.

3.2 DM-DSM Method

For volume extraction purposes we used the DM-DSM (Deformable Model with Dual Surface Minimization) algorithm [12][13]. The surface extraction is reformulated as an energy minimization problem. The energy $E(S;I)$ of the surface S given an image I depends on the image data and the properties of the surface itself. It is a weighted sum of the internal energy penalizing surfaces that are not smooth and the external energy that couples surfaces with the image data. The total energy of the surface S is

$$E(S;I) = \lambda E_{int}(S) + (1-\lambda)E_{ext}(S;I), \qquad (2)$$

where $E_{int}(S)$ is the internal energy, $E_{ext}(S;I)$ is the external energy, and $\lambda \in [0,1]$ is the regularization parameter controlling the tradeoff between external and internal energies.

The internal energy was based on a simple thin-plate shape model [12]. In this study, the external energy values for each voxel were derived from the ICA component images. The external energy value for each voxel was stored in look-up-table, which was called energy image. In the energy image high intensity value corresponded to surface which we were interested in. The energy images were obtained using extended 3D version of varying adaptive window size edge detection method [16][17] to the resulting ICA component images. The starting point of the volume search was defined manually to inside of the object.

4 Results

The values of Jaccard coefficients between the reference volumes and the extracted cardiac tissue volumes are reported in Table 1. In Fig. 2, Fig. 3 and Fig. 4 the results of the ICA separation and the volume extraction are shown. We only present two resulting ICA component images, which contain the ventricles (blood pool) and the myocardium.

Table 1. The Jaccard similarity coefficients between the automatically segmented structures and the ground truth structures. The separation results of the myocardium from the 3D dynamic data were not good enough for the volume extraction

		Blood pool	Myocardium
2D phantom	myocardial flow 500 ml/min*100g	.958	.717
	myocardial flow 300 ml/min*100g	.931	.725
	myocardial flow 100 ml/min*100g	.923	.681
	myocardial flow 40 ml/min*100g	.920	.534
3D phantom	No motion	.652	
	Motion	.607	

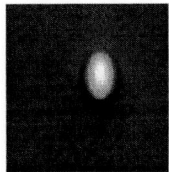 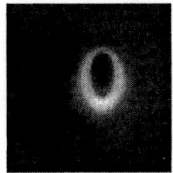 

Fig. 2. In the left the ICA separation of the blood pool and in the middle the separation of the myocardium from the 2D phantom with high myocardial flow (500 ml/min*100g). In the right the sum of the all time frame images from the original 2D phantom image

4.1 The Myocardial Flow Test

The ICA method was able to separate automatically the blood pool and myocardium even with very high myocardial blood flow values from the dynamic 2D phantom data. The separation results of the blood pool and myocardium were visually excellent. The quantitative results showed that the blood pool was extracted with very high accuracy in all cases, but the extraction of the myocardium was more dependent on the perfusion level (Table 1). We used tresholding to define the volume of the blood pool and the myocardium from the resulting ICA component images. Fig. 2. shows the result of the separation

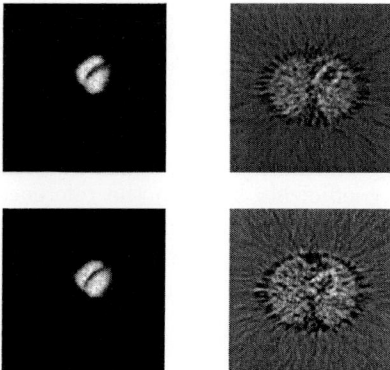

Fig. 3. The ICA separation of the functional components from the MCAT based phantom images. In the top left the blood pool and in the top right the myocardium from phantom image without motion and in the bottom left the blood pool and in the bottom right the myocardium from the phantom image with heart beating and respiratory motion

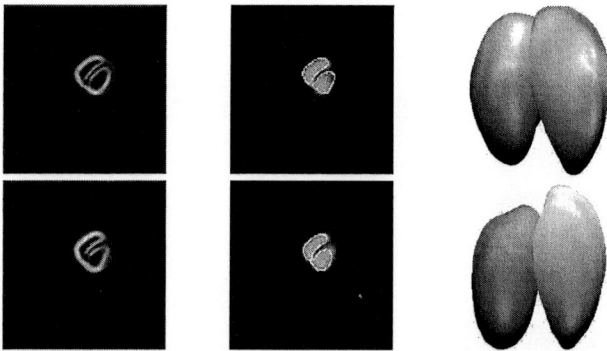

Fig. 4. The volume extraction result of the right and left ventricle from the MCAT based phantom images. In the top left the energy image of the blood pool, in the top middle the extracted volumes of the ventricles and in the top right the 3D visualization of the extracted ventricle volumes from the phantom images without the motion. In the bottom left the energy image of the blood pool, in the bottom middle the extracted volumes of the ventricles and in the bottom right the 3D visualization of the extracted volumes from phantom images with heart beating and respiratory motion. The volumes of extracted ventricles with the static phantom and the motion phantom were different (The Jaccard coefficient between these volumes was 0.7178). The surfaces have been smoothed based on [18]

with myocardial flow 500 ml/min*100g and it is compared to the sum of the time frame images. This illustrates the problem of the initial low contrast of different tissues in $H_2^{15}O$ study.

4.2 The Motion Test

The result of blood pool and myocardium separations from the MCAT based PET phantom images are shown in Fig. 3. With the static phantom and the phantom with the cardiac and respiratory motion the separation of the blood pool was visually good and it could be used for volume extraction. When looking the separation of the myocardium, we could see that the separation was more difficult and the result is not visually so good. The ICA method had problems of separating the myocardium from surrounding tissues. Especially, the separation of the myocardium from the liver with the simulated study including heart and respiratory motions was problematic. One problem with motion for separation was also that ICA caught just some phase of the heart cycle and we could not define, which phase the detected heart cycle phase was.

The separation result of myocardium was not good enough for the volume extraction. Therefore, we concentrated to extract individually the right and left ventricles from the resulting blood pool component images (cf. Fig. 3). Fig. 4 shows the result of the volume extraction with the DM-DSM method. We were able to extract the right and the left ventricles visually with good quality from the ICA component images in both cases. The extracted volumes were compared to the reference phantom corresponding to the end-diastole phase. Table 1 shows the accuracy of the methodology. With the static phantom the extracted volume corresponded more to the reference volume than with the motion phantom, because we do not know which phase of cardiac cycle was detected from the motion phantom.

5 Discussion

We have examined the applicability of the ICA method for the separation of the functional components from the dynamic cardiac perfusion images. Using ICA, it was possible to separate the ventricles (blood pool) with varying myocardial blood flow and even with large motion of the heart and lungs during the dynamic study. The separation of myocardium was more difficult task. However, it was possible to separate the myocardium with very high myocardial blood flow values, but the effect of the cardiac and respiratory motion was more problematic for the ICA method. In addition, we demonstrated the possibility to extract the volumes of the right and left ventricles from resulting ICA component images using the DM-DSM method. The extracted volumes could be used for alignment of two image sets that allows for the quantitative comparison of two studies. The DM-DSM algorithm effectively avoids local minima, which reduces its sensitivity to its initialization. Nevertheless, in this study, we needed manual interaction to generate the initialization.

In this study we created realistic phantom data for testing automatic image analysis methods in the case of dynamic $H_2^{15}O$ cardiac perfusion images. The 3D dynamic phantom contained both dynamic information of the tracer and the motion of heart and respiratory motion. It was generated using the Monte Carlo-based simulation tool and the MCAT phantom. The motion which was included to the phantom described the extreme positions and shapes of the heart and respiratory motion during the cardiac cycle, which is not realistic in true PET imaging, where the motion in one time frame is the average motion over the frame time. The reference phantom was taken from the end-diastole phase, where the heart muscle is thinnest and the ventricles largest. This could explain the problem of separating the myocardium from surrounding tissues. The extracted volumes of the ventricles were different in static and moving case. This result implicated that it is important to construct the phantoms carefully. To achieve comparable results with patient studies also the motion needs to be taken into account. In this study, the patient movement during the acquisition was not simulated. Due to the relative short scanning time (6 minutes) in dynamic $H_2^{15}O$ cardiac perfusion study we could assumed that patient do not move.

Conventionally random matrix has been used for the initialization of the FastICA algorithm. In this study we used fixed initialization for the ICA separation. This made it possible to achieve more reproducible results in automatic way, although the used matrix may not be the optimal solution for the initialization. We have also shown that perhaps the separation should be performed on sinograms [6], because the selection of the image reconstruction method affects the result.

Our long term goal of the research is to find a procedure to correct the motion artifacts between two studies of one patient, so that both visual and quantitative analysis of the images can be performed, at least the comparison of equivalent myocardial segments. Our idea is to first enhance the contrast in $H_2^{15}O$ studies with the ICA method so that it is possible automatically extract from component images the ventricles or the myocardium for movement correction purposes. To reach this goal we evaluated the ICA method for separation of functional components from cardiac perfusion PET phantom images in this study. We studied the effect of varying myocardial flow and motion to the ICA separation results. In addition, we showed that it is possible to extract the volume of the ventricles with the DM-DSM method for movement correction purposes. In the next step we will use patient data and apply the extracted volumes for the alignment of two image sets between two or more studies of one patient.

References

1. Bergmann, SR, Herrero P, Markham J, Weinheimer CJ, Walsh MN.: *Noninvasive quantitation of myocardial blood flow in human subjects with oxygen-15-labeled water and positron emission tomography*, J Am Coll Cardiol; **14**: 639–652, (1989).
2. Ter–Pogossian, M.M., Bergmann S.R., Sobel B.E.: *Influence of Cardiac and Respiratory Motion on Tomographic Reconstructions of the Heart: Implications for quantitative Nuclear Cardiology*. J Comput Assist Tomogr, **6**(6):1148–1155, (1982).
3. Hyvärinen A., Karhunen J., Oja E.: *Independent Component Analysis*, USA, John Wiley Sons, Inc., (2001).
4. Lee, J.S., Lee D.S., Ahn, J.Y., Cheon, G.J., Kim, S-K., Yeo, J.S., Seo, K., Park, K.S., Chung, J-K., Lee, M.C.: *Blind separation of cardiac components and extraction of input function from $H_2^{15}O$ dynamic myocardial PET using Independent Component Analysis*. J Nucl Med, **42**(6): 938–943, (2001).

5. Kalliokoski, K.K., Nuutila, P., Laine, H., Luotolahti, M., Janatuinen, T., Raitakari, O.T., Takala, T.O., Knuuti, J.: *Myocardial perfusion and perfusion reserve in endurance-trained men*, Med Sci Sports Exerc 34:948-953, (2002)
6. Magadán-Méndez, M., Kivimäki, A., Ruotsalainen, U.: *ICA separations of Functional Components from Dynamic Cardiac PET data*. IEEE Nuclear Science Symposium Conference Record, **4**: 2618–2622 (2003)
7. Pretarius, P.H., Xia, W.,King, M.A., Tsui, B.M.W, Lacroix, K.J.: *A mathematical model of motion of the heart for use in generating source and attenuation maps for simulating emission imaging*. Med Phys., **26**(11): 2323–2332, (1999)
8. Segars, W.P., Lalush, D.S., Tsui B.W.M.: *Modelling respiratory mechanics in the MCAT and spline based MCAT phantoms*. IEEE Trans. Nucl. Sci., **48**(1): 89–97, (2001)
9. Reilhac, A., Lartizien, C., Costes, N., Comtat, C., Evans, A.C.: *PET-SORTEO: A Monte Carlo-based simulator with high count rate capabilities*. IEEE Trans. Nucl. Sci., **51**(1): 46–52, (2004)
10. Ruotsalainen, U., Raitakari, M., Nuutila, P., Oikonen, V.,Sipilä, H., Teräs, M., Knuuti, J., Bloomfield, P., Iida, H.: *Quantitative Blood Flow Measurement of Skeletal Muscle Using Oxygen-15 water and PET*. J Nucl Med **38**:314–319, (1997)
11. Furuie, S.S., Herman, G.T., Narayan, T.K., Kinahan, P.E., Karp, J.S., Lewitt, R.M., Matej, S.: *A methodology for testing of statistically significant differences between fully 3D PET reconstruction algorithms*. Phys. Med. Biol. **39**:341–354, (1994)
12. Tohka, J., Mykkänen, J.: *Deformable Mesh for Automated Surface Extraction from Noisy images*. Int. J. Image and Graphics, **4**(3):405–432, (2004)
13. Mykkänen, J., Tohka, J., Luoma, J., Ruotsalainen, U.: *Automatic Extraction of Brain Surface and Mid-sagittal plane from PET images applying deformable Models*. Technical Report of Department of Computer Sciences, University of Tampere, Finland, http://www.cs.uta.fi/reports/r2003.html, (2003)
14. Hyvärinen A., Oja E.,: *A fast fixed-point algorithm for Independent Component Analysis*, Neural Computation, **9**(7): 1483–1492, (1997).
15. Jackson, D.A., Sommers, K.M., Harvey, H.H.: *Similarity coefficients: Measures of co-occurence and association or simply measures of occurrence?* The American Naturalist, **133**(3):436–453, (1989)
16. Albán, E., Katkovnik, V., Egiazarian, K.: *Adaptive window size gradient estimation for image edge detection*. Proceedings of SPIE Electronic Imaging, Image Processing: Algorithms and Systems II, Santa Clara, California, USA, 54–65, (2003)
17. Albán, E., Tohka, J., Ruotsalainen U.: *Adaptive Edge Detection Based on 3D Kernel Functions for Biomedical Image Analysis*. To appear in Proceedings of SPIE Electronic Imaging, Image Processing: Algorithms and Systems IV, San Jose, California, USA, January, (2005)
18. Tohka, J.: *Surface Smoothing Based on a Sphere Shape Model*. In proc. of 6th Nordic Signal Processing Symposium, NORSIG2004, 17–20, (2004)

Detecting and Comparing the Onset of Myocardial Activation and Regional Motion Changes in Tagged MR for XMR-Guided RF Ablation

Gerardo I. Sanchez-Ortiz[1], Maxime Sermesant[2],
Kawal S. Rhode[2], Raghavendra Chandrashekara[1],
Reza Razavi[2], Derek L.G. Hill[2], and Daniel Rueckert[1]

[1] Department of Computing, Imperial College London, U.K
[2] Guy's Hospital, King's College London, U.K

Abstract. Radio-frequency (RF) ablation uses electrode-catheters to destroy abnormally conducting myocardial areas that lead to potentially lethal tachyarrhythmias. The procedure is normally guided with x-rays (2D), leading to errors in location and excessive radiation exposure. One of our goals is to provide pre- and intra-operative 3D MR guidance in XMR systems (combined X-ray and MRI room) by locating myocardial regions with abnormal electrical conduction patterns. We address the inverse electro-mechanical relation by using motion in order to infer electrical propagation. For this purpose we define a probabilistic measure of the onset of regional myocardial activation derived from motion fields. The 3D motion fields are obtained using non-rigid registration of tagged MR sequences to track the heart. The myocardium is subdivided in segments and the derived activation isochrones maps compared. We also compare regional motion between two different image acquisitions, thus assisting in diagnosing arrhythmia, in follow up of treatment, and particularly in determining whether the electro-physiological intervention succeeded. We validate our methods using an electro-mechanical model of the heart, synthetic data from a cardiac motion simulator for tagged MRI, a cardiac MRI atlas of motion and geometry, MRI data from 6 healthy volunteers (one of them subjected to stress), and an MRI study on one patient with tachyarrhythmia, before and after RF ablation. Results seem to corroborate that the ablation had the desired effect of regularising cardiac contraction.

1 Introduction

Advances in non-rigid motion tracking techniques that use tagged MR (SPAMM) now enable us to measure more subtle changes in cardiac motion patterns. One example of disease with associated changes in motion patterns is tachyarrhythmia: a pathological fast heart rhythm originating either in the atria (superventricular) or ventricles (ventricular), often the result of abnormal paths of conduction. Radio-frequency (RF) ablation is the indicated treatment for patients with life threatening arrhythmia as well as for those on whom drug treatment is ineffective. Applying a RF current via an ablation electrode induces hyperthermia and destruction of the abnormally conducting areas. These procedures are typically carried out under x-ray (2D) guidance, leading to errors in the location of the abnormal areas as well as to excessive x-ray exposure for the patient.

One of our goals is to provide pre- and intra-operative 3D MR guidance [1] [2] in XMR systems (combined X-ray and MRI room) by detecting the onset of regional motion and relating it to the electrical activation pattern. For this purpose in this work we define a probabilistic measure of regional motion activation derived from a 3D motion field extracted by using non-rigid 3D registration of tagged MR image sequences. Since we address the inverse electro-mechanical problem, trying to infer time of electrical activation by extracting information from the cardiac motion, we use an electro-mechanical model of the heart to validate these results. Isochrones computed from MR motion are compared between different image acquisitions, and also to those isochrones obtained with the electro-mechanical model. A cardiac MR atlas of motion and geometry is also used to validate results in a relatively noise free case.

The other goal of this work is to detect changes in the regional motion patterns between two different image acquisitions. The purpose of this being the follow up of medical treatment in general, and in particular of patients that have undergone RF ablation. For these patients the method can aid in the identification and localisation of abnormal or changing motion patterns, and also can help determine whether the ablation had the desired effect of regularising cardiac contraction. In order to validate this methodology we use MR images of 6 healthy volunteers (one subjected to stress), synthetic data generated with a cardiac motion simulator of MR images, and pre- and post-intervention MR images on a patient with tachyarrhythmia.

2 Methods

2.1 Registration for Motion Tracking

We use a non-rigid registration algorithm [3] to track the motion and deformation of the heart in a sequence of 3D short- and long-axis tagged MR images. The goal of the non-rigid registration is to align each time frame of the tagged MR image sequence with the end-systolic (ES) time frame of the image sequence by maximising the normalised mutual information of both time frames. To model cardiac motion we use a free-form deformation based on cubic B-splines. The output of the registration is a continuous time varying 3D motion or vector field (see Figure 1a), $\mathbf{F}(\mathbf{p},t)$ where $\mathbf{F} : \Re^4 \to \Re^3$ and $\mathbf{p} \in \Re^3$ is the space coordinate (or voxel (x,y,z) in the discrete implementation).

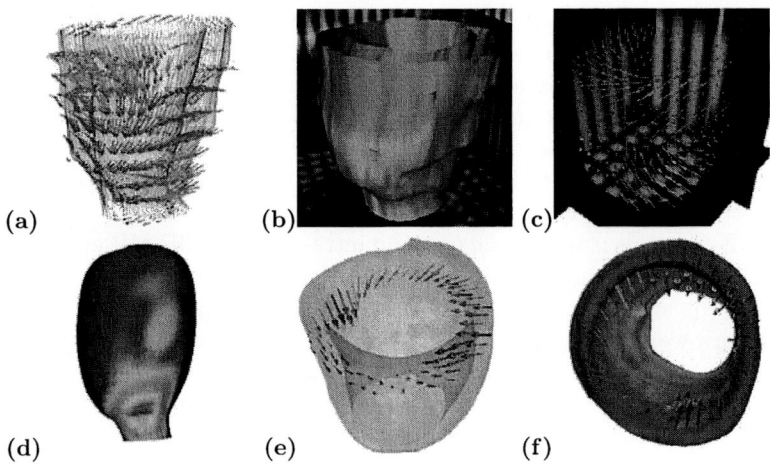

Fig. 1. The reconstructed motion field is shown in (**a**) with displacement vectors and the myocardial surface. The end-diastole myocardial surface ($t = 0$) of a volunteer is shown in (**b**) with the subdivision in 12 segments. In (**c**) the synthetic tagged MR data is displayed with the recovered displacement field while the reconstructed surface in (**d**) is coloured with the magnitude of the difference between the normal and modified parameters. The region where the abnormal motion was produced was accurately identified and can be seen in red and yellow. Two views of the smooth cardiac atlas geometry with a slice of the motion field vectors are shown in (**e**) and (**f**). All colour scales go from blue to red

2.2 Coordinate System and Myocardial Segmentation

A manual segmentation of the myocardium at end-diastole (ED) (see Figure 1b) is used to determine the region of interest (**myo**) for the registration at time $t = 0$. Using **F**, the myocardial region can then be automatically propagated over the entire cardiac cycle (as in Figure 1a).

In order to be able to compare different image acquisitions, a common (cylindrical) coordinate system based on the left ventricle is defined for each subject. In this manner we avoid potential misregistration errors due to subject motion between scans. Using cylindrical coordinates based on the LV allows us to express the non-rigid motion measurements derived from **F** in terms of radial, circumferential and longitudinal directions.

Using this coordinate system, the myocardium **myo** is then subdivided into small meaningful regions or segments s, and the motion derived measurements for each of these myocardial segment is computed during the cardiac cycle. For the purpose of comparing motion between different scans we use $S = 12$ segments, with 4 sections around the z-axis that roughly correspond to septum, lateral, anterior and posterior walls, and 3 sections along the z-axis, corresponding to base, middle region and apex (see Figure 1b).

2.3 Differential Motion Descriptors and Changes in Motion Patterns

Some differential features derived from the motion field $\mathbf{F}(\mathbf{p},t)$ can provide an insight of how a specific region of the myocardium is contracting. We write them as the set of functions

$$F^m = F^m(\mathbf{p},t) \text{ where } m \in \mu = \{D, R, C, Z, \dot{R}, \dot{C}, \dot{Z}, E, r, c, z, \dot{r}, \dot{c}, \dot{z}\} \quad (1)$$

and $F^m : \Re^4 \to \Re$ are defined as the total deformation or displacement $F^D = ||\mathbf{F}||$, the radial, circumferential and longitudinal components of the deformation (F^R, F^C and F^Z) with respect to the a cylindrical coordinate system and their corresponding time derivatives or velocities ($F^{\dot{R}}$, $F^{\dot{C}}$ and $F^{\dot{Z}}$), the magnitude of the strain matrix $F^E = ||E_{i,j}||$, the radial, circumferential and longitudinal components of the strain (F^r, F^c and F^z), and their time derivatives ($F^{\dot{r}}$, $F^{\dot{c}}$ and $F^{\dot{z}}$), all with respect to the the same cylindrical coordinate system. Although F^D and F^E are not linearly independent of their components in the cylindrical coordinate system, in this work we explore the efficiency of them all as motion descriptors and those that turn out to be of less importance are minimized by the use of the confidence weights w_m defined in Section 2.4.

We use a Lagrangian framework where the transform $\mathbf{F}(\mathbf{p},t)$ follows, at time t, the position of the 3D voxels $\mathbf{p} \in \mathbf{myo}$ that correspond to the myocardium at time $t = 0$.

The values of $F^m(\mathbf{p},t)$ are computed for each voxel and the values averaged for each of the myocardial segments s, for all time frames during the cardiac cycle leading to the function

$$F^m(s,t) = \frac{1}{\int_{\mathbf{p} \in s} d\mathbf{p}} \int_{\mathbf{p} \in s} F^m(\mathbf{p},t) d\mathbf{p} \quad \text{for all regions } s \in \mathbf{myo}. \quad (2)$$

In order to evaluate changes in the motion patterns between two data sets $\mathbf{F}_1$ and $\mathbf{F}_2$, for instance those corresponding to pre- and post-ablation scans, the difference between the two functions F_1^m and F_2^m is computed for each segment, integrated over time and normalised using the maximum value of the function for the specific segment. This normalization of the values compensates for the differences in the dynamic behaviour expected in the various regions of the heart (like apex and base for instance). A statistical measure is derived from the above combined quantities [4, 5] and each segment is assigned a measure of motion change and classified as having either no, small or significant changes.

2.4 Activation Detection

Although the study of myocardial electrical phenomena such as excitation-contraction relation, re-entries and patterns occurring inside the myocardium remain open problems for study (see references in [6, 7]), in this work we use the underling assumption that we can relate the onset of regional motion, derived from the images sequences, to the electrical activation. That is, by using the inverse relation of electro-mechanical coupling. Ideally the onset of regional contraction could be inferred from the motion field with a simple measure such as strain. However, because of the limitations imposed by noise, errors and the relatively low space and time resolution of the image acquisition and the extracted motion field, a more robust measure has to be used. For this purpose we investigate the subset of differential descriptors $\mathbf{F}^m$ where $m \in M = \{R, C, Z, \dot{R}, \dot{C}, \dot{Z}, E, \dot{r}, \dot{c}, \dot{z}\}$.

The first step to characterise the regional motion of the heart during the cardiac cycle is to measuring a regional $(T_{ES}(s))$ and global (T_{ES}) end-systolic times, as well as the critical times for each motion descriptor. We therefore define

$$T_{max}^m(s) = t^* \ such\,that \ \ F^m(s,t^*) \geq F^m(s,t)$$
$$\forall t \in [0, T_{ES}(s)]$$

and

$$T_{min}^m(s) = t^* \ such\,that \ \ F^m(s,t^*) \leq F^m(s,t)$$
$$\forall t \in [T_{max}^m(s), T_{ES}(s)].$$

Notice that for T_{min}^m the search interval begins at T_{max}^m, i.e. when the maximum value has been reached (it is the late minimum value of F^m that will help us define the end-systolic time, not those small values at the beginning of the cycle). Because the computation of these values requires a first estimate of the end-systolic time, we use as initialisation the time frame where the heart visually appears to be at end-systole. However, a short iterative process rapidly provides a better estimate for $T_{ES}(s)$.

In the case of displacement and strain, the end-systolic time is linked to their maximum values, while in the case of velocity and rate of change of strain it corresponds to their minimum values (when the heart has paused its contraction). Therefore,

$$T_{ES}^m(s) = \begin{cases} T_{max}^m(s) \ \text{for} \ m \in \{R,C,Z,E\} \\ T_{min}^m(s) \ \text{for} \ m \in \{\dot{R},\dot{C},\dot{Z},\dot{r},\dot{c},\dot{z}\} \end{cases} \quad (3)$$

and combining these times we obtain an estimate that corresponds to the regional time of end-systole:

$$T_{ES}(s) = \sum_{m \in M} w_m T_{ES}^m(s).$$

The weights w_m are normalised (i.e. $\sum_{m \in M} w_m = 1$) and reflect the confidence we have on each of the differential motion descriptors m. Although at present we have assigned their values manually, a statistical measure derived from the data is being developed in order to compute them automatically. In order to obtain a global estimate for end-systolic time for each feature we integrate those values over the entire myocardium: $T_{ES} = \int_{s \in \mathbf{myo}} T_{ES}(s) ds$.

Using the above equations we can now define a probabilistic measure of the activation for every voxel in the myocardium, at anytime time during the cardiac cycle:

$$A(s,t) = \sum_{m \in M} w_m \int_0^t \frac{F^m(s,\tau)}{\int_0^{T^m_{max}(s)} F^m(s,\tau')d\tau'} d\tau \qquad (4)$$

where we impose $F^m(s,t) = 0$ if $t > T^m_{max}(s)$ in order to keep the values normalised (notice that some motion descriptors like the velocities and the time-derivatives of strain reach their maximum values before end-systole).

The value of $A(s,t)$ monotonically increases from zero to one as we expect every voxel to have been activated by the time the motion descriptors reach the maximum value at time $T^m_{max}(s)$. In order to avoid singularities in the equation we excluded from the computation, and labelled as not active, those voxels that might remain relatively static (*i.e.* those for which $F^m(s, T^m_{max}(s)) \approx 0$).

By integrating over time we obtain an accumulated probability and we can therefore set a (percentage) threshold P, between 0 and 1, to define the time t_a at which the activation of a voxel s takes place. That is, if $A(s, t_a) = P$ then s becomes active for $t = t_a$. The activation *isochrones* are then defined, for a given threshold P, as the function $A(s) = t_a$, for all $s \in$ **myo**.

2.5 Cardiac Motion Simulator for Tagged MRI

For the purpose of validating the proposed methodology with a controlled case we also implemented and modified a cardiac motion simulator for tagged MRI [7]. The motion simulator is based on a 13-parameter model of left-ventricular motion developed by Arts *et al.* [8] and is applied to a volume representing the LV that is modeled as a region between two confocal prolate spheres while the imaging process is simulated by a tagged spin-echo imaging equation [9].

A pair of sequences of synthetic tagged LV images was produced in the following manner: first, a 'post-intervention' (normal) sequence was computed using the standard model parameters, and secondly, a 'pre-intervention' (abnormal) sequence for which the motion parameters were modified in a small region of the myocardium. The modification to the parameters consisted mainly in moving the phase of the contraction forward in time and changing the magnitude of the motion. Two such pairs of image sequences were produced, with different abnormal parameters and in different regions of the myocardium. Examples of these synthetic images can be seen in Figure 1.

3 Results and Discussion

3.1 Changes in Regional Motion Patterns

The detection of changes in motion patterns was evaluated on synthetic data as well as real MR data from six subjects. In order to test the algorithm when the ground truth is available, results on the 'pre-' and 'post-intervention' sequences of **synthetic** tagged LV images were compared in two cases, with different parameters and regions of abnormal motion (see one case in Figure 1c). In both cases these regions were accurately located. One segment showed significant changes while the rest were correctly classified as having no change (see Figures 1d and 3).

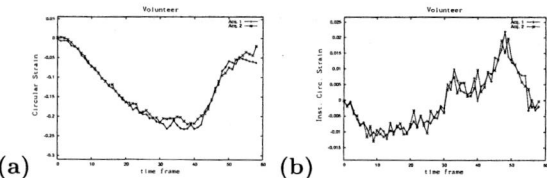

Fig. 2. Reproducibility: Time plots of a typical myocardial segment of a healthy volunteer. The reproducibility of the motion fields is demonstrated with the similar curves obtained for two independent acquisitions of the same subject. The plots show the accumulated (a) and instantaneous (b) circumferential strain, for each of the two image acquisitions

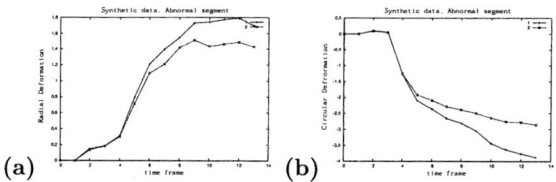

Fig. 3. Synthetic data: Time plots of two segments of the cardiac motion simulator for tagged MRI. Each plot shows results for the normal and modified motion parameters of a segment in the region of abnormal motion, where significant change was correctly detected. The plots show radial (a) and circumferential (b) deformation from end-diastole to end-systole

We also acquired data from four volunteers. For each of them two separate sets of image sequences were acquired with only few minutes between the acquisitions. Since no change is expected in these pairs of image acquisitions, this allowed us to verify the **reproducibility** of the motion fields computed by the algorithm and to test the comparison method against false positive detection. The motion patterns encountered were all very similar and no region was classified as having a significant change (see Figure 2).

With another volunteer we acquired three sets of image sequences. The first two as described above, with only few minutes between the acquisitions. The third data set was acquired few minutes after the second, but while subjecting the volunteer to **stress**. The stress was induced by placing one foot of the subject into a bucket of cold water with ice. This experiment allowed us to compare normal motion patterns with those obtained under stress, and again, to validate the method regarding reproducibility and false positives. No segment showed a significant difference between the first two acquisitions, but when comparing normal motion to that under stress we found that three segments showed no change, four presented small but noticeable changes, and the remaining five showed a significant amount of change (see Figure 4).

Finally, MRI data was acquired from an eight year old patient with acute super-ventricular tachyarrhythmia, before and after **RF ablation**. The image

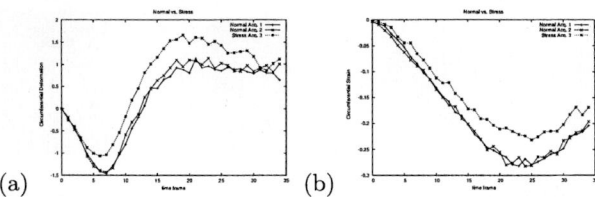

Fig. 4. Stress study: Time plots of a myocardial segment of a healthy volunteer, with and without stress. There are no significant changes in the motion pattern between the first two image acquisitions. In the third image acquisition, during which stress was induced on the subject, a noticeable alteration was detected. The plots show circumferential deformation **(a)** and strain **(b)** for each of the three image acquisitions

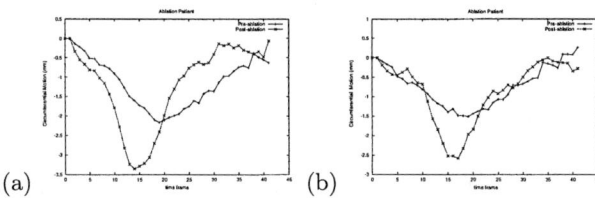

Fig. 5. RF ablation patient: Time plots of circumferential motion of two myocardial segments of a cardiac patient, before and after RF ablation. A significant change can be seen in the post-intervention sequence, when this region of the myocardium exhibits a faster and more pronounced motion, indicating a regularisation of the contraction

acquisition and catheter intervention were carried out with an XMR system [1]. Our results confirmed that the motion pattern changed in most parts of the myocardium (visual inspection of the reconstructed 3D surfaces and displacement vectors also showed pronounced changes in the overall contraction pattern), while the largest changes were found in five segments. Examples of the compared motion also show the corrective effect of the intervention (see Figure 5).

3.2 Activation Detection

Figure 6 shows the results of activation detection (Equation 4) obtained on the MR repetition and stress study described in Section 3.1. The times of activation of different regions of the myocardium are shown as different colours over the end-diastolic myocardial surface (**activation isochrones** maps). The first three images in the figure compare the isochrones obtained from the three MR data acquisitions of the same subject: two repetition scans with no changes in between them, and a third scan acquired while the volunteer was subjected to stress. Results of subtracting pairs of isochrones maps are also shown: the difference between the two normal repetition acquisitions, in Figure 6e, and the difference between a normal and the stress acquisition, in Figure 6f. We can see that the

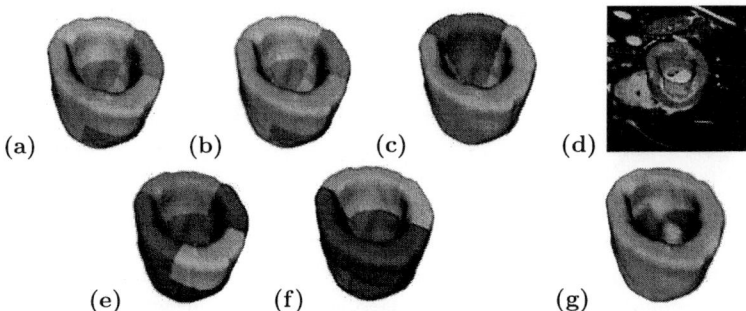

Fig. 6. Isochrones of stress data. The motion-derived activation isochrones computed from the two normal MR acquisitions, **(a)** and **(b)**, and a third one acquired while the volunteer was subjected to stress **(c)**. Two isochrones subtraction maps are also shown below their corresponding images: the difference between the two normal repetition acquisitions in **(e)**, and the difference between a normal and the stress acquisition in **(f)**. The orientation of the myocardium can be seen in **(c)**, where a zoomed-out view of the anatomical MR image is shown with the myocardial surface skeleton. Isochrones computed from the electro-mechanical model are shown in **(g)**. The colour scale for the isochrones maps go from blue to red (0-500ms, with green approx. 200ms), and for the isochrones subtraction maps from blue to red (0-100ms approx.)

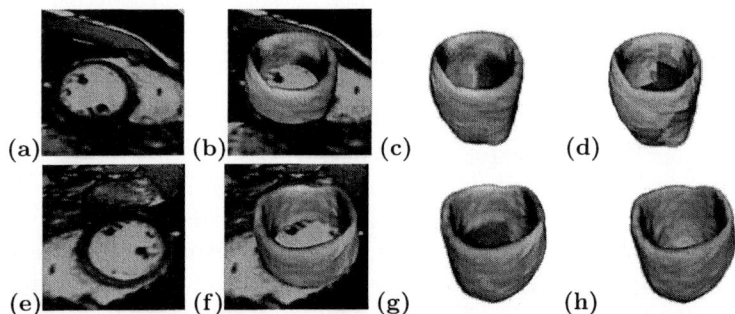

Fig. 7. Isochrones of cardiac atlas. Two views (top and bottom row) of the isochrones were computed for the atlas using both, the electro-mechanical model **(b)**, **(c)**, **(f)** and **(g)**, and the proposed activation measure derived from the motion field **(d)** and **(h)**. The colour scale goes from blue to red, where blue shows the earliest time and red the latest. The orientation of the left and right ventricle can be seen on the MR images of the subject used as a reference for the atlas (**(a)** and **(e)**)

difference between the isochrones of the two normal acquisitions is small, thus validating the method regarding reproducibility, while on the other hand some larger changes can be appreciated between the isochrones of the normal and the stress scans, thus highlighting the regions that were most affected by stress.

Since we are addressing the problem of inverse electro-mechanical coupling, that is, trying to infer the time of electrical activation by extracting information from the cardiac motion images, we have also used a forward 3D **electro-mechanical model** of the heart [6, 10] to validate the activation detection results. The segmentation of the myocardium of a healthy volunteer at end-diastole was used as geometric input for the model. The muscle fiber orientation and the Purkinje network location were fitted to the geometry from a-priori values of the model. Figure 6g shows the isochrones values computed using the electro-mechanical model applied to the subject of the stress study. Good correlation can be seen between these and the isochrones derived from MR motion.

We also used a **cardiac atlas** of geometry and motion generated from 3D MR images sequences of 14 volunteers to test our activation measure in a realistic but smooth and virtually noise-free data set [11] (see Figures 1e and 1f). For the purpose of comparing activation detection results to those obtained with the high-resolution electro-mechanical model, a larger number of smaller segments was used (also, segments can be very small in this case since there is little noise in the data). Figure 7 compares the isochrones for the atlas computed by both, the electro-mechanical model, and the proposed activation measure derived from the motion field. Promising agreement can be seen on these results of activation detection.

4 Conclusions and Future Work

Despite current limitations such as distinguishing between epi- and endo-cardial activation patterns, the methodology seems promising for the assessment of intervention results and could also be used for the detection of arrhythmia, ischaemia, regional disfunction, as well as for follow up studies in general.

Because acquisition of tagged images can be carried out in less than 20 minutes, either immediately before the RF ablation or the day before the intervention [1], the proposed analysis is suitable for clinical practice in guiding and monitoring the effects of the ablation procedure on ventricular arrhythmias [12], with little extra discomfort added to the patient.

In order to account for possible changes in the heart rate between the pre- and post-intervention acquisitions, we intend to re-scale one of the image sequences in the time domain, by using the 4D registration technique described in [11]. Results will be compared to those obtained without rescaling (for instance, in the case of the stress study, where there was a small change in the heart rate).

---

[1] When images are acquired on different days further image alignment has to be carried out in order to register the different acquisitions. We are currently investigating results of our methods in these circumstances.

References

1. K.S. Rhode, D.L.G. Hill, P.J. Edwards, J. Hipwell, D. Rueckert, G.I. Sanchez-Ortiz, S. Hegde, V. Rahunathan, and R. Razavi. Registration and tracking to integrate X-ray and MR images in an XMR facility. *IEEE Transactions on Medical Imaging*, 22(11):1369– 1378, 2003.
2. G.I. Sanchez-Ortiz, M. Sermesant, R. Chandrashekara, K.S. Rhode, R. Razavi, D.L.G. Hill, and D. Rueckert. Detecting the onset of myocardial contraction for establishing inverse electro-mechanical coupling in XMR guided RF ablation. In *IEEE Int. Symposium on Biomedical Imaging (ISBI'04)*, pages 1055–1058, USA, 2004.
3. R. Chandrashekara, R. Mohiaddin, and D. Rueckert. Analysis of 3D myocardial motion in tagged MR images using nonrigid image registration. *IEEE Transactions on Medical Imaging*, 23(10):1245–1250, 2004.
4. G. I. Sanchez-Ortiz, R. Chandrashekara, K.S. Rhode, R. Razavi, D.L.G. Hill, and D. Rueckert. Detecting regional changes in myocardial contraction patterns using MRI. In *Proc. SPIE Medical Imaging: Physiology, Function, and Structure from Medical Images*, pages 710–721, San Diego, USA, 2004.
5. G.I. Sanchez-Ortiz, M. Sermesant, R. Chandrashekara, K.S. Rhode, R.Razavi, D.L.G. Hill, and D.Rueckert. Measuring myocardial motion changes in tagged MR to establish inverse electro-mechanical coupling for XMR-guided RF ablation. In *Int. Workshop on Augmented environments for Medical Imaging and Computer-aided Surgery (AMI-ARCS'04)*, pages 32–40, Rennes, France, 2004.
6. M. Sermesant, K. Rhode, A. Anjorin, S. Hedge, G. Sanchez-Ortiz, D. Rueckert, P. Lambiase, C. Bucknall, D. Hill, and R. Razavi. Simulation of the electromechanical acitvity of the heart using XMR interventional imaging. In *Medical Image Computing and Computer-Assisted Intervention (MICCAI'04)*, LNCS, vol. I, pages 786–794, France, 2004.
7. E. Waks, J. L. Prince, and A. S. Douglas. Cardiac motion simulator for tagged MRI. In *IEEE Workshop on Mathematical Methods in Biomedical Image Analysis*, pages 182–191, San Francisco, CA, 1996.
8. T. Arts, W. C. Hunter, A. Douglas, A. M. M. Muijtjens, and R. S. Reneman. Description of the deformation of the left ventricle by a kinematic model. *Biomechanics*, 25(10):1119–1127, 1992.
9. J. L. Prince and E. R. McVeigh. Motion estimation from tagged MR images. *IEEE Transactions on Medical Imaging*, 11(2):238–249, June 1992.
10. M. Sermesant, K. Rhode, G.I. Sanchez-Ortiz, O. Camara, R. Andriantsmianova, S. Hedge, D.Rueckert, P. Lambiase, C. Bucknall, E. Rosenthal, H. Delingette, D.L.G. Hill, N. Ayache, and R. Razavi. Simulation of cardiac pathologies using an electromechanical biventricular model and XMR interventional imaging. *Medical Image Analysis*, 2005. In Press.
11. D. Perperidis, M. Lorenzo-Valdes, R. Chandrashekara, R. Mohiaddin, G. I. Sanchez-Ortiz, and D. Rueckert. Building a 4D atlas of the cardiac anatomy and motion using MR imaging. In *IEEE Int. Symposium on Biomedical Imaging (ISBI'04)*, pages 412–415, Arlington, USA, Apr 2004.
12. G.I. Sanchez-Ortiz, M. Sermesant, K.S. Rhode, R. Chandrashekara, R.Razavi, D.L.G. Hill, and D.Rueckert. Detecting the onset of myocardial contraction and regional motion changes for XMR guided RF ablation. In *Int. Society for Magnetic Resonance in Medicine (ISMRM'05)*, Miami, USA, May 2005.

Suppression of IVUS Image Rotation.
A Kinematic Approach

Misael Rosales[1,2], Petia Radeva[2], Oriol Rodriguez[3], and Debora Gil[2]

[1] Lab. de Física Aplicada, Fac. de Cs.
Dpto. de Física (ULA), Mérida/Venezuela
misael@cvc.uab.es
[2] CVC, Edifici O, Campus UAB, 08193 Bellaterra, Spain
[3] Universitary Hospital "Germans Tries i Pujol", Badalona, Spain

Abstract. IntraVascular Ultrasound (IVUS) is an exploratory technique used in interventional procedures that shows cross section images of arteries and provides qualitative information about the causes and severity of the arterial lumen narrowing. Cross section analysis as well as visualization of plaque extension in a vessel segment during the catheter imaging pullback are the technique main advantages. However, IVUS sequence exhibits a periodic rotation artifact that makes difficult the longitudinal lesion inspection and hinders any segmentation algorithm. In this paper we propose a new kinematic method to estimate and remove the image rotation of IVUS images sequences. Results on several IVUS sequences show good results and prompt some of the clinical applications to vessel dynamics study, and relation to vessel pathology.

1 Introduction

The introduction of the IntraVascular UltraSound (IVUS) in the field of medical imaging [1, 2] as an exploratory technique has significantly changed the understanding of the arterial diseases and individual patterns of diseases in coronary arteries. Each IVUS plane visualizes the cross-section (Fig. 1 (left)) of the artery allowing extraction of qualitative information about: the causes and severity of the narrowing of the arterial lumen, distinction of thrombus of the arteriosclerotic plaque, recognition of calcium deposits in the arterial wall, determination and location of morpho-geometrics arteries parameters [3, 4, 5], among others. The main role of IVUS is to serve as a guide in the interventional procedures allowing to measure the morphological structures along the vessel. Artifacts caused by the periodic rotation of the image, introduce an error in the measurements precision in tangential direction [6, 8]. The vessel wall follows a periodic oscillatory motion in an image sequence corresponding to the heart cycles. This motion has a rotation center positioned on the vessel wall border in most cases. We can visually evidence this effect by using the mean of an IVUS sequence along its temporal direction, as shown in Fig. 1 (right). This image represent the average grey level of pixels along 25 frames corresponding to approximately one heart

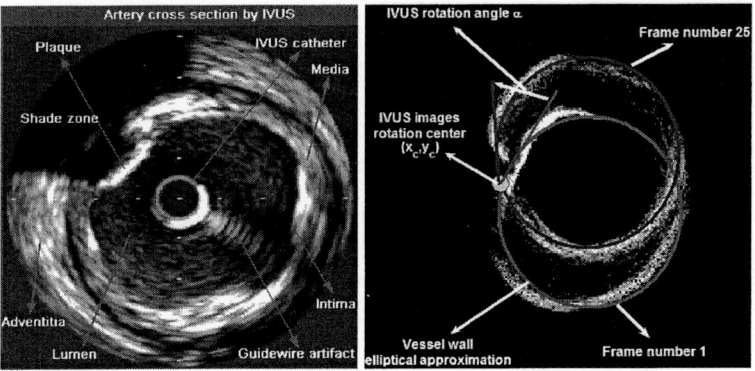

Fig. 1. Morphological arterial structures and artifacts (left). Empirical evidence of IVUS rotation effect (right)

cycle. Since brighter structures correspond to the vessel wall, in this particular case, the center of rotation is the most "brilliant point". The main goal of this work is to estimate and remove the rotation effect, in order to improve the longitudinal IVUS visualization. Instead of using an optical flow scheme, prone to be misled by blood random movement, we suggest using kinematic principles. A good estimation of the rotation images gives the possibility of understanding their mechanical and physiological genesis providing the possibility to study in a robust form the vessel dynamics, and to establish new diagnostic tools. The article is organized as follows: In section 2 we discuss some physical considerations about the rotation effect, the general aspect over the kinematic model is discussed in section 2.1, the neuronal network training procedures are explained in section 2.2 and the estimation and removing procedures of the rotation effect are discussed in sections 2.3 and 2.4, respectively. Finally, the results and conclusions are explained in sections 3 and 4.

2 Physical Considerations of IVUS Image Rotation

The IVUS rotation effect is an image sequence artifact described by various authors [6, 8, 10] as a gray levels shift in the vessel tangential direction, that avoids reliable measurements of distance in longitudinal views. There are two main reasons of this artifact: mechanical and anatomic-physiological factors. The mechanical factor corresponds to the catheter movement during pullback. This motion is locally caused by pulsatile blood flow, the vessel wall dilation and the intrinsic heart muscle dynamics and globally by the catheter trajectory geometry. Anatomic factor is due to the dynamic response of the intrinsic vessel architecture to blood pressure and its mutual interaction with the heart muscle. We model the rotation artifact as follows:

2.1 Kinematic Model for Rotation Estimation

In order to find the rotation center of the sequence, we consider that the vessel wall shape can be modelled as a discrete structure, which temporal evolution depends on the reciprocal interaction between the radial force that comes from the blood pulse and the vessel wall local shape perceived by the catheter. Current IVUS techniques assume that the vessel is circular, the catheter is located in the center of the artery, and the transducer is parallel to the long axis of the vessel. However, both transducer obliquity and vessel curvature can produce an image giving the false impression that the vessel is elliptical. Transducer obliquity is especially important in large vessels and can result in an overestimation of dimensions and reduction of image quality [6]. In general, each image sequence has its own center of rotation. This center can stay at rest or change the spatial position along the sequence. In order to find such center of rotation, we will assume that the vessel wall is a discrete linear elastic oscillating system, with a total energy coming from the pulsatile radial force of the heart blood pulse [11]. Because arterial structures have approximately an elliptical shape, we use polar coordinates to study their temporal evolution. It follows that if the trajectory is given by: $(x, y) = (r(t)cos(\theta(t)), r(t)sin(\theta(t)))$, the total energy of one element (x_i, y_i) of the vessel wall is equal to:

$$E_i = T_i + U_i \qquad (1)$$

where $T_i = \frac{m_i v_i^2}{2} + \frac{m_i}{2}(r_i \omega_i)^2 \quad U_i = \frac{k_i r_i^2}{2}$
$v_i = \sqrt{vx_i^2 + vy_i^2} \quad r_i = \sqrt{x_i^2 + y_i^2} \quad w_i = \frac{\partial \theta_i}{\partial t}$.

The quantities T_i and U_i are the kinetic and elastic energy respectively, m_i, v_i, ω_i and k_i are the mass, tangential velocity, angular velocity and elasticity constant of the i-th element of the vessel wall. In this paper the elastic constant is set to $k_i = 1$. The mass of one element can be estimated considering the minimal "voxel" volume $V_b \approx 6.4 \times 10^{-5}$ mm^3 swept by the ultrasound beam [14]. Using this fact, the element of mass is $m = \rho * V_b \approx 1.09 \frac{grs}{cm^3} * 6.4 \times 10 - 8grs \approx 6.97 \times 10^{-5} kg$, where ρ is the tissue density [12]. Within the above kinematic framework, the rotation center along the IVUS sequence is represented by the region in the vessel wall that has minimal total energy. The steps to compute and suppress the rotation are the following:

2.2 Neuronal Network Training

We find candidate structures in the media and intima, which can follow the vessel wall kinematic during approximately a complete period of one cardiac event ≈ 25 images. In order to find the potential candidate structures, a Perceptron Multilayer Neural Network (60 : 50 : 60 : 30) was trained using a standard Back Propagation Algorithm [9]. The input features were the radial grey level intensity defined as: $I(r) = I_o exp(-\alpha(N_\theta)rf)$, where I_o is the beam intensity at $r = 0$ and the absorbtion coefficient, α gives the rate of diminution of average power

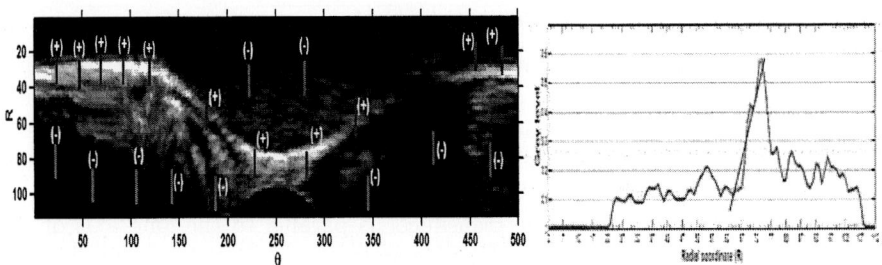

Fig. 2. Positive (+) and negative (-) pattern (left). Positive pattern features (right)

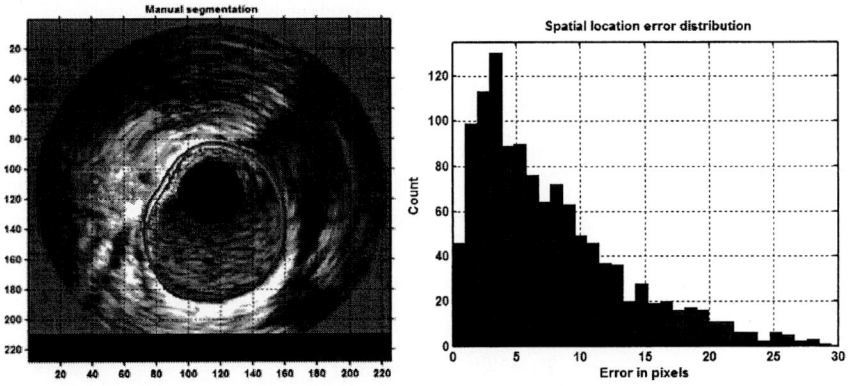

Fig. 3. Intima validated segmentation (left). Error distribution in mm (right)

with respect to the distance along a transmission path [7]. It is composed of two parts, one (absorption) proportional to the frequency f, the other (scattering) dependent on the ratio of grain, particle size or the scatterer number N_θ located along the ultrasound beam path. Absorption coefficient, image pixel grey level, standard deviation and mean of the data were used to train the Neural Network. The absorption coefficient gives local information about the lumen-vessel transition, the grey level distribution gives local and global information about the vessel structure. The global and local vessel wall structure information is given by the low and high frequencies, of the grey level distribution. Standard deviation gives global textural information and the mean of the data is the base line of the global grey level intensity. The training and test data were obtained from IVUS images in polar form. Figure 2 (left) shows an example of a positive (+) corresponding to intima and negative (-) patterns, corresponding to blood, adventitia, shadows and artifact zones. The extraction of features for positive patterns is show in Fig. 2 (right). The local absorption coefficient was obtained from the regression line slope [15] of the image profile, grey level intensity vs. radial penetration. Figure 3 (left) shows an example of the validated data and its error spatial location (center) computed by 4 IVUS sequences of different patients. The error is defined as, the Euclidean distance between the manual location of intima and the spatial location found out by the Neural Network algorithm. In order to improve the rotation center, the spatial location of the selected points should be greater as the vessel wall border. Although the spatial error between validated intima points find out by the neuronal network is as high as 5 pixels $\mp$ 12 (See Fig. 3 (right)), nevertheless with the selected intima points recovery of temporal kinematics is optimal. This fact makes the method very flexible since the main goal of this work is to find the global kinematic of the vessel wall not the vessel segmentation. In this sense we find the global motion of the structure physically connected to the vessel (intima and media) border whose thickness is approximately 10 to 15 pixels. Figure 3 (right) shows an elliptical approximation and the intima point find out by the neuronal network.

2.3 The IVUS Image Rotation Estimation

a. **Center estimation.** We determine the spatial location of the rotation center (x_c, y_c) in frame k, as the position (ij) on the vessel wall that has a minimal total energy given by (Eq. 2.1). The spatial location of the rotation center is put into the image point that reaches the condition:

$$(x_c, y_c) = argmin_{ij} \sum_{k=1}^{f_n} E_{ij}^k$$

where $f_n = 25$ is the image number used to evaluate this condition and (ij) are the row and columns of the average IVUS images. Figure 4 shows the kinematics parameters used by the estimation of the total energy such as required in Eq. 2.1. The temporal evolution of the vessel wall candidates is obtained using their kinematics variables: radial coordinates (a) and angular position (b). The total energy is computed considering that all points having the same mass (See section 2.1). Figure 4 (c) shows the minimal energy distribution for a particular frame and 25 candidates. The spatial evolution of the center of rotation is shown in Fig. 4 (d).

b. **Angle estimation.** Once the rotation center of the IVUS sequence has been determined, the procedure to calculate the rotation angle for each frame is as follows: 1.- An elliptical approximation following the ellipse fitting procedures of [16] is adjusted to the spatial distribution of the points structures (See Fig. 5). The fitting of ellipses is made over the mean of the IVUS sequence in the longitudinal direction. If the ellipse center is noted by (x_e^k, y_e^k), then the rotation angle α_k (See Fig. 5) for a frame k is given by:

$$\alpha_k = \arctan\left(\frac{y_e^k - y_c^k}{x_e^k - x_c^k}\right)$$

where (x_e^k, y_e^k) and (x_c^k, y_c^k) are the rotation center and ellipse center of image k respectively.

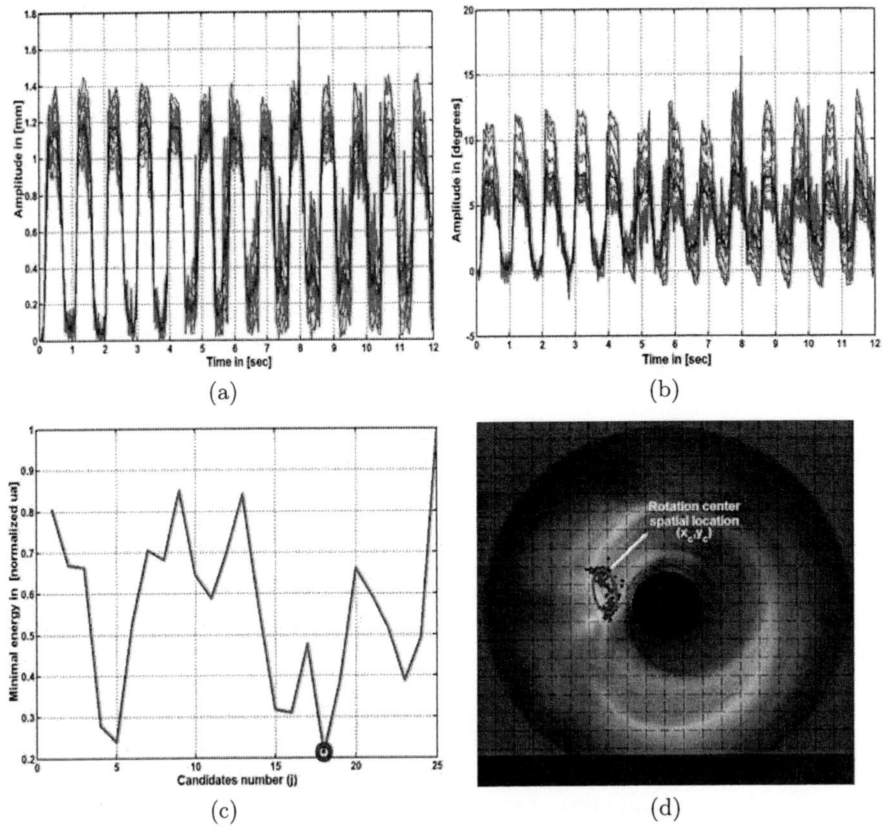

Fig. 4. Kinematics variables: r (a) and θ (b) coordinates evolution. Minimal energy (d) location by 25 candidates points. Rotation center location (d)

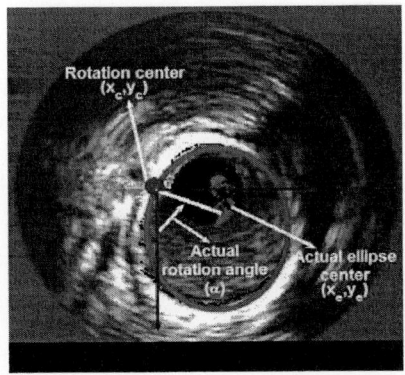

Fig. 5. Parameters to estimate the IVUS image rotation α_k

2.4 Removing the IVUS Rotation Effect

The suppression of the rotation is given by the following linear transformation: The original image $I_o(x,y)$ is translated to the rotation center coordinates (x_c^k, y_c^k) and rotated through angle α_k. The new image $I'_o(x',y')$ is finally located in a new arbitrary center (x_a^k, y_a^k) as follows:

$$\begin{pmatrix} x' \\ y' \end{pmatrix} = \begin{pmatrix} sin(\alpha_k) & cos(\alpha_k) \\ -cos(\alpha_k) & sin(\alpha_k) \end{pmatrix} \begin{pmatrix} x - x_c^k \\ y - y_c^k \end{pmatrix} + \begin{pmatrix} x_a^k \\ y_a^k \end{pmatrix} \quad (2)$$

where (x',y') and (x,y) are the new and old cartesian image coordinates, (x_c^k, y_c^k) is the actual rotation center, α_k is the rotation angle and (x_a^k, y_a^k) is the new image center of image respectively.

3 Results

Our experiments focus on assessment of the rotation suppression and illustration of a possible application to pathology diagnosis.

3.1 Validation of the Rotation Suppression

Validation of rotation removing is done by analyzing the temporal evolution of the rotation angles. We consider that the rotation has been removed if the rotation angle profile after correction is constant to zero. There are several ways of checking the former hypothesis. First, we can compare the temporal evolution of the angle before and after rotation suppression. Lack of rotation is also reflected in average images and longitudinal cuts. After suppression there is no grey level shift so that bright structures stay still and the longitudinal cuts shape is a straight line in contrast to the wavy shape of original cuts. The former analysis is illustrated in Fig. 7. Rotation profiles before and after removing the rotation artifact are shown in Fig. 6 (a) and (b), respectively. In figure 7 (a) and (b) we depict the average sequences image with their corresponding centers of the adjusted ellipses. Note that their spatial variation has significantly reduced after image correction. Finally, longitudinal cuts before and after rotation elimination are exhibited in figure 7 (c) and (d).

3.2 Healthy and Pathological Rotations Profiles

We studied the local rotation profile in order to illustrate differences between healthy and pathological vessel segments. The comparative analysis is based on the period and the amplitude of the profile. For our basic comparative analysis we considered a healthy segment and 28 pathological ones including soft plaque, hard plaque and atheroma. Graphics along 250 frames are shown in Fig.8. The healthy patient presents a regular periodic behavior with an oscillation amplitude of approximately 40 degrees and a period that coincides with the heart beat rate.

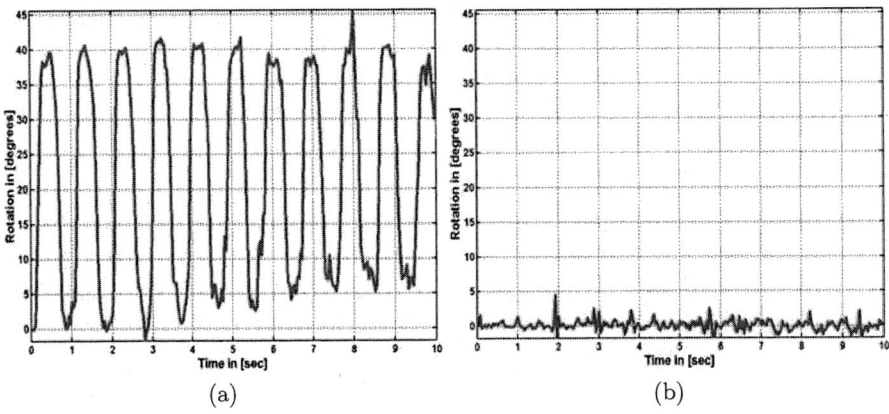

Fig. 6. Angle rotation profiles, before (a) and after (b) rotation suppression

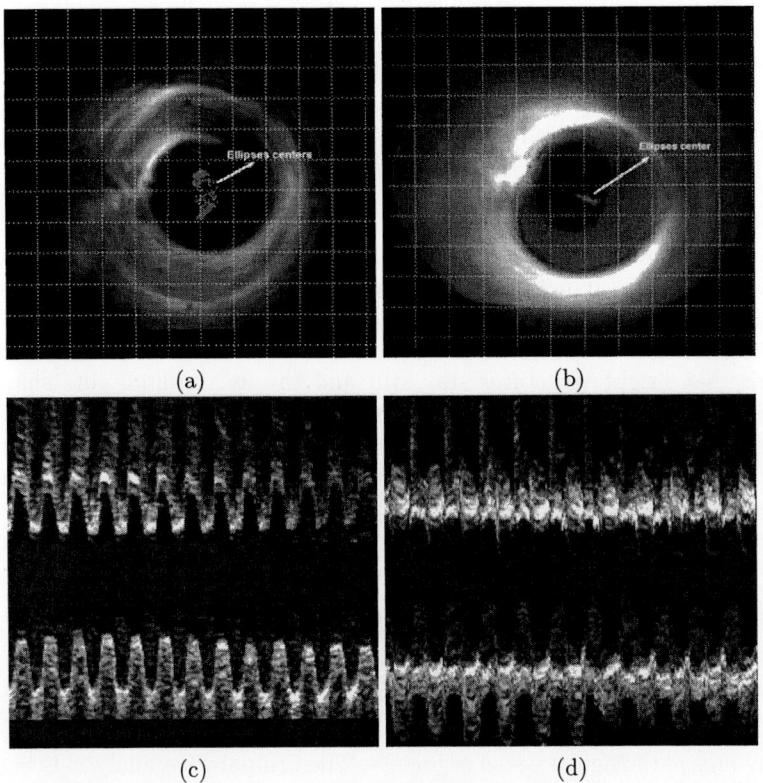

Fig. 7. Ellipses centers spatial location, before (a) and after (b) correction of rotation. Longitudinal cuts before(c) and after (d) rotation suppression

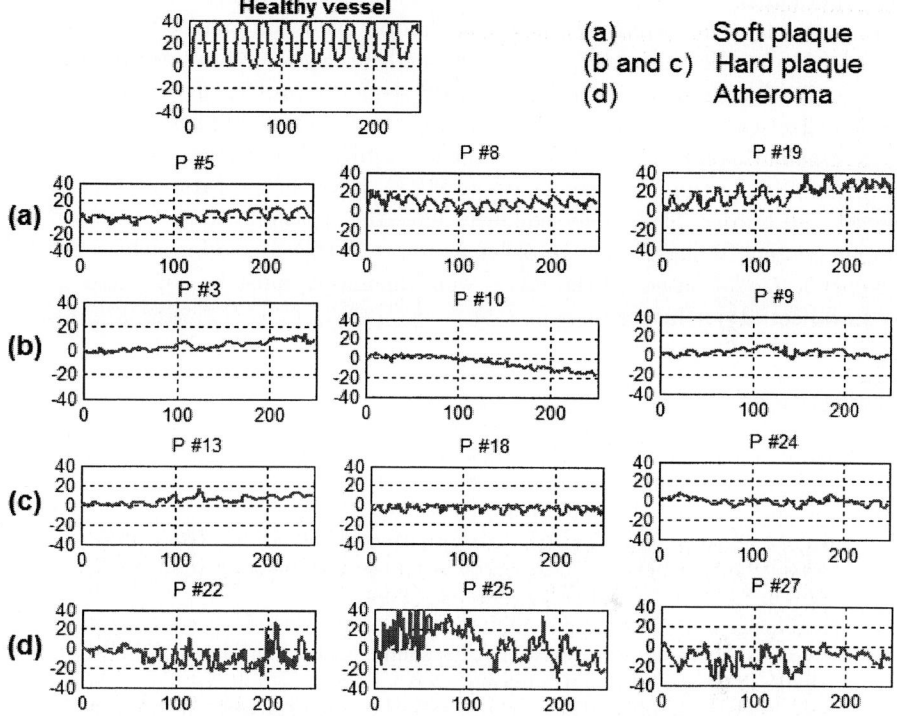

Fig. 8. Differences between healthy and pathological vessel segments of the rotation profiles for selected patients

Soft plaque still follows a periodic pattern synchronized with the heart beat, although its amplitude drops to a range of 5 to 20 degrees, depending on the severity of the lesion. The mechanical properties (elasticity and rigidity) of hard plaque result in a suppression of vessel oscillation, yielding almost flat rotation profiles. Finally, the atheroma profile is the most irregular one lacking of any visual periodicity.

4 Conclusions

The clinical applications of the rotation artifact have not been reported today, since the physiological, physical and geometrical reasons are not well known. We developed a kinematic model that allows to estimate and remove the IVUS rotation images effect. The model is based on the assumptions that the vessel wall shape can be modelled as a discrete structure. The proposed kinematic model can be used two ways: IVUS stabilization and kinematics characterization of the vessel wall. The first aspect is the main goal of this work, therefore we introduced a kinematics method to estimate and remove the rotation of IVUS sequences.

The method is based on the assumption that the vessel wall can be described as a discrete structure which kinematics temporal evolution can be followed by a trained Neural Network, during at least one heart cycle. A first examination of the qualitative shape of the rotation IVUS sequence profiles shows that the rotation effect can be used as a complementary tool, to evaluate vessel pathologies from kinematic point of view. Due to the anatomical distribution of the coronary arteries, most of the IVUS sequences ($> 85\%$), have their rotation center in the vessel wall border. Still some sequences have the rotation center located in the lumen center. In these cases in order to improve this first approach, a general geometric model based on the vessel wall kinematics must be considered. This is an object of our future work.

References

1. Yock P., Linker D., and et al., Intravascular two dimensional catheter ultrasound, Initial clinical studies, Circulations, No. 78, II-21, 1988.
2. Graham S., Brands D., and et al., Assesment of arterial wall morphology using intravascular ultrasound in vitro and in patient, Circulations, II-56, 1989.
3. Metz J., Paul G., and et al Intravascular Ultrasound Imaging, Jonathan M. Tobis y Paul G. Yock. Churchil Livinstone Inc., 1992.
4. Metz Jonas A., Paul G., and et al, Intravascular ultrasound basic interpretation, in Beyond Angiography, Intravascular Ultrasound, State of the art, Vol. XX, Congress of the ESC Viena-Austria, California, USA., 1998.
5. Jumbo G., Raimund E., Novel techniques of coronary artery imaging, in Beyond Angiography, Intra Vascular Ultrasound, state of the art, Vol. XX, Congress of the ESC Viena-Austria, University of Essen,Germany, 1998.
6. Di Mario C., et al. The angle of incidence of the ultrasonic beam a critical factor for the image quality in intravascular ultrasonography. Am Heart J 1993;125:442-448.
7. Arendt Jesen J., Linear Descripcion of Ultrasound Imaging System, Notes for the international Summer School on Advanced Ultrasound Imaging, Tecnical University of Denamark, 2001.
8. Berry E., and et al, Intravascular ultrasound-guided interventions in coronary artery disease, NHS R D HTA Programme, 2000.
9. Patrick H. W. Inteligencia Artificial, Addison Wesley Iberoamericana, 1994.
10. Korte Chris L., Intravascular Ultrasound Elastography, Interuniversity Cardiology Institute of the Netherlands (ICIN), 1999.
11. Mazumdar J., Biofluids Mechanics, World Scientific Publishing, 1992.
12. Young B., and Heath J., Wheater's, Histología Funcional, 4ta edición, Ediciones Hardcourt, S.A, 2000.
13. Boston Scientific Corporation, Scimed division, The ABCs of IVUS, 1998.
14. Misael Rosales, Petia Radeva and et al. Simulation Model of Intravascular Ultrasound Images, Lecture Notes in Computer Science Publisher: Springer-Verlag Heidelberg, Medical Image Computing and Computer-Assisted Intervention MICCAI 2004, Volume 3217/ 2004, pp. 200-207, France, September, 2004.
15. Vogt M., and et al, Strutural analysis of the skin using high frequency broadband ultrasound in the range from 30 to 140 mhz, IEEE Internationalk Ultrasonics Syposium, Sendai, Japan, 1998.
16. Andrew W. and et al, Direct Least Square Fitting of Ellipses, IEEE Transactions on Pattern Analysis and Machine Intelligence, vol 21, number 5, p 476-480, 1999

Computational Modeling and Simulation of Heart Ventricular Mechanics from Tagged MRI

Zhenhua Hu[1], Dimitris Metaxas[2], and Leon Axel[3]

[1] Department of Computer and Information Science,
University of Pennsylvania, Philadelphia, PA 19104, USA
zhhu@seas.upenn.edu
[2] The Center of Computational Biomedical, Imaging and Modeling,
Rutgers University, New Brunswick, NJ 08854-8019, USA
dnm@cs.rutgers.edu
[3] Department of Radiology,
New York University, New York City, NY 10016, USA
leon.axel@med.nyu.edu

Abstract. Heart ventricular mechanics has been investigated intensively in the last four decades. The passive material properties, the ventricular geometry and muscular architecture, and the myocardial activation are among the most important determinants of cardiac mechanics. The heart muscle is anisotropic, inhomogeneous, and highly nonlinear. The heart ventricular geometry is irregular and object dependent. The muscular architecture includes the organization of the fiber and the connective tissues. Studies of the myocardial activation have been carried out at both cell and tissue levels. Previous work from our research group has successfully estimated the in-vivo motion and deformation of both the left and the right ventricles. In this paper, we present an iterative model to estimate the in-vivo myocardium material properties, the active forces generated along fiber orientation, and strain and stress distribution in both ventricles. Compared to the strain energy function approach, our model is more intuitively understandable. Using the model, we have simulated the mechanical events of a few different heart diseases. Noticeable strain and stress differences are found between normal and diseased hearts.

1 Introduction

According to World Health Organization (WHO) estimates, 16.7 million people around the world die of heart disease each year, which contributes to nearly one-third of global deaths [27]. In the United States, heart disease has been the No. 1 killer each year since 1900 except 1918, causing more than 700,000 deaths per year since the 1960s; the estimated direct and indirect cost of heart disease is $368.4 billion in 2004 [1]. However, many aspects of the heart dynamics are still not well understood, although it has been studied for centuries.

The primary function of the heart is mechanical pumping. Myocardial stress and strain are important determinants of various aspects of cardiac physiology, pathophysiology, and clinical factors. Myocardial deformation and the associated stress fields reflect local contractile status and are related to heart wall motion and to the ventricular pressures. The quantitative description of heart ventricular strain and stress is important for the evaluation of cardiac performance in the diagnosis of heart disease.

Many models have been proposed to describe and predict mechanical function of heart in a rational and systematic manner since the late nineteenth century [30]. Most models used knowledge of continuum mechanics, anatomy, physiology, deformation, boundary conditions and material properties of the heart. Since the right ventricle is difficult to approximate with any simple parameterized 3D shape, most models have focused on the left ventricle. In one of the earliest models, a thin-walled spherical geometry was assumed [29]. The stress was derived from cavity pressure and radius using the classical Laplace membrane solution. After more than seventy years, ellipsoidal thin-walled models were proposed [24, 25]. In these models, the assumption that the thickness is much less than the radii of curvature contradicts reality and severely limits their application. Thick-walled ventricular mechanics models were then proposed [28, 19]. However, the geometry was still assumed to be a regular shape such as sphere or axisymmetric ellipsoidal.

With the invention of computers and the finite element method, irregular shapes were used in the modeling [8], which could describe the complex geometry of the ventricles more efficiently and accurately. The deformation was still assumed to be infinitesimal. Along with the development of large elastic deformation theory, finite deformation was incorporated into new models that offered real insight into myocardial stress distributions [20, 15]. Isotropic material properties were assumed in the models mentioned so far because of mathematical simplicity.

Appropriate formulation of constitutive relations for passive myocardium is needed for the analysis of deformation and stress in the heart. To determine the constitutive relations of myocardium, it is necessary to consider its anatomical structure. In this paper, we only give some introduction to heart ventricular anatomy that is related to our modeling. Details of the anatomy of the heart can be found in [3, 7].

Tissue structure studies have revealed that the myocardium muscle cells are bundled into fibers, which have different orientation through the heart wall. Hort, and later Streeter were the first to carry out systematic and quantitative measurements of muscle fiber orientations of canine heart [12, 25]. Their main findings are: (1) fibers predominantly lie in planes parallel to the endocardial and epicardial surfaces, and (2) the fiber directions generally vary in a continuous manner from +60° (i.e., 60° counter-clockwise from the circumferential axis) on the endocardium to -60° on the epicardium. This discovery has important implications for the mechanical properties of the myocardium. Hunter proposed the first non-axisymmetric large deformation finite element model of the left ventricle with anisotropic material properties [14]. In his model, the myocardium was represented as an incompressible, transversely isotropic material and transmural fiber orientation distribution was incorporated. Transversely isotropic material properties were then incorporated in many other works [13, 4, 9, and 5]. According to biaxial tissue testing, the stress-strain relations for the fiber and cross-fiber directions are different, which verified that heart material is anisotropic.

More recent studies reported that myocardial muscle fibers are tightly bound into branching laminar sheets that are approximately four cells thick [16]. The sheets are generally oriented normal to the ventricular surfaces. The adjacent sheets are loosely coupled and can slide over each other relatively easily, which may be the mechanism responsible for large ventricular wall thickening at end systole [5, 26].

Accurate representation of cardiac geometry is needed for the realistic modeling of heart ventricular mechanics. Cardiac geometry was traditionally given only qualitative descriptions. In the last three decades, a variety of imaging methods have been used to reconstruct three-dimensional geometry of heart ventricles [21]. However, most imaging methods are either invasive or cannot give information on the intramural motion due to the lack of suitable landmarks.

Magnetic resonance imaging (MRI) provides a new method to study heart function. Coupled with magnetization tagging, MRI not only provides tomographic images of the heart wall, it can also provide a means to noninvasively track material points within the wall [2, 31]. Two-dimensional deformation of the heart wall can be calculated by analyzing sequential tagged images, without considering through-plane motion components. To reconstruct three-dimensional trajectory of material points, we must combine tag displacement data in three orthogonal directions (two short axis and one long axis) [22, 10].

In this paper, we propose a new model of myocardium stiffness. The active stress and material properties are estimated iteratively. Using experimental strain data, we predict in-vivo stress distribution in the heart wall. Moreover, we simulate the mechanical events of a few different heart diseases.

2 Background

2.1 Cardiac Cycle

The cardiac cycle is divided into two major phases: systole (contraction) and diastole (relaxation). The pressure-volume loop of the left ventricle is shown in Figure 1. The mitral valve closes at A and the left ventricle undergoes isovolumic contraction with rapidly rising pressure until B, when the left ventricular pressure exceeds the aortic pressure and the aortic valve opens, blood is ejected and the left ventricle's volume begins to decrease. The aortic valve closes at the end of systole at C, due to the decreasing intraventricular pressure, which falls below the aortic pressure. The left ventricle then undergoes isovolumic relaxation from C to D. The mitral valve reopens at D when the pressure of left ventricle is lower than that of left atrium. Blood pressure in the right ventricle cavity changes similarly over the heart contraction cycle though its magnitude is much smaller than that of the left ventricle [3, 7].

2.2 Passive Myocardium Material Properties

Various stretching tests have been performed on isolated myocardium and many constitutive formulations have been derived based on these experimental results. The most commonly performed experiment on excised cardiac tissue is the uniaxial test. In uniaxial test, the specimens are mounted in a uniaxial testing apparatus and stretched in muscle-fiber direction; the stress-extension relations are measured during the stretching. Papillary muscle and ventricular trabeculae have been used in most uniaxial testing due to their geometrical simplicity. The data demonstrated that cardiac muscle is a nonlinear pseudo-elastic material with time-dependent mechanical properties. Fig. 2 shows schematic stress-extension relations for left ventricle midwall undergoing uniaxial loading in muscle fiber direction.

However, uniaxial data are not sufficient for the constitutive relations because of the anisotropy of cardiac tissue. Researchers have obtained more complete information using biaxial testing procedures. Biaxial specimens are usually rectangular slices of ventricular free-walls tangential to the epicardial surface such that the muscle fibers lie within the plane of the specimens. Different from uniaxial testing, the specimens are stretched in two in-plane orthogonal directions and the force-length data are measured in both directions. The data suggested that passive myocardium is a nonlinear, anisotropic, nearly pseudoelastic and perhaps regionally heterogeneous material, which is consistent with both structural and uniaxial data. The fiber axis is two to three times stiffer than the cross-fiber direction in the LV midwall.

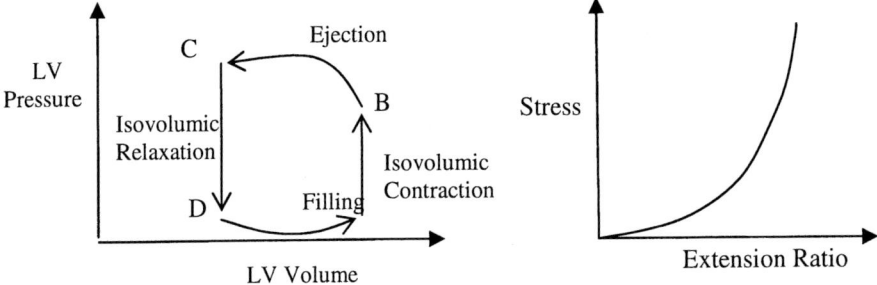

Fig 1. Schematic pressure-volume loop of left ventricle

Fig. 2. Schematic stress-extension relations for left ventricle midwall undergoing uniaxial loading in muscle fiber direction

Although biaxial testing is useful to identify two-dimensional constitutive relations, it is not sufficient for three-dimensional models. Three-dimensional tests have been used for this purpose [11]. The only difference from biaxial testing is that a punch indents a small portion of the top surface of the specimen in the third orthogonal direction; the indentation force and punch penetration depth are measured along with biaxial force-length relations. The slope of the indentation force-depth relationship is called a measure of transverse stiffness.

3 Estimation of Material Properties and Active Forces

Since the stresses cannot be measured directly, there is a need for a reliable model for estimating the state of stress in a beating heart. One important approach is using the classical balance relations of continuum mechanics and computational techniques such as finite element method. With knowledge of cardiac anatomy, muscle fiber and sheet orientation, and boundary conditions, the most significant problems in developing such a model are estimating the stress-strain relationship of the materials of the heart and the laws governing the active contraction of the heart muscle. In this section, we present our methods to estimate the material properties and active forces for a beating heart.

3.1 Boundary Conditions

Since our reconstructed data do not include the atrium, we need to assume some reasonable boundary conditions at the base area. We assume that each basal node of ventricular wall is attached to a spring from above. The spring is at its stress-free state at end-diastole. Since base area moves down towards the apex area during systole, the springs will be stretched and generate forces pulling basal nodes. The elasticity of the spring is assumed to be equal to the Young's modulus of ventricular wall.

During contraction, the blood pressure acting on the endocardial wall is P_{LV}, which changes with time. We assume the blood pressure is uniform in the whole ventricle and follow empirical results [6].

From the reconstructed motion, we observe that the apex area has relatively small displacements. Moreover, there is no other organ attached to the ventricle at the apex. Therefore, it is reasonable to assume that the displacements of the nodes around the apex are given, either from reconstructed data for real heart models or zero for regular geometric model.

3.2 Nodal Forces from Active Contraction

For simplicity, we assume that cardiac muscle only generates forces along the fiber orientation. To be implemented in the finite element method, these active forces need to be converted into the equivalent node forces. Consider a 2D square in Fig. 3; suppose the fiber is located at the center of the square and the fiber angle with respect to the horizontal axis is θ. Assume the active force within fibers is F, which is along the fiber direction. We need to calculate the equivalent forces on the four nodes named 1, 2, 3, and 4. We assume all four nodal forces are parallel to the fiber force and the magnitudes of diagonal forces are equal to each other because of symmetry:

$$f_1 = f_3 \quad f_2 = f_4 \tag{3.1}$$

To calculate the equivalent nodal forces, we cut the plane in the middle virtually by a plane perpendicular to the fiber orientation. From force balance, it is easy to find that the summation of forces 1 and 4 is equal to the active force:

$$f_1 + f_4 = F \tag{3.2}$$

To determine the magnitudes of forces 1 and 4, we use torque balance with respect to node 1:

$$F \cdot L_F - f_4 \cdot L_4 = 0 \tag{3.3}$$

where:

$$L_F = \frac{1}{2} L (1 - \tan \theta)$$
$$L_4 = \frac{L}{\cos \theta} \tag{3.4}$$

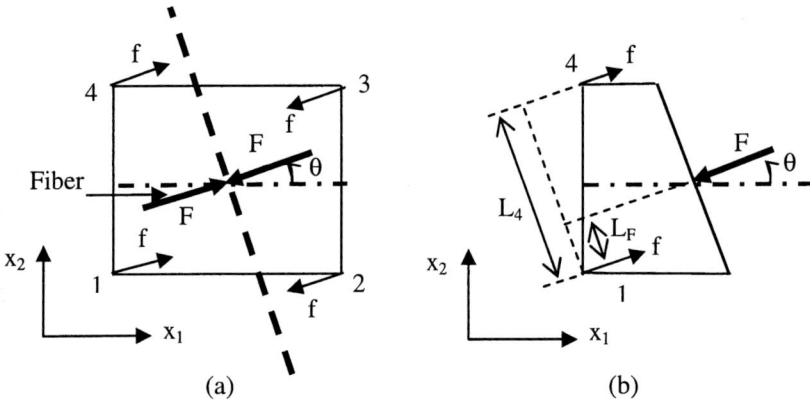

Fig. 3. Calculation of equivalent node forces in a square element: (a) one square element with active forces along fiber orientation; (b) one half of the same element

where L is the length of the side of the square and θ is assumed to be less than or equal to 45° (similarly we can derive the forces for other values of θ). Substitute (3.2) and (3.4) into (3.3), we get:

$$f_1 = f_3 = \left(1 - \frac{1}{2}\cos\theta + \frac{1}{2}\sin\theta\right)F \qquad \theta \in [0,45] \qquad (3.5)$$

$$f_2 = f_4 = \frac{1}{2}(\cos\theta - \sin\theta)F$$

To compute active force, we need to multiply active stress by the corresponding cross section area. Since we assume fibers lie in the plane parallel to the surface, one simple way to calculate the cross section area is taking the average of four face areas that fibers cross through. The active force for one element is:

$$F = \left(\frac{1}{4}\sum_{i=1}^{4} S_i\right) f_{af} \qquad (3.6)$$

where f_{af} is the element's active fiber stress that we will estimate, and S_i is area of the i th face that fiber crosses through.

In above computation, the directions of nodal forces are relative to the local element coordinates and parallel to the fiber direction. We need to express them in global coordinates before they can be used in finite element computing. Suppose fiber direction in global coordinates is given by:

$$\vec{r} = r_1\vec{e}_x + r_2\vec{e}_y + r_3\vec{e}_z \qquad (3.7)$$

The equivalent force at node 1 will be:

$$\vec{f}_1 = \frac{1}{2}F\left(r_1\vec{e}_x + r_2\vec{e}_y + r_3\vec{e}_z\right) \qquad (3.8)$$

In order to get the net active forces on each node, we go through the same procedure for each element of the heart ventricles. For one node i, suppose it is one of the corner nodes of ni elements, the net active force on node i is:

$$\vec{f}_i = \sum_{k=1}^{ni} \vec{f}_{i,k} \qquad (3.9)$$

where $f_{i,k}$ is the nodal force acting on node i generated from neighbor element k. Combing equations (3.5-3.9), we derive the relationship between equivalent nodal forces and active forces in matrix form as follows:

$$P_{af} = A_{af} f_{af} \qquad (3.10)$$

where P_{af} is the nodal force vector with $3n$ entries (n is the number of nodes, 3 entries for each node in 3D problems), f_{af} is the finite element active stress vector with m entries (m is the number of elements), and A_{af} is the matrix ($3n$ by m) that maps the element active stresses onto the nodal forces. The mapping matrix A_{af} depends on the geometry and fiber angle only.

3.3 Singular Value Decomposition

If we neglect body forces and residual stresses, the finite element equation will be:

$$K \cdot a = P_{cavity} + P_{af} \qquad (3.11)$$

where K is the stiffness matrix, P_{cavity} is the pressure generated by blood in the ventricular cavity, P_{af} is the equivalent nodal force given by equation (3.10). Since displacement is given by the reconstructed motion data and we use empirical blood pressure, the active stresses can be determined once K is known. Substituting equation (3.10) into (3.11), we have:

$$A_{af} \cdot f_{af} = K \cdot a - P_{cavity} \qquad (3.12)$$

Observing that the coefficient matrix A_{af} is non-square and equation (3.12) is an over-determined set of linear equations, we use singular value decomposition (SVD) to solve these equations. Supposing A_{af} is a $M \times N$ matrix, it can be written as the product of an $M \times N$ column-orthogonal matrix U, an $N \times N$ diagonal matrix W with positive or zero elements, and the transpose of an $N \times N$ orthogonal matrix V [23].

$$A_{af} = U \cdot W \cdot V^T \qquad (3.13)$$

The solution of equation (3.12) is then given by:

$$f_{af} = V \cdot [diag(1/\omega_j)] \cdot \left(U^T \cdot (K \cdot a - P_{cavity})\right) \qquad (3.14)$$

where ω_j is the j th diagonal element of matrix W, and $1/\omega_j$ is replaced by zero if $\omega_j = 0$. The solution given by equation (3.14) will not exactly solve equation (3.12), however, it minimizes:

$$r = \left| A_{af} \cdot f_{af} - \left(K \cdot a - P_{cavity} \right) \right| \qquad (3.15)$$

where r is called the residual of the solution.

3.4 Assumptions on Stiffness

One of our major tasks is to identify the myocardial material properties and active stress during systole. Passive myocardial material properties change greatly when strains are large. Conventional models will predict that passive myocardial material properties vary only a small amount during iso-volumic contraction since there is very little deformation at this phase. From this observation, we cannot use the passive material properties only to compute the stiffness matrix K in equation (3.11). Otherwise, it will be in contradiction with the real underlying mechanics; since the active stress goes up rapidly during iso-volumic contraction and the cavity pressure goes up quickly as well; if the stiffness does not change much, the deformation of myocardium would be quite large, which is different from the real measurements. Therefore, we propose a new model, that myocardial stiffness consists of two components: one component depends on active forces only while another component depends on strains only:

$$Y_{s,\,total} = Y_{s,\,af} + Y_{s,\,\varepsilon} \qquad (3.16)$$

where $Y_{s,total}$ is the overall Young's modulus of the myocardium, $Y_{s,af}$ is the active Young's modulus that depends on active forces, and $Y_{s,\varepsilon}$ is the passive Young's modulus that varies with strains.

For simplicity, we assume the active Young's modulus is a linear function of active force along the fiber:

$$Y_{s,\,af} = C_1 f_{af} \qquad (3.17)$$

where C_1 is a coefficient, and f_{af} is the active stress generated by the fiber. Since the Young's modulus and active stress have the same units, C_1 is unit-less. Note that $Y_{s,af}$ is equal to zero when there is no active stress. Since myocardium has different Young's modulus along different directions, to make it reasonably agree with experimental results, we need to use at least three functions to represent active Young's modulus along fiber, sheet, and sheet normal orientation, respectively. There are three parameters to be determined in these functions, and we express them in one vector as $(C_{1,f},\ C_{1,s},\ C_{1,sn})$. They are the constant coefficients corresponding to fiber, sheet, and normal sheet, respectively. The corresponding passive Young's modulus is derived from published experimental data.

3.5 Estimation Algorithm

Although both sides of equation (3.12) have unknown variables, we can estimate them based on more constraints. The active force generated by the fibers acts to increase the cavity pressure during iso-volumic contraction and then push the blood out of ventricles during ejection. The magnitude of blood pressure mainly depends on how much active force has been generated, while the deformation of the ventricles depends on active force, blood pressure, and the material properties. The deformation itself determines how much the volume of the cavity will change. Since the volume of the cavity does not change during isovolumic contraction, this imposes one more constraint on the model.

We use the binary search method to estimate the approximate amount of active force needed to generate the given blood pressure during isovolumic contraction. We then use a generalized EM algorithm [18] to estimate the active Young's moduli and active stress. The procedure is as follows:

1. Let $i = 0$, Assume initial active stress $f_{af}^{(i)}$ is 0,
2. Use mean active stress $E(f_{af}^{(i)})$, active Young's modulus coefficients, and the passive Young's modulus from experimental data to calculate K in equation (3.15),
3. Use K from step 2, the reconstructed motion data, and empirical blood pressure to calculate active stress $f_{af}^{(i+1)}$ using equation (3.15),
4. Take the mean active stress $E(f_{af}^{(i+1)})$ from all elements, if

$$\frac{\left|E(f_{af}^{(i+1)}) - E(f_{af}^{(i)})\right|}{E(f_{af}^{(i+1)})} < \delta,$$ stop the iteration (δ is a small number), else let $i = i + 1$, go to step 2.

The step 2 is the E-step since we use the mean value of active stress to compute the active Young's modulus. The step 3 is the M-step since we use the singular value decomposition (SVD), which is fundamentally a general least-square method.

4 Simulation of Patho-Physiological Ventricles

The goal of our modeling and simulation is to provide quantitative data on the strain and stress distribution within the heart ventricles to clinicians who may find this information useful for treating cardiovascular diseases (CVD). Since our research is focused on heart ventricles, we will study two kinds of heart disease - ischemia and conduction abnormalities.

Fig. 4. Simulation of ischemia with different size: small ischemia (yellow area) (left); 33% of the left ventricle's free wall is ischemia (yellow area) (right)

4.1 Ischemia

Heart muscle needs a steady supply of oxygen and nutrients to function properly. The coronary arteries supply oxygenized blood flow to the heart. These arteries normally have smooth inner lining. However, they can become clogged with fats, cholesterol and plaques. As a result, a blocked coronary artery prevents sufficient oxygen and nutrients from reaching a section of the heart. This is called ischemia. There is an imbalance between oxygen supply and oxygen demand during myocardial ischemia and the myocardium becomes hypoxic. When ischemia lasts for more than a few minutes, heart muscle can begin to die and lead to infarction of the tissue, causing a heart attack.

In our model, we consider acute ischemia. We assume the acute ischemia region does not generate active force and is less stiff than normal regions. We control the size of ischemia from a small area to 33% of the free wall of left ventricle (Fig. 4).

4.2 Conduction Delay

The heart muscle needs electric stimulation for every contraction. Cardiac conduction system refers to the system of electrical signaling that instructs these muscle cells to contract. The Sinoatrial (SA) node is the electric impulse station of the heart. It is located in the upper corner of the right atrium. The action potential from the SA node spreads to neighboring cardiac muscle cells and triggers atrial systole. The SA node eventually activates the atrioventricular (AV) node that is located between right atrium and right ventricle. A group of fast-conducting fibers called the His bundle carry the AV node activity to the interventricular septum very quickly. The His bundle divides into two branches when it reaches the interventricular septum. The left branch is for left ventricle and the right branch is for right ventricle. The bundle branches run all the way down to apex of the heart through the septum and then go back up along the ventricles. The bundle branches running up the outer edges are called Purkinje fibers [6].

The conduction within the bundle branches may be delayed in diseased hearts. This results in a delay of the depolarization for part of the ventricular muscle. It is useful to study the impact of such delay on the function of heart ventricles. We use a simple model to simulate conduction delay within left ventricle. We assume 33% of the free wall of left ventricle is activated later than other regions. The region with delayed activation will be initially stretched a little bit. From the Frank-Starling law, this region will generate greater active force than other regions when it gets activated. The results are shown in Section 5.3.2.

5 Results

In this section, we present experimental results of our modeling and corresponding simulation. First, we will show the reconstructed 3D strain of heart ventricles. Next, we will present the material properties and active force estimation results and the stress distribution in heart ventricles. Finally, we demonstrate the application of our model to different heart disease simulations.

5.1 Clinical Strain Data

The 3D motion reconstruction technique was applied to data from 5 normal volunteers. The motion of heart ventricles from end-diastole to end-systole was extracted. Motion data from all 5 normal volunteers had similar behavior. There are 5 frames from end-diastole to end-systole in this data set. There are two layers of elements in the radial direction in the left ventricle. The right ventricle has one layer of elements along the radial direction. We use this motion data to compute the strain of the ventricles. The strain components are presented in local wall coordinates. The normal and shear strain components of left and right ventricles in *RCL* (radial, circumferential, and longitudinal) coordinates at end-systole are shown in Fig. 5. We observe that the ventricles become thicker radially, and shorter circumferentially and longitudinally during systole.

5.2 Material Properties and Stress

The passive material properties of heart ventricles are assumed to take some average values from the experimental data reported by other researchers [26]. We use the passive Young's modulus given in Table 1 in our model. Estimated passive stress distribution in one normal bi-ventricle heart at end-systole is shown in Fig. 6.

Table. 1. Passive Young's modulus of heart ventricles: $Y_{sf,\varepsilon}$, $Y_{ss,\varepsilon}$, and $Y_{snf,\varepsilon}$ are passive Young's modulus along fiber, sheet, and sheet-normal directions, respectively

	Axial strain			
	0.0	0.1	0.2	0.3
$Y_{sf,\varepsilon}$ (KPa)	40	90	140	200
$Y_{ss,\varepsilon}$ (KPa)	10	40	80	125
$Y_{snf,\varepsilon}$ (KPa)	6	20	40	70

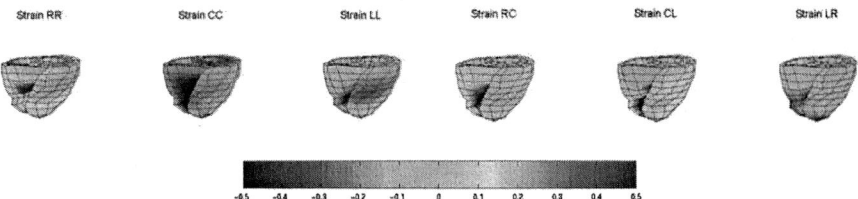

Fig. 5. Normal heart strain components at end-systole

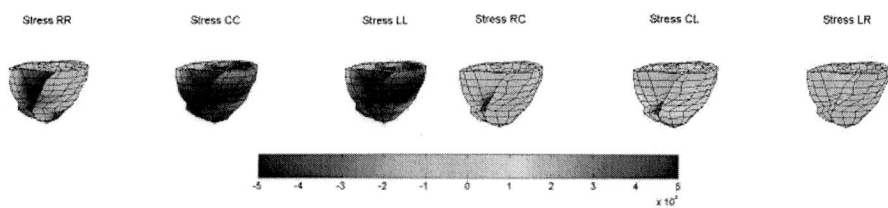

Fig. 6. Normal heart stress components at end-systole

5.3 Patho-Physiological Ventricles

5.3.1 Ischemia

We used regular geometry in the simulation of heart diseases. In this simulation experiment, ischemia region occupies 33% of the left ventricle's free wall (see Fig. 4). The strain components are shown in Fig. 7. As we can see, the ischemia region has different strain components from normal region, especially the radial and circumferential components. From the simulation results, we observe that a small area of ischemia does not alter the systolic function much and the strain and stress pattern of the ventricle is very similar to the normal heart. However, the strain and stress are quite different from those of normal heart when the ischemia size becomes large. In addition, a large ischemia alters not only the function of itself but also the border zone.

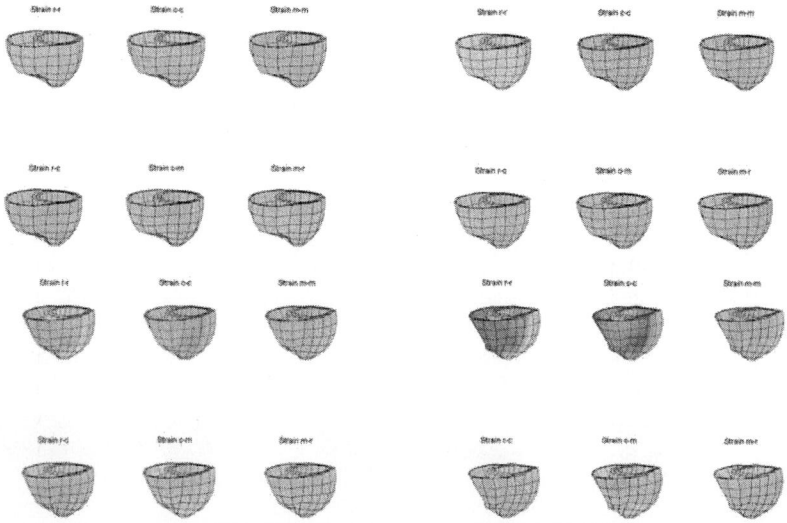

Fig. 7. The strain pattern in diseased heart with ischemia at frame 2 (top left), 3 (top right), 4 (bottom left), and 5 (bottom right), respectively

5.3.2 Conduction Abnormalities

In this simulation experiment, 33% of the left ventricle's free wall (the same area as Fig. 4) is activated later than other regions. The Frank-Starling law tells us that this region will contract more rigorously due to the initial stretch. We can see from Fig. 8 that the greater contraction force compensates the delay and the final strain within this area is similar to other areas.

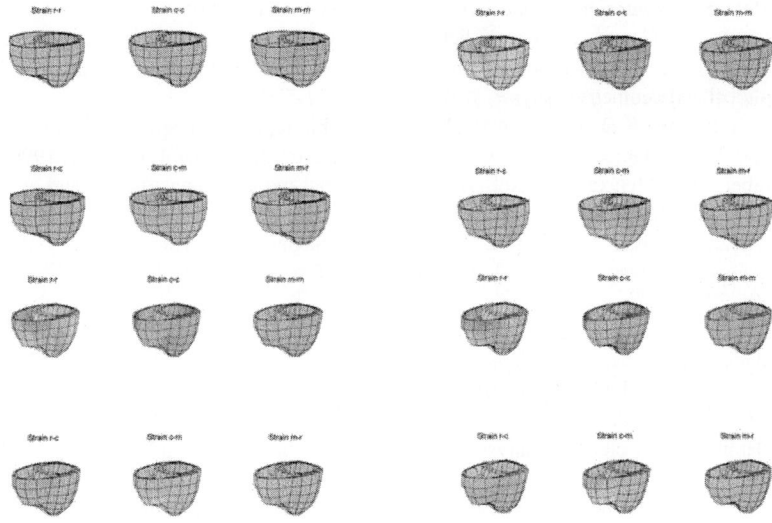

Fig. 8. The strain pattern in diseased heart with conduction abnormality at frame 2 (top left), 3 (top right), 4 (bottom left), and 5 (bottom right), respectively

6 Conclusion

We have developed an iterative model to estimate the strain and stress of both cardiac ventricles by using accurate displacements reconstructed from MRI-SPAMM tagging and a deformable model. Compared to the traditional strain energy function method, our model gives more intuitive and understandable parameters. In addition, we have used our models to simulate two different heart diseases, which have different strain and stress patterns from normal heart. The modeling and simulation results may be clinically useful for heart disease treatment in the future. In future work, we will extend our model to the full cardiac cycle by including the relaxation phase and carry out sensitivity analysis on parameter estimation.

Reference

[1] American Heart Association. *2004 Heart and Stroke Statistical Update*. Dallas, TX: Am. Heart Assoc.
[2] L. Axel, L. Dougherty. Heart wall motion: Improved method of spatial modulation of magnetization for MR imaging. *Radiology*, 272:349-50, 1989.

[3] R.M. Berne, M.N. Levy. *Principles of Physiology*. Mosby-Year Book, 1996.
[4] P.H.M. Bovendeerd, T. Arts, J.M. Huyghe, D.H. van Campen, R.S. Reneman. Dependence of local left ventricular wall mechanics on myocardial fiber orientation: A model study. J. Biomech. 25(10), 1129-1140, 1992.
[5] K. Costa. *The Structural Basis of Three-Dimensional Ventricular Mechanics*, Ph.D. Dissertation, University of California, San Diego, CA, 1996.
[6] Y.C. Fung. *Biomechanics: Circulation*, 2nd Edition. Springer-Verlag, New York, 1997.
[7] L. Glass, P. Hunter, A. McCulloch. *Theory of Heart: Biomechanics, Biophysics, and Nonlinear Dynamics of Cardiac Function*. Springer-Verlag, 1991.
[8] P. Gould, D. Ghista, L. Brombolich, I. Mirsky. In vivo stresses in the human left ventricle: Analysis accounting for the irregular 3- dimensional geometry and comparison with idealized geometry analyses. J. Biomech. 5, 521-539, 1972.
[9] J.M. Guccione, K.D. Costa, A.D. McCulloch. Finite element stress analysis of left ventricular mechanics in the beating dog heart. J. Biomech. 28(10), 1167-1177, 1995.
[10] I. Haber, D.N. Metaxas, L. Axel. Three-dimensional motion reconstruction and analysis of the right ventricle using tagged MRI. *Medical Image Analysis*, 4:335-355, 2000.
[11] H.R. Halperin, P.H. Chew, M.L. Weisfeldt, K. Sagawa, J.D. Humphrey, F.C.P. Yin. Transverse stiffness: A method for estimation of myocardial wall stress. Circ. Res., 61:695-703, 1987.
[12] W. Hort. Mikrometrische Untersuchungen an verschieden weiten Meerschweinchenherzen, *Verhandl. Deut. Ges. Kreislaufforsch*, 23: 343-346, 1957.
[13] J.D. Humphrey, F.C.P. Yin. Biomechanical experiments on excised myocardium: Theoretical considerations. *Journal of Biomechanics*, 22:377-383, 1989.
[14] P.J. Hunter. Finite element analysis of cardiac muscle mechanics. Ph.D. thesis, University of Oxford. 1975.
[15] R.F. Janz, A.F. Grimm. Finite element model for the mechanical behavior of the left ventricle. Circ. Res. 30, 244-252, 1972.
[16] I.J. LeGrice, B.H. Smaill, L.Z. Chai, S.G. Edgar, J.B. Gavin, P.J. Hunter. Laminar structure of the heart: ventricular myocyte arrangement and connective tissue architecture in the dog. Am. J. Physiol. 269: H571-582, 1995.
[17] A.D. McCulloch. Cardiac mechanics. In: J.D. Bronzino (ed.), *The Biomechanical Engineering Handbook*. CRC Press, Boca Raton, FL. Chapter 31, pp. 418-439, 1995.
[18] G.J. McLachnan, T. Krishnan. The EM algorithms and extensions. John Wiley & Sons, Inc. 1997.
[19] I. Mirsky. Effects of anisotropy and nonhomogeneity on left ventricular stresses in the intact heart. Bull. Math. Biophys. 32(2), 197-213, 1970.
[20] I. Mirsky. Ventricular and arterial wall stresses based on large deformations analysis. Biophys. J. 13(11), 1141-59, 1973.
[21] W.G. O'Dell, A.D. McCulloch. Imaging three-dimensional cardiac function. Annu. Rev. Biomed. Eng. 2:431-56, 2000.
[22] J. Park, D. N. Metaxas, L. Axel. Analysis of left ventricular wall motion based on volumetric deformable models and MRI-SPAMM. *Medical Image Analysis*, 1:53-71, 1996.
[23] W.H. Press, S.A. Teukolsky, W.T. Vetterling, B.P. Flannery. Numerical recipes in C++, the Art of Scientific Computing. 2nd Edition. Cambridge University Press, 2002.
[24] H. Sandler, H.T. Dodge. Left ventricular tension and stress in man. *Circ. Res.* 13(2), 91-104, 1963.

[25] D. D. Streeter Jr., W. T. Hanna. Engineering mechanics for successive states in canine left ventricular myocardium: I. Cavity and wall geometry. *Circulation Research*, 33:639-655, 1973.
[26] T. P. Usyk, R. Mazhari, A. D. McCulloch. Effect of laminar orthotropic myofiber architecture on regional stress and strain in the canine left ventricle. *Journal of Elasticity*, 61: 143-164, 2000.
[27] WHO *World Health Report, 2003*; WHO website: www.who.int/ncd/cvd.
[28] A.Y.K. Wang, P.M. Rautaharju. Stress distribution within the left ventricular wall approximated as a thick ellipsoidal shell. Am. Heart J. 75(5), 649-662, 1968.
[29] R.H. Woods. A few applications of a physical theorem to membranes in the human body in a state of tension. *J. Anat. Physiol.* (26) 362-370, 1892.
[30] F.C.P. Yin. Ventricular wall stress. *Circ. Res.* 49 (4), 829-842, 1981.
[31] A.A. Young, L. Axel. Three-dimensional motion and deformation of the heart wall: estimation with spatial modulation of magnetization – a model based approach. Radiology, 185: 241-247, 1992.

A Realistic Anthropomorphic Numerical Model of the Beating Heart

Rana Haddad[1], Patrick Clarysse, Maciej Orkisz, Pierre Croisille, Didier Revel, and Isabelle E. Magnin

Creatis, CNRS Unit #5515, INSERM U630, F-69621 Lyon Cedex, France
rana.haddad@creatis.insa-lyon.fr
http://www.creatis.insa-lyon.fr/

Abstract. A realistic anthropomorphic numerical model of the beating heart is presented. It includes the main cardiac anatomical structures, vessels junctions and part of the coronary network. Its main feature is that it is based on an imaging study on the same human subject from which both structural and motion information are retrieved. This confers to the model a remarkable consistency. Heart's deformation is assessed through successive 3D non rigid registrations in cine MR sequences. The resulting model can be used for the evaluation of cardiac image processing algorithms such as myocardium segmentation and cardiac image registration.

1 Introduction

Up-to-date medical imaging modalities, such as Magnetic Resonance Imaging (MRI) are now able to provide very accurate pictures of the heart's anatomy in 3D and through the cardiac cycle. Image acquisitions generally relies on triggering systems (i. e. ECG gating) to reconstruct the anatomy from data recording over a certain time interval (currently several cardiac cycles). For diagnostic purposes, clinicians need image post-processing techniques in order obtain quantitative parameters, currently cavity volumes, ejection fraction or local motion indexes such as myocardial wall thickening. Therefore, a significant number of methods have been developed for the segmentation of the heart and the estimation of its motion [13, 4, 16, 9]. Some of them may surely obtain good results in specific cases but the novel user usually lacks objective performance comparison. A *beating heart model* could serve as a reference to evaluate the performances of segmentation methods [3], [8]. Such a model could also be of great interest for the evaluation of image reconstruction [6] and registration [12] algorithms. In these cases, the model would include not only a computer description of the main cardiac structures but also the associated medical images in mono or even multi-modalities. Another interest of such a model would be the training for minimally invasive beating heart surgeries [18]. If the model is doted of some tuning parameters, various normal and pathological conditions could then be simulated.

This paper reports the current state of development of a beating heart model which includes the heart's cavities, the junction of the main vessels and the coronaries. One of the main characteristics of this model is that it issues from the same single anatomical and dynamic imaging study on a healthy subject. Also, the model comprises both a surface description of the anatomical structures and the associated MR images.

2 Method

As described in Figure 1, the model's development first relies on the construction of a static anatomical model which will be animated in a second phase. These two steps are detailed hereafter.

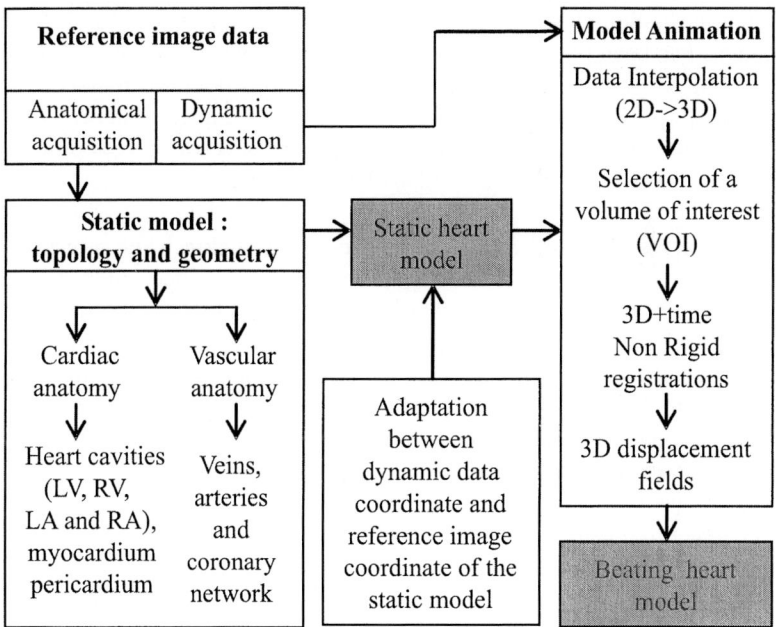

Fig. 1. Overview of the anthropomorphic beating heart model development

2.1 Static Model Construction

The anatomy of the heart and the main vascular structures, including the coronaries, is obtained from the acquisition of a specific 3D image set in MRI examination of a healthy subject. The static model construction relies of the following four successive steps:

1. Acquisition of the reference data set
2. Extraction of the contours and 3D surface reconstruction of anatomical structures
3. Semi-automatic extraction and reconstruction of the coronary network
4. Assembling the whole model

Reference Data Set. A 3D MRI acquisition has been performed on a healthy subject on a INTERA 1.5T, Philips Medical Systems at the Cardiological Hospital, Lyon, France. 3D sense sequence has been used with ECG gating and echo navigation (free respiration) to acquire a dense set of 80 transverse slices (slice thickness=2mm, spacing between slices=1mm, image resolution=$0.53 \times 0.53 mm^2$, matrix=512x512) to cover the whole heart and peripheral vessels at mid-systole (Figure 2(a)). Cine images has been acquired on the same volunteer to be used for the animation of the model (Section 2.2). Seven slice levels cover the whole heart. For each slice, thirty frames sample the cardiac cycle. At total, 210 short axis images has been obtained. The parameters are: slice thickness=7mm, spacing between slices=10mm, image resolution=$1.05 \times 1.05 mm^2$, matrix=256x256 (Figure 2(b)(c)). 2 and 4 chambers long axis cine images (30 frames) complete the reference data set. Their parameters are : slice thickness=6mm, image resolution=$1.13 \times 1.13 mm^2$, matrix=256x256 (Figure 2(d)).

Contour Extraction and 3D Surface Reconstruction. The contours of the left and right ventricles and atria, the pericardium have been manually traced by a radiologist. Each contour, which has been individually checked, is labeled by the corresponding anatomical structure it is part of. In the same way, the junctions of the four pulmonary veins, vena cava and the aorta have been segmented. Figure 3(a) shows some of the extracted contours. 3D reconstruction of the cardiac structures is based on the method developed by B. Geiger for surface reconstruction from planar contours and implemented in the *Nuages*[1] software. Surfaces are available as 3D triangle meshes. A smoothed version of the surfaces has been generated using Laplacian smoothing (vtkSmoothPolyDataFilter of the VTK library[2]). Figure 3(b) shows some of the reconstructed surfaces.

Semi-automatic Extraction and Reconstruction of the Coronary Network. Semi-automatic extraction and reconstruction of the coronary network perceptible in the data set was achieved in following steps: 1) image filtering aiming at the enhancement of the coronary arteries, while attenuating other structures, 2) segment-by-segment extraction of arterial centerlines, along with estimation of local diameters, 3) extraction of cross-sectional lumen boundaries of each arterial segment in image planes locally perpendicular to the centerline, 4) connection and smoothing of the centerline segments, 5) generation of smoothed lumen surfaces.

---

[1] http://www-sop.inria.fr/prisme/
[2] http://public.kitware.com/VTK/

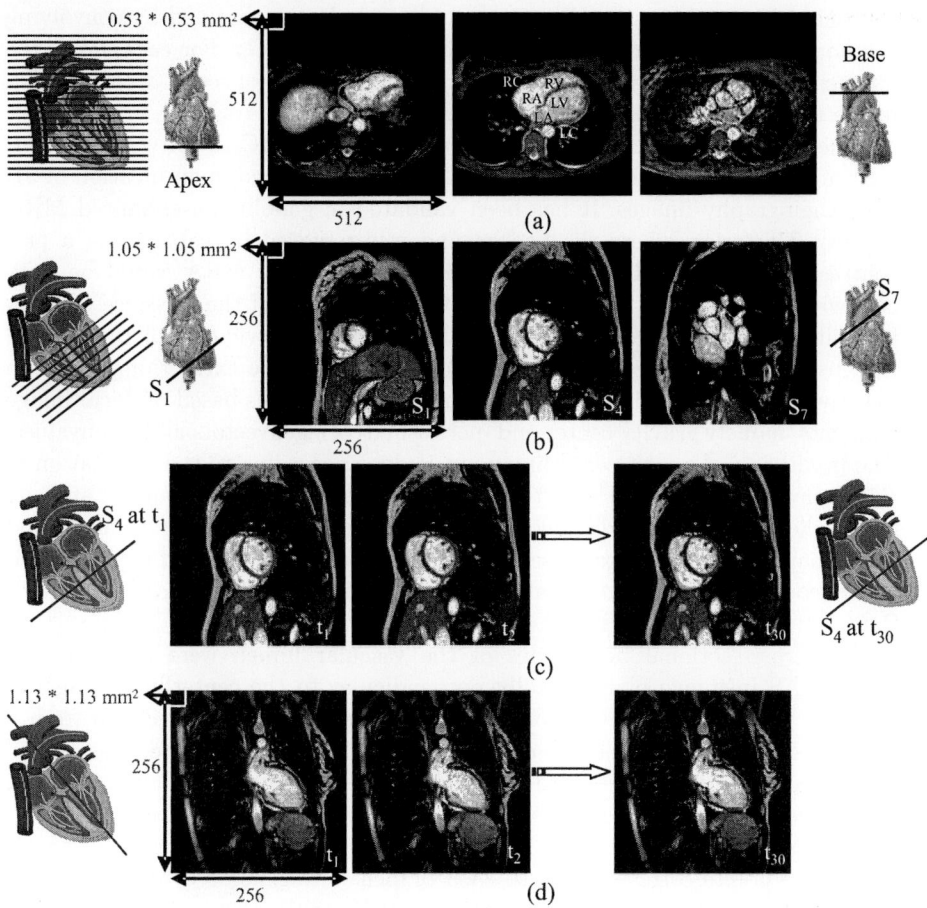

Fig. 2. (a) Reference MRI data set and the annotated anatomical structures (LV and RV are left and right ventricles, LA and RA are left and right atria, respectively). Note that right (RC) and left (LC) coronaries are also visible. (b) 3 levels of short axis slices. (c) 3 over the 30 frames acquired at one short axis slice level. (d) 3 over the 30 frames acquired at one long axis slice (2 chambers)

The first step was carried out using Frangi's multi-scale filter based on Hessian matrix eigenvalues $\lambda_3 < \lambda_2 < \lambda_1$ [5]. As the arteries appear bright on darker background, and their expected shape is cylindrical, the filter enhances the points that meet the following criterion: $\lambda_1 \approx 0$, $|\lambda_1| << |\lambda_2|$ and $\lambda_2 \approx \lambda_3$. This criterion at a given scale σ is computed as follows: $P_\sigma = \left[1 - \exp\left(-R_A^2/2a^2\right)\right] \exp\left(-R_B^2/2b^2\right) \left[1 - \exp\left(-S^2/2c^2\right)\right]$, where the parameters $a = b = 0.25$, and c depends on image dynamics. With $R_A = |\lambda_2/\lambda_3|$, the first term is a measure of cross-sectional circularity. With $R_B = |\lambda_1|/\sqrt{\lambda_2/\lambda_3}$, the second term is a measure of elongation. The last term uses the Frobenius norm S of the Hessian matrix to reject unstructured regions. The second derivatives used to construct the Hessian at a given scale are obtained by convolving the image with appropriate derivatives of the Gaussian G_σ. For each voxel, the filter chooses the strongest response across a discrete set of scales σ corresponding to the expected radii of the arteries.

The filtered images were used as input to the MARACAS software [7]. This software was designed for the purpose of segmentation and quantification of 3D MR Angiography images. It has been validated in gadolinium-enhanced MRA of carotid arteries, where the background was removed by subtracting a pre-contrast mask image. Its direct application in coronary MRA would be difficult without the above-described pre-filtering, because of the presence of high-intensity large structures, namely cavities, in the close vicinity of the vessels. In a user-selected vessel, MARACAS performs centerline extraction within an iterative prediction/estimation framework. This process is based on local image moments, namely gravity center and inertia-matrix eigenvectors and eigenvalues. The prediction of a next centerline point is done according to the orientation of the eigenvector associated with the smallest eigenvalue. The estimation attracts the predicted point toward the local gravity center, while using first and second order shape constraints like those used in snakes, *i.e.* elasticity and flexibility. Simultaneously, an approximate value of the local diameter of the vessel is deduced from a multi-scale analysis of the eigenvalues [11].

The cross-sectional boundaries of the vascular lumen were extracted by MARACAS in image planes locally perpendicular to the centerline, using iso-contours with an adaptive iso-value. The local iso-values were deduced from a low-pass filtered intensity curve of the centerline points. The centerlines of separate arterial segments provided by MARACAS were then connected. Namely, the segments belonging to the same branch were concatenated and smoothed with B-spline interpolation.

Surface points corresponding to each branch were generated by placing discrete circular contours along the centerline, with constant spacing. These contours were locally orthogonal to the centerline and their radii were calculated as follows. In each branch, refined local radii were inferred from the previously extracted cross-sectional boundaries. Thus obtained curve representing the radius evolution along the artery was smoothed by low-pass filtering. The resulting set of points representing the discrete circles along all branches, was used as input into a triangular-mesh generation process.

Assembling the Model. The overall model is the combination of all the anatomical and vascular structures (Figure 3(d)) that are perfectly superimposable onto the original MR data set (Figure 4(b)). Figure 3 illustrates the different steps of the reconstruction process. Two views of the resulting 3D model are shown in Figure 4 as compared with anatomical sheets.

2.2 Animating the Model

Overview. Motion estimation of the heart was achieved by non rigid image registration between frames in sequence of cine MR images (30 frames) acquired

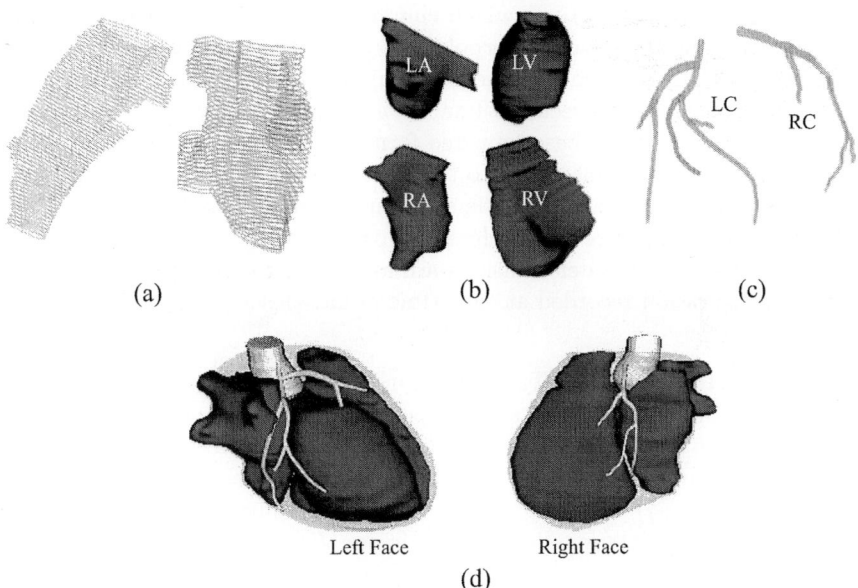

Fig. 3. Static model construction : (a) Extracted contours, (b) Reconstructed cardiac surfaces, (c) Reconstructed coronary network, (d) Assembled model

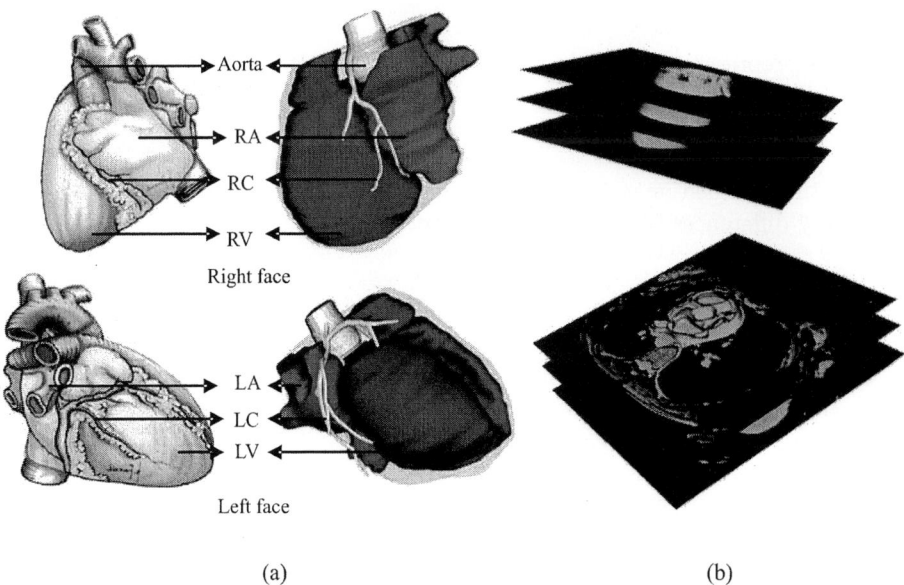

Fig. 4. (a) 3D static heart model as compared with anatomical sheets. (b) Views of the correspondence between transverse reference MR images and the 3D static model

on the same volunteer (Figure 2(b)(c)). Such a technique allows to recover global as well as local motion features which characterize the heart dynamics. Images at each time point are linearly interpolated to an isotropic 3D volume. A volume of interest (VOI) has been selected to focus on the heart area and to reduce the computing time. Consecutive image pairs are then matched using a non rigid registration method based on free form deformations (FFD) (see below). Deformation from one time frame to the next time frame is available through a set of grid point displacements. The static heart model has to be put into the same coordinate reference as for the short axis cine images. Once in the same coordinate system, FFD deformations are applied iteratively to the model and its new configuration recorded at each time point.

Non Rigid Registration. The 3D non-rigid registration technique is an iterative algorithm based on the spline-based free form deformation principle. A regular grid $\Phi(n_x \times n_y \times n_z)$ of control points $\Phi_{i,j,k}$ is overlaid to the floating volumetric image (which domain is denoted as $\Omega = \{(x,y,z)|0 \leq x \leq X, 0 \leq y \leq Y, 0 \leq z \leq Z\}$) and deformed until the minimum of an intensity based similarity criterion between the floating image B and a reference image A is reached. Equation (1) defines the 3D Free-Form deformation T_{FFD} used in the algorithm.

$$T_{FFD}(x,y,z) = \sum_{l=0}^{3}\sum_{m=0}^{3}\sum_{n=0}^{3} B_l(u)B_m(v)B_n(w)\phi_{i+l,j+m,k+n} \qquad (1)$$

where B_l represents the lth basis function of the B-spline, and $i = \lfloor x/n_x \rfloor - 1$, $j = \lfloor y/n_y \rfloor - 1$, $k = \lfloor z/n_z \rfloor - 1$, $u = x/n_x - \lfloor x/n_x \rfloor$, $v = y/n_y - \lfloor y/n_y \rfloor$, $w = z/n_z - \lfloor z/n_z \rfloor$, with $\lfloor \rfloor$ the floor operator [14].

As the matched images are of the same nature, we chose the sum of squared intensity differences (SSD) as the registration metric (equation 2).

$$SSD = \frac{1}{N}\sum |A(i) - B(i)|^2 \qquad (2)$$

where the intensities in the reference and transformed floating images are $A(i)$ and $B(i)$, respectively, and N is equal to the number of pairs of pixels in the intersection region of the two images. The minimum of the criterion is searched for using a gradient descent technique. The algorithm stops when the similarity stabilizes or when the maximum iteration number is reached. In practice, the parameters were fixed to the following values : the size of grid is $20 \times 20 \times 20$ pixels, the maximum number of iteration is 30, the maximum step length of the optimizer is 15 and VOI dimension is $128 \times 128 \times 70$ pixels. We designed the algorithm from software components of the ITK[3] library (Figure 5).

---

[3] http://www.itk.org

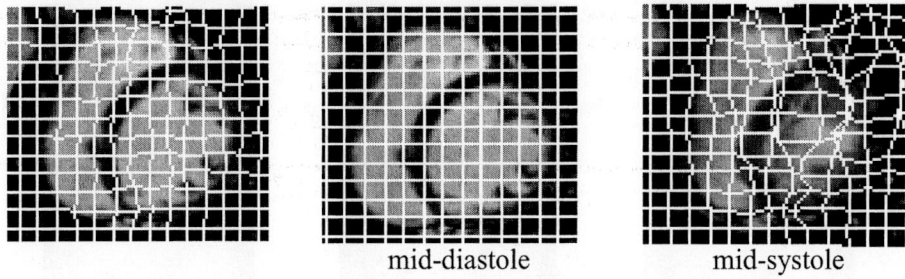
mid-diastole   mid-systole

Fig. 5. FFD deformed grid overlaid to the floating image in 2D

Model Animation. Non Rigid registration is repeated for each successive VOI pairs of the sequence. The result is a set of 30 steps of 3D control point displacements. The model animation consists in applying the retrieved FFD's to the static model. To this end, the model is transformed to the same coordinate system as the cine short axis images. This is performed using image position and orientation information available in image file headers, assuming that the subject has not moved between the two acquisitions. This has been usually checked. Then, the successive FFD warpings are applied to the reoriented static model to simulate the heart motion. Each deformed model configuration is recorded in a separate file. Figure (6) illustrates some of the computed displacement fields using the non rigid FFD based registration algorithm. One can observe the global extension of the LV myocardium close to the end systole and the opposite effect during diastole.

3 Discussion and Conclusion

In this paper, a novel 3D realistic beating model of the heart and main vessels is presented. A few heart models have been previously reported in the literature. Some of them issue from the biomechanical community [2, 17, 10] but the geometry of the heart structures are quite idealized and usually restricted to the LV. To our knowledge, the models of Segars and colleagues [15] and Wierzbicki and colleagues [18] are similar to the one proposed in this paper. The main differences relies on the fact that, in our model, all the imaging data, including heart and vessels anatomy and heart's motion, have been recorded from the same human patient, with the same MRI modality and during the same examination. This insures a very good consistency of the resulting model. Moreover, the model includes the native images that can be transformed according to the estimated motion model. Therefore, the overall model (structure surfaces + images + motion model) constitutes a unique basis for an accurate realistic anthropomorphic imaging model of the beating heart and vessels. Still, the present model has to be improved on a certain number of points. Long axis cine MR images should be accounted for in the motion estimation process to better approach the motion

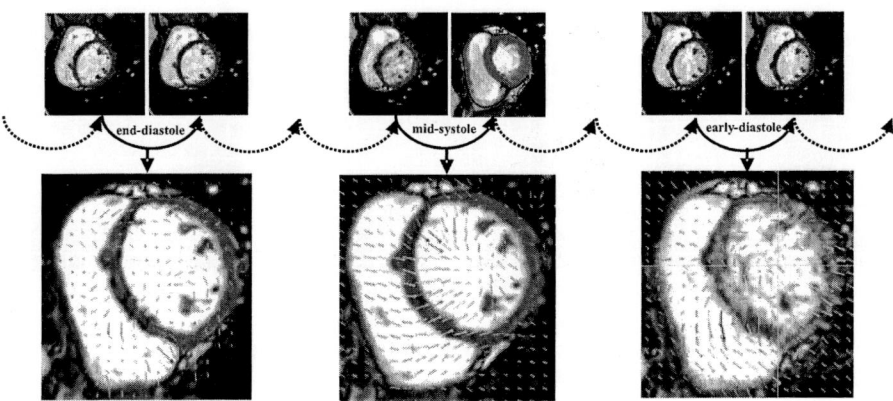

Fig. 6. 2D displays of some computed displacement fields from 3D non rigid registration

in true 3D. Then, the resulting motion model should be carefully evaluated. It is also envisaged to investigate the integration of motion estimations from MR tagging which is known as the reference for non invasive cardiac deformation assessment [1]. Simulating various heart dynamics could be achieved either by adequately tuning the model's parameters (control points) or processing additional examinations with healthy volunteers as well as pathological subjects. We then truly hope to deliver these realistic models for various purposes such as evaluation of cardiac image processing algorithms.

References

1. Amini A. and L.Prince J. *Computational Imaging and vision: Measurement of Cardiac Deformations from MRI: Physical and Mathematical Models*, volume 23. Kluwer Academic Publishers, Netherlands, 2001.
2. T. Arts, W.C. Hunter, A. Douglas, A.M.M. Muijtjens, and R. S. Reneman. Description of the deformation of the left ventricle by a kinematic model. *Journal of Biomechanics*, 25(10):1119–1127, 1992.
3. V. Chalana and K. Yongmin. A methodology for evaluation of boundary detection algorithms on medical images. *IEEE Transactions on Medical Imaging*, 16(5):642–652, 1997.
4. A. F. Frangi, W. J. Niessen, and M. A. Viergever. Three-dimensional modeling for functional analysis of cardiac images, a review. *IEEE Transactions on Medical Imaging*, 20(1):2–25, 2001.
5. A.F. Frangi, W.J. Niessen, and K.L. Vincken. Multiscale vessel enhancement filtering. In S. Delp W.M. Wells, A. Colchester, editor, *MICCAI'98 - Medical Image Computing and Computer-Assisted Intervention*, pages 130–137, Cambridge, MA, USA, 1998. Berlin : Springer.
6. P. Grangeat, A. Koenig, T. Rodet, and S. Bonnet. Theoretical framework for a dynamic cone-beam reconstruction algorithm based on a dynamic particle model. *Physics in Medicine and Biology*, 47:2611–2625, 2002.

7. M. Hernández-Hoyos, M. Orkisz, P. Puech, C. Mansard, and P.C. Douek. Computer-assisted analysis of three-dimensional angiograms. *RadioGraphics*, (22):421–436, 2002.
8. P. Jannin, J. M. Fitzpatrick, D. J. Hawkes, X. Pennec, R. Shahidi, and M.W. Vannier. Validation of medical image processing in image-guided therapy. *IEEE Transactions on Medical Imaging*, 21(12):1445–1449, 2002.
9. M. Lorenzo-Valdés, G. I. Sanchez-Ortiz, Andrew G. Elkington, Raad H. Mohiaddin, and D. Rueckert. Segmentation of 4D cardiac MR images using a probabilistic atlas and the EM algorithm. *Medical Image Analysis*, 8:255–265, 2004.
10. P.M.F. Nielsen, I.L. LeGrice, B.H. Smaill, and P.J. Hunter. Mathematical model of geometry and fibrous structure of the heart. *American Journal Physiology*, 260:1365–1378, 1991.
11. M. Orkisz and M. Hernández-Hoyos. From inertia matrix to the analysis of vascular pathologies. In *Int. Conf. Computer Vision and Graphics*, pages 6–15, Zakopane, Poland, 2002.
12. N. Pauna, P. Croisille, N. Costes, A. Reilhac, T. Mäkelä, O. Cozar, M. Janier, and P. Clarysse. A strategy to quantitatively evaluate MRI/PET cardiac rigid registration methods using a monte carlo simulator. In Springer, editor, *Functional Imaging and Modeling of the Heart (FIMH'03)*, volume 2674 of *Lecture Notes in Computer Science*, pages 194–204, Lyon, France, 2003.
13. Q-C. Pham, F. Vincent, P. Clarysse, P. Croisille, and I.E. Magnin. A FEM-based deformable model for the 3D segmentation and tracking of the heart in cardiac MRI. In *2nd International Symposium on Image and Signal Processing and Analysis (ISPA 2001)*, pages 250–254, Pula, Croatia, 2001.
14. D. Rueckert, L. I. Sonoda, C. Hayes, D. L. G. Hill, M. O. Leach, and D. J. Hawkes. Nonrigid registration using free-form deformations: Application to breast MR images. *IEEE Transactions on Medical Imaging*, 18(8):712–721, 1999.
15. W.-P. Segars, D.-S. Lalush, and B.-M.-W. Tsui. A realistic spline-based dynamic heart phantom. In *IEEE Nuclear Science Symposium and Medical Imaging Conference*, Seattle, Washington, USA, 1999.
16. Jasjit S. Suri. Computer vision, pattern recognition and image processing in left ventricle segmentation: the last 50 years. *Pattern Analysis and Applications*, 3:209–242, 2000.
17. E. Waks, J. L. Prince, and A. S. Douglas. Cardiac motion simulator for tagged MRI. In *IEEE Workshop Mathematical Methods in Biomedical Image Analysis*, pages 182–191, San Francisco, CA, USA, 1996.
18. M. Wierzbicki, M. Drangova, G. Guiraudon, and T. Peters. Validation of dynamic heart models obtained using non-linear registration for virtual reality training, planning, and guidance of minimally invasive cardiac surgeries. *Medical Image Analysis*, 8:387–401, 2004.

Multi-formalism Modelling of Cardiac Tissue

Antoine Defontaine, Alfredo Hernández, and Guy Carrault

LTSI - INSERM U642, Université de Rennes 1,
Campus de Beaulieu, Bât 22,
F-35042 Rennes Cedex, France
http://www.ltsi.univ-rennes1.fr

Abstract. Many models of the cardiovascular system (e.g. cardiac electrical activity, autonomous nervous system, ...) have been proposed for the last decades. Research is now focusing on the integration of these different models, in order to study more complicated physiopathological states in clinical applications context. To get round the practical limitations of existing models, multi-formalism modelling appears as a way to ease the integration of these different models together.

This paper presents an original methodology allowing to combine different types of description formalisms. This method has been applied to define a multi-formalism model of cardiac action potential propagation on a 2D grid of endocardial cells, combining cellular automata and a set of cells defined by the Beeler-Reuter model. Results, obtained under physiologic and ischemic conditions, highlight the improvements in term of computing compared with mono-formalism systems, while keeping the necessary explanatory strength for a practical clinical use.

1 Introduction

Cardiac modelling and simulation have been the subject of important research during the last three decades. Different models of the electrical activity have been proposed for the main types of cardiac myocites in normal or pathological conditions [1, 2]. These models are defined at different levels of detail (i.e. taking into account more or less independant ionic currents) and different formalisms (usually ordinary differential equations for cellular defined models and cellular automata for models defined at a wider scale).

Typically, individual models defined at a *same level* of detail and under the *same formalism* are coupled in the form of 1D, 2D or 3D objects to represent a given part of cardiac tissue, or to reproduce the whole cardiac anatomy. Applications range from the understanding of the cardiac function, in normal or pathologic conditions (e.g. ischemia [3]), to the assistance in the definition of new therapies [1]. However, none of the existing approaches allows a complete consideration of whole cardiac activity and, choice and compromise have to be done depending on the expected simulation.

After a short presentation of the main current cardiac modelling approaches, this paper proposes an original modelling and simulation method based on a generic multi-formalism approach. Relevant results obtained under normal or pathologic (ischemia) conditions, highlighting the interest of the method in terms of computational needs and clinical interpretation, are presented and discussed on the remaining parts.

2 Current Views of Cardiac Modelling Problem

Two different approaches can be identified in the definition of computational cardiac models, depending on the level of detail employed for their definition: whole cardiac models at a cellular level and complete heart models developed at the tissue or organ level [4, 5]. Both views still suffer from difficulties that reduce their clinical application: the former approach requires heavy computational resources while the later one is not able to reproduce certain pathologies defined at different scales. A hybrid approach combining the two previous types of description is now emerging.

2.1 Cellular Level

A number of cardiac models have been proposed at a cellular level [6, 7, 1]. In this type of approach, systems are defined by a network of many 'atomic' cells whose description is usually implemented by means of models representing different physiological aspects [8, 9, 10, 11].

In general, a system of such cells is defined as follows [12, 13, 5]:

$$\frac{dV_i}{dt} = G(P_C) + K \cdot \nabla^2 V \quad (1)$$

where V_i is the membrane potential of cell i, G is a function that depends on a set of parameters P_C, K is a diffusion coefficient and $\nabla^2 V$, the Laplacian of the membrane voltages of the neighbouring cells.

Usually, thousands of cells are coupled in a predefined geometry to represent one or more cavities of the heart. Due to this extensive definition, models defined at the cellular level require massive computing resources. Moreover, their coupling with other models remains tricky and even with high performance calculating resources, computational time limits their clinical application.

2.2 Tissue Level

Models developed at the tissue level are based on a coupled network of macrostructures, often using a cellular automata (CA) approach, which represent specific anatomical structures of the heart [14, 15, 16].

The state behaviour of each automaton of such an event-based approach can be defined by [17, 5]:

$$E = H(P_A) \quad (2)$$

where E is the state of the cellular automaton and H is the function governing internal state transitions, depending on parameters P_A. When a given macrostructure reaches the depolarisation state, neighbouring tissues are activated by the transmission of a flag (external state transition). Particular properties of cardiac cells, such as the dependance of the depolarisation slope to the stimulation frequency, have also been included in some CA models [17].

Due to their low computational costs, this kind of models has been used in different clinical setups. Although some major cardiac rhythms can be reproduced and explained by these models, some difficulties remain when dealing with complex rhythms and when simulating pathologies implying modifications at a cellular or molecular level such as myocardial ischemia [17]. These difficulties are inherent to the definition of the models at a macroscopic scale and, consequently, to the inability of considering a physiopathological process at a cellular level.

2.3 Multi-formalism Approach

In this context, one can easily think that a way to take advantage from the benefits of each approach would be to selectively define different regions of the modelled heart at different scale levels, depending on its physiological or pathological state. Such a consideration is also legitimated by the practical clinical diagnosis performed by the physician, which aims at refining progressively the investigated region of the heart, going from a global consideration of healthy parts to a precise analysis of pathological sources.

A similar problem of hybrid approach has been identified in other applications [18] and has led to specific researches and developments on multi-formalism modelling (DEVS++, AToM^3, Modelica, ...). This type of modelling consists in gathering components described in different ways (known as description formalisms), which can be basically summarised as discrete or continuous specifications. Although these approaches reveal efficient in traditional engineering fields, few works have been done on specific modelling of natural processes, including cardiac modelling.

This approach parallels recent works by Poole et al [19] which presented preliminary results of a multi-formalism approach on 1D segments of cardiac tissue, rather than an exhaustive use of supercalculation. They expanded the general theory of synchronous concurrent algorithms (SCA) which consists in unifying the different types of models on a global clock measuring discrete time. Indeed, the advantage of a multi-formalism method lies in the partial use of discrete models (cellular automata) that require less computational resources than corresponding continuous models (based on ODE definition). Nevertheless, Poole's work suffer from certain limitations: i) their use of SCA can be considered as a cosimulation approach [18] limiting the gains in term of computing time, ii) the way the different models are coupled is unclear and iii) used cellular automata lack of dynamical properties.

3 Proposed Methodology

The proposed approach tries to go deeper in the way of dealing with a multi-formalism definition. Based on Zeigler's work [20], our main goal has been to define a tool as generic as possible with ease of use in other fields than cardiology and ease of implementing new types of models or new simulation algorithms. The main difficulty associated with such a multi-formalism approach concerns the definition of a unique but generalisable coupling criteria, particularly at the interfaces between atomic models of different formalisms. Moreover, the proposition of such a method should be accompanied by a quantitative method to evaluate the differences between the multi-formalism approach and a monoformalism used as gold standard.

In our work, we have chosen to use a coupling function of the neighbouring potential as performed for continuous models [5]. In our concern for developing as generic as possible a system, using the same manner of coupling (same method in the tissue model) for all the types of tissues allows to define a unique standard coupling procedure. Adaptations of the methods will be done in each model definition as follows:

Let $C_{i,j,k}^{F}$ be an atomic cell component of a cardiac tissue, defined by a formalism F (where F can be continuous F_c or discrete F_d). The generic coupling behaviour can be extended from (1) as follows:

$$C_{i,j,k}^{F} = G_{F}(P) + Coup_{F}(K \cdot \nabla^2 V) \qquad (3)$$

where G_F is the function of parameters P, $Coup_F$ the coupling method and K as defined in (1). The coupling method can be defined as follows:

$$Coup_{F} = \begin{cases} thres & \text{if the cell model is discrete}(F = F_d) \\ id & \text{if the cell model is continuous}(F = F_c) \end{cases} \qquad (4)$$

where id is the identity function and $thres$ a threshold function setting external activation for the cellular automata if the input is greater than the limit value necessary for depolarisation of an equivalent continuous model (Fig. 1).

With this approach, the coupling between a set of cells of a tissue will always be defined by the generic definition (3), whatever their description formalisms are. This allows to keep into account the influence of the neighbouring cells during the whole activation. Consequently, a minimum of information will be lost during the propagation of depolarisation fronts, allowing not to alter the clinical interpretation. Each specification of the methods will be done for each model definition, in the sense of an object oriented approach.

4 Implementation Considerations

Traditional processing of the 'cable equation' (1) is usually done using a centralised approach (Fig. 3(a)) where the whole simulation is done at the same level and, usually, inside a unique simulation loop which can solve only one modelling formalism. An alternative to this approach, which is particularly adapted to the multi-formalism case, has been proposed by Zeigler [20]. It is based on a distributed simulator structure that parallels the model architecture (Fig. 3(b)). The introduction of coordinator objects grouping different sub-models, eases the use of a multi-formalism approach and can facilitate a parallel implementation of the simulator.

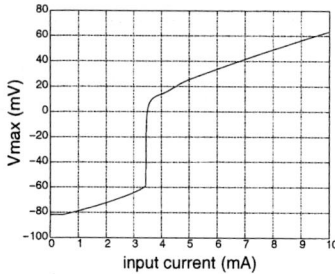

Fig. 1. Peak membrane voltages for a Beeler-Reuter model, as a function of a variable input current. The value retained for the activation threshold of cellular automata is $3.5\ mA$

Fig. 2. Output of the proposed CA model, representing a piece-wise linear fitting of the BR action potential

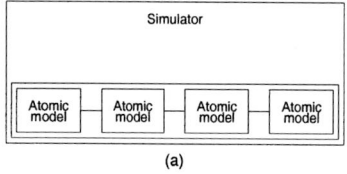
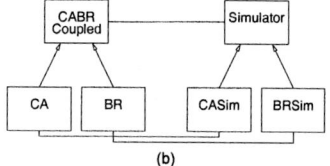

Fig. 3. Simulation approaches: a. classical mono-formalism centralised approach: the link between the different components and the whole simulation are done in a unique level; b. distributed multi-formalism approach: the coupled model represents the coordinator with an associated simulator

We have developed a generic library of modelling and simulation based on this architecture. It consists of a list of different types of models inherited from a standard mother class. By this definition, inherited classes can represent any combination of elements (atomic element, structure of atomic elements, complex structure mixing atomic elements and pre-existing structures, ...). Based on Zeigler's introduction of coordinator objects, simulation is done deepening the level of the considered structure up to reaching only atomics elements and specific simulators, adapted to each atomic model's formalism, are used. The simulation of the global system is performed at the coordinator level whereas each component is simulated at the model level.

5 Results

5.1 Experimental Settings

The proposed generic method has been implemented on a 256 x 256 square tissue of endocardial cells, corresponding to an average size of $10\ mm$ x $10\ mm$.

The coupling coefficient K has been set in order to maintain a conduction velocity of $0.5\ m.s^{-1}$ in the case of healthy tissues.

Different types of tissues have been simulated: mono-formalism healthy (discrete or continuous), multi-formalism healthy (discrete and continuous), mono-formalism ischemic and multi-formalism ischemic. In the case of multi-formalism tissues, peripheral cells are defined by a discrete approach while the 64×64 central cells are continuous.

Cellular models fall in the two previous categories:

- Continuous models: Beeler Reuter (BR) model [9] is used in the case of healthy tissues. Ischemic model used has been adapted from the Beeler Reuter model by Sahakian [21] to take into account membrane current modifications. A different coupling coefficient is also used with gradual transition from normal to pathologic cells [21, 3, 13].
- Discrete models: Cellular automata (CA) traditionally used are composed of main action potential states (i) idle, ii) rapid depolarisation, iii) absolute refractive period and iv) relative refractive period) but suffer from a lack of dynamical properties. Contrary to those static models, our automata possess two main dynamical properties: refractory period dependance to the stimulation frequency as well as the response to premature activations [17]. The CA output is defined by means of a piece-wise linear function fitting the Beeler Reuter action potential, where each linear segment is associated to a different state of the CA (Fig. 2).

5.2 Simulation Results

Depolarisation fronts obtained for healthy tissues are presented in Fig. 4(a-c). The differences between BR defined mono-formalism tissue and the two others are quantified in Fig. 4(d-e).

Depolarisation fronts are coherent with the known behaviour of electrical propagation on cardiac tissue. Slight differences that appear between BR tissue and the two others (Fig. 4(d-e)) are due to the atomic behaviour of each model and, especially, to the difference on the depolarisation slopes (Fig. 2). Even if the previous results are interesting only for validation purposes, it is important to note that the clinical interpretation won't be altered, from a qualitative point of view, for discrete or hybrid tissues, presenting the same mean propagation properties (i.e. conduction velocities). Moreover, experimental results show a reduction of the computing time in the case of multi-formalism approach compared to mono-formalism one, highlighting the clinical interest of such a method.

Such a multi-formalism approach takes sense when dealing with ischemic tissues. Results obtained for ischemic tissues are presented in Fig. 5(a-b) and the difference between mono and multi-formalism simulation are quantified in Fig. 5(c).

Even if the spatial depolarisation fronts simulated by these two apporaches are not strictly identical, similar behaviour can be observed in both tissues:

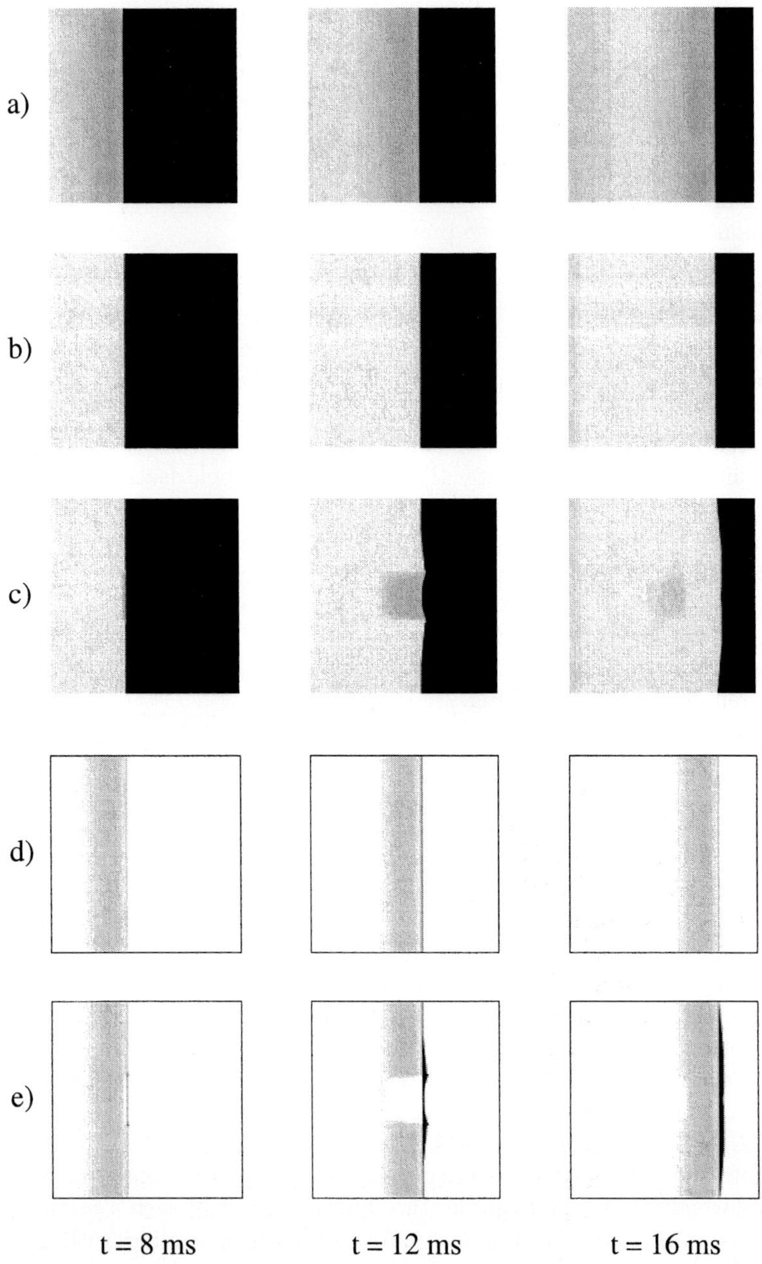

t = 8 ms　　　　t = 12 ms　　　　t = 16 ms

Fig. 4. Depolarisation fronts for healthy tissues: a. BR tissue, b. CA tissue, c. CABR tissue. Greyscale ranges from black for resting potential ($-84.5\ mV$) to white for a depolarised potential ($22\ mV$).
Differences in the depolarisation fronts for healthy tissues: a. between BR and CA, b. between BR and CABR. Greyscale ranges from white where no difference appears to black for the greatest differences

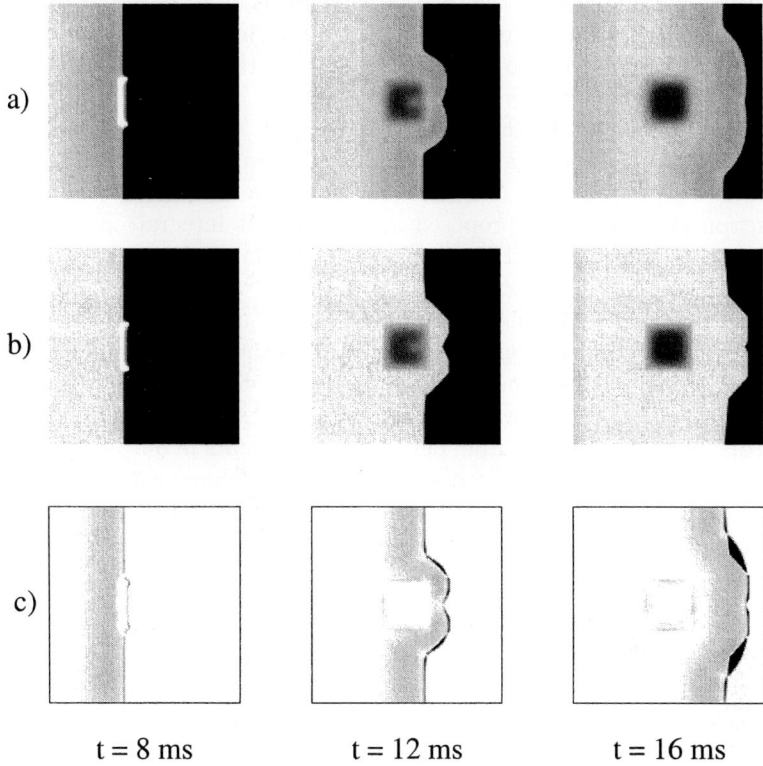

Fig. 5. Depolarisation fronts for ischemic tissues: a. BRIsch tissue, b. CAIsch tissue. c. Differences in the depolarisation fronts for ischemic tissues

- Alteration of the propagation front linked with the presence of ischemic area.
- Abnormal and incomplete depolarisation of pathologic cells (note the small difference in the *64* x *64* centre square of Fig. 5(c)): i) quicker depolarisation in the border of the ischemia, ii) depolarisation block at the center of the ischemic area, iii) modification of the depolarisation front in the shadow of the ischemia.
- Temporal correspondance of the simulated events.

These experimental results have brought out improvements in terms of computing time compared with mono-formalism systems, while keeping the necessary qualitative explanatory strength for a practical clinical use.

6 Conclusion and Perspectives

An original simulation method based on a multi-formalism approach has been presented in this paper. Depolarisation fronts obtained from healthy or ischemic tissues have highlighted that the qualitative clinical interpretations are not altered despite the partial use of simpler description models (CA). Experiments have also shown a reduction of the computing time that would ease a practical use of such models.

One great advantage of this method is the possibility of simulating physiopathological states even in a hybrid approach. Compared to other projects in the same field [19], which use Aliev-Panfilov model (a morphological description of the action potential), the proposed method lets us integrate physiologically detailed models, such as the BR model, while still minimizing the global computing expenses.

Current development consist in improvements of our simulation library: i) integration of more detailed models such as Luo and Rudy [10, 11] and ii) improvement of our CA simulation method, by taking into account their specific state transition properties. Future work will deal with the extension of these results to a 3D cardiac volume obtained from current developments in our laboratory [22]. It will be based on a multi-scale description of the volume combined with the presented multi-formalism approach, with the constant aim of not altering the clinical interpretations by being able to reproduce the important physiological markers of cardiac electrical activity.

Acknowledgements

This work has been partly supported by the ECOS-NORD cooperation program, action number V03S03.

References

1. Noble, D.: Modeling the heart – from genes to cells to the whole organ. Science. **295** (2003) 1678–1682
2. Bardou, A.L., Auger, P.M., Birkui, P.J. and Chasse, J.L.: Modeling of cardiac electrophysiological mechanisms: from action potential genesis to its propagation in myocardium. Critical Review of Biomedical Engineering. **24** (1996) 141–221
3. Wilders, R., Verheijck, E.E., Joyner, R.W., Golod, D.A., Kumar, R., van Ginneken, A.C.G., Bouman, L.N. and Jongsma H.J.: Effects of Ischemia on Discontinuous Action Potential Conduction in Hybrid Pairs of Ventricular Cells. Circulation. **99** (1999) 1623–1629
4. Trudel, M.-C., Dubé, B., Potse, M., Gulrajani, R.M. and Leon, L.J.: Simulation of QRST Integral Maps With a Membrane-Based Computer Heart Model Employing Parallel Processing. IEEE Transactions on Biomedical Engineering. **51** (2004) 1319–1329
5. Defontaine, A., Hernández, A. and Carrault, G.: Modelling and Simulation: Application to Cardiac Modelling. Acta Biotheoretica. **52** (2004) 273–290
6. Porman, J.B.: A Modular Simulation System for the Bidomain Equations. Department of Electrical and Computer Engineering, Duke University. (1999)
7. McCulloch, A. Bassingthwaighte, J., Hunter, P. and Noble, D.: Computational bilogy of the heart: from structure to function. Progess in Biophysics and Molecular Biology. **69** (1998) 153–155

8. Aliev, R.R. and Panfilov, A.V.: A Simple Two-variable Model of Cardiac Excitation. Chaos, Solitons and fractals. **7** (1996) 293–301
9. Beeler, G.W. and Reuter, H.: Reconstruction of the action potential of ventricular myocardial fibres. Journal Of Physiology. **268** (1977) 177–210
10. Luo, C.H. and Rudy, Y.: A model of the ventricular cardiac action potential. Depolarization, repolarization, and their interaction. Circulation Research. **68** (1991) 1501–1526
11. Luo, C.H. and Rudy, Y.: A dynamic model of the cardiac ventricular action potential (I & II). Circulation Research. **74** (1994) 1071–1113
12. Keener, J. and Sneyd, J.: Mathematical Physiology. Springer-Verlag. (1998)
13. Clayton, R.H., Parkinson, K. and Holden, A.V.: Re-entry in computational models of ischaemic myocardium. Chaos, Solitons and Fractals. **13** (2002) 1671–1683
14. Malik, M., Cochrane, T., Davies, D.W. and Camm, A.J.: Clinically relevant computer model of cardiac rhythm and pacemaker/heart interaction. Medical and Biological Engineering and Computing. **25** (1987) 504–512
15. Ahlfeldt, H., Tanaka, H., Nygards, M.E., Furukawa, T. and Wigertz, O.: Computer simulation of cardiac pacing. Pacing Clinical Electrophysiology. **11** (1988) 174–184
16. Virag, N., Vesin, J.M. and Kappenberger, L.: A computer model of cardiac electrical activity for the simulation of arrhythmias. Pacing Clinical Electrophysiology. **21** (1998) 2366–2371
17. Hernández, A.I., Carrault, G. and Mora, F.: Model-based interpretation of cardiac beats by evolutionary algorithms: signal and model interaction. Artificial Intelligence in Medicine. **26** (2002) 211–235
18. de Lara, J. and Vangheluwe, H.: Computer aided multi-paradigm modelling to process petri-nets and statecharts. In International Conference on Graph Transformations (ICGT), Lecture Notes in Computer Science. **2505** (2002) 239–253
19. Poole, M.J., Holden, A.V. and Tucker, J.V.: Hierarchical reconstructions of cardiac tissue. Chaos, Solitons and Fractals. **13** (2002) 1581–1612
20. Zeigler, B.P., Praehofer, H. and Kim, T.G.: Theory of Modeling and Simulation, Second Edition, Integrating Discrete Event and Continuous Complex Dynamic Systems. Academic Press. (2000)
21. Sahakian, A.V., Myers, G.A. and Maglaveras, N.: Unideirectional Block in Cardiac Fibers: Effects of Discontinuities in Coupling Resistance and Spatial Changes in Resting Membrane Potential in a Computer Simulation Study. IEEE Transactions on Biomedical Engineering. **39** (1992) 510–522
22. Garreau, M., Simon, A., Boulmier, D. and Guillaume H.: Cardiac Motion Extraction in Multislice Computed Tomography by using a 3D Hierarchical Surface Matching process. IEEE Computers in Cardiology Conference, Chicago, USA. (2004)

Analysis of Tagged Cardiac MRI Sequences

Aymeric Histace[1], Christine Cavaro-Ménard[1], Vincent Courboulay[2], and Michel Ménard[2]

[1] LISA, Université d'Angers, 62 avenue Notre Dame du Lac 49000 Angers
aymeric.histace@univ-angers.fr
[2] L3i, Université de La Rochelle, L3i, Pôle science et technologie, 17 000 La Rochelle
vincent.courboulay@univ-lr.fr

Abstract. The non invasive evaluation of the cardiac function presents a great interest for the diagnosis of cardiovascular diseases. Tagged cardiac MRI allows the measurement of anatomical and functional myocardial parameters. This protocol generates a dark grid which is deformed with the myocardium displacement on both Short-Axis (SA) and Long-Axis (LA) frames in a time sequence. Tracking the grid allows the estimation of the displacement inside the myocardium. The work described in this paper aims to make the automatic tracking of the grid of tags on cardiac MRI sequences robust and reliable, thanks to an informational formalism based on Extreme Physical Informational (EPI). This approach leads to the development of an original diffusion pre-processing allowing us to increase significantly the robustness of the detection and the follow-up of the grid of tags.

1 Introduction

The non invasive assessment of the cardiac function is of major interest for the diagnosis and the treatment of cardiovascular pathologies. Whereas classical cardiac MRI only allows to measure anatomical and functional parameters of the myocardium (mass, volume...) tagged cardiac MRI makes the evaluation of the intra-myocardial displacement possible. For instance, this type of information can lead to a precise characterization of the myocardium viability after an infarction. Moreover, data concerning myocardium viability allows to decide of the therapeutic : medical treatment, angiopathy, or coronary surgery and to follow the amelioration of the ventricular function after reperfusion.

The SPAMM (Space Modulation of Magnetization) acquisition protocol [22] we used for the tagging of MRI data, displays a deformable 45°-oriented dark grid which describes the contraction of myocardium (Figure 1) on the images of temporal Short-Axis (SA) and Long-Axis (LA) sequences. The 3D+T follow-up of this grid makes the evaluation of the intra-myocardial displacement possible.

Nevertheless, tagged cardiac images present particular characteristics which make their analysis difficult. More precisely, images are of low contrast compared with classical MRI, and their resolution is only of approximately one centimeter.

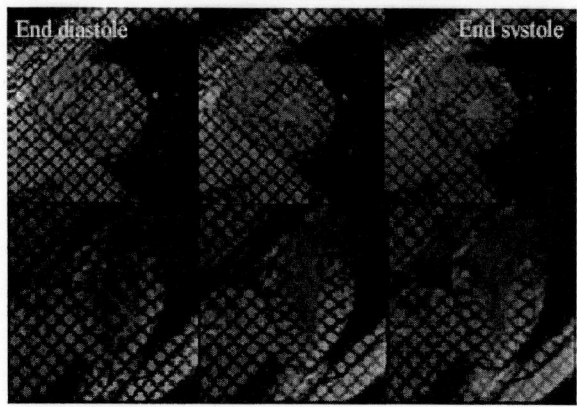

Fig. 1. SA and LA tagged MRI of the Left Ventricle

Numerous studies were carried out concerning the analysis of the deformations of the grid of tag on both SA and LA sequences: These methods can be divided into two major families:

- direct estimation of the displacement field of the myocardium (optical flow [3], analysis of the Harmonic Images [13, 7], image registration [15]);
- undirect estimation of the displacement field (active contours [11, 21, 6, 1, 2], use of the spectral information [23, 5]).

The common disadvantages of those approaches are their sensibility to noise and to the fading of the grids of tags, their poor adaptation when tags are close to myocardial boundaries and their bad adaptation to important deformations of the grid between two consecutive instants. Moreover, manual interventions are often needed to obtain precise results and execution time can sometimes reach high values [10].

Moreover, the clinical validation of the different methods often shows a lack of robustness and reproducibility which is incompatible with a medical application.

In order to avoid these problems, we propose in this article, an original method for the detection and the follow-up of the grid of tags, based on active contours and image diffusion. More particularly, we will show that the integration of an adapted external energy in a simple contour active model allows to avoid the usual problems encountered by the different technics presented in literature. We will also show that our approach allows to obtain precise and robust results of detection.

We present in a first part the principle of the detection and follow-up method, to continue with the description of our diffusion process based on a recent theory developed in [4] called Extreme Physical Information (EPI). In a second part, we present the application of the resulting diffusion process to our particular topic. In a last part, we present results of detection and follow-up of the grid of tags on tagged cardiac MRI showing the robustness of the technic, and illustrating it with examples of quantification and representation of the extracted data.

2 Method

2.1 Principle

The general principle of the technic we present here has been already described in a past article [9] : To detect and to follow-up the grid of tags on SA and LA sequences, we deform a virtual grid, modeled by B-splines and controlled by 44 nods P (the intersections of the grid), each one characterized by a particular energy, noted E, to minimize (Eq. (1)):

$$E = w_{internal} \cdot E_{internal} + w_{external} \cdot E_{external} \qquad (1)$$

The internal energy imposes the regularity of the whole grid to obtain thus a coherent result. For our application, we chose the weighted sum of two terms defined by [18] :

- the energy E_{esp}^{int} ensures a regular spacing between each intersection point (i,j) of the grid of tags.

$$E_{esp}^{int} = \sum_{i,j} \left[\left(\frac{1 - r_1^2(i,j)}{1 + r_1^2(i,j)} \right)^2 + \left(\frac{1 - r_2^2(i,j)}{1 + r_2^2(i,j)} \right)^2 \right] \qquad (2)$$

where $r_k(i,j)$ is the ratio among the distances which separate the intersection point (i,j) and its two related intersections in the k direction (k=1 for 45° and k=2 for 135°). We can note that this expression is equal to zero when $r_1(i,j) = r_2(i,j) = 1$ for all (i,j), i.e. when intersection points are regularly spaced.

- the energy E_{align}^{int} ensures the alignement of the related intersection points on each lines of the grid of splines :

$$E_{align}^{int} = \sum_{i,j} \left[\cos^2 \left(\frac{\theta_1(i,j)}{2} \right) + \cos^2 \left(\frac{\theta_2(i,j)}{2} \right) \right] \qquad (3)$$

where $\theta_k(i,j)$ is the angle of the intersection point (i,j) and its two related intersection points in the k direction (k=1 for 45° and k=2 for 135°). We can note once again that this expression is equal to zero when $\theta_1(i,j) = \theta_2(i,j) = 180°$ for all (i,j), i.e. when two intersection points on a same line are aligned.

Concerning $E_{external}$, this term takes into account the tag information of the studied tagged MR image. Numerous proposition have been made since 1988 to extract the tag information (gradient of the image, extraction in the Fourier's domain), but no one of them appears to be really adapted to the problematic in terms of robustness regarding small variations of $w_{internal}$ and $w_{external}$.

As a consequence, it appears that an original definition of $E_{external}$ is essential for the implementation of a robust detection method of the grid of tags.

Thus, we propose to increase this robustness thanks to a method based on the enhancement of the tag information using a diffusion process which takes into account *a priori* data characterizing the grid.

2.2 Anisotropic Diffusion

Regarding the existing fundamental anisotropic diffusion method presented in the literature [14,19], it appears that their leading differential equations were not adapted to our application because of the impossibility to take into account particular characteristics of the information to enhance.

This is confirmed by the tests presented Figure (2) where we can see that, whereas an optimal parameterization of the diffusion process is implemented, the grid of tags is too much altered even for a small number of iterations.

To integrate in the diffusion process the local orientations of the grid of tags, which will allow to preserve it from alteration, an enrichment of the fundamental diffusion equation (*i.e.* the heat equation (Eq. (4)), can be seen as a solution.

$$\begin{cases} \psi(r,0) = \psi_0(r) \\ \frac{\partial \psi}{\partial t} = \triangle u = div(\nabla \psi) \end{cases} \quad (4)$$

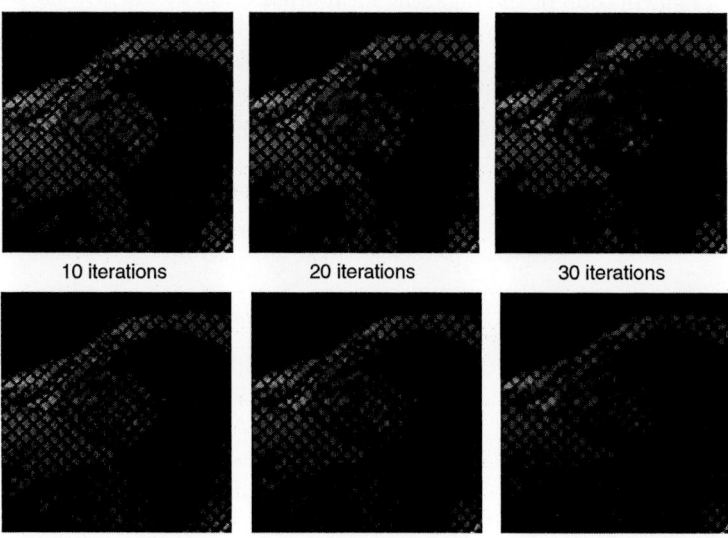

Fig. 2. Top sequence: Perona-Malik's Diffusion $dt = 0, 2$, bottom sequence: Weickert's Diffusion $dt = 0, 2$

Thus, we propose to introduce in equation (4) a new parameter noted **A** which is a potential vector :

$$\frac{\partial \psi}{\partial t} = (\nabla - \mathbf{A}).(\nabla - \mathbf{A})\psi \quad . \tag{5}$$

This integration allows to take into account particular properties of the structure to be enhanced through a judicious choice for **A**.

2.3 About A

As we have just said, the **A** potential allows to control the diffusion process and introduce some prior knowledge about the image evolution.

The choice we do for **A** is based on the fact that equation (5) allows to weight the diffusion process with the difference of orientation between the local gradient and **A**.

To explain the way we implement **A**, let us consider Figure (3):

We can notice on this Figure that when angle θ is null (*i.e.* **A** and $\nabla \psi$ are colinear), the studied pixel will not be diffused. Thus, a precise local estimation of this angle can lead us to preserve particular patterns in the processed image for a given vector **A**.

Thus, a solution to the problem of enhancement of the grid would have been to impose particular orientations for **A**, considering the fact that the gradients to preserve are well known and correspond to the orientations of the grid-of-tag ones (45°, 135°, 225°, 315°). However, because the contraction of the LV induces a deformation of the tags, the local orientation of the grid for an instant of acquisition different from the initial one, can be no more characterized by imposed particular orientations. Moreover, because of the poor quality of MRI sequences, it appears that a calculation of the local orientation of **A** directly made on cardiac tagged images, would be strongly deteriorated by noise.

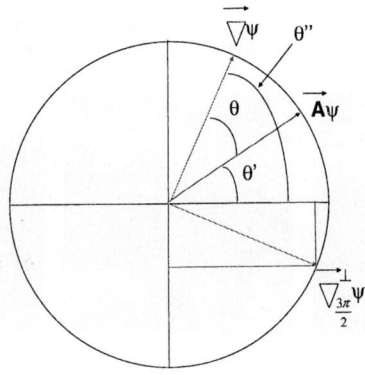

Fig. 3. Local geometrical implementation of **A** in terms of the local gradient $\nabla \psi$

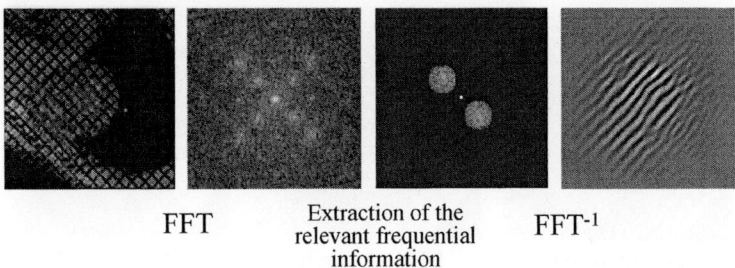

Fig. 4. Extraction of the tag information in the Fourier's representation

As a consequence, we propose to make a local estimation of the direction of **A** in the Fourier area, using the principle developed by Zhang *and al* [23] (Figure 4).

This approach allows a denoising of the tag information which leads us to a more precise estimation of **A** and allows to take into account the deformations of the grid due to the contraction of the LV. Moreover, in order to compute a precise estimation of θ, we propose a method for its calculation based on the work of Rao [16] and Terebes [17] using the analysis of the eigen vectors of a particular neighborhood of the studied pixel.

3 Results

The result presented in Figure (5.a), shows the restoration of the 45°-oriented tag on the first image of a tagged cardiac sequence by the diffusion approach.

As we can see in Figure 5.a, the diffusion process makes possible the fading of noisy artifacts, and non-45°-oriented lines.

Moreover, because the orientation of **A** is locally calculated taking into account a particular neighborhood, the diffusion process remains efficient even if the tag is locally deformed due to myocardial contraction (Figure 5.b).

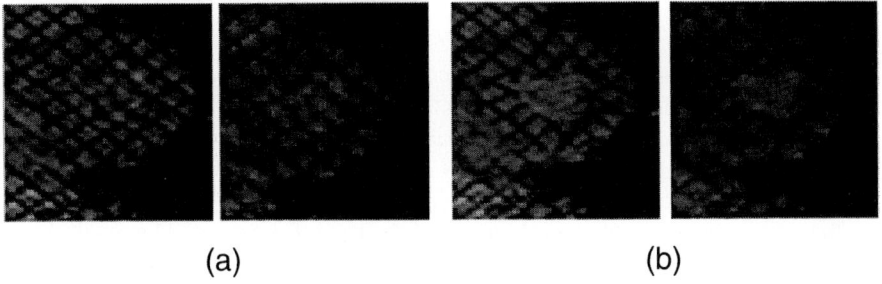

(a) (b)

Fig. 5. Preservation of the 45°-oriented tag on (a) the initial image of a tagged sequence and (b) for $t \neq t_0$

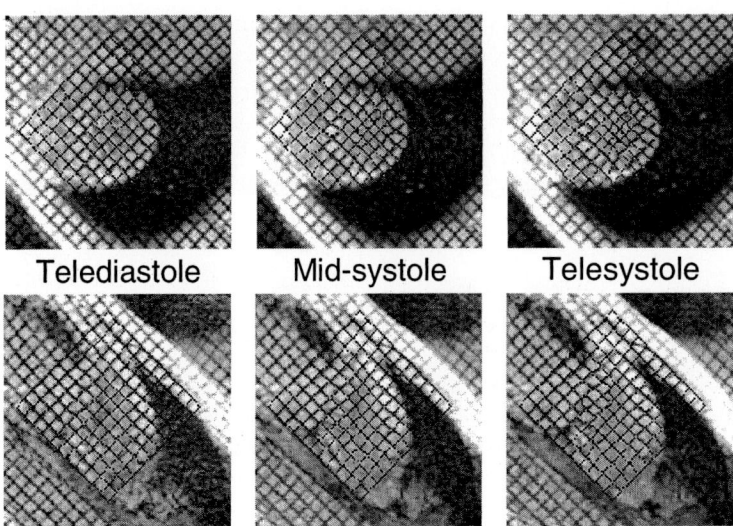

Fig. 6. Detection and following of the grid of tags on both SA and LA sequences

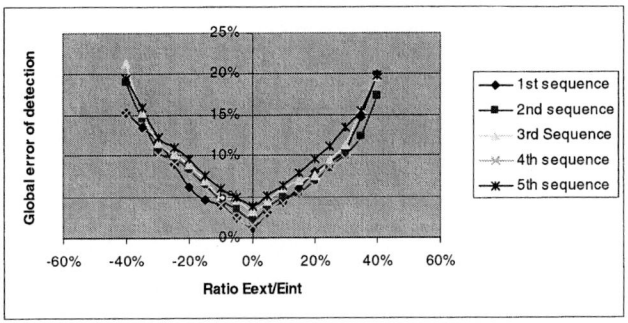

Fig. 7. Variation of the made error (expressed in number of false detected pixels) in terms of the ratio $\frac{w_{external}}{w_{internal}}$

The different preserved images (45° and 135° ones) are then integrated as a new external energy in our active contour model for the detection and the following of the grid of tags as follows :

Each intersection point of the initial grid represents a node for which the global energy E is computed on a $N \times N$ neighborhood. A research for the minimum of E on the considered neighborhood allows to displace the studied node to a new position in accordance with the tag information. The resulting grid obtained for a particular instant t is used for the initialisation of the detection at the successive instant $t+1$ time. The $N \times N$ neighborhood has been empirically fixed to 5. For $N = 3$ the research window does not contain enough information. The value $N = 7$ gives the same results as those obtained for $N = 5$, but increase the calculation time. The same minimization of energy presented in [9] being implemented, the results presented Figure (6) are then obtained.

These results are characterized by a good precision compared with usual methods, but also by a good robustness regarding the necessary weighting of $E_{external}$ and $E_{internal}$. Indeed, a quantitative study of the variation of the committed error (expressed in number of false detected pixel) in terms of the ratio $\frac{w_{external}}{w_{internal}}$ shows that a 20% variation of it does not alter significantly the precision of the detected grid.

In addition, the detection has been tested on 10 different sequences without changing any parameters and the obtained results have been judged satisfying by medical experts on all images.

4 Quantification

In order to make a first validation of the method, we have also quantified classical cardiac parameters on the 10 studied sequences as radial, circumferential, longitudinal displacements, torsion or deformations. We present in Tab.1 a comparison between our obtained results for the quantification of the radial displacements and two studied of the medical literature.

Table 1. Comparison between our quantification and two studies of the medical literature concerning the estimation of the radial displacements (expressed in millimeters) for healthy volunteers

	Base	Mdian	Apex
[20] (12 patients)	5.9 ± 0.4	6 ± 0.3	4.65 ± 0.2
[12] (31 patients)	5.0 ± 1.3	4.3 ± 1.1	4.2 ± 1.6
Our estimation (10 patients)	5.7 ± 0.5	4.9 ± 0.7	4.3 ± 0.9

5 Conclusion

The method presented in this article, based on both active contours and diffusion process, finally allows to (i) smooth the image with a preservation of the tag patterns, (ii) to ensure the robustness and the precision of the grid detection and (iii) to completely automate the detection and follow-up process.

Moreover, by associating the detection method with an original automatic detection of the myocardial boundaries (epical and endocardial ones) [8], it is possible to realize a two-dimensional temporal map (according the recommandations of the American Hospital Association) characterizing the local displacements and local deformations of the myocardium (Figure 8).

The results presented are very interesting for radiologists to evaluate torsion, shearing, longitudinal and radial displacements of the LV and then to draw early diagnoses of particular cardiopathies.

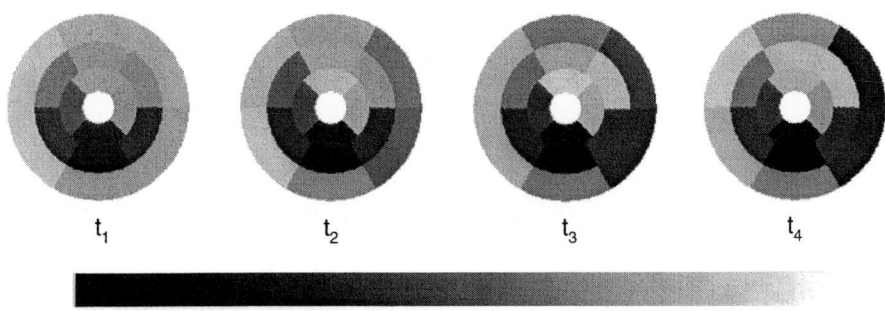

←Contraction-Dilatation→

Fig. 8. Radial contraction of the heart

References

1. A.A. Amini, Y. Chen, R.W. Curwen, V. Mani, and J. Sun. Coupled B-snake grids and constrained thin-plate splines for analysis of 2D tissue deformations from tagged MRI. *IEEE Transaction on Medical Imaging*, 17(3):344–356, 1998.
2. T. Denney. Estimation and detection of myocardial tags in MR images without user-defined myocardial contours. *IEEE Transactions on Medical Imaging*, 18(4):330–344, 1999.
3. L. Dougherty, J. Asmuth, A. Blom, L. Axel, and R. Kumar. Validation of an optical flow method for tag displacement estimation. *IEEE Transactions on Medical Imaging*, 18(4):359–363, 1999.
4. B.R. Frieden. *Physics from Fisher Information*. Cambridge University Press, 1998.
5. M. Groot-Koerkamp, G. Snoep, A. Muijtjens, and G. Kemerink. Improving contrast and tracking of tags in cardiac magnetic resonance images. *Magnetic Resonance in Medicine*, 41:973–982, 1999.
6. M. Guttman, E. Zerhouni, and E. McVeigh. Analysis of cardiac function from MR images. *IEEE Computer Graphics and Applications*, 17(1):30–38, 1997.
7. I. Haber, R. Kikinis, and C.F. Westin. Phase-driven finite elements model for spatio-temporal tracking in tagged mri. In *Proceedings of Fourth International Conference On Medical Image Computing and Computer Assisted Intervention (MICCAI'01)*, pages 1352–1353, 2001.
8. A. Histace, C. Cavaro-Mnard, and B. Vigouroux. Tagged cardiac mri : Detection of myocardial boudaries by texture analysis. In *ICIP 2003*, volume 2, pages 1061–1064, September 2003.
9. A. Histace, L. Hermand, and C. Cavaro-Mnard. Tagged cardiac mr images analysis. In *Biosignal 2002*, pages 313–315, June 2002.
10. Z. Hu, D. Metaxas, and L. Axel. In vivo strain and stress estimation of the heart left and right ventricles from MRI images. *Medical Image Analysis*, 7:435–444, 2003.
11. S. Kumar and D. Goldgof. Automatic tracking of SPAMM grid and the estimation of deformation parameters from cardiac MR images. *IEEE Transactions on Medical Imaging*, 13(1):122–132, 1994.

12. C. Moore, C. Lugo-Olivieri, E. McVeigh, and E. Zerhouni. Three dimensional systolic strain patterns in the normal human left ventricle: Characterization with tagged mr imaging. *Radiology*, 214(2):453–466, 2000.
13. N.F. Osman, E.R. Mc Veigh, and J.L. Prince. Imaging heart motion using Harmonic Phased MRI (HARP). *IEEE Transactions on Medical Imaging*, 19:186–202, 2000.
14. P. Perona and J. Malik. Scale-space and edge detection using anistropic diffusion. *IEEE Transcations on Pattern Analysis and Machine Intelligence*, 12(7):629–639, 1990.
15. C. Petitjean, N. Rougon, F. Prêteux, Ph. Cluzel, and Ph. Grenier. A non rigid registration approach for measuring myocardial contraction in tagged mri using exclusive f-information. In *Proceedings International Conference on Image and Signal Processing (ICISP'2003), Agadir, Morocco*, 25-27 June 2003.
16. A. Rao and R.C. Jain. Computerized flow field analysis: Oriented texture fields. *Transactions on pattern analysis and machine intelligence*, 14(7), 1992.
17. O. Baylou P. Borda M. Terebes, R. Lavialle. Mixed anisotropic diffusion. In *Proceedings of the 16th International Conference on Pattern Recognition*, volume 3, pages 1051–4651, 2002.
18. S. Urayama, T. Matsuda, N. Sugimoto, S. Mizuta, N. Yamada, and C. Uyama. Detailed motion analysis of the left ventricular myocardium using an MR tagging method with a dense grid. *Magnetic Resonance in Medicine*, 44(73-82), 2000.
19. J. Weickert. *Anisotropic Diffusion in image processing*. Teubner-Verlag, Stuttgart, 1998.
20. A. Young, C. Kramer, V. Ferrari, L. Axel, and N. Reichek. Three-dimensional left ventricular deformation in hypertrophic cardiomyopathy. *Circulation*, 90(854-867), 1994.
21. A.A. Young, D.L. Kraitchmann, L. Dougherty, and L. Axel. Tracking an finite element analysis of stripe deformation in magnetic resonance tagging. *IEEE Transactions on Medical Imaging*, 14(3):413–421, 1995.
22. E. Zerhouni, D. Parish, W. Rogers, A. Yang, and E. Shapiro. Human heart : tagging with MR imaging - a method for noninvasive assessment of myocardial motion. *Radiology*, 169(1):59–63, 1988.
23. S. Zhang, M. Douglas, L. Yaroslavsky, R. Summers, V. Dilsizian, L. Fananapazir, and S. Bacharach. A fourier based algorithm for tracking SPAMM tags in gated magnetic resonance cardiac images. *Medical Physics*, 32(8):1359–1369, 1996.

Fast Spatio-temporal Free-Form Registration of Cardiac MR Image Sequences

Dimitrios Perperidis[1], Raad Mohiaddin[2], and Daniel Rueckert[1]

[1] Visual Information Processing Group,
Department of Computing, Imperial College, London,
180 Queen's Gate, London SW7 2BZ, United Kingdom
[2] Royal Brompton and Harefield NHS Trust,
Sydney Street, London, United Kingdom

Abstract. In this paper we present a novel approach to the problem of spatio-temporal alignment of cardiac MR image sequences. This novel method has the ability to correct spatial misalignment caused by global acquisition and local shape differences as well as temporal misalignment caused by differences in the length of the cardiac cycles or by differences in the motion patterns of the hearts. In contrast to our previous approach [1], the algorithm optimizes the spatial and temporal transformation components separately, thus significantly speeding up the registration process. To achieve this we have developed a novel approach for the calculation of the temporal registration which does not require the spatial alignment of the image sequences. The method was evaluated using fifteen cardiac MR image sequences from healthy volunteers and the results were compared to the previously presented method. The results indicate that the performance of the method is similar to the previously presented method [1] while the the computational complexity has been significantly reduced.

1 Introduction

Cardiovascular diseases are a very important cause of death in the developed world [2]. Their early diagnosis and treatment is crucial in order to reduce mortality and to improve patient's quality of life. MR imaging plays an increasingly important role for the high resolution imaging of the cardiovascular system since it allows the acquisition of 4D cardiac image sequences which describe the cardiac anatomy and function.

The recent advantages in the development of cardiac imaging modalities have led to an increased need for cardiac registration methods (for recent reviews of cardiac image registration methods see [3] and for a general review of image registration methods see [4]). In general, cardiac image registration is a very complex problem due to the complicated non-rigid motion of the heart and the thorax as well as the anisotropic resolution with which cardiac images are usually acquired. In the recent years cardiac image registration has emerged as an important tool for a large number of applications. It has a fundamental role in the construction of anatomical atlases of the heart [5, 6]. It has also been employed for the analysis of the myocardial motion [7] and for the segmentation of cardiac images [8]. Image registration has been also used for the fusion of information from a number of different modalities such as CT, MR, PET, and SPECT [9, 10]. In addition, cardiac image registration is crucial for the comparison of images of the same subject, e.g. before and after pharmacological treatment or surgical intervention. Furthermore, inter-subject alignment of cardiac image sequences to the same coordinate space (anatomical reference) enables direct comparisons between the cardiac anatomy and function of different subjects to be made.

While a large number of registration techniques exist for cardiac imaging, most of these techniques focus on 3D images ignoring any temporal misalignment between the two image sequences. In an earlier publication, [1], we developed a novel approach for the spatio-temporal alignment of cardiac MR image sequences. It uses a deformable spatio-temporal transformation which has been decoupled into temporal and spatial components. This method will not only bring a number of sequences of cardiac images acquired from different subjects or the same subject (for example short and long axis cardiac image sequences) into the same spatial coordinate system but also into the same temporal coordinate system. This allows direct comparison between both the cardiac anatomy of different subjects and the cardiac function to be made. The registration approach corrects temporal misalignment caused by different acquisition parameters, different length of cardiac cycles and different motion patterns of the hearts. It also corrects any spatial misalignment caused by global shape differences in the image sequences and local shape differences.

Due to the coupling of the spatial and temporal registration, the computational complexity of the previously presented spatio-temporal deformable registration method, [1], is very high. In this paper we present a novel approach for the spatio-temporal deformable registration of cardiac MR image sequences addressing the issues regarding computational complexity. The approach is based on the same transformation model as the one in [1]. However, it optimizes each transformation component separately. In particular, the temporal registration algorithm does not require the image sequences to be spatially aligned in order to calculate the optimal temporal registration.

2 Spatio-temporal Registration

Since the heart is undergoing a spatially and temporally varying degree motion during the cardiac cycle, 4D cardiac image registration algorithms are required when registering two cardiac MR image sequences. Spatial alignment of corresponding frames of the image sequences (e.g. the second frame of one image sequence with the second frame of the other) is not sufficient since these frames may not correspond to the same temporal position in the cardiac cycle of the hearts. This is due to differences in the acquisition parameters (the initial offset or trigger delay in the acquisition of the first time frame and different frequency in the acquisition of consecutive time frames), differences in the length of cardiac cycles (e.g. one cardiac cycle maybe longer than the other) and differences in the dynamic properties of the hearts (e.g. one heart may have a longer contraction phase and shorter relaxation phase). Spatio-temporal alignment will enable comparison between corresponding anatomical positions and corresponding positions in the cardiac cycle of the hearts. It will also resolve spatial ambiguities which occur when there is not sufficient common appearance in the two 3D MR cardiac images.

A 4D cardiac image sequence can be represented as a sequence of n 3D images $I_k(x, y, z)$ with a fixed field of view Ω_I and an acquisition time t_k with $t_k < t_{k+1}$, in the temporal direction. The resulting image sequence can be viewed as a 4D image $I(x, y, z, t)$ defined on the spatio-temporal domain $\Omega_I \times [t_1, t_n]$. The goal of 4D image registration described in this paper is to relate each point of one image sequence to its corresponding point of the reference image sequence. In this case the transformation $\mathbf{T} : (x, y, z, t) \rightarrow (x', y', z', t')$ maps any point of one image sequence $I(x, y, z, t)$ into its corresponding point in the reference image sequence $I(x', y', z', t')$. The mapping used in this paper is of the following form:

$$\mathbf{T}(x, y, z, t) = (x'(x, y, z), y'(x, y, z), z'(x, y, z), t'(t)) \qquad (1)$$

and can be of a subvoxel displacement in the spatial domain and of a sub-frame displacement in the temporal domain. The 4D mapping can be resolved into decoupled spatial and temporal components $\mathbf{T}_{spatial}$ and $\mathbf{T}_{temporal}$ respectively where

$$\mathbf{T}_{spatial}(x, y, z) = (x'(x, y, z), y'(x, y, z), z'(x, y, z)), \mathbf{T}_{temporal}(t) = t'(t)$$

One consequence of this decoupling is that each temporal frame t in image sequence I will map to another temporal frame t' in image sequence I', ensuring causality and preventing different regions in a 3D image $I_t(x, y, z)$ from being warped differently in the temporal direction by $\mathbf{T}_{temporal}$.

2.1 Spatial Alignment

The aim of the spatial part of the transformation is to relate each spatial point of an image to a point of the reference image, i.e. $\mathbf{T}_{spatial} : (x, y, z) \rightarrow (x', y', z')$ maps any point (x, y, z) of a particular time frame t in one image sequence into its corresponding point (x', y', z') of another particular time frame t' of the reference image sequence. The transformation $\mathbf{T}_{spatial}$ consists of a global transformation and a local transformation:

$$\mathbf{T}_{spatial}(x, y, z) = \mathbf{T}^{global}_{spatial}(x, y, z) + \mathbf{T}^{local}_{spatial}(x, y, z) \qquad (2)$$

The global transformation addresses differences in the size, orientation and alignment of the hearts while the local part addresses differences in the shape of the hearts. An affine transformation with 12 degrees of freedom utilizing scaling and shearing in addition to translation and rotation is used as $\mathbf{T}^{global}_{spatial}$.

A free-form deformation (FFD) model based on B-splines is used in order to describe the differences in the local shape of the hearts. To define a spline-based FFD we denote the spatial domain of the image volume as $\Omega_I = \{(x, y, z) \mid 0 \leq x < X, 0 \leq y < Y, 0 \leq z < Z\}$. Let Φ denote a $n_x \times n_y \times n_z$ mesh of control points $\phi_{i,j,k}$ with uniform spacing δ. Then, the FFD can be written as the 3D tensor product of the familiar 1D cubic B-splines [11]:

$$\mathbf{T}^{local}_{spatial}(x, y, z) = \sum_{l=0}^{3} \sum_{m=0}^{3} \sum_{n=0}^{3} B_l(u) B_m(v) B_n(w) \phi_{i+l, j+m, k+n} \qquad (3)$$

where $i = \lfloor \frac{x}{n_x} \rfloor - 1, j = \lfloor \frac{y}{n_y} \rfloor - 1, k = \lfloor \frac{z}{n_z} \rfloor - 1, u = \frac{x}{n_x} - \lfloor \frac{x}{n_x} \rfloor, v = \frac{y}{n_y} - \lfloor \frac{y}{n_y} \rfloor, w = \frac{z}{n_z} - \lfloor \frac{z}{n_z} \rfloor$ and where B_l represents the l-th basis function of the B-spline. One advantage of B-Splines is that they are locally controlled which makes them computationally efficient even for a large number of control points. In particular, the basis functions of cubic B-Splines have a limited support, i.e. changing a control point affects the transformation only in the local neighborhood of that control point.

2.2 Temporal Alignment

The temporal part of the transformation consists of a temporal global part, $\mathbf{T}_{temporal}^{global}$, and a temporal local part, $\mathbf{T}_{temporal}^{local}$:

$$\mathbf{T}_{temporal}(t) = \mathbf{T}_{temporal}^{global}(t') + \mathbf{T}_{temporal}^{local}(t')$$

$\mathbf{T}_{temporal}^{global}$ is an affine transformation which corrects for differences in the length of the cardiac cycles and differences in the acquisition parameters. $\mathbf{T}_{temporal}^{local}$ is modeled by a free-form deformation using a 1D B-spline and corrects for temporal misalignment caused by different cardiac dynamic properties (difference in the length of contraction and relaxation phases, different motion patterns, etc). To define a spline-based temporal free-form deformation we denote the temporal domain of the image sequence as $\Omega_t = \{(t) \mid 0 \leq x < T\}$. Let Φ denote a set of n_t control points ϕ_t with a temporal spacing δ_t. Then, the temporal free-form deformation can be defined as a 1D cubic B-spline:

$$\mathbf{T}_{temporal}^{local}(t) = \sum_{l=0}^{3} B_l(u)\phi_{t_{i+l}} \tag{4}$$

where $i = \lfloor \frac{t}{n_t} \rfloor - 1, u = \frac{t}{n_t} - \lfloor \frac{t}{n_t} \rfloor$ and B_l represents the l-th basis function of the B-spline.

$\mathbf{T}_{temporal}^{local}$ deforms the temporal characteristics of each image sequence in order to follow the same motion pattern with the reference image sequence. The combined 4D transformation model (eq. 1) is the spatio-temporal free-form deformation (STFFD)[1].

2.3 Separate Optimization of the Spatial and Temporal Component

The computational complexity of the previously presented spatio-temporal deformable registration method, [1], is very high. We can reduce the computational complexity by optimizing each transformation component (the temporal and the spatial one) separately. We first optimize the temporal component, $\mathbf{T}_{temporal}$, of the transformation $\mathbf{T}$ and then the spatial component, $\mathbf{T}_{spatial}$. In this method the calculation of the temporal transformation does not require the hearts to be aligned in the spatial domain.

Optimization of the Temporal Component. The global temporal component, $\mathbf{T}_{temporal}^{global}$, is calculated by aligning the start and end frames of the image sequences while the local temporal component, $\mathbf{T}_{temporal}^{local}$, is a temporal free-form deformation (eq. 4) which aligns temporal feature points which have been detected. In particular,

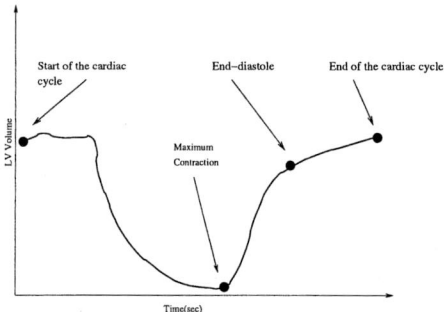

Fig. 1. The temporal feature points used for the temporal alignment between two image sequences

$T_{temporal}^{local}$ aligns the beginning of the cardiac cycles, the position of maximum contraction, the end diastolic time point of the left ventricle and the end of the cardiac cycles (as in figure 1). In order to detect these temporal positions in each image sequence we calculate the normalised cross-correlation coefficient between each frame with the first frame:

$$CC = \frac{\sum_x \sum_y \sum_z (I_0(x,y,z) - \bar{I}_0) \cdot (I_i(x,y,z) - \bar{I}_i)}{\sqrt{\sum_x \sum_y \sum_z (I_0(x,y,z) - \bar{I}_0)^2} \cdot \sqrt{\sum_x \sum_y \sum_z (I_i(x,y,z) - \bar{I}_i)^2}} \quad (5)$$

were I_0 is the first frame, $\bar{I}_0$ the mean intensity of the first frame, I_i each frame and $\bar{I}_i$ the mean intensity of each frame.

The idea behind this approach is that during the contraction phase of the cardiac cycle, in which the volume and shape of the heart changes significantly, each time frame will be less similar to the end-diastolic time frame (first frame of the sequence) and during the relaxation phase of the cardiac cycle each time frame will be more and more similar to the end-diastolic time frame. The end-systolic image, in which the heart has reached its maximum contraction, should have the highest degree of dissimilarity with the first image since the heart has different shape and size due to the contraction.

Figure 2 (a) shows the plot of the calculated normalised cross-correlation and its second derivative for a particular subject and 2 (b) shows the volume of the left ventricle of the same subject over time. We can see the similarity of the cross-correlation and the volume curves. The maximum contraction position is found by the minimum cross-correlation value. In order to find the end-diastolic position we need to find the minimum value of the second derivative after the location of the maximum contraction 2 (b). The second derivative is calculated using the finite differences method.

Optimization of the Spatial Component. The optimal spatial transformation, $T_{spatial}$, is calculated using non-rigid 3D registration (equation 2) of the first frames of the image sequences [11]. $T_{spatial}^{global}$ is an affine transformation correcting translation, rotation, shearing and scaling differences between the end-systolic time frames, while $T_{spatial}^{local}$ is a free-form deformation (eq. 3) deforming each sequence's end-systolic

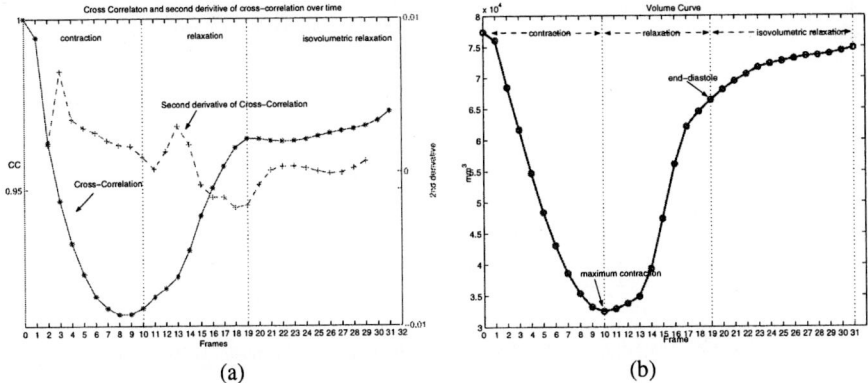

Fig. 2. The cross-correlation and the second derivative of the cross-correlation between the first frame and each consecutively frame (a) and the volume of the left ventricle of the same subject over time (b)

time frame to map the reference sequence's end-systolic time frame. Both $T^{global}_{spatial}$ and $T^{local}_{spatial}$ are optimized using normalised mutual information [12] based on the intensity histogram of the spatial domain of overlap of the two images.

3 Results

To evaluate the spatio-temporal deformable registration algorithm we have acquired fifteen cardiac MR image sequences from healthy volunteers. All image sequences used for our experiments were acquired on a Siemens Sonata 1.5 T scanner using TrueFisp pulse sequence. For the reference subject 32 different time frames were acquired (cardiac cycle of length 950msec). Each 3D image of the sequence had a resolution of $256 \times 192 \times 46$ with a pixel size of $0.97mm \times 0.97mm$ and a slice thickness of 3mm. Fourteen 4D cardiac MR images were registered to the reference subject. The length of the cardiac cycle of these images sequences varied from 300msec to 800msec. An initial estimate of the global spatial transformation was provided due to the large variety in the position and orientation of the hearts. Since all the image sequences contained almost entire cardiac cycles, the global temporal transformation was calculated in order to compensate the differences in length of the cardiac cycles of the subjects (by matching the first and the last frames of the image sequences). The spacing of the control points of the local transformation were 10mm in the spatial domain and 90msec in the temporal domain.

Figure 3 (a) shows the volume curves of the left ventricle after the optimization of the spatio-temporal global transformation and 3 (b) after optimization of the spatio-temporal local transformation by optimizing the transformation components separately. The volume of the left ventricles were calculated after segmenting the images using the EM-algorithm developed by Lorenzo-Valdés et al. [8]. We can clearly see that with the

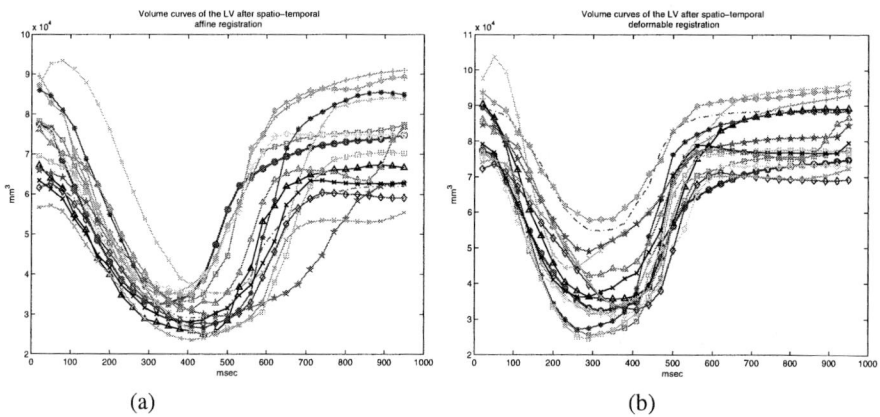

Fig. 3. The volume curves of the left ventricle for all subjects after optimisation of the global spatio spatio-temporal transformation (a) and after local spatio-temporal registration using the fast registration method (c)

introduction of the deformable components the hearts are significantly better aligned in the temporal domain.

Figure 4 provides an example of the spatio-temporal free-form registration. The images in the top row (a-c) are the short-axis, the long-axis and the temporal views of a frame in the middle of the image sequence (in the temporal views the vertical direction corresponds to time) after the optimization of the global transformations (affine spatio-temporal registration). The lines in the images represent the contours of the reference image sequence. The images in the middle row of figure 4 are the same images after spatio-temporal free-form registration by combined optimization of the transformation components [1]. The images in the bottom row of figure 4 are the same images after spatio-temporal free-form registration by optimizing the transformation components separately. We can clearly see with the introduction of the deformable temporal and spatial transformation there is a significant improvement in the alignment of the image sequences both in the spatial and in the temporal domain.

We calculated the error in the estimation of the maximum contraction and end-diastolic positions in the cardiac cycle by manually determining the time frames in which the maximum contraction and the end-diastole appears in each image sequence and comparing them with positions identified by the algorithm. The mean error in the detection of maximum contraction is 1.2 frames while the mean error of the end-diastole detection is 0.93 frames. Furthermore, The quality of the registration in the spatial domain was measured by calculating the volume overlap for the left and right ventricles as well as for the myocardium. The volume overlap for an object O is defined as:

$$\Delta(T, S) = \frac{2 \times |T \cap S|}{|T| + |S|} \times 100\% \qquad (6)$$

Here T denotes the voxels in the reference (target) image part of object O and S denotes the voxels in the other image part of object O. We have also calculated the mean

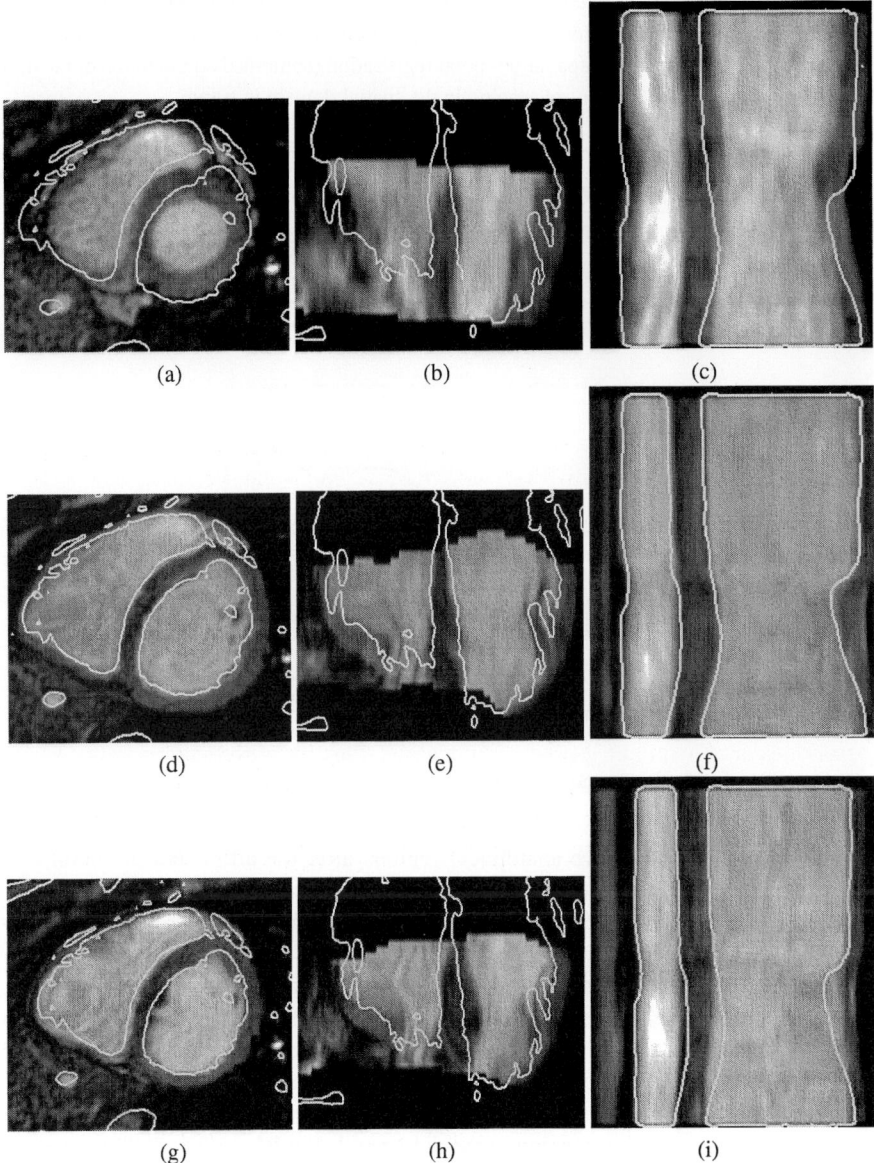

Fig. 4. Results of the 4D cardiac MR registration algorithm. The figure shows the short axis, the long axis and the temporal views after (a-c) affine alignment, (d-f) after the spatio-temporal free-form registration using combined optimization of the transformation components [1], and (g-i) after the spatio-temporal free-form registration using separate optimization of the transformation components. Animations of the registrations can be found at http://www.doc.ic.ac.uk/~dp1/Conferences/FIMH05/

Table 1. The mean volume overlap and surface distance after affine spatio-temporal registration, after spatio-temporal free-form registration by combined optimization of the transformation components [1] and after affine spatial-temporal registration (combined optimization of the transformation components). The control spacing in the spatial domain is 10mm

Anatomical region	Volume Overlap			Surface Distance in mm		
	Affine	Combined	Separate	Affine	Combined	Separate
Left ventricle	76.16%	85.57%	82.38%	4.16	2.96	3.41
Right ventricle	77.39%	84.67%	83.56%	4.95	3.60	3.93
Myocardium	70.57%	73.18%	71.62%	4.77	4.16	4.21

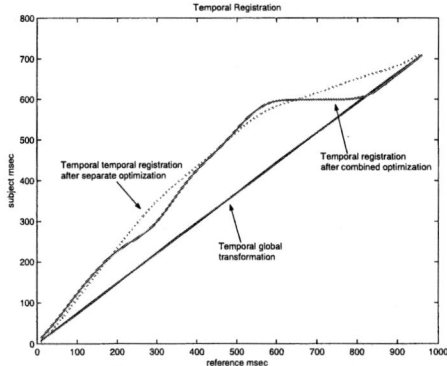

Fig. 5. Temporal alignment between two image sequences

surface distance of the above anatomical regions after the affine and the deformable 4D registration. The mean surface distance has been measured by comparing for each subject the distance between each surface point and the closest point on the surface of the reference subject.

Table 1 shows the mean overlap and the mean surface distance (in mm) for each anatomical region after affine spatio-temporal (similar to the previously presented spatio-temporal affine registration method [13]), after spatio-temporal free-form registration (simultaneous optimization of the transformation components) and after spatio-temporal free-form registration (separate optimization of the transformation components). We can clearly see that the introduction of the deformable models significantly improve the overlap of the anatomical features.

The results reported in table 1 and in figures 3 and 4 indicate that calculating the temporal transformation separately and using the combined approximation approach provides similar results. Obviously, the separate optimization of the temporal transformation aligns a limited number of positions in the cardiac cycle, while the combined optimization provides a better temporal alignment between these cardiac positions. This can been seen in figure 5, in the volume curves of figure 3 and in the temporal views of figure 4. However, the computational complexity of the combined optimization approach is very high. A typical combined optimization takes more than 24 hours while a typical registration using this method takes around an hour on a modern PC. Furthermore, an advantage of this method is that we can generate a temporal alignment between two image sequences without having to perform image registration. There is no need for the image sequences to be registered in the spatial domain in order to calculate the temporal alignment.

4 Conclusions

We have presented an alternative approach to our earlier spatio-temporal free-form registration for cardiac MR image sequences [1]. This contribution addresses the issues related to the computational complexity of the previously presented method. The fast spatio-temporal free-form registration method uses the same transformation model based on B-Splines as the previously presented method. However, this approach calculates the optimal spatial and temporal components of the transformation separately. It first calculates the temporal component and then the spatial component. For the calculation of the temporal component we used a novel approach based on normalised cross-correlation. An additional advantage of this method is that the temporal alignment can be calculated without the sequences to be registered in the spatial domain. The method has been evaluated using fifteen cardiac MR image sequences and the results have been compared to the previous registration method [1]. The experiments indicated performance of this method is similar to the performance of the previously presented method. The combined optimization of the transformation components results to marginally better spatio-temporal registrations. However, the computational complexity of this method is significantly lower than the previous method.

References

1. D. Perperidis, R. Mohiaddin, and R. Rueckert. Spatio-temporal free-form registration of cardiac MR image sequences. In *7th International Conference on medical image computing and computer assisted intervention (MICCAI 04)*, Lecture Notes In Computer Science. Springer, 2004.
2. American Heart Association. Heart and stroke statistical update. http://www.americanheart.org/, 2002.
3. T. Mäkelä, P. Clarysse, N. Sipila, O.and Pauna, and Q. C. Pham. A review of cardiac image registration methods. *IEEE Transactions on Medical Imaging*, 21(9), 2002.
4. B. Zitová and J. Flusser. Image registration methods: a survey. *Image and Vision Computing*, 21:977–1000, 2003.
5. D. Perperidis, M. Lorenzo-Valdés, R. Chandrashekara, A. Rao, R. Mohiaddin, G.I Sanchez-Ortiz, and D. Rueckert. Building a 4D atlas of the cardiac anatomy and motion using MR imaging. In *2004 IEEE International Symposium on Biomedical Imaging: From Nano to Macro*, 2004.
6. A.F. Frangi, D. Rueckert, J. A. Schnabel, and W.J. Niessen. Automatic construction of multiple-object three-dimensional statistical shape models: Application to cardiac modeling. *IEEE Transaction on Medical Imaging*, 21(9):1151–1165, 2002.

7. R. Chandrashekara, Mohiaddin R., and Rueckert D. Analysis of 3D myocardial motion in tagged MR images using nonrigid image registration. *IEEE Transactions on Medical Imaging*, 3(10):1245–1250, 2004.
8. M. Lorenzo-Valdés, G. I Sanchez-Ortiz, A. Elkington, Mohiaddin R., and D. Rueckert. Segmentation of 4D cardiac MR images using a probabilistic atlas and the EM algorithm. *Medical Image Analysis*, 8(3):255–265, 2004.
9. M.C. Gilardi, G. Rizzo, A. Savi, C. Landoni, V. Bettinardi, C. Rosseti, G. Striano, and F. Fazio. Correlation of SPECT and PET cardiac images by a surface matching registration technique. *Computerized Medical Imaging and Graphics*, 22(5):391–398, 1998.
10. T. G. Turkington, T. R. DeGradom, M. W. Hanson, and E. R. Coleman. Alignment of dynamic cardiac PET images for correction of motion. *IEEE Transaction on Nuclear Science*, 44(2):235–242, 1997.
11. D. Rueckert, L. I. Sonoda, C. Hayes, D.L.G Hill, M.O. Leach, and D.J Hawkes. Non-rigid registration using free-form deformations: Application to breast MR images. *IEEE Transactions on Medical Imaging*, 18(8):712–721, 1999.
12. C. Studholme. *Measures of 3D Medical Image Alignment*. PhD thesis, United Medcial and Dental Scools of Guy's and St. Thomas Hospitals,, 1997.
13. D. Perperidis, A. Rao, M. Lorenzo-Valdés, R. Mohiaddin, and D. Rueckert. Spatio-temporal alignment of 4D cardiac MR images. In *Lecture Notes in Computer Science: Functional Imaging and Modeling of the heart, FIMH'03*, Lyon, France, June 5-6, 2003. Springer.

Comparison of Cardiac Motion Fields from Tagged and Untagged MR Images Using Nonrigid Registration

Raghavendra Chandrashekara[1], Raad H. Mohiaddin[2], and Daniel Rueckert[1]

[1] Visual Information Processing Group, Department of Computing,
Imperial College, London, SW7 2BZ, UK
{rc3, dr}@doc.ic.ac.uk
[2] Cardiovascular Magnetic Resonance Unit,
Royal Brompton and Harefield NHS Trust,
Sydney Street, London, SW3 6NP, UK
R.Mohiaddin@rbh.nthames.nhs.uk

Abstract. This paper presents a comparison of the motion fields computed from TrueFISP untagged and SPAMM tagged magnetic resonance (MR) images using a 4D nonrigid registration algorithm that we have developed for cardiac motion tracking [3]. Our results, which were obtained from a group of 7 normal volunteers, indicate that although there is a good correlation between the motion fields computed from the tagged and untagged MR images, some of the twisting motion is not captured in the motion fields derived from the TrueFISP MR images.

1 Introduction

Magnetic resonance (MR) tagging [2] has been widely used as a means of measuring local deformation fields in the myocardium of the left ventricle (LV) which are useful indicators of cardiovascular diseases. Although a number of different techniques have been developed including B-spline surface models [6, 8, 1], optical flow [5], and harmonic phase (HARP) MRI [7] there is still no agreed gold standard. The main difficulties encountered are the need to estimate through-plane motion as well as the process of tag fading.

We have recently proposed an algorithm [3] based on a 4D B-spline motion model and nonrigid image registration as a means of tracking the motion of the myocardium in the left ventricle (LV) from tagged magnetic resonance (MR) images. The 4D motion field was reconstructed by registering sequences of short-axis (SA) and long-axis (LA) images of the LV simultaneously to the corresponding set of segmented images of the myocardium taken at end-diastole. The motion field obtained from the registration algorithm was spatially and temporally smooth and allowed displacements and velocities to be computed at arbitrary time instants between end-diastole and end-systole; and because the registration algorithm makes no assumptions about the type of tag pattern in the images acquired it may even be possible to compute motion fields from untagged

MR images. Such a possibility would mean that segmentation and motion field analysis could be done using a single set of images which would have important implications for patient diagnosis. Since there are no clear landmarks or features in the myocardium of the LV it is possible that not all of the deformation can be measured in the untagged MR images. Clearly, it is important to determine what types of motion can be captured from the untagged MR images. So, in this paper we present a comparison of the motion fields computed from TrueFISP and SPAMM tagged MR images using nonrigid image registration.

This paper is organized as follows: In section 2 we describe the registration algorithm developed in [3]. This algorithm is then used to compute the motion fields in the myocardium of the left ventricle (LV) from a group of normal volunteers. Two sets of images were acquired during the same scanning session for each volunteer, TrueFISP untagged and SPAMM tagged MR images. The motion fields computed for each volunteer using the two different sets of images are compared in section 3. Finally, in section 4 we make conclusions on the work presented in this paper.

2 Method

To track the motion of the heart we need to find a transformation, $\mathbf{T}(\mathbf{x}, t)$, which describes how a particular material point, $\mathbf{x} = [x, y, z]^T$, in the myocardium moves over time. Since the tag pattern used in a cardiac examination is a two-dimensional one, deformations can only be measured parallel to the imaging planes. So, it is usual to acquire image volumes with tag-planes applied in different directions to measure the deformation of the heart in all three directions. Suppose we acquire two volume image sequences:

$$S = \{S_0, S_1, \ldots, S_{T-1}\} \quad (1)$$
$$L = \{L_0, L_1, \ldots, L_{T-1}\} \quad (2)$$

where S_t is the image volume taken at time t with the tag-planes applied perpendicular to the short-axis image planes, L_t is the image volume taken at time t with the tag-planes applied perpendicular to the long-axis planes, and T is the total number of time frames.

The method proposed in [3] was to pose the problem of tracking the motion of the heart as a 4D registration problem in which the image sequences S and L are registered to the following image sequences respectively:

$$S' = \{S'_0, S'_1, \ldots, S'_{T-1}\}, \text{where } \forall t, S'_t = S_0 \quad (3)$$
$$L' = \{L'_0, L'_1, \ldots, L'_{T-1}\}, \text{where } \forall t, L'_t = L_0 \quad (4)$$

By registering the image sequences S and L to S' and L' respectively we can relate all points in the myocardium at $t = 0$ to their corresponding positions for all other times and thus reconstruct the deformation field in the myocardium.

The coordinate system, X, used to register the image sequences and hence perform the motion tracking is defined with respect to the reference image, S_0, taken at end-diastole ($t = 0$). The x-, y- and z-axes of X correspond to the x-, y-, and z-axes of the reference image and the transformation we use to perform the image sequence registration, $\mathbf{T}(\mathbf{x}, t)$, is defined by a 4D linear combination of the cubic B-spline basis functions,

$$\mathbf{T}(\mathbf{x},t) = \mathbf{x} + \sum_{k=0}^{3}\sum_{l=0}^{3}\sum_{m=0}^{3}\sum_{n=0}^{3} B_k(s)B_l(u)B_m(v)B_n(w)\phi_{g+k,h+l,i+m,j+n} \quad (5)$$

where the B_i's are the cubic B-spline basis functions and the ϕ's are the $n_x n_y n_z n_t$ three dimensional displacement vectors defined at the control point positions of the 4D B-spline. $s, u, v, w \in [0, 1]$, and g, h, i, j are given by

$$g = \lfloor x/\delta_x \rfloor - 1, s = x - g + 1 \quad (6)$$
$$h = \lfloor y/\delta_y \rfloor - 1, u = y - h + 1 \quad (7)$$
$$i = \lfloor z/\delta_z \rfloor - 1, v = z - i + 1 \quad (8)$$
$$j = \lfloor t/\delta_t \rfloor - 1, w = t - j + 1 \quad (9)$$

where $\delta_x, \delta_y, \delta_z$, and δ_t are the spacings of the control points in the x-, y-, z- and t-directions respectively. This is similar to the approach taken by Huang et al [6] except that in our case we are using nonrigid image registration to perform the motion tracking.

The image sequences S and L are registered to S' and L' by finding the optimal set of displacement vectors, $\phi_{g,h,i,j}$, such that the cost function

$$\mathcal{C}_{[1,T-1]} = \sum_{t=1}^{T-1} \mathcal{C}_t \quad (10)$$

is maximized, where

$$\mathcal{C}_t = w_S \frac{H(S_0) + H(\mathbf{T}_t(S_t))}{H(S_0, \mathbf{T}_t(S_t))} + w_L \frac{H(L_0) + H(\mathbf{T}_t(L_t))}{H(L_0, \mathbf{T}_t(L_t))} \quad (11)$$

is the normalized mutual information between the SA and LA images being registered. Here $H(A)$ represents the marginal entropy of the intensity distribution in image A, $H(A, B)$ is the joint entropy of the intensity distributions in images A, and B, and $\mathbf{T}_t(A)$ represents the image A after it has been transformed into the coordinate system X by $\mathbf{T}(\mathbf{x}, t)$. w_S and w_L are weighting factors which depend on the number of voxels in the myocardium in the end-diastolic time frame

$$w_S = \frac{N(S_0)}{N(S_0) + N(L_0)} \quad (12)$$

$$w_L = \frac{N(L_0)}{N(S_0) + N(L_0)} \quad (13)$$

where $N(I)$ is equal to the number of voxels in the myocardium in image I.

Initially, the displacement vectors, $\phi_{g,h,i,j}$ are set to $[0,0,0]^T$ for all g, h, i, and j, so that $\mathbf{T}(\mathbf{x},t) = \mathbf{x}$ for all $\mathbf{x}$ and t. The $\phi_{g,h,i,j}$ are then optimized by considering each subset:

$$\Phi_j = \{\phi_{g,h,i,j}\}_{g=0,h=0,i=0}^{n_x,n_y,n_z} \tag{14}$$

of displacement vectors in turn. Φ_j contains the displacement vectors defined at one particular point in time but at different positions in space. Since changing the displacement vectors in Φ_j only affects the computation of the similarity metric at those time frames which are within the region of support of the B-spline functions, we need only compute the similarity metric between those time frames. The Φ_j are then optimized by using a gradient descent optimization procedure so that $\mathcal{C}_{[t_a,t_b]}$ is maximized, where t_a and t_b are the minimum and maximum time frames between which the similarity metric in equation 10 is affected. Before optimizing Φ_{j+1} we reinterpolate the motion field so that

$$\forall k > j, \mathbf{T}(x_i, y_i, z_i, t(k)) = \mathbf{T}(x_i, y_i, z_i, t(j)) \tag{15}$$

by using the algorithm of Unser et al [9]. Doing this enables a good initial estimate for the motion field at $t(j+1)$. Once the Φ_j are optimized for each j individually, we consider the displacement vectors in all the Φ_j at once. Gradient descent optimization is now used to maximize the metric $\mathcal{C}_{[1,T-1]}$ for all time frames. The final results of the optimization is a transformation which describes the motion of the myocardium for all times between 0 and $T-1$.

3 Results

Data was acquired from 7 normal volunteers consisting of a series of SA and LA volume images of the heart. Both TrueFISP and tagged MR images were acquired for each volunteer so that the motion fields computed from the two sets of images could be compared with each other. A typical set of images from one volunteer is shown in figure 1.

For each volunteer we computed the motion fields using the algorithm described in section 2: once using the TrueFISP MR images and once using the tagged MR images. We label the two transformations $\mathbf{T}_{\text{TrueFISP}}$ and $\mathbf{T}_{\text{SPAMM}}$ respectively.

A cylindrical coordinate system was defined whose longitudinal axis passed through the center of the LV and was perpendicular to the SA imaging planes. The myocardium in each SA slice was divided into 16 sectors around the center of the LV and the average radial, circumferential, and longitudinal displacements were computed in each sector using the two transformations, $\mathbf{T}_{\text{TrueFISP}}$ and $\mathbf{T}_{\text{SPAMM}}$. Scatter plots of the radial, circumferential, and longitudinal displacements were drawn to see how well $\mathbf{T}_{\text{TrueFISP}}$ and $\mathbf{T}_{\text{SPAMM}}$ were correlated. The results are shown in figure 2.

Linear regression analysis was then performed on the scatter plots of the motion fields, the results of which are presented in table 1. As can be seen there is

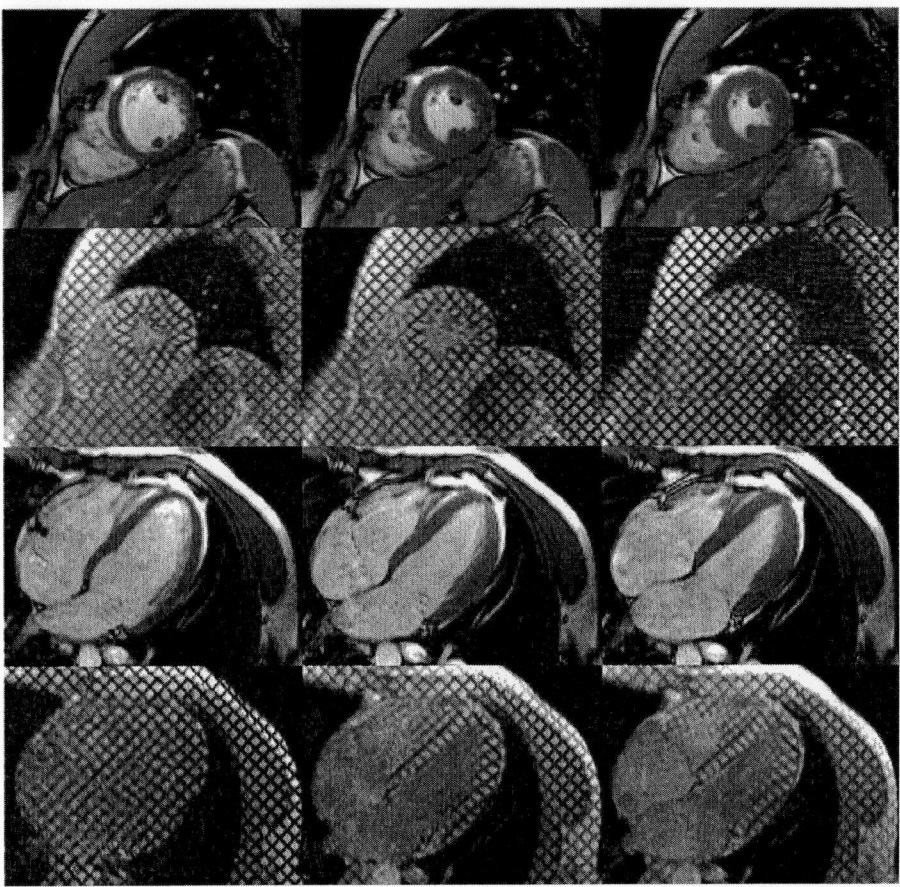

Fig. 1. This figure shows the two sets of MR images acquired for one of the seven volunteers. Row 1 shows a SA view of the LV using the TrueFISP imaging sequence, while row 2 shows the same view using a SPAMM imaging sequence. Rows 3 and 4 show a LA view of the LV. The first column corresponds to end-diastole, the second to mid-systole, and the last to end-systole

a good correlation between the radial and longitudinal displacements computed from $\mathbf{T}_{\text{TrueFISP}}$ and $\mathbf{T}_{\text{SPAMM}}$, but there is less of an agreement between the circumferential displacements. In figure 3 we show the motion fields computed from $\mathbf{T}_{\text{TrueFISP}}$ and $\mathbf{T}_{\text{SPAMM}}$. The first and second rows show arrow plots of the displacement fields in a mid-ventricular SA slice for one of the volunteers using $\mathbf{T}_{\text{TrueFISP}}$ and $\mathbf{T}_{\text{SPAMM}}$ respectively. As can be seen the motion fields are remarkably similar but there are regions where not all of the twisting motion has been captured in $\mathbf{T}_{\text{TrueFISP}}$. These regions are indicated by the circles in the third column. In the third and fourth rows, virtual tag grids have been placed on the SA tagged MR image sequences and have been deformed over time by $\mathbf{T}_{\text{TrueFISP}}$ and $\mathbf{T}_{\text{SPAMM}}$ respectively. From the third row we see that there is a

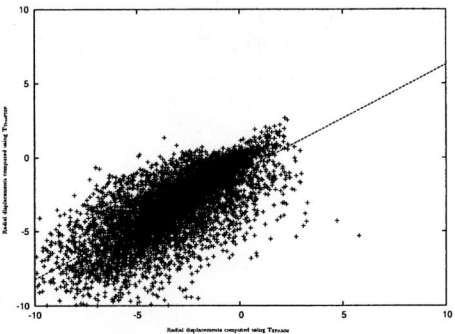

(a) Radial displacement scatter plot

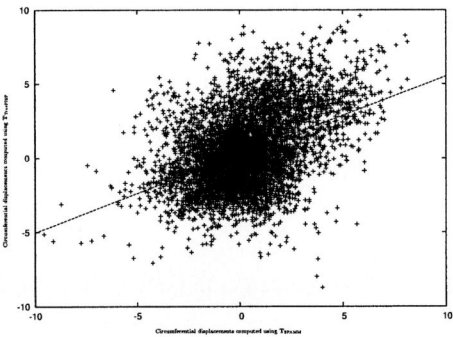

(b) Circumferential displacement scatter plot

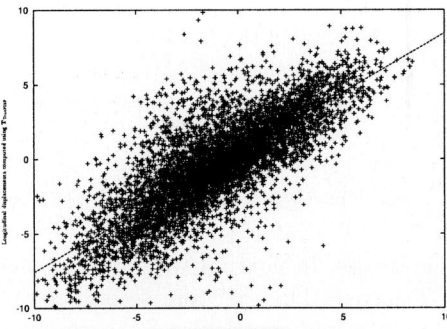

(c) Longitudinal displacement scatter plot

Fig. 2. Scatter plots showing the correlation between the radial, circumferential, and longitudinal displacements computed using $\mathbf{T}_{\text{TrueFISP}}$ and $\mathbf{T}_{\text{SPAMM}}$. Results of linear regression analysis on these plots are given in table 1

Table 1. The results of linear least squares fitting for the radial, circumferential, and longitudinal displacements computed using T_{TrueFISP} and T_{SPAMM}

	Line of Best Fit	Correlation Coefficient
Radial	$y = 0.73x - 0.97$	0.74
Circumferential	$y = 0.53x + 0.24$	0.43
Longitudinal	$y = 0.80x + 0.42$	0.78

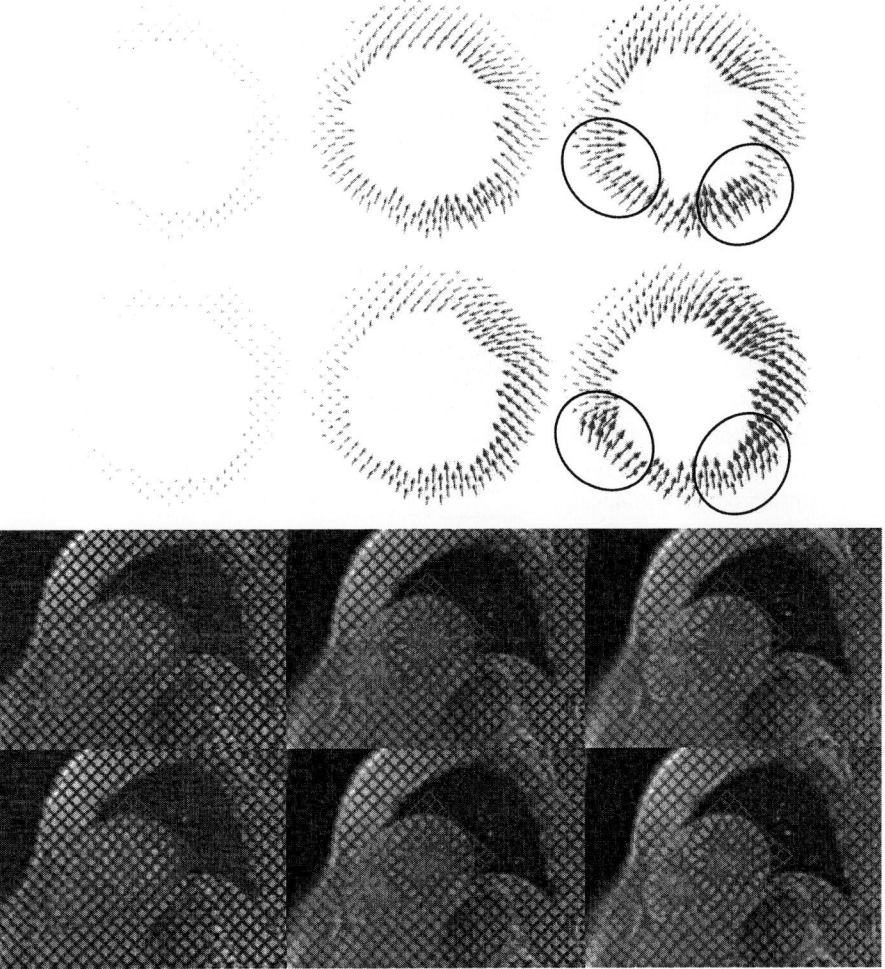

Fig. 3. The first and second rows show the computed motion fields in a mid-ventricular SA slice for one of the volunteers using T_{TrueFISP} and T_{SPAMM} respectively. The columns, from left to right, correspond to end-diastole, mid-systole, and end-systole. Regions of the myocardium which show a discrepancy in the computed motion fields are indicated by the circles in the third column. In the third and fourth rows, virtual tag grids have been placed on the SA tagged MR image sequences and have been deformed over time by T_{TrueFISP} and T_{SPAMM} respectively

good agreement in the motion fields computed from the two sets of images since the virtual tag grid follows the tag pattern in the images.

4 Discussion and Conclusions

This paper presented a comparison of the motion fields computed using a 4D nonrigid registration algorithm from TrueFISP and SPAMM tagged MR images. The correlation coefficients were found to be 0.74, 0.43, and 0.78 for the radial, circumferential, and longitudinal displacements respectively computed from the tagged and untagged motion fields. These results indicate that although not all of the twisting motion is captured in the motion fields derived from the untagged images, the motion fields may still be used for obtaining radial and longitudinal displacements. Further work still needs to be done on how the image acquisition method and the quality of the images acquired affect the correlation between the computed deformation fields. In particular the heart rate of a volunteer may not be constant throughout the duration of a cardiac examination and methods for compensating for the variation in heart rate still need to be developed.

In future work we intend to build a 4D statistical model of the motion of heart as in [4]. Such a model would restrict the transformations obtained from the registration algorithm to statistically likely types of motion predicted by the motion model and should improve the estimation of motion and strain fields from untagged MR images.

References

1. A. A. Amini, Y. Chen, M. Elayyadi, and P. Radeva. Tag surface reconstruction and tracking of myocardial beads from SPAMM-MRI with parametric B-spline surfaces. *IEEE Transactions on Medical Imaging*, 20(2):94–103, February 2001.
2. L. Axel and L. Dougherty. Heart wall motion: Improved method of spatial modulation of magnetization for MR imaging. *Radiology*, 172(2):349–360, 1989.
3. R. Chandrashekara, R. H. Mohiaddin, and D. Rueckert. Cardiac motion tracking in tagged MR images using a 4D B-spline motion model and nonrigid image registration. In *Proceedings of the 2004 IEEE International Symposium on Biomedical Imaging: From Nano to Macro*, Arlington, VA, USA, April 15–18 2004. IEEE.
4. R. Chandrashekara, A. Rao, G. I. Sanchez-Ortiz, R. H. Mohiaddin, and D. Rueckert. Construction of a statistical model for cardiac motion analysis using nonrigid image registration. In C. Taylor and J. Alison Noble, editors, *Information Processing in Medical Imaging*, pages 599–610, Ambleside, UK, July 2003. Springer.
5. S. N. Gupta and J. L. Prince. On variable brightness optical flow for tagged MRI. In *Information Processing in Medical Imaging*, pages 323–334, June 1995.
6. J. Huang, D. Abendschein, V. G. Dávila-Román, and A. A. Amini. Spatio-temporal tracking of myocardial deformations with a 4-D B-spline model from tagged MRI. *IEEE Transactions on Medical Imaging*, 18(10):957–972, October 1999.

7. N. F. Osman, E. R. McVeigh, and J. L. Prince. Imaging heart motion using harmonic phase MRI. *IEEE Transactions on Medical Imaging*, 19(3):186–202, March 2000.
8. C. Ozturk and E. R. McVeigh. Four-dimensional B-spline based motion analysis of tagged MR images: Introduction and *in vivo* validation. *Physics in Medicine and Biology*, 45:1683–1702, 2000.
9. M. Unser, A. Aldroubi, and Murray Eden. Fast B-spline transforms for continuous image representation and interpolation. *IEEE Transactions on Pattern Analysis and Machine Intelligence*, 13(3):277–285, March 1991.

Tracking of LV Endocardial Surface on Real-Time Three-Dimensional Ultrasound with Optical Flow

Qi Duan[1], Elsa D. Angelini[2], Susan L. Herz[1], Olivier Gerard[3], Pascal Allain[3], Christopher M. Ingrassia[1], Kevin D. Costa[1], Jeffrey W. Holmes[1], Shunichi Homma[1], and Andrew F. Laine[1]

[1] Columbia University, New York, NY, USA
{qd2002, slh2002, cmi2001, kdc17, jh553, laine, sh23}@columbia.edu
[2] Ecole Nationale Supérieure des Télécommunications, 46 rue Barrault, 75 013 Paris France
elsa.angelini@enst.fr
[3] Philips Medical Systems Research Paris, 51, rue Carnot, BP 301, 92156 Suresnes, France
{Olivier.Gerard, pascal.allain}@philips.com

Abstract. Matrix-phased array transducers for real-time three-dimensional ultrasound enable fast, non-invasive visualization of cardiac ventricles. Segmentation of 3D ultrasound is typically performed at end diastole and end systole with challenges for automation of the process and propagation of segmentation in time. In this context, given the position of the endocardial surface at certain instants in the cardiac cycle, automated tracking of the surface over the remaining time frames could reduce the workload of cardiologists and optimize analysis of volume ultrasound data. In this paper, we applied optical flow to track the endocardial surface between frames of reference, segmented via manual tracing or manual editing of the output from a deformable model. To evaluate optical-flow tracking of the endocardium, quantitative comparison of ventricular geometry and dynamic cardiac function are reported on two open-chest dog data sets and a clinical data set. Results showed excellent agreement between optical flow tracking and segmented surfaces at reference frames, suggesting that optical flow can provide dynamic "interpolation" of a segmented endocardial surface.

1 Introduction

Segmentation and dynamic tracking of the endocardial surface are critical for quantitative assessment of an echocardiographic exam and diagnosis of pathologies such as myocardium ischemia. Ultrasound is the cardiac screening modality with the highest temporal resolution, but is still limited to two-dimensions in most medical centers. Developments in 3D echocardiography started in the late 1980s with the introduction of off-line 3D medical ultrasound imaging systems. Many review articles have been published over the past decade, assessing the progress and limitations of 3D ultrasound technology for clinical screening [1-4]. These articles reflect the diversity of 3D systems developed for both image acquisition and reconstruction.

Although 2D transducers can be configured to assemble a 3D volume from a series of planar views, only matrix phased array transducers can scan true three-dimensional volumes [5]. This technology fundamentally differs from former generations of 3D systems as a true volume of data is acquired in real-time (1 scan per second), enabling the cardiologist to view moving cardiac structures from any given plane in real-time. A first generation of real-time three-dimensional ultrasound (RT3D) scanners was introduced in early 1990s by Volumetrics [5] but acquisitions artifacts prevented the technology from meeting its initial expectation and reaching its full potential. A new generation of RT3D transducers was introduced by Philips Medical Systems (Best, The Netherlands) in 2000s with the SONOS 7500 transducer that can acquire a fully sampled cardiac volume in four cardiac cycles. Each scan produces ¼ of the cardiac volume so that 4 scans are performed and spatially aggregated to generate one ultrasound volume over one cardiac cycle. This technical design enabled to dramatically increase spatial resolution and image quality.

Clinical evaluation of 3D ultrasound data for assessment of cardiac function is performed via interactive inspection of animated data, along selected projection planes. Facing the difficulty of inspecting a 3D data set with 2D visualization tools, it is highly desirable to assist the cardiologist with quantitative analysis tools of the ventricular function. Complex and abnormal ventricular wall motion, for example, can be detected, at a high frame rate, via quantitative four-dimensional analysis of the endocardial surface and computation of local fractional shortening [6]. Such preliminary study showed that RT3D ultrasound provides unique and valuable quantitative information about cardiac motion, when derived from manually traced endocardial contours. Philips Medical Systems recently introduced a new RT3D ultrasound machine, the iE33 [7] that incorporates a ventricular analysis tool named QLAB quantification software. This tool includes an interactive segmentation capability for the endocardium using a 3D deformable model that alleviates the need for full manual tracing of the endocardial border. To assist the segmentation process over the entire cardiac cycle, we evaluated the use of optical flow (OF) tracking between segmented frames and tried to answer the following questions in this study: *Can OF track the endocardial surface between ED and ES with reliable positioning accuracy? How does dynamic information derived from OF tracking on RT3D ultrasound compare to a single segmentation method, given the high inter and intra variability of segmentation by experts? Can OF be used as a dynamic interpolation tool of the endocardial surface?*

Cardiac motion analysis from images has been an active research area over the past decade. However, most research efforts were based on CT and MRI data. Previous efforts using ultrasound data for motion analysis include intensity-based OF tracking, strain-imaging, and elastography. Intensity-based OF tracking methods described in [8-13] combine local intensity correlation with specific regularizing constraints (e.g. continuity). For strain-imaging or elastography, strain calculation and motion estimation are typically derived from auto-correlation and cross-correlation on RF data. The commercialized strain imaging package, "2D Strain" from General Electric [14] uses such a paradigm. Most published works on strain-imaging or elastography [14-18] are limited to 1D or 2D images. Early works [19] used simple simulated phantoms while recent work [20] used 3D ultrasound data sequence for LV volume estimation. The presence of speckle noise in ultrasound prevents the use of gradient-based methods while relatively large region-matching methods are relatively robust to the presence of noise. In this study, we propose a surface tracking application for a 4D correlation-based OF method on 3D volumetric ultrasound intensity data.

2 Method

2.1 Correlation-Based Optical Flow

Optical flow (OF) tracking refers to the computation of the displacement field of objects in an image, based on the assumption that the intensity of the object remains constant. In this context, motion of the object is characterized by a flow of pixels with constant intensity. The assumption of intensity conservation is typically unrealistic for natural movies and medical imaging applications, motivating the argument that OF can only provide *qualitative* estimation of object motions. There are two global families of OF computation techniques: (1) Differential techniques [21-23] that compute velocity from spatio-temporal derivatives of pixel intensities; (2) Region-based matching techniques [24, 25], which compute the OF via identification of local displacements that provide optimal correlation of two consecutive image frames. Compared to differential OF approaches, region-based methods using correlation measures are less sensitive to noisy conditions and fast motion [26] but assume that displacements in small neighborhoods are similar. In three-dimensional ultrasound, this later approach appeared more appropriate and was selected for this study. Given two data sets from consecutive time frames: $(I(\mathbf{x},t), I(\mathbf{x},t+\Delta t))$, the displacement vector $\Delta \mathbf{x}$ for each pixel in a small neighborhoods Ω around a pixel x is estimated via maximization of the cross-correlation coefficient defined as:

$$r = \frac{\sum_{\mathbf{x} \in \Omega}(I(\mathbf{x},t) * I(\mathbf{x}+\Delta\mathbf{x},t+\Delta t))}{\sqrt{\sum_{\mathbf{x} \in \Omega} I^2(\mathbf{x},t) \sum_{\mathbf{x} \in \Omega} I^2(\mathbf{x}+\Delta\mathbf{x},t+\Delta t)}} \quad (1)$$

In this study, correlation-based OF was applied to estimate the displacement of selected voxels between two consecutive ultrasound volumes in the cardiac cycle. The search window Ω was centered about every (5×5×5) pixel volume and was set to size (7×7×7).

2.2 Three-Dimensional Ultrasound Data Sets

The tracking approach was tested on three data sets acquired with a SONOS 7500 3D ultrasound machine (Philips Medical Systems, Best, The Netherlands):

- Two data sets on an anesthetized open chest dog were acquired before (baseline) and 2 minutes after induction of ischemia via occlusion of the proximal left anterior descending coronary artery. These data sets were obtained by

positioning the transducer directly on the apex of the heart, providing high image quality and a small field of view. Spatial resolution of the analyzed data was $(0.56mm^3)$ and 16 frames were acquired for one cardiac cycle.
- One transthoracic clinical data set from a heart-transplant patient. Spatial resolution of the analyzed data was $(0.8mm^3)$ and 16 frames were acquired for one cardiac cycle.

Because of the smaller field of view used to acquire the open-chest dog data and the positioning of the transducer directly on the dog's heart, image quality was significantly higher in this data set, with some fine anatomical structures visible. Cross-section views at end-diastole (ED) from the open-chest baseline data set, and the patient data set are shown in Fig. 1.

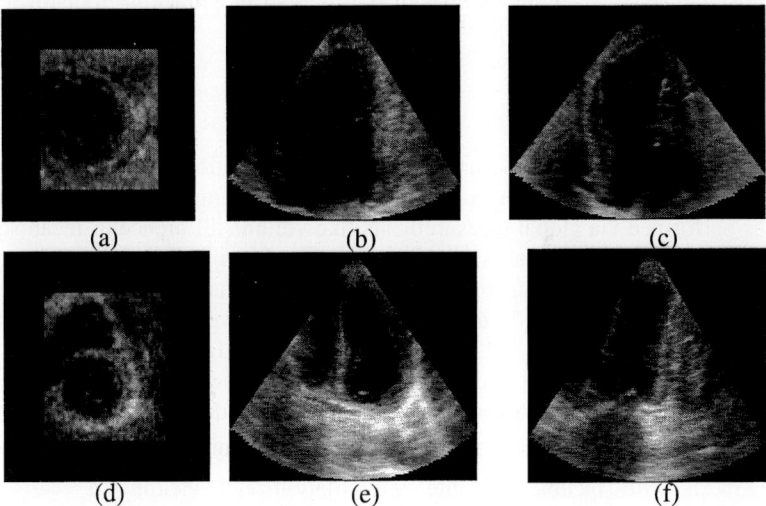

Fig. 1. Cross-sectional views at ED for (a-c) Open-chest dog data, prior to ischemia, (d-f) Patient with transplanted heart. (a, d) axial, (b, e) elevation and (c, f) azimuth views

2.3 Segmentation

The endocardial surface of the left ventricle (LV) was segmented with two methods.

Manual Tracing. An expert performed manual tracing of all time frames in the data sets, on rotating B-scan views (long-axis views rotating around the central axis of the ventricle) and C-scan views (short-axis views at different depths).

QLAB Segmentation. The QLAB software, (Philips Medical Systems), was used to segment the endocardial surface. Initialization was performed by a human expert and a parametric deformable model was fit to the data at each time frame. Segmentation results were reviewed by the expert and adjusted manually for final corrections. We emphasize here that QLAB is used as a semi-automated segmentation tool. The QLAB software was designed to process human clinical data sets. Significant anatomical differences between canine and human hearts could lead to misbehavior of the segmentation software we decided to only apply the software tool to clinical data sets.

2.4 Tracking with Optical Flow

Tracking of the endocardial surface with OF was applied with initialization using the manually traced surfaces (for dog data and clinical data) and the QLAB segmented surfaces (for clinical data). Starting with a set of endocardial surface points (about three thousand points, roughly 1 mm apart for manual tracing and about eight hundred points, roughly 3 mm apart for QLAB) defined at ED, the OF algorithm was used to *track the surface in time through the whole cardiac cycle*. Since OF with correlation-based method is very sensitive to speckle noise, all data sets were pre-smoothed with edge-preserving anisotropic diffusion as developed in [27]. We emphasize here that OF is not applied as a segmentation tool but as a surface tracking tool for a given segmentation method.

2.5 Evaluation

We evaluated OF tracking via visualization and quantification of dynamic ventricular geometry compared to segmented surfaces. Usually comparison of segmentation results is performed via global measurements like volume difference or mean-squared error. In order to provide local comparison, we propose a novel comparison method based on a parameterization of the endocardial surface in prolate spheroidal coordinates [28] and previously used for comparison of ventricular geometries from two 3D ultrasound machines in [29]. The endocardial surfaces were registered using three manually selected anatomical landmarks: the center of the mitral orifice, the endocardial apex, and the equatorial mid-septum. The data were fitted in prolate spheroidal coordinates (λ, μ, θ), projecting the radial coordinate λ to a 64-element surface mesh with bicubic Hermite interpolation and yielding a realistic 3D endocardial surface. The fitting process (illustrated in Fig. 2 for a single endocardial surface) was performed using the custom finite element package Continuity 5.5 developed at the University of California San Diego (http://cmrg.ucsd.edu). In this figure, we can observe the initial positioning of the data points and the surface mesh, and the fitted surface after fitting with very high agreement between the data and the mesh. A zoom is provided on a small region where the fitted surface and the points do not match exactly, due to region-based global optimization of radial projections.

The fitted nodal values and spatial derivatives of the radial coordinate, λ, were then used to map relative differences between two surfaces, $\varepsilon = (\lambda_{seg} - \lambda_{OF}) / \lambda_{seg}$ using custom software. The Hammer mapping was used to preserve relative areas of the flattened endocardial surface [30].

For each time frame, root mean squared errors (RMSE) of difference in λ, over all nodes on the endocardial surface were computed, between OF and individual segmentation methods. Ventricular volumes were also computed from the segmented and the tracked endocardial surfaces. Finally relative λ differences maps were generated for end-systole (ES), providing a direct quantitative comparison of ventricular geometry. These maps are visualized with iso-level lines, quantified in percentage values of radial differences.

3 Results

3.1 Open-Chest Dog Data

Quantitative results, comparing OF and manual tracing, are plotted in Fig. 3 while three-dimensional rendering of the endocardial surfaces and difference maps at ES are shown in Fig. 4.

RMSE results reported a maximum difference on radial absolute difference of 0.19 (average radial coordinate value was 0.7±0.2 at ED and 0.6±0.3 at ES) at frame 11 (start of diastole) on the baseline data set and 0.08 (average radial coordinate value was 0.7±0.3 at ED and 0.6±0.2 at ES) at frame 12 on the post-ischemia data set. Maximum LV volume differences were less than 7 ml on baseline data and 5ml on the post-ischemia data set. RMSE values were smaller for OF tracking on larger volumes.

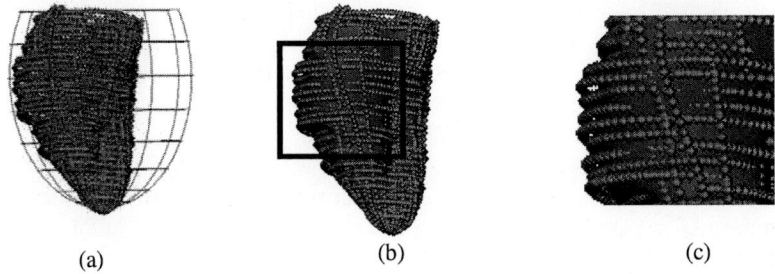

Fig. 2. Fitting process of the endocardial surface. (a) Initial FEM mesh and data points. (b) Fitted FEM surface and data points. (c) Zoom on a small region with the FEM fitted surface and the data points

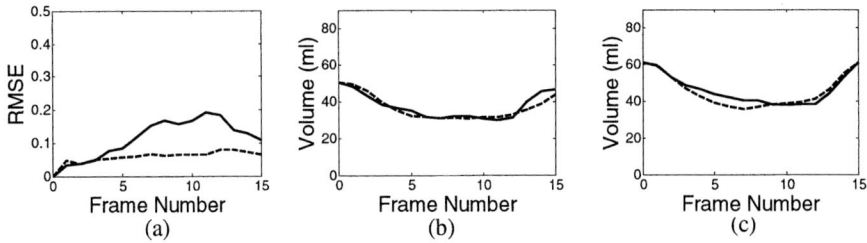

Fig. 3. Results on open-chest data sets. (a) RMSE of radial absolute differences: baseline data (solid line), post-ischemia data (dashed line); (b-c) LV volumes from manual tracing (solid line), and OF tracking (dashed line) for: (b) baseline, post-ischemia (c)

On the radial difference maps in Fig. 4, we observe similar difference patterns in the baseline and the post-infarct data, demonstrating repeatability of the OF tracking performance on a same ventricular geometry but with different contractility patterns. An area with large error in the baseline comparison localized on the anterior-lateral wall disappeared in post-ischemia tracking. This error is caused by a small portion of tracked points that were confused by acquisition artifacts at the boundary between the first and second quadrants. This result suggests that the OF algorithm has difficulty to track large motion in the unconstrained healthy heart. Errors were well distributed over the entire surface with overall shape agreement. Similar maps can be used to examine local fractional shortening using the technique developed by the Cardiac Mechanics Group at Columbia University [6] and revealed similar patterns of abnormal wall motion after ischemia using of tracked surface or manual tracing, and corroborates the accuracy of OF tracking to provide dynamic functional information.

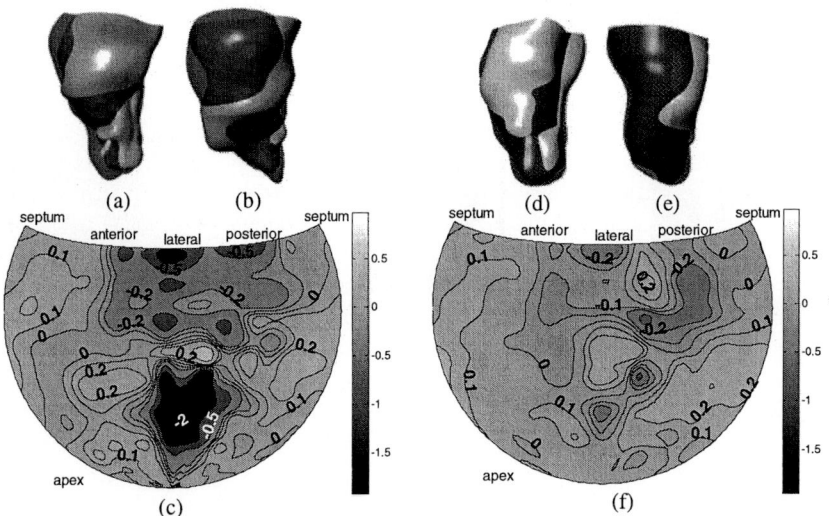

Fig. 4. Endocardial surfaces from open-chest dog data sets at ES. (a-c) Results on baseline data. (d-f) Results on post-ischemia data. Three-dimensional rendering of endocardial surfaces were generated from manual tracing (dark gray) and OF tracking (light gray) for (a, d) lateral views and (b, e) anterior views. (c, f) Relative difference maps between OF and manual tracing surfaces

3.2 Clinical Data

OF tracking was run with initialized surfaces provided by either manual tracing or the QLAB segmentation tool. Because of lower image quality on the clinical data set, compared to the open-chest dog data, we performed two sets of additional experiments. First, we checked if the time frame selected for initialization had an influence on the tracking quality.

Based on manual tracing, we initialized OF tracking for the whole cardiac cycle with ED (forward tracking) or ES (backward tracking) and compared RMSE over the entire cycle. Results, plotted in Fig. 5a show very comparable performance, confirming that the OF seems to be repeatable and insensitive to initialization set up. We therefore selected the first volume in the sequences, triggered by EKG to correspond to ED. A second experiment tested the agreement between QLAB and OF tracking when increasing the number of reference surfaces used to re-initialize OF over the cardiac cycle. Results, plotted in Fig. 5b, show that agreement of OF tracking and QLAB segmentation increases with re-initialization frequency and reaches RMSE levels similar to the experiment with manual tracing for re-initialization every other frame. We point out that strong smoothing constraints, applied by the deformable model of the QLAB segmentation, lead to surface positioning that did not always correspond to the apparent high contrast interface. Finally, we compared RMSE values from forward tracking and from averaging forward and backward tracked shapes. We observed a large increase in agreement with the QLAB smooth segmentation when averaging tracked surfaces.

Given these results, we modified our protocol for tracking based on QLAB and re-initialized the OF every other frame, using the segmented surface. We did not need to use re-initialization for tracking based on manual tracing, since the RMSE was already very small without re-initialization.

RMS error of radial relative differences and LV volumes were computed and these results are plotted in Figure 6 Maximum RMSE, over the whole endocardial surface, was equal to 0.09 (average radial coordinate value was 0.6±0.2 at ED and 0.5±0.2 at ES) at frame 09 (after ES) with manual-tracing initialization and 0.08 (average radial coordinate value is about 0.6±0.2 at ED and 0.5±0.2 at ES) at frame 10 (after ES) with QLAB initialization (with re-initialization). Three-dimensional rendered results and relative difference maps at ES are shown in Figure 7.

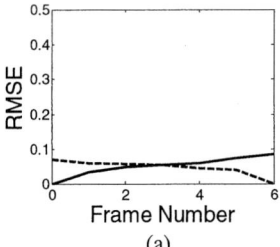 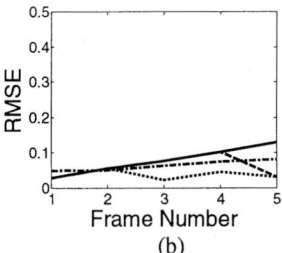

(a)   (b)

Fig. 5. (a) RMSE between OF tracking and manual tracing: forward (solid line) and backward (dashed line). (b) RMSE between OF tracking and QLAB segmentation: forward tracking without re-initialization (solid line), forward tracking with re-initialization every four frame (dashed line), forward tracking with re-initialization every two frame (dotted line), and average result from forward and backward tracking without re-initialization (dashdot linc)

These experiments showed that OF tracking initialized with manual tracing provides ventricular endocardial surfaces similar to this segmentation method, with less than 0.1 maximum absolute differences in RMSE and maximum LV volume differences below 10 ml. When initialized with QLAB, OF tracking with re-initialization shows results with less than 0.08 maximum RMSE difference and less than 13 ml for LV volume differences. These differences are similar to inter and intra-observer variability for measurement of LV volume from studies in 1979 [31, 32].

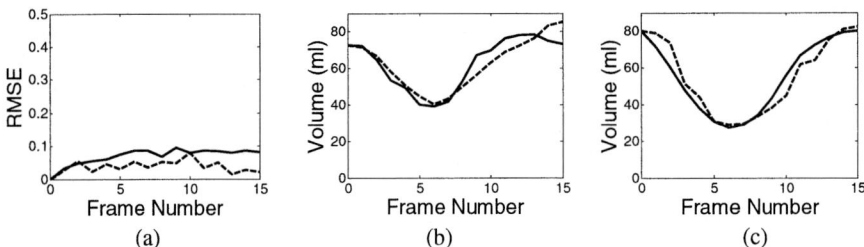

Fig. 6. Results on clinical data. (a) RMSE of radial difference for OF initialized with manual tracing (solid line) and QLAB segmentation with 2-frame re-initialization (dashed line); (b-c) LV volumes over one cardiac cycle: (b) Manual tracing (solid line) and OF initialized with manual tracing (dashed line); (c) QLAB segmentation (solid line) and OF initialized with QLAB segmentation (dashed line)

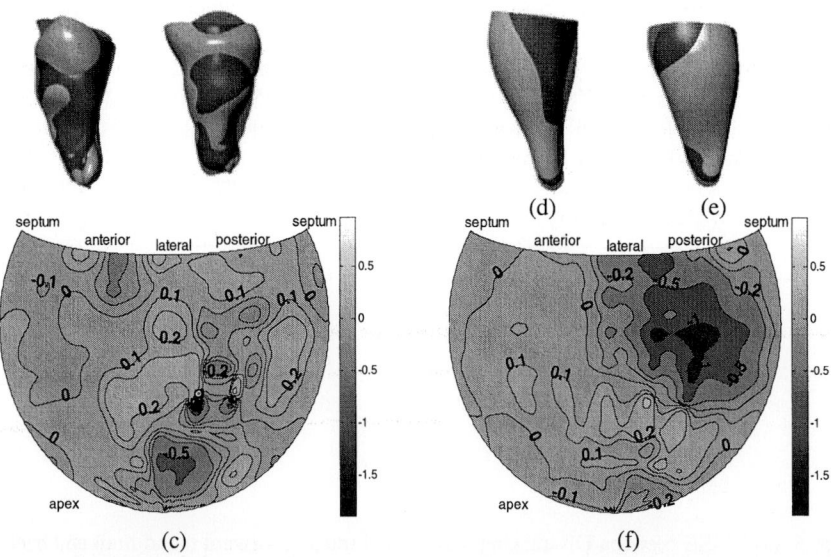

Fig. 7. Endocardial surfaces from clinical data at ES. (a-c) Manual tracing; (d-f) QLAB segmentation. Three-dimensional rendering of endocardial surfaces from segmentation method (dark gray) and OF tracking (light gray): (a, d) lateral view; (b, e) anterior view. Relative radial difference maps between OF tracking and the segmentation method

Ventricular geometries are illustrated in Fig. 7. We again observed high overall agreement between endocardial geometries provided by manual tracing and OF tracking. Radial differences were distributed over the entire surface, with higher values on the lateral-posterior wall, which corresponded to poor image quality (due to scanning set up). The QLAB segmentation provided very smooth surfaces, well tracked by the OF. Larger errors were again observed on the lateral posterior wall. Comparison of the two experiments shows that OF over one time-frame can preserve the smoothness of the surface but will tend towards more convoluted surfaces during temporal propagation of the tracking process.

Based on the high agreement with manual tracing, we can infer that OF might be a good candidate method to guide a deformable model with high smoothness constraints to better adapt to the ultrasound data and incorporate temporal information in the segmentation process. On the other hand, OF tracking could be adapted to these smoothness constraints, better ranked by cardiologists, to track larger spatial windows around the endocardial surface.

4 Conclusion

A correlation-based optical flow (OF) method was evaluated on RT3D ultrasound for tracking of the LV endocardial surface based on segmentations of the ED frame via manual tracing and QLAB segmentation from Philips Medical Systems. Geometries of the endocardial surface over the whole cardiac cycle from individual segmentation methods or tracked via OF were compared via quantitative mapping of relative radial differences of a fitted finite-element mesh. Results showed a promising ability of OF tracking to follow the endocardial surface on experimental and clinical data sets. Very high accuracy was achieved with two open-chest dog heart data sets and clinical data set using manual tracing. Similar patterns of abnormal motion appeared on the fractional shortening maps for the abnormal dog data set. When initialized with QLAB segmentation and re-initializated during the cardiac cycle, OF tracking also showed good tracking ability. Our experiments showed that OF tracking performance was not affected by tracking direction or initial time frame selection. Introducing re-initialization and averaging shape results from forward and backward tracking improved agreement between OF tracking and QLAB segmentation. This study showed encouraging results regarding OF performance to accurately track the LV endocardial surface, yield dynamic information from RT3D ultrasound and provide automated dynamic interpolation of segmented endocardial surfaces.

Acknowledgements

This work was funded by National Science Foundation grant BES-02-01617, American Heart Association #0151250T, Philips Medical Systems, New York State NYSTAR/CAT Technology Program, and the Louis Morin Fellowship program. Dr. Andrew McCulloch at the University of California San Diego provided the finite element software "Continuity" through the National Biomedical Computation Resource (NIH P41RR08605). The authors also would like to thank Dr. Pulerwitz (Department of Medicine, Columbia University).

References

1. E. O. Ofili and N. C. Nanda, "Three-Dimensional and Four-Dimensional Echocardiography," Ultrasound Medical Biology, vol. 20, 1994.
2. A. Fenster and D. B. Downey, "Three-Dimensional Ultrasound Imaging," in Handbook of Medical Imaging. Volume1. Physics and Psychophysics, vol. 1, 2000, pp. 463-510.
3. R. N. Rankin, A. Fenster, et al., "Three-Dimensional Sonographic Reconstruction: Technique and Diagnostic Applications," American Journal of Radiology, vol. 161, pp. 695-702, 1993.
4. M. Belohlavek, D. A. Foley, et al., "Three- and Four-Dimensional Cardiovascular Ultrasound Imaging: A New Era for Echocardiography," Mayo Clinic Proceedings, vol. 68, pp. 221-240, 1993.
5. O. T. V. Ramm and S. W. Smith, "Real Time Volumetric Ultrasound Imaging System," Journal of Digital Imaging, vol. 3, pp. 261-266, 1990.
6. S. Herz, T. Pulerwitz, et al., "Novel Technique for Quantitative Wall Motion Analysis Using Real-Time Three-Dimensional Echocardiography," Proceedings of the 15th Annual Scientific Sessions of the American Society of Echocardiography, 2004.
7. "Philips Ultrasound - Ie33," 2004.
8. S. Tsuruoka, M. Umehara, et al., "Regional Wall Motion Tracking System for High-Frame Rate Ultrasound Echocardiography," Proceedings of the 1996 4th International Workshop on Advanced Motion Control, AMC'96. Part 1, Tsu, Jpn, 1996.
9. I. Mikic, S. Krucinski, et al., "Segmentation and Tracking in Echocardiographic Sequences: Active Contours Guided by Optical Flow Estimates," IEEE Trans Med Imaging, vol. 17, pp. 274-284, 1998.
10. D. Boukerroui, J. A. Noble, et al., "Velocity Estimation in Ultrasound Images: A Block Matching Approach," in Lecture Notes in Computer Science, 2003, pp. 586-598.
11. W. Yu, N. Lin, et al., "Motion Analysis of 3d Ultrasound Texture Patterns," in Lecture Notes in Computer Science, 2003, pp. 252-261.
12. N. Paragios, "A Level Set Approach for Shape-Driven Segmentation and Tracking of the Left Ventricle," IEEE Trans Med Imaging, vol. 22, pp. 773-776, 2003.
13. E. Bardinet, L. D. Cohen, et al., "Tracking and Motion Analysis of the Left Ventricle with Deformable Superquadratics," Med Image Analysis, vol. 1, pp. 129-149, 1996.
14. V. Behar, D. Adam, et al., "The Combined Effect of Nonlinear Filtration and Window Size on the Accuracy of Tissue Displacement Estimation Using Detected Echo Signals," Ultrasonics, vol. 41, pp. 743-753, 2004.
15. J. Bang, T. Dahl, et al., "A New Method for Analysis of Motion of Carotid Plaques from Rf Ultrasound Images," Ultrasound Med Biol, vol. 29, pp. 967-976, 2003.
16. S. I. Rabben, S. Bjaerum, et al., "Ultrasound-Based Vessel Wall Tracking: An Auto-Correlation Technique with Rf Center Frequency Estimation," Ultrasound Med Biol, vol. 28, pp. 507-517, 2002.
17. J. D'Hooge, P. Claus, et al., "Deformation Imaging by Ultrasound for the Assessment of Regional Myocardial Function," 2003 IEEE Ultrasonics Symposium,, Honolulu, HI, USA, 2003.
18. E. E. Konofagou, W. Manning, et al., "Myocardial Elastography - Comparison to Results Using Mr Cardiac Tagging," 2003 IEEE Ultrasonics Symposium,, Honolulu, HI, United States, 2003.
19. M. A. Gutierrez, L. Moura, et al., "Computing Optical Flow in Cardiac Images for 3d Motion Analysis," Proceedings of the 1993 Conference on Computers in Cardiology, London, UK, 1993.

20. I.-S. Shin, P. A. Kelly, et al., "Left Ventricular Volume Estimation from Three-Dimensional Echocardiography," Proceedings of SPIE, Medical Imaging 2004 - Ultrasonic Imaging and Signal Processing, San Diego, CA, United States, 2004.
21. B. D. Lucas and T. Kanade, "An Iterative Image Registration Technique with an Application to Stereo Vision," International Joint Conference on Artificial Intelligence (IJCAI), 1981.
22. B. K. P. Horn and B. G. Schunck, "Determining Optical Flow," Artificial Intelligence, vol. 17, 1981.
23. H. Nagel, "Displacement Vectors Derived from Second-Order Intensity Variations in Image Sequences," Computer Vision Graphics Image Processing, vol. 21, pp. 85-117, 1983.
24. P. Anandan, "A Computational Framework and an Algorithm for the Measurement of Visual Motion," International Journal of Computer Vision, vol. 2, pp. 283-310, 1989.
25. A. Singh, "An Estimation-Theoretic Framework for Image-Flow Computation," International Conference on Computer Vision, 1990.
26. J. L. Barron, D. Fleet, et al., "Performance of Optical Flow Techniques," Int. J of Computer Vision, vol. 12, pp. 43-77, 1994.
27. Q. Duan, E. D. Angelini, et al., "Assessment of Visual Quality and Spatial Accuracy of Fast Anisotropic Diffusion and Scan Conversion Algorithms for Real-Time Three-Dimensional Spherical Ultrasound," SPIE International Symposium Medical Imaging, San Diego, CA, USA, 2004.
28. C. M. Ingrassia, S. L. Herz, et al., "Impact of Ischemic Region Size on Regional Wall Motion," Proceedings of the 2003 Annual Fall Meeting of the Biomedical Engineering Society, 2003.
29. E. D. Angelini, D. Hamming, et al., "Comparison of Segmentation Methods for Analysis of Endocardial Wall Motion with Real-Time Three-Dimensional Ultrasound," Computers in Cardiology, Memphis TN, USA, 2002.
30. S. Herz, T. Pulerwitz, et al., "Novel Technique for Quantitative Wall Motion Analysis Using Real-Time Three-Dimensional Echocardiography," Annual Scientific Sessions of the American Society of Echocardiography, 2004.
31. N. B. Schiller, H. Acquatella, et al., "Left Ventricular Volume from Paired Biplane Two-Dimensional Echocardiography," Circulation, vol. 60, pp. 547-555, 1979.
32. E. D. Folland, A. F. Parisi, et al., "Assessment of Left Ventricular Ejection Fraction and Volumes by Real-Time, Two-Dimensional Echocardiography. A Comparison of Cineangiographic and Radionuclide Techniques," Circulation, vol. 60, pp. 760-766, 1979.

Dense Myocardium Deformation Estimation for 2D Tagged MRI

Leon Axel, Ting Chen, and Tushar Manglik

Radiology Dept, NYU, 600A, 650, 1st Ave, New York, NY, 10016, USA
{Leon.axel, ting.chen, tushar.manglik}@med.nyu.edu

Abstract. Magnetic resonance tagging technique measures the deformation of the heart wall by overlying darker tag lines onto the brighter myocardium and tracking their motion during the heart cycle. In this paper, we propose a new spline-based methodology for constructing a dense cardiac displacement map based on the tag tracking result. In this new approach, the deformed tags are tracked using a Gabor filter-based technique and smoothed using implicit splines. Then we measure the displacement in the myocardium of both ventricles using a new spline interpolation model. This model uses rough segmentation results to set up break points along tag tracking spline so that the local myocardium deformation will not be influenced by the tag information in the blood or the deformation in other parts of the myocardium. The displacements in x- and y- directions are calculated separately and are combined later to form the final displacement map. This method accepts either a tag grid or separate horizontal and vertical tag lines as its input by adjusting the offsets of images taken at different breath hold. The method can compute dense displacement maps of the myocardium for time phases during systole and diastole. The approach has been quantatively validated on phantom images and been tested on more than 20 sets of in-vivo heart data.

1 Introduction

Tagged MRI [1] data has been extensively used in clinical applications to extract the myocardial motion during the heart cycle. The deformation of tag lines reflects the deformation of the underlying heart wall so that the clinician has a more direct way to view the cardiac motion. Of more importance, quantative analysis based on the tag tracking results, i.e., strain analysis, helps us to understand the cardiac motion pattern better and can play an important role in the diagnosis of various kinds of heart failure and/or malfunctioning.

To accurately track the myocardial motion and calculate the strain, one must create a dense displacement map over the heart cycle. In [2], a B-spline solid model has been used to trace the intersection of 3D tag planes and the image plane in different time phases during the heart cycle. However, the model suffers from the over smoothness caused by the limited number of the control points. The model also has the defect that the motion of the free wall and septum may have an influence on each other in case they share a common control point. Besides, the B-spline model in [2] did not consider the displacements in the myocardium of the right ventricle.

In [3] tag lines are automatically detected by using the image intensity profile along a tag's perpendicular direction as a *priori* information in a MAP estimation framework. This method avoids the use of segmentation results during the tag tracking process and has a strong performance on 2D image tag tracking. However, no details are given on how to create a dense displacement map based on the tag tracking results.

In [5] spatial-temporal filters have been used together with an interpolator approach to track tags. The method succeeded in tracking tag lines with their counterparts in different slices. However, the tags are smoothed using a quadratic approximation in each 2D slice and tag lines are assumed to be continuous during the tracking process. Both factors will introduce errors in the calculation of dense displacement map.

In [9], a new segmentation framework, called metamorphs, that combines regional image information, prior shape information, and deformable models has been developed to segment 2D and 3D clinical objects such as the endocardial and epicardial surfaces of the myocardium, based on the work in [7, 8]. In [6], Gabor filters have been used to remove the tag lines in tagged MRI images. The magnitude of the local response to a bank of Gabor filters is used to replace the intensity of the pixel in the output image of this method. These results together with the boundary information in the original image can be used to provide us an estimation of the cardiac surfaces using the metamorphs.

In this paper, we propose a new methodology based on the integration of spline interpolation, metamorph segmentation, and registration. This method solves the problem of over smoothness in tag tracking, wrong influences on tag fitting between different parts of the heart, and is capable of constructing a dense displacement map in the whole myocardium instead of only in the left ventricle wall. We tried our method on numerical phantoms, heart MRI of healthy volunteers, and heart MRI of patients to validate its effectiveness.

The rest of the paper will be arranged as follows: in section 2, we will describe our method and explain the way it works. Related information will be given on Gabor filters, metamorphs, and the spline model. In section 3, we will demonstrate the effectiveness of the method by showing the dense displacement map and strains of a numerical phantom and validate the result using the ground truth of displacements and strains. More figures of experiment results will also be shown in this section to illustrate the combination of the spline model and metamorph segmentation. In section 4, we discuss the advantage of using our method to compute the dense displacement map and its effect on the computation of cardiac strains. Finally, we will list some possible directions for future researches.

2 Method

This section describes the metamorphs segmentation and the spline model. We will use 2 subsections to introduce these two models. In the subsection on metamorph segmentation, we will also introduce the Gabor filters. In the third subsection, we will also explain the integration of these two models. Finally, we will give details about the computation of the dense displacement map.

2.1 Gabor Filters and Metamorphs

To segment the myocardium in a tagged MR image, we first use a bank of Gabor filters to remove the tags. The Gabor filter acts as a band-pass filter with the central spatial frequency of the filter set equal to the frequency of tags in the image. The filter that we use to remove the tags in the myocardium has the form of:

$$h(x, y) = g(x, y)\sin(\omega y + \theta) \tag{1}$$

where $g(x, y) = \dfrac{1}{2\pi\sigma_x\sigma_y}\exp(-\dfrac{(x/\sigma_x)^2 + (y/\sigma_y)^2}{2})$, ω is the sine wave frequency in the range $(0.8\omega_{tag}, 1.2\omega_{tag})$, where ω_{tag} is the spatial frequency of the tags, θ is the sine wave phase $(0, 2\pi)$, and σ_x σ_y are the standard deviations along the x- and y-axis. Equation (1) can be used to remove tags that are horizontal to the y-axis. A similar equation can be used to remove tags that are horizontal to the x- axis. This filter has a higher response in regions with the presence of tags and suppresses the low spatial frequency component of the image. Since the tags in the blood will be washed out soon after the initial tagging while the tags in the myocardium fade more slowly over time, the filter responds to the myocardium and suppresses the blood in the cardiac chambers. A bank of Gabor filters consisting filters with different phase shifts are applied on each pixel in the image and the maximum response will be used as the final response, thus the tag-attenuated and non-tag-attenuated myocardium will have the same response and the tags are removed. In figure 1 we show two 1D Gabor filters, the one with a 0 shift (blue) responds to the minimum tagged region (the region between tags) and the one with a 90 degree shift (pink) responds to the maximum tag attenuated region (the tags).

The metamorphs method is a hybrid segmentation method that combines the shape and interior texture information. A hierarchy of both global and local deformations parameterizes the model geometry. During the global deformation, a global energy is minimized by optimizing the global parameters using the unified gradient descent method. In each step of the deformation, we calculate the partial derivatives of the image energy with respect to each of the global parameters, and then change the values for these parameters in the opposite direction to achieve the global minimum solution of parameters (Due to the restriction on the length of the paper, please refer to reference [9] for details). The local deformations are computed using the external force formulation, in which the model evolves based on Lagrangian dynamics:

$$\dot{\mathbf{q}}_l + \mathbf{K}\mathbf{q}_l = f_{ext} \tag{2}$$

where $\mathbf{K}$ is the stiffness matrix, $\mathbf{q}_l$ is the local displacement vector, and f_{ext} is the external force. In metamorphs, we use a combination of the edge map distance gradients, the region edge distance gradients, and the second order derivative of the original image as the external force. The form of the external force is shown below:

$$f_{ext} = -(a\nabla(M(I)) + b\nabla M(S(I)) + c\nabla(\nabla G(I))) \tag{3}$$

where I is the original image, $M(.)$ is the function that extracts the edge from the image and computes the distance transform of the edge map by calculating at every pixel in the image the distance to the nearest edge, ∇ is the gradient operator, $S(.)$ is the function that computes the binary mask of the object of interest, $G(.)$ is the Gaussian operator, and a, b, c are the weights for the edge map distance gradient, the region edge distance gradient, and the second order derivative gradient flow respectively.

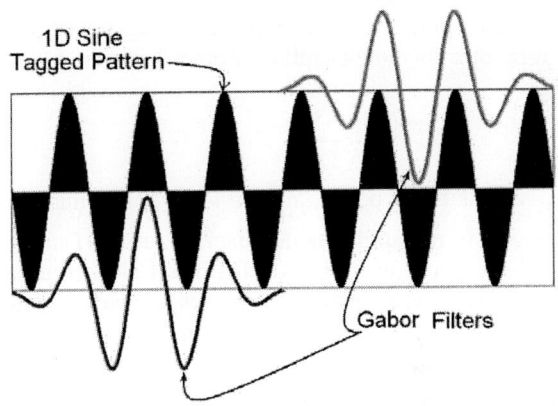

Fig. 1. 1D Gabor Filter tag filling

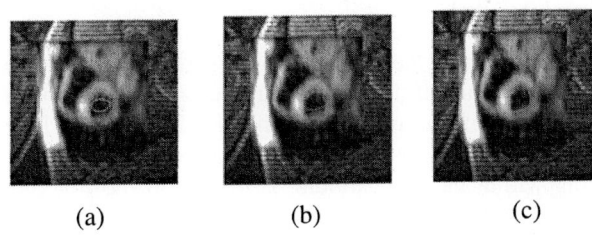

Fig. 2. The integration of Gabor filters and metamorphs. a) a metamorph is initialized inside the left ventricular region in the output image of Gabor filters; b) the global deformable fitting results (intermediate); c) the local and final segmentation results

For endo-cardiac segmentation in this paper, the metamorph is initialized manually on one slice and then propagate to other slices. We use both the Gabor filter results (for the tagging texture) and the original image (for edge information) to guide the deformation of metamorphs. For epi-surface segmentation, the metamorph is either initialized based on the segmentation results of endo-cardiac surfaces or just manually segmented.

In figure 2 we show how to segment the MR image using the Gabor filters and metamorphs. In a) there is an image that has been preprocessed using the Gabor filters. The tags have been removed and the myocardium and blood region are not any longer of similar intensity. A deformable model is initialized in the left ventricle (in green) and deforms to fit to the myocardium boundaries using both global transformation (b) and local deformation (c).

2.2 Spline Model

The spline model is initialized using the tag tracking result. We use a snake-based technique to track the tags. The snakes are driven by the gradient and they converge to the local minimum of image intensity. However, the results may not converge to the intensity minimum because of snake's internal constraints and local noise.

A band of pp-form splines are then fitted to these snakes to refine the tracking results. The pp-form of a polynomial spline of order k has the following form:

$$p_j(x) = \sum_{i=1}^{k}(x-\xi_j)^{k-i} c_{ji} \qquad (4)$$

where $\xi_1,\ldots,\xi_{l+1}$ are its break points, c_{ji} are local polynomial coefficients, and l is the number of pieces. In our spline model, the k is set to 4, and the break points are densely and evenly (every 2 to 3 pixels) distributed along the tag line. This spline model is used to refit the snakes to the tags. We resample each snake at the interval of 1 pixel and calculate the breaks for the spline using the local intensity information and the discretized snake location. The local information may also include the local gradient and also other nearby splines. The resampled snakes are more smooth and closer to the local image intensity minimum after the spline fitting.

2.3 Integration

The segmentation results of the metamorphs have been used to provide break points on the spline model. Since the tags in the background (beyond the epi-surface or interior to the endo-surface of the myocardium) do not deform in the same pattern as those in the myocardium, we assume that the tags break at the boundary when they begin to deform. In previous work, researchers also noticed this so that they limited spline-tag fitting within the myocardium. However, in those approaches, the spline itself is still continuous. The continuous spline has a bad fitting performance in the regions that are close to the myocardium boundary and bring unnecessary mutual influence between different parts of the heart when there are limited control points. In our model, the spline is separated into several portions, each for a distinct part of the myocardium. The new model has a better fitting performance and prevents unnecessary influence between different parts of the myocardium.

2.4 Computing the Dense Displacement Map

After we have the refined spline-fitted tags in both the horizontal and the vertical direction, we compute the dense displacement map. Since the displacements in the x- and y- direction are independent of each other, we can compute the displacement map in each direction separately and then combine them to form the final displacement map. We first create the separate dense displacement map in the horizontal and the vertical directions. We use a spline-based technique to interpolate between the existing spline-fitted tags to form a series of virtual tags with the interval of 1 pixel in the undeformed image. Because of the sparsity of the tags, sometimes there are missing part in these virtual tags. In such a case, we use a spline model that is the same as those we use to fit the real tags to complete the virtual tag. More than one loop may be needed in cases there are not enough tags to interpolate with. The interpolation result can be used as a close approximation to the real displacement field. The algorithm can be expressed as:

```
Repeat
    virtual_tags=Interpolate(existing_tags);
    for each tag in the set of virtual tags
        If isincomplete(current_tag)
            then splineinterpolate(current_tag);
    End;
        existing_tags = virtual_tags;
until there is no change in existing_tags
```

In figure 3, we show an image with the spline model (in red) and the virtual tags (in green).

After the computation of the separate displacement maps, we combine them together. Since we have all those virtual tags that can be treated as straight lines that are parallel to the axis in the undeformed image, we can compute the displacement at each pixel by calculating the distance between the pixel and the cross points of the corresponding horizontal and vertical virtual tags in the deformed image.

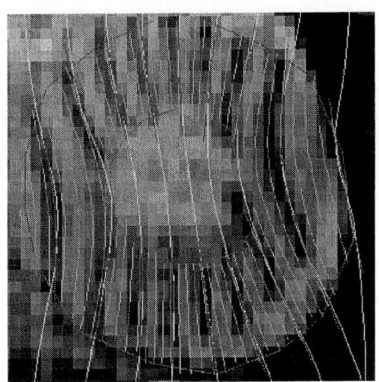 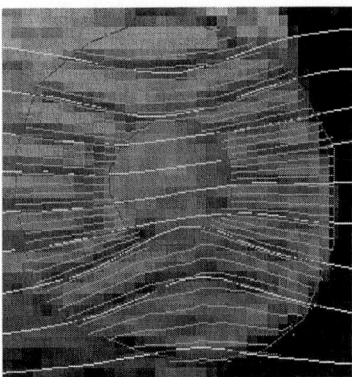

Fig. 3. The spline model and the virtual tags. The spline model is in red, the virtual tags are in green. For interpolation purposes, the spline model is extended a little beyond the myocardium boundary. The extended part is estimated using the information inside the myocardium

3 Results

We first show the result of our method on a numerical phantom. The phantom underwent a series of displacements and deformations under the influence of an underlying displacement field. In the phantom, the horizontal and the vertical tags are in separate images. In figure 4 we show the location of the spline model and the virtual tags. In figure 5, we show the dense displacement map created by our method and its difference to the ground truth of displacement. We also calculated the strain based on the dense displacement map, and the result can be seen in figure 6.

We also show the dense displacement map in other 2 real heart MR images and computed the strain based on the displacement map. In one image, the horizontal and the vertical tags were imaged separately. In the other image, there was a tag grid combining both the horizontal and the vertical tags. Our method output smooth and accurate dense displacement maps for both data sets.

We also include the result of an experiment that we calculate the displacement map around both ventricles in figure 9.

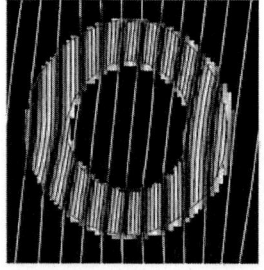 

Fig. 4. The vertical and horizontal spline model (red lines) and virtual tags (blue) in the phantom

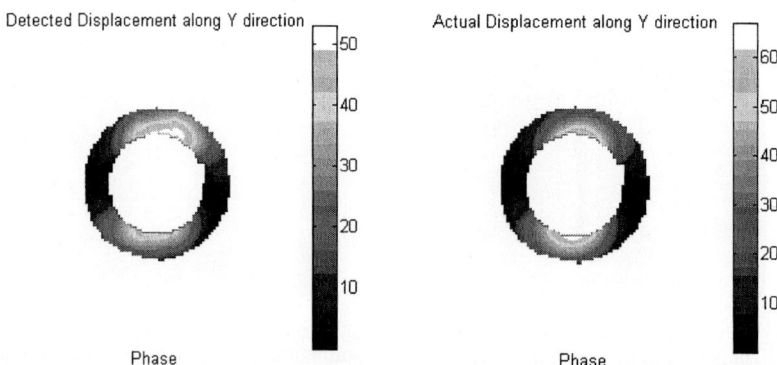

Fig. 5. Comparison between the displacement map created by our method and the ground truth. The plot shows the arbitrary value of the displacement in millimeters (same in all the following plots with color bars). These two displacement map have a similar distribution. However, the magnitude of the real displacement map is a little bigger than that of the computed one (within 10% error at most pixels) and the difference does not have effect on the computation of the strain

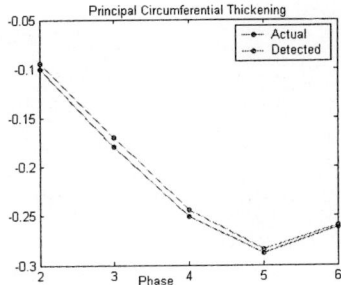

Fig. 6. The comparison between the strain derived from the calculated displacement map and the real strain. There is very little difference in the strain in all 6 phases of deformation. The plot shows the ratio between the changed length and the original length of the myocardium in the circumferential direction

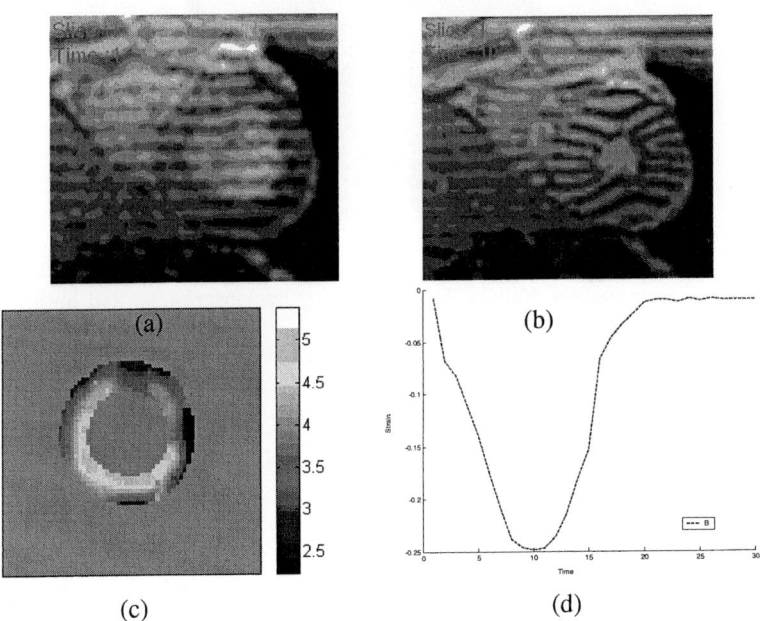

Fig. 7. The dense displacement map of a healthy volunteer. a) is the undeformed heart at time phase 1, b) is the deformed heart at time phase 10, c) is the displacement from phase 10 to 1, d) is the average strain magnitude (the ratio between the changed and the original length) during the heart cycle calculated based on the displacement map

Finally, multiple patient data have been unified and combined into one plot the show the difference between healthy hearts and abnormal hearts.

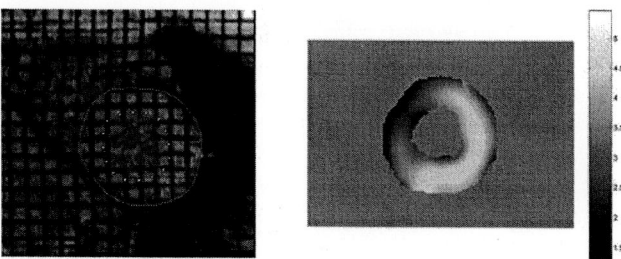

Fig. 8. The dense displacement map of a patient's MRI with tag grid. The patient suffers severe heart dysfunction that his heart cannot beat synchronously

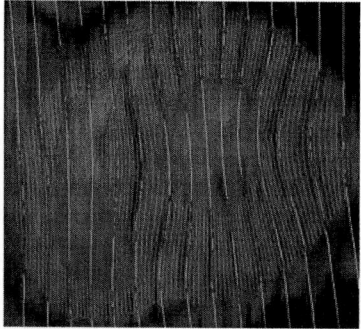

Fig. 9. The virtue tags in around left and right ventricles. Red lines are the endo- and epicardiac segmentation results. Yellow lines are the tag tracking results. Red lines are spline models. Green lines are virtue tags

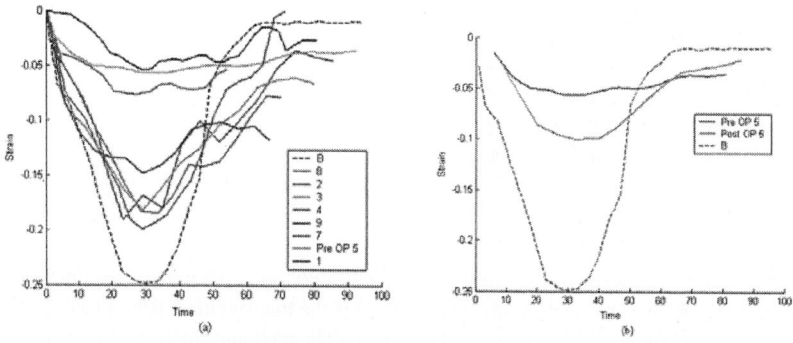

Fig. 10. (a) shows Strain developing in a human heart as a function of time. In both the images, plot "B" belongs to healthy volunteer where as rest of the plots belong to patients. In image (b) a patent underwent a heart surgery and improvement in his heart condition is quantified on this graph and compared to a healthy volunteer. The y-axis corresponding to the ratio of the changed length to the original length of the myocardium

4 Discussions

Our approach of computing the dense displacement map has a strong performance by combining the segmentation module and the spline model. By inserting the break points into the spline model, the wrong tag information in the blood and the background does not have any influence on the myocardium displacement map computation. The pp-form spline with break points at the myocardial boundary also prevents the mutual influence of displacement between different parts of the myocardium. In addition, the spline model is capable of calculating the displacement map around the right ventricle.

In the future, we plan to extend the method in 3D. By using information from both long axis and short axis heart MRs, we will be able to construct virtual tag planes and calculate the displacement in 3D spatial space.

Acknowledgement

The MR images we use in our experiment are scanned at the NYU, school of medicine. We should thank Vinay Pai and Dan Kim for scanning. We also receive help from Professor Dimitris Metaxas and his research group at Rutgers University.

Reference

1. L. Axel and L. Dougherty. MR imaging of motion with spatial modulation of magnetization. Radiology, 171(3):841-845, 1989.
2. A. A. Amini, Y. Chen, M. Elayyadi, and P. Radeva. Tag surface reconstruction and tracking of myocardial beads from SPAMM-MRI with parametric B-spline surfaces, IEEE Transactions on Medical Imaging, Vol. 20, No. 2, February 2001.
3. Y.Chen and A. A. Amini. A MAP framework for tag line detection in SPARMM data using Markov random fields on the B-spline solid. Mathematical Methods in Biomedical Image Analysis, 2001:131-138
4. J. Park, D. Metxas, and L. Axel. Volumetric deformable models with parameter functions: A new approach to the 3D motion analysis of the LV from MRI-SPAMM. In *Proc.* International Conference on Computer Vision:700-705, 1995
5. Denney, T.S. and J.L. Prince, Frequency domain performance analysis of Horn and Schnuck's optical flow algorithm for deformable motion, IEEE transaction on Image Processing, Vol. 4 No. 9 pp:1324-1328, 1995
6. T. Manglik, L. Axel, V. M. Pai, D. Kim, P. Dugal, A. Montillo, and Q. Zhen. Use of bandpass Gabor filters for enhancing blood-myocardium contrast and filling-in tags in tagged MR images. In Proc. of ISMRM vol.11:1793, 2004.
7. T. Chen and D. Metaxas: Gibbs prior models, marching cubes, and deformable models: A hybrid framework for 3D medical image segmentation. In Proc. of MICCAI 2003, Montreal, vol 2, 703-710.
8. T. Chen and D. Metaxas: Image segmentation based on the integration of Markov random fields and deformable models. In Proc. of MICCAI 2000

9. X. Huang, D. Metaxas, and T. Chen. MetaMorphs: deformable shape and texture models, In Proc. of the IEEE Computer Society Conf. on Computer Vision and Pattern Recognition, CVPR'04, Oral Presentation, Washington, D.C., June, 2004.
10. D. Gabor. Theory of communication. Journal of Institution of Electrical Engineers 93,429-457, 1946.

Appendix

Deformation refers to change in shape of an object between an initial (undeformed) state and a subsequent (deformed) state.

$$x_i = x_i(X_1, X_2) \tag{A.1}$$

The description of position (x_1 and x_2), motion or deformation of a point in space as a function of it initial undeformed position (X_1 and X_2) is known as a Lagrangian formulation (equation A.1).

$$F = x\nabla X \tag{A.2}$$

$$\nabla X = \partial/\partial X_i \hat{e}_i \tag{A.3}$$

$$L = \frac{1}{2}(F'F - 1) \tag{A.4}$$

The deformation gradient F (equation A.2) is given as a partial differentiation of equation A.1.

The Lagrangian (or Green's) finite strain tensor can be calculated from equation A.4.

A Surface-Volume Matching Process Using a Markov Random Field Model for Cardiac Motion Extraction in MSCT Imaging

Antoine Simon[1], Mireille Garreau[1], Dominique Boulmier[2],
Jean-Louis Coatrieux[1], and Hervé Le Breton[1,2]

[1] Laboratoire Traitement du Signal et de l'Image, INSERM U642,
Université de Rennes 1, Campus de Beaulieu, 35042 Rennes, France
http://www.ltsi.univ-rennes1.fr/
[2] Centre Cardio-Pneumologique, CHU Pontchaillou, 35033 Rennes, France

Abstract. Multislice Computed Tomography (MSCT) scanners offers new perspectives for cardiac kinetics evaluation with 3D time image sequences of high contrast and spatio-temporal resolutions. A new method is proposed for cardiac motion extraction in Multislice CT. Based on a 3D surface-volume matching process, it provides the detection of the heart left cavities along the acquired sequence and the estimation of their 3D surface velocity fields. A 3D segmentation step and surface reconstruction process are first applied on only one image of the sequence to obtain a 3D mesh representation for one t time. A Markov Random Field model is defined to find best correspondences between 3D mesh nodes at t time and voxels in the next volume at $t+1$ time. A simulated annealing is used to perform a global optimization of the correspondences. First results obtained on simulated and real data show the good behaviour of this method.

1 Introduction

Cardiovascular diseases are a major cause of mortality, being responsible for about 30% of registered adult deaths in industrialized countries. Because a more sensitive measurement of myocardial function might result in earlier diagnosis, more effective treatment of heart disease may be possible. Technological improvements in cardiac imaging provide rich opportunities for such a progress.

The minimally invasive assessment of heart motion has been studied from four-dimensional (4D) data sets with magnetic resonance (MR) imaging (especially MR imaging tagging [1,2]), ultra-sound images [3] and recently with transthoracic 3D echocardiography [4], electron-beam computed tomography and the Dynamic Spatial Reconstructor (DSR) [5,6]. The recent significant advances of spiral computed tomography, with the introduction of ultra-fast rotating gantries (0.5 s/tr), multi-rows detectors and retrospective ECG-gated reconstructions, provide higher contrast and spatio-temporal resolutions and allow a huge progress toward the imaging of moving organs. The technical advances of MSCT imaging allow the reconstruction of all cardiac structures on one 3D image, for successive instants of the cardiac cycle, under one single breathhold. Some first reference studies have been conducted in MSCT for the detection of coronary diseases [7, 8], but very few for the quantitative 3D cardiac motion estimation [9].

The issue of nonrigid motion estimation from 3D images is one of the most important challenges of computer vision. Methods which have been proposed for this purpose are most often classified into three kinds of approaches: geometric deformable models, optical flow estimation and feature matching methods. In geometric model-based approaches, parametric models [10, 11, 12] involve the parametric formulation of the object and/or of the movement. This kind of methods is interesting to extract global motion and to represent it with few parameters. Non parametric models [13, 14], using mainly mass-spring and finite element methods, extract local motion using differential constraints. Optical flow methods [15, 5] are mainly based on intensity conservation and motion smoothing constraints. That constraint of intensity conservation with time is difficult to advance with MSCT data because of the contrast agent diffusion combined with the retrospective reconstruction of the sequence. Furthermore, these methods providing dense motion fields are difficult to handle with big data volumes in which the study deals only with few objects. Feature matching methods [16, 17, 18] are based on the search of correspondences between entities (considered at t and $t + 1$ times) according to descriptive parameters. These methods enable to focus the study on the objects of interest, and to extract local and global motion. But most of them are highly dependent from the segmentation quality because they need an accurate segmentation for each instant of the sequence.

We propose a new method to extract ventricular shapes and their motion from cardiac MSCT images. This approach provides, with one unique process and from the segmentation of only one moment, the detection of the object of interest along the time sequence of 3D images and its motion. This problem of dual 3D shape and motion estimation is viewed by a statistical approach as it has been used in other works to estimate dense 2D motion fields in outdoor scene analysis and video indexing [19, 20]. In this paper, the 3D sparse non-rigid motion field to estimate at each instant is formulated as a Markov Random Field model under spatio-temporal regularity hypotheses. This motion field is provided by a matching method based on features of different types which are surface 3D mesh nodes for one part and image voxels for the other part. These extracted motion fields can then be used for global and local motion quantification and interpretation. First results obtained on simulated and real data give promising results.

In the remainder of this paper, we describe in Sect. 2 the surface-volume matching method we have developed including the pre-processing step, the definition of the Markov Random Field model and of the associated energy function, and the optimization stage. In Sect. 3 we present the results obtained on simulated and real data. In Sect. 4 we conclude and discuss about the issues to develop in further research.

2 Surface-Volume Matching Based Method

From a time sequence of 3D MSCT cardiac images, our approach allows the spatio-temporal detection of the left heart cavities and the quantification of their deformations. This is achieved by a matching method which provides, for each instant of the sequence, the correspondences between a 3D surface mesh extracted at one instant and the 3D volume image available at the next instant. The overall process description of this method can be decomposed in such a way: a) a 3D segmentation step and surface reconstruction process are first applied on only one 3D image of the time sequence (at t_0 time) to provide the first surface in the sequence ; b) the surface-volume matching process is applied to estimate a 3D velocity field between a surface (at t_0 time) and the next volume image (at t_1 time). An iterative procedure is used to provide the best estimated motion field. c) A new 3D surface can be extracted at t_1 time from the estimated motion field; d) steps b) and c) are repeated until all images of the sequence have been processed.

2.1 Pre-processing Stage: 3D Segmentation and Surface Reconstruction

A segmentation process is applied to detect in the first image of the time sequence the object of interest. The method is based on a region growing process combined with contour detection and mathematical morphology operators [21]. It has been developed to provide a tool adapted to cardiac MSCT images and allows to extract a 3D object or a subset of objects in an isotropic 3D volume with minimal user interaction. The extracted object is then submitted to a Marching Cubes algorithm, giving access to a 3D mesh representation. This mesh is finally adjusted to provide a mesh such as edges have a length similar to distances between voxels in 3D images. This constraint has been chosen to be as close as possible of the available spatial details. This first stage provides the 3D surface mesh for one t time to be used in the matching process. The next temporal 3D image (at $t + 1$ time) is submitted to a contour detection operator labelling voxels as "borders" or not.

2.2 Surface-Volume Matching Process

A feature matching problem implies to choose the entities to match and to define local energies which can be combined to provide a distance measure between entities. The best correspondences of selected entities can be finally obtained from the minimization of a global energy.

Definition of Entities and Attached Local Energies. One original contribution of this method is to establish correspondences between spatial entities which are not of the same type. The matching process is conducted between 3D mesh nodes on one part and image voxels on the other part. A local energy is defined to compare a 3D mesh node with a voxel candidate for correspondence.

A first energy term $E_{1(f_i,d)}(N_i, V_k)$ models the local correspondence between one node N_i ($i = 1, ..., N_S$) (with N_S the number of nodes) at t time and one voxel V_k ($k = 1, ..., N_V$) (with N_V the number of voxels) at $t+1$ time. It provides a data conformity term and, in such a way, a distance between the observation d (surface at t and 3D image at $t+1$) and the estimated motion f_i from t to $t+1$ times. For each (N_i, V_k), this term is given by the following equation:

$$E_{1(f_i,d)}(N_i, V_k) = \alpha_c.E_{contour}(V_k) + \alpha_t.E_{topol}(N_i, V_k) + \alpha_d.E_{dist}(N_i, V_k) \quad (1)$$

where the energy $E_{contour}$ takes into account the probability that the voxel V_k belongs to a border; the term E_{topol} expresses a topological correspondence between the node N_i and the voxel V_k: the topology is described, for the surface, in terms of coordinates of the neighbouring nodes and compared, in the volume, to the corresponding voxel contour values. The energy E_{dist} takes into account the distance measured between the node N_i and the voxel V_k. α_c, α_t, α_d are weighting coefficients of the energy term.

A second energy $E_2(f_i)$ term which models spatio-temporal regularity constraints is introduced. It is defined from the Markov Random Field model described in the next paragraph.

Estimation of 3D Velocity Fields Based on a Markov Random Field Model. The issue is to extract the 3D motion field corresponding to the evolution of the object from one instant of the sequence to the next instant. This field is obtained by a matching process between a 3D surface mesh previously extracted at t time and the volume image available at $t+1$ time. The 3D motion field to compute between two successive instants is considered as a realization $f = \{f_i / i = 1, ..., N_S\}$ of a 3D Markov Random Field $F = \{F_i / i = 1, ..., N_S\}$, N_S being the number of considered sites in the field.

The sites of the MRF are given by all the 3D surface mesh nodes at t time. This surface corresponds to a graph $G = [S, U]$, with $S = \{i, i = 1, ..., N_S\}$ (N_S being the number of 3D nodes in the mesh) and with U the edges of the graph G. The labels assigned to these sites express the f_i estimations and are given by the voxels found (at $t+1$ time) in best correspondence with the 3D nodes: $f_i \in L$, with L the lattice of 3D voxels ($Nx \times Ny \times Nz$) in the next image.

The field F is considered as a MRF in relation to the neighbourhood of each site i:

$$\mu_i = \{j \in S / \{i, j\} \in U\} . \quad (2)$$

The neighbourhood system μ results from the set of sites and verifies:

$$\forall i \in S, i \notin \mu_i \quad (3)$$

$$\forall \{i, j\} \in S, j \in \mu_i \Leftrightarrow i \in \mu_j \quad (4)$$

According to these definitions, the neighbourhood associated to one node i is given by all nodes j which have a common edge with node i.

The MRF conditional probability is given by:

$$P(f_i/f_{S-i}) = P(f_i/f_{\mu_i}) \ . \qquad (5)$$

According to the Hammersley-Clifford theorem [22], the random field F defined in S is a Gibbs Random Field in relation to a neighbourhood system μ if and only if the probability distribution function is given by:

$$P(f) = \frac{1}{Z} \exp[U(f)] \qquad (6)$$

with $U(f)$ an energy function defining the interactions between the sites. Z is a normalization constant. The most probable realization f is provided by the minimization of this energy function $U(f)$. The function $U(f)$ can be more precisely noted $U(f,d)$, where f is related to the estimation and d to the observations at t and $t+1$ times. The total energy is defined by the linear combination of two terms:

$$U(f,d) = \alpha_1 U_d(f,d) + \alpha_2 U_r(f) \qquad (7)$$

where $U_d(f,d)$ models the estimation error. It is defined as a quadratic error computed on the set of sites from the local energy terms which have been previously described.

$$U_d(f,d) = \sum_{\substack{i \in S \\ k \in L}} (E_{1(f_i,d)}(N_i, V_k))^2 \ . \qquad (8)$$

$U_r(f)$ represents the internal energy of the Markov Field and has a regularization effect. It is defined as:

$$U_r(f) = \sum_{i \in S} E_2(f_i) \qquad (9)$$

with $E_2(f_i) = |f_i - f_{\mu_i}|$ and $f_{\mu_i} = \frac{1}{n_i} \sum_{j \in \mu_i} f_j$, n_i being the number of neighbours of site i.

Optimization Stage. To perform a global optimization of the correspondences, the stochastic relaxation Metropolis algorithm is used, combined with a simulated annealing process. A first level iterative process is defined: a random scanning of nodes included in the 3D mesh at t time is realized and, for each node, a new candidate voxel at $t+1$ time, which corresponds to one node transition, is randomly chosen. The acceptance criteria for the transition of these nodes is based on the comparison between the energy terms $U(f,d)$ computed for the current configuration and for the proposed new configuration resulted from the tested transition. The number of transitions accepted with an increasing energy is controlled by a temperature parameter. The stopping criteria used for this first iteration is related to the number of accepted transitions with an increasing energy and to the number of scanned nodes.

This first loop is included in a second iterative procedure in which the temperature parameter is linearly decreased at each step. The initial temperature is set to a high value in order to allow an important number of accepted transitions at the beginning of the process. The field of correspondences is then re-adjusted until a stopping criteria is reached.

3 Results

3.1 Tests on Simulated Data

Numerical simulations have been used to test the motion extraction process between two successive instants. These simulations are based on a model including one surface representing the endocardium at t time and one data volume including another surface (the first mesh after deformation) at $t+1$ time. Because of the great complexity to create a whole data volume with various grey levels, the simulated data volume, generated by the transposition of the deformed surface into an empty data volume, corresponds to the volume resulting from the application of the edge operator.

In order to create the surface corresponding to the first instant, a superellipsoidal shape is locally deformed to obtain local topological features comparable to those of the real endocardial surface (cf. Fig. 1). To simulate the motion between two successive instants, this shape is successively deformed using five kinds of motion (translation, twisting, rotation, global expansion/compression, and local deformations) resulting in the mesh corresponding to the second instant (cf Fig. 2). Because of the complexity to model cardiac motion with precision and in order to test the algorithm in the worst situation, the simulated motion has been applied with a greater amplitude than the motion potentially observed between two successive volumes of MSCT databases. Results obtained with this situation are here described. Figure 3 represents one axial slice of these two simulated shapes.

The mesh corresponding to $t+1$ time is then transposed into an empty data volume (75 slices of 128×128 pixels) to obtain a volume corresponding to the result of the edge operator application. In this volume, to simulate the presence of other objects (such as the epicardium) and of noise, another surface (previous mesh after an expansion process) is transposed in the volume and a binary noise is finally added (cf Fig. 4 representing one axial slice of the simulated volume at $t+1$ time). Using this process, the real correspondences are known. It enables to measure the error of matching of the proposed method and to study the evolution of the matching process along iterations. The impact of the different

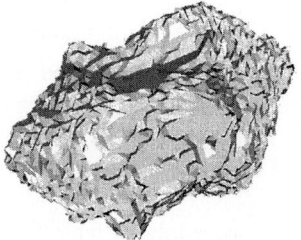
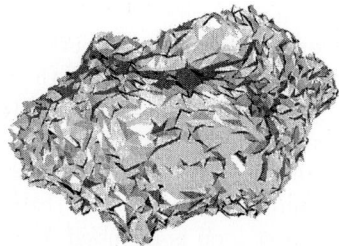

Fig. 1. Simulated surface corresponding to t time

Fig. 2. Simulated surface corresponding to $t+1$ time

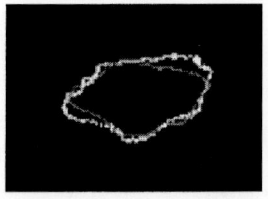 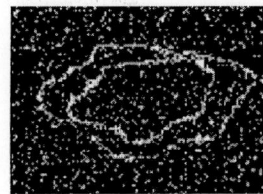 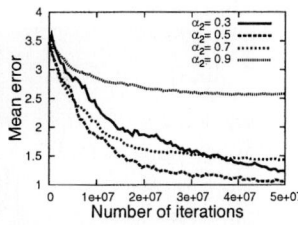

Fig. 3. Axial slice representing two contours corresponding to the simulated endocardium at t time (in grey colour) and $t + 1$ time (in white colour)

Fig. 4. Axial slice of the simulated data volume corresponding to $t + 1$ time (simulated endocardium and epicardium with noise)

Fig. 5. Evolution of the error of matching according to the number of iterations for different values of α_2 in the energy function

parameters involved in the computation of the energies or in the optimization process can also be evaluated.

For instance, Fig. 5 shows the evolution of the matching error according to the number of iterations (one iteration corresponding to the test of one node's transition) for different values of the weighting factor corresponding to the regularization term in the energy computation (cf. Eq. 7), highlighting better results for an optimal value of 0.5.

Tests conducted with different shapes and motions have enable to find, for each parameter involved in the matching process, its value corresponding to the minimal final error. With this set of parameters ($\alpha_1 = 0.50$, $\alpha_2 = 0.30$, $\alpha_c = 0.10$, $\alpha_d = 0.05$, $\alpha_t = 0.30$), a final mean 3D error of matching of 1.0 voxel (minimum: 0 voxel, maximum: 3.7 voxels) has been obtained.

3.2 Results on Real Data

The algorithm has been applied on real human heart data with a temporal database acquired by a Siemens SOMATOM Sensation 16 with ten volume images representing a whole cardiac cycle. Each volume contains about 300 slices of 512×512 pixels, giving a resolution for each voxel of $0.3 \times 0.3 \times 0.6$ mm (cf Fig. 6(a) illustrating one CT axial slice).

The segmentation pre-process has been applied to the first volume of the sequence resulting in the extraction of the heart left cavities and of the beginning of the aorta [21]. In order to highlight main motion and to reduce computational time, the surface-volume matching method has been applied at a lower resolution obtained after three filtering and down-sampling processes. To obtain the surface mesh corresponding to t_0 time, the low resolution segmented volume has been processed by the Marching Cubes algorithm (cf Fig. 6(c)). Using the optimal set of parameters found with the numerical simulations, the motion extraction process has been applied between this mesh (corresponding to t_0 time) and the volume corresponding to t_1 time after an edge operator application. After the test of different detection operators, the Canny operator has been chosen in

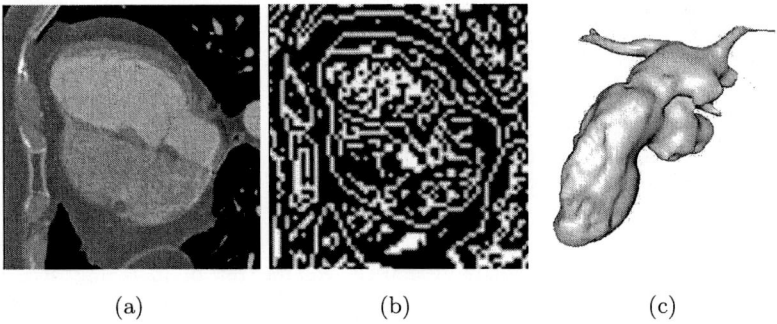

Fig. 6. (a) One axial slice of one data volume of the sequence, (b) corresponding axial slice (at lower resolution) after the Canny filtering process and (c) 3D surface representation of the extracted shape

order to promote false positive more than false negative results (cf Fig. 6(b)). The algorithm has then been iteratively applied to the successive volume images, providing a set of estimated surfaces and motion fields for each instant of the sequence.

To evaluate the precision of the method, the resulting estimated surfaces (one for each instant from t_1 to the end of the sequence) have been compared to surfaces corresponding to the same instant, but obtained with the segmentation tool used in the pre-processing stage. The mean difference between the two surfaces increases at each estimation process (from 0 mm to 0.5 mm), resulting to 1.5 mm after six estimation processes. These first results show that the segmentation part of the algorithm provides satisfying results.

As example of motion extraction results, the estimated motion amplitude is visualized in Fig. 7(a) and 7(b) for the end of the ventricular diastole (systole of

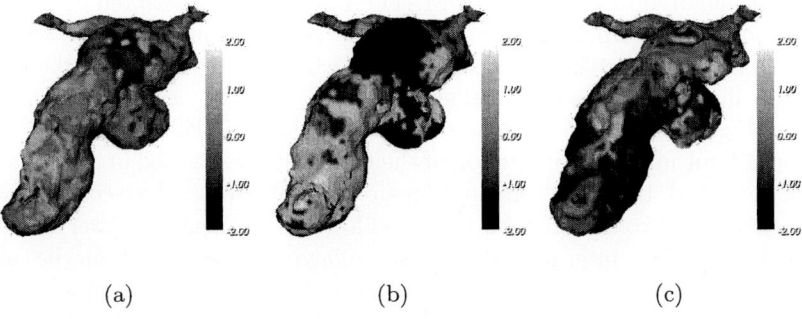

Fig. 7. Estimated motion amplitude at three following instants (from the end of the ventricular diastole (a,b) to the beginning of the ventricular systole (c)) (colours: in white: expansion, black: contraction)

the atrium) and in Fig. 7(c) for the beginning of the ventricular systole, these three surfaces corresponding to successive instants in the sequence. The contraction movements (represented with black colour), characteristic of systole, are as well identified as the expansion movements (represented with white colour). Moreover, movements extracted on the whole sequence are coherent with cardiac phases.

This kind of representations enables to highlight functional abnormalities. For instance, Fig. 7 highlights the pathological situation where the apical area suffers from akinesia.

4   Conclusions

A new solution of motion extraction combined with surface estimation has been introduced and applied to the left ventricle in 4D cardiac MSCT imaging. This approach is based on a statistical surface-volume feature matching method formulated with a Markov Random Field and provides, with one unique process, left cavity surfaces and associated 3D motion vector fields. The algorithm has been tested with simulated and real MSCT dynamic data, highlighting encouraging results and confirming the great potential of MSCT imaging for quantitative clinical measure assessment in cardiac applications. Further works will carry on algorithm optimization and on extensive evaluation.

Acknowledgements

This work is supported by Brittany region and the French Medical Research and Health National Institute (INSERM). The authors express their thanks to Siemens Medical Division France and especially to Brigitte Lebruno for having provided us real MSCT image data.

References

1. Prince, J.L., McWeigh, E.R.: Motion estimation from tagged MR image sequences. IEEE Transactions on Medical Imaging **11** (1992) 238–249
2. Clarysse, P., Han, M., Croisille, P., Magnin, I.E.: Exploratory analysis of the spatio-temporal deformation of the myocardium during systole from tagged MRI. IEEE Transactions on Medical Imaging **149** (2002) 1328–1339.
3. Mulet-Parada, M., Noble, J.A.: 2D+T acoustic boundery detection in echocardiography. Medical Image Analysis **4** (2000) 21–30
4. Papademetris, X., Sinusas, A. J., Dione, D. P., Duncan, J. S.: Estimation of 3D left ventricular deformation from echocardiography. Medical Image Analysis **5** (2001) 17–28
5. Gorce, J. M., Friboulet, D., Magnin, I. E.: Estimation of three-dimensional cardiac velocity fields: assessment of a differential method. Medical Image Analysis **1** (1997) 245–261

6. Eusemann, C.D., Ritman, E.L., Robb, R.A..: Parametric visualization methods for the quantitative assessment of myocardial motion. Acad Radiol. **10** (2003) 66–76
7. Schroeder, S., Kopp, A.F., Ohnesorge, B., Floh, T., Baumbach, A., Kuettner, A., Herdeg, C., Karsch, K., Claussen, C.D.: Accuracy and reliability of quantitative measurements in coronary arteries by multi-slice computed tomography: Experimental and initial clinical results. Clinical Radiology **56** (2001) 466–474
8. Larralde, A., Boldak, C., Garreau, M., Toumoulin, C., Boulmier, D., Rolland, Y.: Evaluation of a 3D Segmentation Software for the Coronary Characterization in Multi-slice Computed Tomography. Lecture Notes in Computer Science **2674** (2003) 39–51
9. Garreau, M., Simon, A., Boulmier, D., Guillaume, H.: Cardiac Motion Extraction in Multislice Computed Tomography by using a 3D Hierarchical Surface Matching process. IEEE Computers in Cardiology Conference; Chicago, USA, (2004)
10. Frangi, A. F., Niessen, W. J., Viergever, M. A.: Three-Dimensionnal Modeling for Functionnal Analysis of Cardiac Images: A Review. IEEE Transactions on Medical Imaging **20** (2001) 2–25
11. Bardinet, E., Cohen, L. D., Ayache, N.: Tracking and motion analysis of the left ventricle with deformable superquadrics. Medical Image Analysis **1** (1996) 129–149
12. Chen, C.W., Huang, T.S., Arrott, M.: Modeling, analysis and visualization of left ventricle shape and motion by hierarchical decomposition. IEEE Transactions on Pattern Analysis and Machine Intelligence **16** (1994) 342–356
13. Choi, S.-M., Kim, M.-H.: Motion visualization of human left ventricle with a time-varying deformable model for cardiac diagnosis. The journal of visualization and computer animation **12** (2001) 55–66
14. Benayoun, S., Ayache, N.: Dense non-rigid motion estimation in sequences of medical images using differential constraints. Int. Journal of Computer Vision **26** (1998) 25–40
15. Song, S. M., Leahy, R. M.: Computation of 3-D velocity fields from 3-D cine ct images of the human heart. IEEE Transactions on Medical Imaging **10** (1991) 295–306
16. Amini, A. A., Duncan, J. S.: Bending and stretching models for LV wall motion analysis from curves and surfaces. Image and Vision Computing **10** (1992) 418–430
17. Kambhamettu, C., Goldgof, D., He, M., Laskov, P.: 3D nonrigid motion analysis under small deformations. Image and Vision Computing **21** (2003) 229–245
18. Shi, P., Sinusas, A.J., Constable, R.T., Ritman, E., Duncan, J. S.: Point-tracked quantitative analysis of left ventricular motion from 3D image sequences. IEEE Transactions on Medical Imaging **19** (2000) 36–50
19. Heitz, F., Bouthemy, P.: Multimodal Estimation of Discontinuous Optical Flow Using Markov Random Fields. IEEE Transactions on Pattern Analysis and Machine Intelligence **15** (1993) 1217–1232
20. Keng Pang Lim, Das, A., Man Nang Chong: Estimation of Occlusion and Dense Motion Fields in a Bidirectional Bayesian Framework. IEEE Transactions on Pattern Analysis and Machine Intelligence **24** (2002) 712–718
21. Guillaume, H., Garreau, M.: Segmentation de cavités cardiaques en imagerie scanner multi-barettes. Forum des Jeunes Chercheurs en Génie Biologique et Médical (2003)
22. Besag, J.: Spatial interaction and the Statistical Analysis of Lattice Systems. Journal of the Royal Statistical Society **36** (1974) 192–236

Evaluation of Two Free Form Deformation Based Motion Estimators in Cardiac and Chest Imaging

Bertrand Delhay[1], Patrick Clarysse[1], Jyrki Lötjönen[2], Toivo Katila[3], and Isabelle E. Magnin[1]

[1] Creatis, CNRS UMR 5515, Inserm U630, 69621 Villeurbanne, France
delhay@creatis.insa-lyon.fr
[2] VTT Information Technology, P.O.B. 1206, FIN-33101 Tampere, Finland
[3] Laboratory of Biomedical Engineering, Helsinki University of Technology, P.O. Box 2200, FIN-02015 HUT, Finland

Abstract. In the context of motion estimation of the heart and thoracic structures from tomographic imaging, we investigated two free form deformations (FFD) based non linear registration methods as motion estimators. Standard and cylindrical FFD (CFFD) methods are evaluated in 2D, both on simulated and in vivo cardiac and thoracic images. Results tend to show that CFFD based method achieves the same accuracy with less parameters. However, the fast convergence of this model is hamped by a higher computing time with a straightforward implantation.

1 Introduction

Due to the great progress of image acquisition devices, it is now possible to better explore the dynamics of moving organs such as the heart and the lungs. Magnetic resonance imaging (MRI) and X-ray computed tomography (CT) can provide information about the anatomy of the heart over the cardiac cycle. The general context of our study is the motion analysis of organs within the ribcage (in particular lungs and heart) for spatio-temporal segmentation of anatomical structures and motion compensation in image sequences. Clinical applications could be the tracking of anatomical structures and tumor during radiotherapy treatments, or the motion compensation of the heart during minimally invasive heart surgeries from preoperating tomographic acquisitions [1]. In the former case, organ's motion is, in current practice, not directly taken into account in the radiation of tumors, except through error margins defined in the radiotherapy treatment planning. Integration of motion information would therefore result in a much more accurate irradiation without the need for breath control of the patient.

In this paper, we focus on the motion estimation of rounded like anatomical structures, such as the whole thorax and the heart from dynamic tomographic image acquisitions. In this context, some author [2, 3, 4] have proposed cylindrical free form deformation (CFFD) as a potentially better suited transformation for the reconstruction of the heart motion from cardiac image sequences. In this paper, a CFFD based registration method is compared to the more classical free form deformation (FFD) based registration both on simulated and in-vivo images of a beating heart and breathing thorax.

2 Material and Method

2.1 Image Registration as Motion Estimator

Image registration algorithms are used to estimate correspondences between two image data sets. The image registration problem can be formally defined by the search of a multidimensional function within a space of admissible function, called *warp space* such that:

$$I^f(\mathbf{g}(\mathbf{x})) \cong I^r(\mathbf{x}), \tag{1}$$

where I^f is the floating image to be warped, thanks to the transformation $\mathbf{g}$, for any coordinates $\mathbf{x}(x,y)$ in 2D, and I^r is the reference image. Thus, the task is to find $\mathbf{g}$ so that the warped floating image matches as well as possible the reference image. In the literature, there exists a lot of methods, based on different characteristics, to seek after this transformation (See [5] and [6] for detailed reviews). In medical imaging, non-rigid registration is usually required to put into correspondence images from two patients or to estimate the motion between two images of the same patient acquired at different time points.

The selected *warp space*, characterizes the registration algorithm. The parametric and global methods can handle coarse motions (limited *warp space*) with a small number of parameters. On the other hand, non parametric approaches like variational methods, pioneered by Horn and Shunck [7], can handle a wide *warp space*. In the regularization framework, the registration problem can state as the search for a transformation $\hat{\mathbf{g}}(\mathbf{x})$ which maximizes the similarity between I^f and I^r with given smoothness properties . Therefore, thin-plates [8] and elastic [9] transformations have been proposed as regularizing functionals. Christensen et al, [10], enforce the consistency by taken into account forward and reverse transformations. Piecewise polynomial functions, like splines [11], present the advantage of a local control of the deformation while being piloted by a limited number of parameters.

2.2 FFD-Based Registration

FFD algorithms [12, 13] aim to recover local smooth deformations. Basically, the principle of a free-form deformation is to deform an object by manipulating an underlying mesh of control points. The transformation is defined by a geometrical regular control point grid [14](Fig. 1). According to [11, 15], in 2D, the FFD function $\mathbf{g}$ can be expressed by the tensor product of B-Splines:

$$\mathbf{g}(\mathbf{x},\mathbf{c}) = \mathbf{g}(x,y,\mathbf{c}) = \sum_{(k_x,k_y) \in K} c_{k_x,k_y} \beta(x/h_x - k_x) \beta(y/h_y - k_y), \tag{2}$$

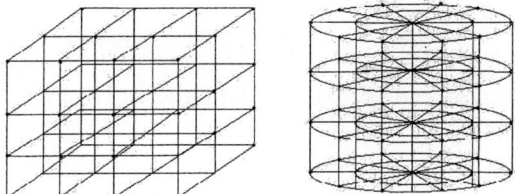

Fig. 1. (left) Cartesian grid for FFD (right) cylindrical grid for CFFD

where K stands for the set of all the points inside the control grid. The more points, the greater the dimension of the *warp space*. If there is M nodes and data lies in a N dimensional space, dimension of the vector of parameters **c** is defined by $M \cdot N$. Within a 2D deformation grid, the nodes are indexed by k_x and k_y. The displacement at node $[k_x, k_y]$ is defined by the parameter c_{k_x, k_y} and, h_x and h_y are the grid spacings which define the distance between two cubic B-Spline functions β.

2.3 CFFD-Based Registration

Without any prior knowledge about the structures and their motion, the standard FFD-based registration method is a reasonable method to establish the transformation. In cardiac image analysis, the standard shape of the heart and its motion are, approximately known at least qualitatively. Recently, CFFD transformation models (Fig. 1), have been proposed to more closely fit some specific motions (ie. motion of the left ventricle) with less parameters than with cartesian grids [2, 3, 4]. They may be more appropriate when the motion to be estimated has a central tendency. In short axis acquisition plane, the left ventricular (LV) myocardium looks like a ring which thickens during the systole. Therefore, it is reasonable to expect that CFFD would allow a better reconstruction of the LV's motion. CFFD have been introduced in the 90's [16]. Chandrashekara *et al* [2] and Deng *et al* [4] used this model for tracking the motion of the myocardium in tagged magnetic resonance images and Lin adapted a cylindrical topological regular grid on manually segmented left ventricular walls in order to perform the segmentation through image sequences [3].

The CFFD function is summarized in polar coordinates as:

$$\mathbf{g}(r, \theta, \mathbf{c}) = \sum_{(k_r, k_\theta) \in K} c_{k_r, k_\theta} \beta^o(r/h_r - k_r) \beta^c(\theta/h_\theta - k_\theta). \qquad (3)$$

β^o stands for open cubic spline function, and β^c stands for closed cubic spline function to handle the periodicity in the circumferential direction. Because of the local support of the B-Spline for a given point $\mathbf{p}(r, \theta)$, equation (3) can be written as:

$$\mathbf{g}(\mathbf{p}, \mathbf{c}) = \mathbf{g}(r, \theta, \mathbf{c}) = \sum_{k=-1}^{2} \sum_{l=-1}^{2} c_{k_{r_{i+k}}, k_{\theta_{j+l}}} \beta^o(r/h_r - k_{r_{i+k}}) \beta^c(\theta/h_\theta - k_{\theta_{j+l}}), \qquad (4)$$

with
$$k_{r_i} = \lfloor r/h_r \rfloor, k_{\theta_j} = \lfloor \theta/h_\theta \rfloor. \tag{5}$$

$\lfloor \rfloor$ is the floor operator and (k_{r_i}, k_{θ_j}) is the index of the first point of the lattice in which **p** is situated.

2.4 Similarity Metrics

In [2], normalized mutual information is used because tag intensity changes with time (tag fading). With classical MRI and CT modalities we assume that the intensity of a tissue does not change much and we use the *sum of square differences* (*SSD*) criterion. The objective function E to be minimized in the registration process is:

$$E(\mathbf{c}) = \sum_{\mathbf{p} \in ROI} (I^f(\mathbf{g}(\mathbf{p},\mathbf{c})) - I^r(\mathbf{p},\mathbf{c}))^2. \tag{6}$$

Most of the time, the intensity value $I^f(\mathbf{g}(\mathbf{p},\mathbf{c}))$ does not correspond to the value available at an exact pixel location. Therefore an interpolation is required to map the floating image. Linear interpolation was used. Minimization of the objective function E, is achieved through a gradient descent search. Kybic [15] developed the computation of the derivative of the objective function. If q is the linear application that transforms a point $\mathbf{x}(x,y)$ into $\mathbf{p}(r,\theta)$, we have:

$$\nabla_c E(\mathbf{c}) = \frac{\partial E}{\partial \mathbf{c}} = 2 \sum_{x \in ROI} e_\mathbf{x} \frac{\partial e_\mathbf{x}}{\partial \mathbf{c}} \quad \text{where} \quad e_\mathbf{x} = I^f(\mathbf{g}(q(\mathbf{x}),\mathbf{c})) - I^r(q(\mathbf{x})), \tag{7}$$

$$\text{and} \quad \frac{\partial e_\mathbf{x}}{\partial \mathbf{c}} = \frac{\partial I^f}{\partial \mathbf{p}}\bigg|_{\mathbf{p}=g(q(\mathbf{x}),\mathbf{c})} \cdot \frac{\partial \mathbf{g}}{\partial \mathbf{c}}\bigg|_{\mathbf{p}=q(\mathbf{x})} \tag{8}$$

In Equation (8), $\frac{\partial I^f}{\partial \mathbf{p}}$ is the gradient of I^f at point **p**, characterized by its two polar coordinates. Thanks to the basis change matrix **Q**:

$$\frac{\partial I^f}{\partial \mathbf{p}} = \mathbf{Q}^{-1} \frac{\partial I^f}{\partial \mathbf{x}} \quad \text{where} \quad \mathbf{Q}^{-1} = \begin{bmatrix} \cos\theta & \sin\theta \\ -\sin\theta & \cos\theta \end{bmatrix}, \tag{9}$$

where $\frac{\partial I^f}{\partial \mathbf{x}}$ is the gradient of I^f at point **x**, characterized by its two cartesian coordinates. The iterative process to derive the optimal parameters is defined by:

$$\mathbf{c}_{n+1} = \mathbf{c}_n + \lambda \nabla_c E(\mathbf{c}), \tag{10}$$

where λ is halved when two consecutive gradient directions are different. In this framework, no regularization is applied, but piecewise B-Spline polynomial functions implicitly favor smooth deformations.

2.5 Test Data Sets

Simulated Thoracic Data. Image sequence (Sim_1) of a breathing thorax, consisting in a two time points sequence, was simulated from a real 2D X-ray dynamic CT scan (Fig. 2a). Dimensions of the images are 512×512 pixels and spatial resolutions are $1mm \times 1mm$. As the true motion is known, rectangular and cylindrical models can be compared through a reference. The theoretical motion field is used to emphasize advantages and drawbacks of the respective methods.

In Vivo Thoracic Data. Fig. 2b shows images which have been acquired during a radiotherapy treatment (Centre Léon Bérard, Lyon, France). A breath control device (Active Breath Control) was used to monitor the patient breathing (See [17] for more details). For two patients, three different volumes are available which corresponds to three steps of breathing. Corresponding 2D slices were extracted from 3D volumes to provide 2D test sets ($ABC_{1,2}$) whose dimensions are 512×512 voxels and spatial resolutions are $0.94mm \times 0.94mm$.

In Vivo Cardiac Data. Both models were tested on heart cine-MR images (Fig. 2c). Short Axis (SA) scans were obtained from a 1.5T Siemens Magnetom Vision at the Helsinki Helsinki Medical Imaging Center, Helsinki University, Finland. The temporal resolution is 30-40 ms, leading to about 20 time points over the cardiac cycle. Dimensions of the data are 224×256 pixels and spatial resolutions are $1.44mm \times 1.44mm$ or $1.31mm \times 1.31mm$. For evaluation purpose, five 2D mid-ventricular sequences ($S_{1,...,5}$) of 20 time points were selected from 5 patient's acquisitions.

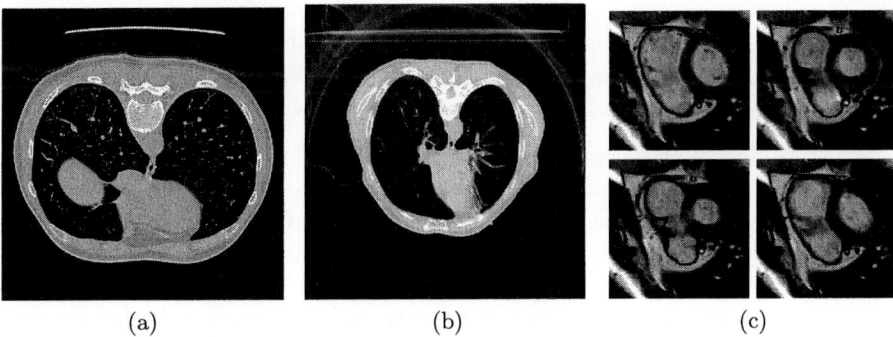

(a)   (b)   (c)

Fig. 2. Data set used to test the two models *a)* 2D Dynamic CT slice used to generate the sequence of a breathing thorax, *b)* one of the three breathing time points from the classical CT data set acquired during radiotherapy protocol (3 corresponding 2D slices has been selected), *c)* 4 of the 20 time points available from S_2 sequence

2.6 Evaluation of the Results

For tests performed on simulated data, the gold standard is available to estimate the accuracy of the results. Dense motion field can be assessed in terms of quadratic (QE) or angular error (AE):

$$QE = \|\boldsymbol{u} - \boldsymbol{u}'\| \quad \text{and} \quad AE = \frac{\boldsymbol{u} \cdot \boldsymbol{u}'}{\|\boldsymbol{u}\| \cdot \|\boldsymbol{u}'\|}, \tag{11}$$

where $\boldsymbol{u}$ is the estimated displacement and $\boldsymbol{u}'$ is the theoretical displacement. We also computed the image difference after registration.

When no ground truth is available, measurement of consistency evaluates the estimated displacements. It can be assessed with the protocol described by Wierzbicki in [18]. Basically, it consists in tacking into account the periodicity of cardiac and breathing motions. Once all the transformations of an image sequence have been estimated, we apply consecutively those transformations on the initial image of the sequence. The difference between the initial image and the loop-warped initial image is computed to assess the accuracy of the models. The error metrics are computed over the same spatial support defined by a manually segmented binary mask.

3 Results and Discussion

The CFFD method has been implemented in C++ language using the ITK library[1] on a Pentium 4 processor, 1.7 GHz. We arbitrarily chose to stop all the registration tests at a maximum of 150 iterations in order to compare the computing times. For both models and for each data set, the evolution of the SSD criterion was less than 0.0001 times its initial value within the 150th iteration. For thoracic data sets, the cylindrical grid central point was put on the spine.

3.1 Grid Resolution Impact with Simulated Data

To evaluate the behavior of the proposed method, we estimated the minimum, the maximum, the mean and the standard deviation of the criterions given in Eq. (11) and of the image difference after registration. The results given in Table 1 and Fig. 3 lead to the following remarks:

- The tuning of the grid size is different for standard FFD and CFFD. The grid points are related to the cartesian coordinate system with standard FFD and to the polar coordinate system with CFFD. The QE criterion clearly decreases when the radial number of points increases. Similarly, AE significantly decreases as a function of the number of points in the circumferential domain. In those tests, the center point was manually defined. The random initialization of this center in a limited window did not change significantly the results. With a sensible initialization, SSD criterion is usually better with CFFD methods than with FFD methods. However, respiration has been simulated by warping initial data with a cylindrical grid; therefore the requested motion might be naturally better estimated with the CFFD than with the classical FFD. This can partially biases the results.
- The same final values can be reached with half the parameters using CFFD.

---

[1] http://www.itk.org/

Table 1. Angular, quadratic error between estimated and theoretical motion fields and image intensity difference between the *reference image* and motion compensated *floating image*. For CFFD, the first term in the field "*N. of Par.*" stands for the number of points in the radial direction, the second one, stands for the number of points in the circumferential direction

N. of Par.	AE (degree)		QE (mm)		Diff. (intensity value)	
	Min-Max	Mean±Std.Dev.	Min-Max	Mean±Std.Dev.	Min-Max	Mean±Std.Dev.
CFFD						
2 × 4	0-179	8.50 ± 30.74	0-24.49	4.47± 4.13	0-2861	122.76 ± 225.36
3 × 4	0-179	10.04 ± 31.88	0-14.18	3.52 ± 2.63	0-2597	103.63 ± 194.60
4 × 4	0-179	10.73 ± 31.82	0-9.55	2.99 ± 2.25	0-2526	89.87 ± 173.42
4 × 8	0-179	7.27 ± 22.07	0-10.78	2.94 ± 2.74	0-2529	81.02 ± 161.23
5 × 8	0-179	5.41 ± 20.23	0-4.71	0.61 ± 0.67	0-1100	24.97 ± 41.16
7 × 8	0-179	5.62 ± 20.85	0-2.64	0.32 ± 0.32	0-800	14.52 ± 21.1254
FFD						
4 × 4	0-179	21.57 ± 27.48	0-24.82	7.96 ± 4.97	0-2731	143.21 ± 250.69
6 × 6	0-179	18.58 ± 32.37	0-13.53	3.06±2.43	0-1923	85.83 ± 153.54
7 × 7	0-179	15.06 ± 28.85	0-10.74	1.74 ± 1.40	0-1749	62.39 ± 111.55
8 × 8	0-179	12.47 ± 34.47	0-9.53	1.07 ± 1.14	0-1146	43.57 ± 72.03
10 × 10	0-179	12.24 ± 30.76	0-9.35	0.8 ± 1.19	0-285	19.55 ± 28.67

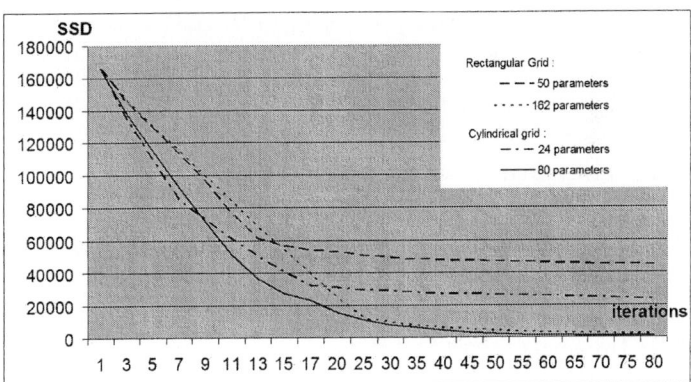

Fig. 3. Evolution of the SSD dissimilarity criterion for rectangular and cylindrical grids in Sim_1 data set

- The dynamics of the curves, in Fig. 3, are indeed different between the two models. CFFD converges faster (especially in the first iterations). Reasonable transformations are provided in a small number of iterations (this tendency will be confirmed by tests on true data). However, the CFFD model suffers from a higher computing time as compared to the standard FFD. The basis change (Eq. 9) introduces a matrix product for each point in the ROI. In that case, the computing time took more than 1.5 times the FFD computing time (this ratio increases with the ROI dimensions). For a good approximation of the final value ($\leqslant 5\%$), the effective time was more important for CFFD ($\approx 1min$ for 150 iterations). However it seems possible to take advantage of the fast evolution of the cylindrical grid during the first iterations in a specially adapted CFFD mutiresolution process.

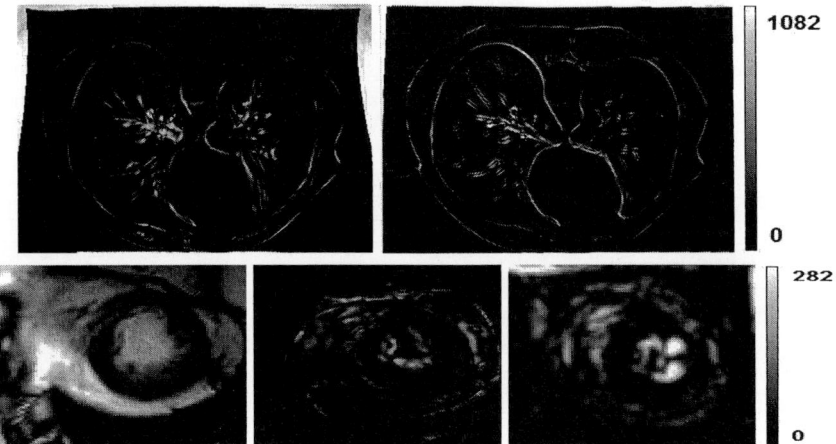

Fig. 4. Difference images provided by registration loop on ABC_2 (*Up row*). Cylindrical grid *(left)* with 96 parameters (6 × 8 × 2) *vs.* Rectangular grid with 98 parameters (7 × 7 × 2)*(right)*. *Bottom row*, from left to right, initial MR image on S_1, difference image with CFFD based registration and with FFD based registration. Cylindrical grid *(middle)* with 64 parameters (4 × 8 × 2) *vs.* Rectangular grid with 72 parameters (6 × 6 × 2)*(right)*

3.2 Consistency Study with in Vivo Data

Once all the transformations have been found between heart and thorax frames, we consecutively applied the motion fields on the initial reference images. In Fig. 4, the difference between an initial reference CT and MR image and the resulting loop-warped initial image on ABC_2 and S_1 are shown. The difference between these two images show that, for a small number of parameters, the mean error is significantly reduced for CFFD, especially for area far from the center.

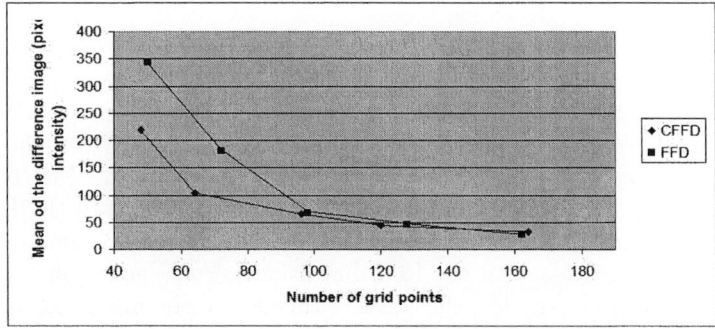

Fig. 5. Evolution of the mean of the image difference for the slice S_3 after registration. Similar behavior can be observed for the other slices

Moreover, with FFD model, between two time points on $S_{1,...,5}$, the computing time was only 2 *min* (150 iterations), whereas CFFD took 3.5 *min*. As shown in Fig. 5, this difference between the two models is important at coarse grid resolutions in favor of CFFD. However, when the grid resolution increases this difference tends to collapse. Then, the usefulness of the CFFD is not as clear as with reduce number of parameters.

4 Conclusion

The objective of this paper was to investigate the efficiency of the CFFD-based non linear registration as motion estimator compared with standard FFD based method in cardiac and thoracic imaging. This evaluation has been carried out in 2D and on both simulated and in vivo image sequences. The obtained results on simulated data set show a good accuracy of the CFFD with a small number of grid points. With a reasonable initialization, CFFD converges in less iterations. Nevertheless, for comparable accuracies, the computing time remains higher with the CFFD method than with the FFD method. So, the advantage for using this new model is not clearly demonstrated on in vivo thorax CT images and heart cine-MR images, especially when considering a high number of grid points. However, specific time-optimized implementations and multiresolution grids might reduce the computing time and the sensitivity of the initialization. This study has also to be re-edited in 3D as the motion of the heart and the thorax structures are truly three-dimensional.

Acknowledgments

This work is supported by the "Région Rhônes-Alpes", France, through the EURODOC Program. We also thank D. Sarrut for providing us the CT scans of the thorax.

References

1. S. Bonnet, A. Koenig, S. Roux P. Hugonnard, R. Guillemaud and P. Grangeat. Dynamic X-ray computed tomography. *Proceedings of the IEEE*, 91:1574–87, 2003.
2. R. Chandrashekara, R.H. Mohiaddin, and D. Rueckert. Analysis of myocardial motion and strain patterns using a cylindrical B-Spline transformation model. In Springer-Verlag, editor, *Proc. IS4TM*, pages 88–99, Berlin Heidelberg, 2003.
3. N. Lin and J.S. Duncan. Generalized robust point matching using an extended free-form deformation model: Application to cardiac images. In *Proc. ISBI*, pages 320–323, 2004.
4. X. Deng and T.S. Denney. Three-Dimensional myocardial strain reconstruction from tagged MRI using a cylindrical B-spline model. *IEEE Transactions on Medical Imaging*, 23(7):371–386, July 2004.
5. M.A. Viergever, J.B.A. Maintz. A survey of medical image registration. *Medical Image Analysis*, 2:1–36, 1998.

6. T. Mäkelä, P. Clarysse, O. Sipila, N. Pauna, Q.C. Pham , T. Katila, I.E. Magnin. A review of cardiac image registration methods. *IEEE Transactions On Medical Imaging*, 21:1011–21, 2002.
7. B. Horn and B. Shunck. Determining Optical Flow. *Artificial Intelligence*, 17:185–203, 1981.
8. F.L. Bookstein. Principal warps: Thin plate splines and decomposition of deformation. *IEEE Transactions on Pattern Analysis and Machine Intelligence*, 11:567–585, 1989.
9. T. Rohlfing, C.R. Maurer, J. Bluemke, D. Jacobs, M. Volume Preserving non rigid registration of MR Breast images using Free Form deformation with an imcompressibility constraint. *IEEE Transactions on Medical Imaging*, 22:730–741, 2003.
10. H.J. Johnson and G.E Christenen. Consistent landmark and intensity-based image registration. *IEEE Transactions on Medical Imaging*, 21:450–461, 2002.
11. M. Unser, A. Albroudi,and M. Eden. B-Spline signal processing: PartI-Theory. *IEEE Transactions On Signal Processing*, 41:821–832, 1993.
12. J. Lötjönen and I.E. Magnin and P. Reissman and J. Nenonen and T. Katila. Segmentation of magnetic resonance images using 3D deformable model. *MICCAI 98, Cambridge, USA*, pages 1211–1221, 1998.
13. D. Rueckert, L.I. Sonoda,C. Hayes, D.L.G Hill. Nonrigid registration using free form deformation: Application to breast MR images. *IEEE Transactions on Medical Imaging*, 11:712–721, 1999.
14. T.W. Sederberg and S.R. Parry. Free-form deformation of solid geometric models. *SIGGRAPH*, pages 151–160, 1986.
15. J. Kybic. *Elastic Image Registration using Parametric Deformations Models*. PhD thesis, Ecole Polytechnique Fédérale de Lausanne, 2001.
16. S. Coquillart. Extended free-form deformation: a sculpturing tool for 3D geometric modeling. *International Conference on Computer Graphics and Interactive Techniques*, pages 187–196, 1990.
17. V. Boldea, D. Sarrut, and S. Clippe. Lung deformation estimation with non-rigid registration for radiotherapy treatment. In *MICCAI 03, Montreal, Canada*, pages 770–777, 2003.
18. M. Wierzbicki, M. Drangova, G. Guiraudon, T. Peters. Validation of dynamic heart models obtained using non-linear registration for virtual reality trainig planning and guidance of minimally invasive cardiac surgeries. *Medical Image Analysis*, 8:387–401, 2004.

Classification of Segmental Wall Motion in Echocardiography Using Quantified Parametric Images

Cinta Ruiz Dominguez[1,2], Nadjia Kachenoura[1], Sébastien Mulé[1],
Arthur Tenenhaus[1], Annie Delouche[1], Olivier Nardi[3], Olivier Gérard[2],
Benoît Diebold[1,3], Alain Herment[1], and Frédérique Frouin[1]

[1] INSERM, U678, CHU Pitié-Salpêtrière, Paris, France
ruiz@imed.jussieu.fr
[2] Philips Medical Systems Research Paris, Suresnes, France
[3] Service d'échocardiographie, HEGP, Paris, France

Abstract. The interpretation of cine-loops and parametric images to assess regional wall motion in echocardiography requires to acquire an expertise, which is based on training. To overcome the training phase for the interpretation of new parametric images, a quantification based on profiles in the parametric images was attempted. The classification of motion was performed on a training set including 362 segments and tested on a second database including 238 segments. The consensual visual interpretation of two-dimensional sequences by two experienced readers were used as the "gold standard". Mono- and multi-parametric classification approaches were undertaken. Results show an accuracy of 74% for training and 68% for test in case of mono-parametric approach. They are 80% and 67% in case of multi-parametric approach. Moreover, the evaluation protocol enables to understand the limitations of this approach. The in-depth study shows that a large part of false-positive segments are apical segments. This suggests that taking into account the segment location could improve the performances.

1 Introduction

Echocardiography is the modality of choice for the detection and the follow-up of wall motion abnormalities. The global wall motion index which is assessed as the sum of regional wall motion scores (RWMS) has a high predictive value. At the present time, the analysis of the contraction is mainly visual and requires a long training to acquire the necessary expertise. Among the new techniques which provide an additional information to clinicians to evaluate the RWMS, the most cited are: "Color Kinesis", which displays the timing and magnitude of endocardial wall motion [1], the "Tissue Doppler Imaging", which shows the instant velocity of the myocardium [2], and the "Strain-rate Imaging", which displays the radial and longitudinal deformation of the myocardium [3]. These images can be evaluated visually but this evaluation remains subjective. Several indices are currently studied to quantify these images [4][5][6].

The methods of parametric imaging provide images that represent parameters estimated on the temporal variation of intensity curve of each pixel [7] [8] [9]. These methods do not require specific acquisition or software. In [8], a qualitative validation of parametric images obtained by the Factor Analysis of the Left Ventricle in Echocardiography (FALVE) has already been proposed. In this paper, quantification indices of parametric FALVE images are proposed. A classification methodology based on ROC curves is applied in order to evaluate the power detection of these indices for regional wall motion abnormalities of the left ventricle. Cut-off values are optimized on the training data set and diagnostic performances are studied on a test database.

2 Patients and Methods

2.1 Patients' Databases

Eighty-six apical two dimensional harmonic gray scale sequences were acquired using an HDI system (Philips Medical Systems, Best, The Netherlands) and digitally recorded with the use of HDI Lab software. Four-chamber and two-chamber views were acquired during routine examinations in order to be representative of in-hospital patients in terms of pathology and echogenicity. Forty-nine patients were enrolled in the study. No patient was excluded. The etiology of the left ventricule dysfunction was coronary artery disease in 28, cardiomyopathy in 7, valvular disease in 5, and other in 9.

Series of three or four cycles were acquired and separated cycles were identified by selecting the onset of QRS complex of ECG, and the associated end-diastolic images. The cycle giving the best superimposed initial and final images was selected automatically, in order to minimize the global motion.

The consensual visual interpretation of two-dimensional sequences by two experienced readers were used as the "gold standard" for comparisons. Each view was segmented as recommended by the American Society of Echocardiography [10]. For each patient, endocardial motion in each segment was examined visually and judged as normal or pathological (hypokinetic, akinetic or dyskinetic).

The patient's database was divided into two groups : a training database and a test database. The training database was constituted by 52 sequences in order to have an equivalent number of normal (n=185) and pathological ($\bar{n}$=177) segments. The test database, constituted by 34 sequences, had normal ($n = 160$) and pathological ($\bar{n}$=78) segments. Only 2 segments were unclassified, due to their extremely bad quality.

2.2 Methods

Parametric Imaging. FALVE is a method used to extract the myocardium contraction information of an image sequence which corresponds to a cardiac cycle [8]. It expresses the time signal amplitude $p(x,y,t)$ of each pixel (x,y) at time t as:

$$p(x,y,t) = f_1(t)I_1(x,y) + f_2(t)I_2(x,y) + e(x,y,t) \quad x = 1,...,M \quad y = 1,...,N \quad (1)$$

The time functions $f_1(t)$ and $f_2(t)$ are called the factors, and the weighting coefficients $I_1(x,y)$ and $I_2(x,y)$ the factor images; M and N are the row and column numbers and $e(x,y,t)$ is the residual error.

The first factor $f_1(t)$ estimates the continuous component of the curves. The pixels (x,y) which show only small variations of the signal during the cardiac cycle (those which stay inside the myocardium or the left cavity during the whole cycle) present an intensity $I_1(x,y)$ larger than $I_2(x,y)$. The second factor $f_2(t)$ estimates the contraction-relaxation component: it increases during systole, then decreases during diastole. The pixels which have a significant variation in intensity, for example the points located in the cavity and close to the endocardium in the initial image of the cycle, present a larger intensity $I_2(x,y)$. While $I_1(x,y)$ is always positive, $I_2(x,y)$ is either positive or negative.

The three-color superposition of these images (green color for I_1, red color for the positive values and blue color for the negative values of I_2), called parametric FALVE image (see Fig. 1(b)), was interpreted by the clinicians in order to detect the contraction abnormalities [8].

Left Ventricle Segmentation. Parametric FALVE images were partitioned according to the guidelines of the American Heart Association [10]. A fast method of myocardial segmentation was implemented. Three anatomic landmarks P1, P2 and P3 (apical, left and right mitral valves points) and a distance d were manually located on an image of the sequence (see Fig. 1(a)). In the apical two- and four-chamber views, the image was divided into two regions separated by a line (defined as the long axis) connecting the apical point P1 to the the mid-point P4 between the mitral valves points. The long axis was divided into three thirds using two orthogonal lines, dividing the image into apical, medium and basal sections. Intersection points were noted P5 and P6. Apical section was divided into three equiangled regions by two radial lines. Outside points were defined on the orthogonal and radial lines at the distance d.

These points delimit a global mask (ROI_g) located on the left ventricle excluding the mitral valves. This mask was divided into seven regions of interest corresponding to the seven segments of the left ventricle (see Fig. 1(b)). Global mask was applied on images sequence to reduce the influence of the mitral valves motion in the estimation of the factors (see Fig. 1(a)).

Long Axis Distance Map. An image where intensity represented the Euclidian distance between the pixel location and the long axis [P5P6] of the left ventricle was computed (see Fig. 2). Using this coordinate system, inspired by [11], it was assumed that the decomposition of the local motion depended on the pixel location : a pixel belonging to the medium wall contracts towards the long axis, a pixel close to the apex contracts towards the point P5, a pixel close to the base contracts towards the point P6.

Let P5=(x_{P5}, y_{P5}) and P6=(x_{P6}, y_{P6}) be the extreme points on the long axis and $\boldsymbol{n} = (x_n, y_n)$ the perpendicular vector to the $[P5P6]$ segment. The long axis distance map was expressed as follows :

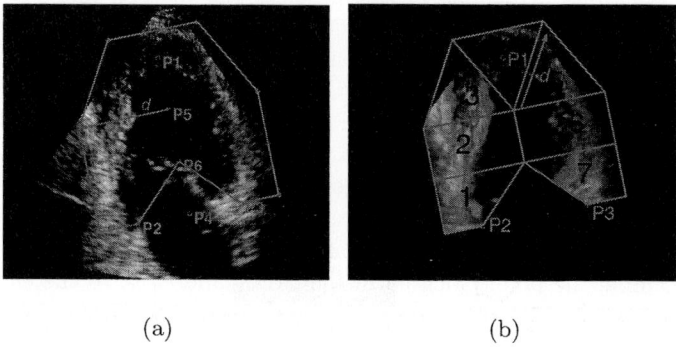

Fig. 1. Points of interest and global mask superimposed on one image of the sequence (a); manual landmarks (P1, P2, P3 and d) and seven segments on the corresponding parametric FALVE image (b)

$$R(x,y) = \begin{cases} \sqrt{(x-x_{P5})^2+(y-y_{P5})^2} & if\ y < \frac{x_n}{y_n}x + \left(y_{P5} - \frac{x_n}{y_n}x_{P5}\right) \\ \sqrt{(x-x_{P6})^2+(y-y_{P6})^2} & if\ y > \frac{x_n}{y_n}x + \left(y_{P6} - \frac{x_n}{y_n}x_{P6}\right) \\ x.x_n + y.y_n & else \end{cases} \quad (2)$$

Finally, the distance values were rounded to integer values.

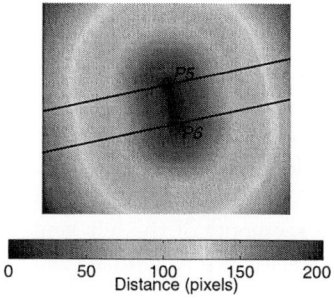

Fig. 2. Long axis distance map

Extraction of Segmental Profiles in Parametric Images. In [8], it has been shown that for a given segment, the color and the width of the band oriented towards the interior of the cavity in the parametric FALVE images were related to the RWMS. The values of the pixels in each parametric FALVE image corresponding to a segmental ROI were transformed into two mean profiles averaging the intensity of pixels located at the same distance. Thus, mean profiles $p_i^1(r)$ and $p_i^2(r)$ corresponding to segment i on factor images $I_1(x,y)$ and $I_2(x,y)$ were computed by formula:

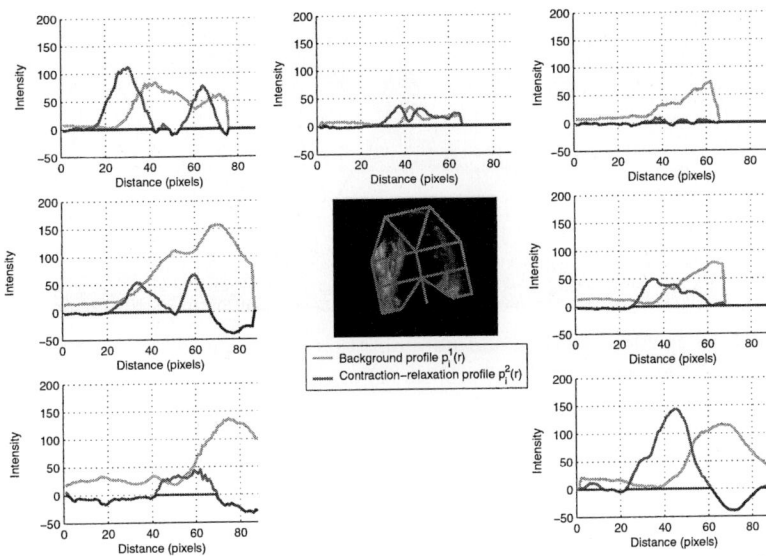

Fig. 3. Parametric FALVE image and contraction-relaxation (red-blue curves) and background (green curves) profiles corresponding to the seven segments

$$p_i^1(r) = \frac{1}{N_r} \sum_{\substack{(x,y)/R(x,y)=r \\ (x,y) \in ROI_i}} I_1(x,y), \quad p_i^2(r) = \frac{1}{N_r} \sum_{\substack{(x,y)/R(x,y)=r \\ (x,y) \in ROI_i}} I_2(x,y), \quad (3)$$

where $r = 1, ..., R_{max}$, N_r being the number of pixels such as $R(x,y) = r$ and R_{max} the maximal distance in ROI_i. The mean profiles $p_i^1(r)$ and $p_i^2(r)$ were called respectively background and contraction-relaxation profiles. The figure 3 shows the background and the contraction-relaxation profiles associated with a parametric FALVE image.

Two types of parameters per segment were proposed for the classification task:

- A normalized signed area (A_n) was estimated from the profile $p_i^2(r)$ to quantify color and width of the band oriented towards the interior of the cavity from the image corresponding to the contraction-relaxation factor. Distance r_{max} was defined as the distance corresponding to the maximum value in $p_i^1(r)$ profile. Positive and negative areas of contraction-relaxation profile from 0 until r_{max} distance were then computed. The maximum of these areas was normalized by the difference of the maximum value and the cavity value of the background profile to take into account the echogenicity of the segment: $(p_i^1(r_{max}) - p_i^1(r_0))$ (see Fig. 4(b)).
- A composite profile was estimated from the profiles $p_i^1(r)$ and $p_i^2(r)$. As the length of profiles R_{max} depended on the distance d that was different for

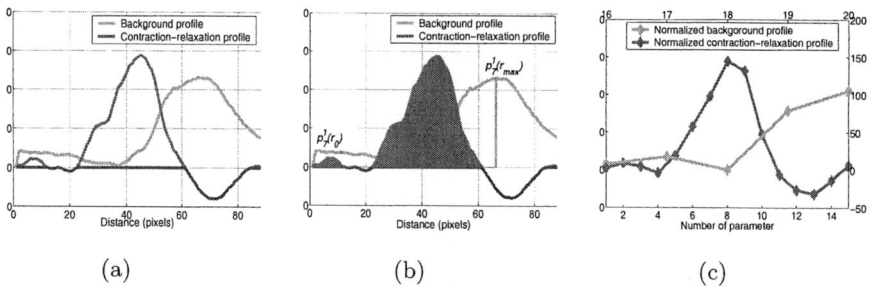

Fig. 4. Mean profiles (a), area and normalization coefficient of contraction-relaxation profile (b) and length-normalized profiles (c) of the segment 7 from the figure 2

each view, two length normalized profiles were computed by changing the sampling rate of the profiles. This was performed in the spectral domain by applying a cascade of three operators: up-sampling (zero padding) by integer factor q_1, filtering by an anti-aliasing (low-pass) FIR filter, and down-sampling by integer factor q_2. A study was performed to determinate the minimal sampling rate : spatial resolution of the factor images $I_1(x,y)$ and $I_2(x,y)$ was reduced similarly in order to keep visible the useful information for the diagnosis of contraction. Five points were retained for the profile $p_i^1(r)$ and fifteen points for the profile $p_i^2(r)$ (see Fig. 4(c)). A composite profile was constructed by juxtaposing the 5 values and the 15 values.

Classification. Two types of classification of segments were then proposed:

- A mono-parametric approach based on A_n index: it was computed for the normal and pathological segments of the training database. The ROC (Receiver Operating Characteristic) curve corresponding to this index was traced (Cut-off, Sensitivity,1-Specificity) [12]. Optimal cut-off was defined as the value of A_n that minimized the difference between the sensitivity and the specificity. Optimal cut-off was finally applied to the classification of the segments of the test database.
- A multi-parametric approach based on logistic regression applied to composite profiles. Logistic regression parameters were estimated from the training database. For each segment, the probability of belonging to the normal (P_N) and pathological class (1-P_N) was computed. The parameters of the logistic regression model were finally applied to the composite profiles of the segments of the test database in order to infer the classification of these segments.

Comparisons of mono-parametric and multi-parametric approaches were performed for the training and the test databases. A non-parametric comparison of the ROC curves was carried out. The sensitivity and the specificity in test of both approaches were compared by McNemar's statistics.

3 Results

The empirical ROC curves corresponding to A_n and P_N indices are shown in Fig. 5(c). The empirical areas under ROC curve (AUC) were respectively 0.82 and 0.89. The difference of AUCs was statistically significantly different from 0 (p=0.001).

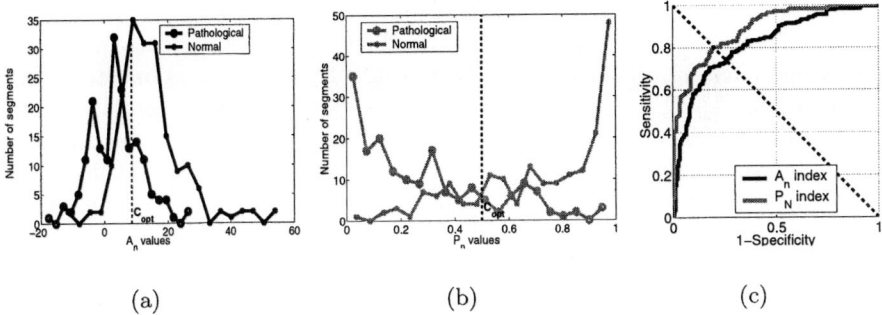

Fig. 5. Distributions of the (a)normalized area and (b) probability of belonging to the normal class for both normal and pathological segments. Corresponding ROC curves (c)

For the optimal cut-offs (A_n=8.6 and P_N=0.5), the sensitivity and the specificity of A_n and P_N indices were respectively 74%, 74% and 80%, 80% on the training database (see Table 1 and Table 2).

Table 1. Training results for A_n

Number of segments	Gold Standard	
Classification	Normal	Pathological
Normal	137	46
Pathological	48	131

Table 2. Training results for P_N

Number of segments	Gold Standard	
Classification	Normal	Pathological
Normal	149	35
Pathological	36	142

For the test database, the sensitivity and the specificity were respectively 67%, 69% for the A_n index and 62%, 70% for the P_N index (see Table 3 and Table 4).

Table 3. Diagnostic test results for A_n

Number of segments	Gold Standard	
Classification	Normal	Pathological
Normal	111	26
Pathological	49	52

Table 4. Diagnostic test results for P_N

Number of segments	Gold Standard	
Classification	Normal	Pathological
Normal	112	30
Pathological	48	48

4 Discussion

Echocardiography is the most widely used imaging modality to assess regional wall motion. The evaluation of RWMS is commonly performed by visual inspection, which is subjective and experience dependent. The use of a consensus of two expert readers seems to be the best "gold standard". This gold standard is conventionally used by clinical evaluation studies [4].

A new quantification method of wall regional left ventricular motion based on profiles derived from parametric FALVE images was presented. To achieve a correct classification, a crucial step is the reduction of the information. This was performed successively as follows : first, using Principal Component Analysis based techniques : FALVE reduces the sequence of images corresponding to one cardiac cycle to 2 factors images; then, by the extraction of 14 mean regional profiles, 2 per segment. Finally, two types of parameters were estimated per segment: one normalized area index, that represented magnitude of the regional wall motion, and two length normalized profiles (20 parameters per segment). The length reduction of the profiles was necessary for the logistic regression because of the reduced number of segments in the training database. Such a reduction avoids learning by heart.

To reduce the inter-patient variability of the area parameter, a normalization was required. Different normalization coefficients were tested : normalization by the length of the profiles, normalization by the maximum value of the contraction-relaxation profile, and normalization by the difference between the maximum value and the cavity value of the background profile. The latter was retained because it gave the best results.

Some regional indices to quantify regional wall motion in echocardiography have been proposed and tested in the literature [4]. Validation of the "color kinesis" index was performed by cut-off determination on a training database. In the cited study, the training database was exclusively composed of normal subjects and the cut-off was defined as one standard deviation around the mean of the normal control group. This provides a specificity greater than the sensibility. In our study, healthy subjects and patients composed training database, presenting a similar number of normal and pathologic segments, and the optimal cut-off was estimated to minimize the sensitivity and specificity difference.

Using the training database, diagnostic performances are significantly better with the index P_N than with the index A_n. But diagnostic performances of P_N and A_n are no more significantly different for the test database. The large difference between training and test performances that is observed for P_N suggests a case of "overtraining" : this could be solved either by reducing the number of parameters or by increasing the number of segments in the training database.

Moreover a large part of the misclassified segments concern the apical segments. These results can be explained by the regional heterogeneity of wall motion amplitude, which has not yet been introduced in the learning phase. Indeed, some major variations of mean indices values could be observed according to the localization of the segment (see Table 5 and Table 6), showing that the motion magnitude is considerably reduced at the apex.

Table 5. Localization effect on A_n

	Mean Values of A_n	
Region	Normal	Pathological
Basal	15.94	4.98
Medium	16.51	7.40
Apical	14.78	3.39
Apex	8.74	-1.38

Table 6. Localization effect on P_N

	Mean Values of P_N	
Region	Normal	Pathological
Basal	0.73	0.35
Medium	0.75	0.31
Apical	0.78	0.26
Apex	0.56	0.16

The modelling of the localization as a covariate factor could improve the performances of the diagnostic test largely [12], but this would require a larger valued database. This is currently under construction, using the same criteria for patients' selection as those presented here. Indeed, the construction of an appropriate database and the collection of medical expertise is a key point of any evaluation approach. Using acquisitions from in-hospital patients enables us to have a good estimate of difficult cases and to be strict with tested indices.

The classification of contraction into four classes is also under study in order to have an approach similar to the clinicians' evaluation. This requires to generalize the ROC approach to 4 classes (normal, hypokinetic, akinetic, dyskinetic) in order to optimize the estimation of thresholds.

5 Conclusion

A methodological approach was developed to test the discrimination power of any quantitative method, aiming at detecting regional wall motion abnormalities. Some encouraging results have been observed for indices derived from a regional analysis based on parametric FALVE images. However the location of the segments being classified should be introduced as an complementary information to improve the performances of the classification. Moreover, several other methods of parametric imaging, such as [7][13] are already planed to be evaluated in the future, using this protocol and an augmented database. Some others methods [1][2][3] could be tested using the same protocol, but would require a modification of the acquisition.

References

1. Mor-Avi V., Vignon P., Koch R., et al.: Segmental analysis of color kinesis images: new method for quantification of the magnitude and timing of endocardial motion during left ventricular systole and diastole. Circulation **95** (1997) 2082-2097.
2. Sutherland G.R., Stewart M.J., Groundstroem K. W. et al.: Color Doppler myocardial imaging: a new technique for the assessment of myocardial function. J. Am. Soc. Echocardiogr. **7** (1994) 441-458.
3. Heimdal A., Stylen A., Torp H., Skjaerpe T.: Real-time strain rate imaging of the left ventricle by ultrasound. J. Am. Soc. Echocardiogr. **11** (1998) 1013-1019.

4. Koch R., Lang R.M., Garcia M.J., et al.: Objective evaluation of regional left ventricular wall motion during dobutamine stress echocardiographic studies using segmental analysis of Color Kinesis Images. J. Am. Coll. of Cardiology, **34**(2) (1999) 409-419.
5. Desco M., Ledesma-Carbayo M.J., Perez E., Santos A., et al: Assessment of normal and ischaemic myocardium by quantitative M-mode tissue Doppler imaging. Ultrasound Med Biol. **28**(5) (2002) 561-569.
6. Edvardsen T, Skulstad H, Aakhus S, Urheim S, Ihlen H: Regional myocardial systolic function during acute myocardial ischemia assessed by strain Doppler echocardiography. J. Am. Coll. of Cardiology, **37**(3) (2001) 726-730.
7. Ruiz Dominguez C., Frouin F., Lim P., Gerard O., et al: Parametric Analysis of Main Motion to study the regional motion of the left ventricle echocardiography in Proc FIMH'03, Lyon (2003) 173-183.
8. Frouin F., Delouche A., Raffoul H., Diebold H. et al: Factor analysis of the left ventricle by echocardiography (FALVE): a new tool for detecting regional wall motion abnormalities Eur J Echocardiogr. **5**(5) (2004) 335-346.
9. Caiani E.G., Lang R.M., Korcarz C.E., DeCara J.M. et al: Improvement in echocardiographic evaluation of left ventricular wall motion using still-frame parametric imaging. J. Am. Soc. Echocardiogr. **6**(3) (2002) 926-934.
10. Cerqueira M.D., Weissman N.J., Dilsizian V., et al.: Standardized myocardial segmentation and nomenclature for tomographic imaging of the heart: a statement for healthcare professionals from the Cardiac Imaging Committee of the Council on Clinical Cardiology of the American Heart Association. Circulation **105**(4) (2002) 539-542.
11. Declerk J., Feldmar J., Ayache N.: Definition of a four continuous planispheric transformation for the tracking and the analysis of the left-ventricle motion. Medical Image Analysis **2**(2) (2002) 197-213.
12. Sullivan Pepe M. : The statistical evaluation of medical tests for classification and prediction. Oxford University Press (Ed) (2003).
13. Tilmant C. *Estimation d'indices de contractilité myocardique par analyse d'images échocardiographiques.* PhD Thesis, University of Auvergne, France, 2004.

Author Index

Admiraal-Behloul, Faiza 54
Albán, Edisson 338
Allain, Pascal 434
Andriantsimiavona, R. 325
Angelini, Elsa D. 434
Antila, K. 92
Arts, Theo 314
Ashikaga, Hiroshi 256
Aslanidi, Oleg V. 162
Axel, Leon 369, 446
Ayache, Nicholas 256

Baeyens, Enrique 75
Benson, Alan P. 304
Biktashev, Vadim N. 283, 293
Biktasheva, Irina V. 283, 293
Blok, Mark 54
Boulmier, Dominique 457
Bovendeerd, Peter 314
Briscoe, Chris 85
Brisinda, Donatella 143, 205
Brooks, Dana H. 195

Calderero, Felipe 195
Calvani, Menotti 205
Camara, O. 325
Carol, Antoni 65
Carrault, Guy 394
Cavaro-Menard, Christine 404
Chandran, Krishnan B. 12
Chandrashekara, Raghavendra 348, 425
Chapelle, D. 325
Chen, Ting 446
Cimrman, R. 325
Clarysse, Patrick 384, 467
Clayton, R.C. 153
Clayton, Richard 246
Coatrieux, Jean Louis 236, 457
Cootes, T. 92
Costa, Kevin D. 434
Courboulay, Vincent 404
Croisille, Pierre 384

Danilouchkine, Mikhail G. 33, 54
Defontaine, Antoine 394

Delhaas, Tammo 314
Delhay, Bertrand 467
Delingette, Hervé 256
Delouche, Annie 477
Deng, J. 123
Diebold, Benoît 477
Dindoyal, I. 123
Dössel, Olaf 172
Duan, Qi 434

Faris, Owen 256
Fenici, Peter 143
Fenici, Riccardo 143, 205
Fischer, Gerald 44, 183
Fraile, Juan C. 75
Frangi, Alejandro F. 33
Franzone, Piero Colli 267
Fritscher, Karl D. 113
Frouin, Frédérique 477

Garreau, Mireille 236, 457
Gérard, Olivier 434, 477
Ghodrati, Alireza 195
Gil, Debora 65, 359
González, Lorena 75

Haddad, Rana 384
Haigron, Pascal 236
Hänninen, Helena 278
Hanser, Friedrich 44
Hayek, Giuseppe 205
Hendriks, Emile A. 54
Herment, Alain 477
Hernández, Alfredo I. 236, 394
Hernandez, Aura 65
Herz, Susan L. 434
Hill, Derek L.G. 325, 348
Hintermüller, Christoph 44
Hintringer, F. 183
Histace, Aymeric 404
Ho, S.Y. 183
Holden, Arun V. 153, 162, 226, 293, 304
Holmes, Jeffrey W. 434
Homma, Shunichi 434

Hu, Zhenhua 369
Hurmusiadis, Vassilios 85
Husa, Terhi 278

Ingrassia, Christopher M. 434

Juslin, Anu 338

Kachenoura, Nadjia 477
Katila, Toivo 467
Kharche, S. 153
Koikkalainen, J. 92
Koning, Gerhard 23
Korhonen, Petri 278

Laine, Andrew F. 434
Lambert, Jennifer L. 162
Lambrou, T. 123
Lamminmäki, E. 92
Le Breton, Hervé, 457
Lelieveldt, Boudewijn P.F. 23, 33, 54
Lilja, M. 92
Linney, A.D. 123
Lopez, John J. 12
Lorenz, Cristian 1, 102
Lötjönen, Jyrki 92, 467

Mabo, Philippe 236
MacLeod, Rob 195
Magadán-Méndez, Margarita 338
Magnin, Isabelle E. 384, 467
Manglik, Tushar 446
McVeigh, Elliot 256
Meloni, Anna Maria 143, 205
Ménard, Michel 404
Metaxas, Dimitris 369
Modre, Robert 44
Mohiaddin, Raad 414
Mohiaddin, Raad H. 425
Moireau, P. 325
Moreau-Villéger, Valérie 256
Mulé, Sébastien 477

Nardi, Olivier 477

Olszewski, Mark E. 12
Oost, Elco 23
Ordás, Sebastián 33
Orkisz, Maciej 384

Pavarino, Luca F. 267
Perán, Jose R. 75
Perperidis, Dimitrios 414
Pfeifer, Bernhard 44
Pilgram, Roland 113

Radeva, Petia 65, 359
Razavi, Reza 325, 348
Reiber, Johan H.C. 23, 33, 54
Reilhac, Anthonin 338
Revel, Didier 384
Rhode, Kawal S. 348
Rodeck, C.H. 123
Rodriguez, Oriol 65, 359
Rosales, Misael 359
Rossen, James D. 12
Rubio, Jerónimo J. 75
Rueckert, Daniel 348, 414, 425
Ruff, C.F. 123
Ruiz Dominguez, Cinta 477
Ruotsalainen, Ulla 338

Sachse, Frank B. 172, 216
Sainte-Marie, J. 325
Sanchez-Ortiz, Gerardo I. 348
Schubert, Rainer 113
Seemann, Gunnar 172, 216
Seger, Michael 44
Sermesant, Maxime 256, 325, 348
Simon, Antoine 457
Sonka, Milan 12, 23
Srinivasan, Neil T. 162
Stenroos, Matti 278
Sternickel, Karsten 143

Taccardi, Bruno 216, 267
Tadmor, Gilead 195
Tenenhaus, Arthur 477
Tierala, Ilkka 278
Tilg, Bernhard 44, 183
Todd-Pokropek, A. 123
Tohka, Jussi 338
Toivonen, Lauri 278
Tong, Wing Chiu 226
Trieb, Thomas 44

Ubbink, Sander 314

Väänänen, Heikki 278
van Assen, Hans C. 33
van de Vosse, Frans 314

Veenman, Cor J. 54
Vesterinen, Paula 278
Vigmostad, Sarah C. 12
von Berg, Jens 1, 102

Wahle, Andreas 12
Weiß, Daniel L. 172

Wesarg, Stefan 133
Westenberg, Jos J.M. 33
Wieser, L. 183

Zhang, H. 153
Zhuchkova, Ekaterina 246

Lecture Notes in Computer Science

For information about Vols. 1–3436

please contact your bookseller or Springer

Vol. 3556: H. Baumeister, M. Marchesi, M. Holcombe (Eds.), Extreme Programming and Agile Processes in Software Engineering. XIV, 332 pages. 2005.

Vol. 3543: L. Kutvonen, N. Alonistioti (Eds.), Distributed Applications and Interoperable Systems. XI, 235 pages. 2005.

Vol. 3537: A. Apostolico, M. Crochemore, K. Park (Eds.), Combinatorial Pattern Matching. XI, 444 pages. 2005.

Vol. 3535: M. Steffen, G. Zavattaro (Eds.), Formal Methods for Open Object-Based Distributed Systems. X, 323 pages. 2005.

Vol. 3532: A. Gómez-Pérez, J. Euzenat (Eds.), The Semantic Web: Research and Applications. XV, 728 pages. 2005.

Vol. 3531: J. Ioannidis, A. Keromytis, M. Yung (Eds.), Applied Cryptography and Network Security. XI, 530 pages. 2005.

Vol. 3528: P.S. Szczepaniak, J. Kacprzyk, A. Niewiadomski (Eds.), Advances in Web Intelligence. XVII, 513 pages. 2005. (Subseries LNAI).

Vol. 3526: S.B. Cooper, B. Löwe, L. Torenvliet (Eds.), New Computational Paradigms. XVII, 574 pages. 2005.

Vol. 3525: A.E. Abdallah, C.B. Jones, J.W. Sanders (Eds.), Communicating Sequential Processes. XIV, 321 pages. 2005.

Vol. 3524: R. Barták, M. Milano (Eds.), Integration of AI and OR Techniques in Constraint Programming for Combinatorial Optimization Problems. XI, 320 pages. 2005.

Vol. 3523: J.S. Marques, N.P. de la Blanca, P. Pina (Eds.), Pattern Recognition and Image Analysis, Part II. XXVI, 733 pages. 2005.

Vol. 3522: J.S. Marques, N.P. de la Blanca, P. Pina (Eds.), Pattern Recognition and Image Analysis, Part I. XXVI, 703 pages. 2005.

Vol. 3521: N. Megiddo, Y. Xu, B. Zhu (Eds.), Algorithmic Applications in Management. XIII, 484 pages. 2005.

Vol. 3520: O. Pastor, J. Falcão e Cunha (Eds.), Advanced Information Systems Engineering. XVI, 584 pages. 2005.

Vol. 3518: T.B. Ho, D. Cheung, H. Li (Eds.), Advances in Knowledge Discovery and Data Mining. XXI, 864 pages. 2005. (Subseries LNAI).

Vol. 3517: H.S. Baird, D.P. Lopresti (Eds.), Human Interactive Proofs. IX, 143 pages. 2005.

Vol. 3516: V.S. Sunderam, G.D. van Albada, P.M.A. Sloot, J.J. Dongarra (Eds.), Computational Science – ICCS 2005, Part III. LXIII, 1143 pages. 2005.

Vol. 3515: V.S. Sunderam, G.D. van Albada, P.M.A. Sloot, J.J. Dongarra (Eds.), Computational Science – ICCS 2005, Part II. LXIII, 1101 pages. 2005.

Vol. 3514: V.S. Sunderam, G.D. van Albada, P.M.A. Sloot, J.J. Dongarra (Eds.), Computational Science – ICCS 2005, Part I. LXIII, 1089 pages. 2005.

Vol. 3513: A. Montoyo, R. Mu\~noz, E. Métais (Eds.), Natural Language Processing and Information Systems. XII, 408 pages. 2005.

Vol. 3510: T. Braun, G. Carle, Y. Koucheryavy, V. Tsaousidis (Eds.), Wired/Wireless Internet Communications. XIV, 366 pages. 2005.

Vol. 3509: M. Jünger, V. Kaibel (Eds.), Integer Programming and Combinatorial Optimization. XI, 484 pages. 2005.

Vol. 3508: P. Bresciani, P. Giorgini, B. Henderson-Sellers, G. Low, M. Winikoff (Eds.), Agent-Oriented Information Systems II. X, 227 pages. 2005. (Subseries LNAI).

Vol. 3507: F. Crestani, I. Ruthven (Eds.), Information Context: Nature, Impact, and Role. XIII, 253 pages. 2005.

Vol. 3506: C. Park, S. Chee (Eds.), Information Security and Cryptology – ICISC 2004. XIV, 490 pages. 2005.

Vol. 3505: V. Gorodetsky, J. Liu, V. A. Skormin (Eds.), Autonomous Intelligent Systems: Agents and Data Mining. XIII, 303 pages. 2005. (Subseries LNAI).

Vol. 3504: A.F. Frangi, P. I. Radeva, A. Santos, M. Hernandez (Eds.), Functional Imaging and Modeling of the Heart. XV, 489 pages. 2005.

Vol. 3503: S.E. Nikoletseas (Ed.), Experimental and Efficient Algorithms. XV, 624 pages. 2005.

Vol. 3502: F. Khendek, R. Dssouli (Eds.), Testing of Communicating Systems. X, 381 pages. 2005.

Vol. 3501: B. Kégl, G. Lapalme (Eds.), Advances in Artificial Intelligence. XV, 458 pages. 2005. (Subseries LNAI).

Vol. 3500: S. Miyano, J. Mesirov, S. Kasif, S. Istrail, P. Pevzner, M. Waterman (Eds.), Research in Computational Molecular Biology. XVII, 632 pages. 2005. (Subseries LNBI).

Vol. 3499: A. Pelc, M. Raynal (Eds.), Structural Information and Communication Complexity. X, 323 pages. 2005.

Vol. 3498: J. Wang, X. Liao, Z. Yi (Eds.), Advances in Neural Networks – ISNN 2005, Part III. L, 1077 pages. 2005.

Vol. 3497: J. Wang, X. Liao, Z. Yi (Eds.), Advances in Neural Networks – ISNN 2005, Part II. L, 947 pages. 2005.

Vol. 3496: J. Wang, X. Liao, Z. Yi (Eds.), Advances in Neural Networks – ISNN 2005, Part II. L, 1055 pages. 2005.

Vol. 3495: P. Kantor, G. Muresan, F. Roberts, D.D. Zeng, F.-Y. Wang, H. Chen, R.C. Merkle (Eds.), Intelligence and Security Informatics. XVIII, 674 pages. 2005.

Vol. 3494: R. Cramer (Ed.), Advances in Cryptology – EUROCRYPT 2005. XIV, 576 pages. 2005.

Vol. 3493: N. Fuhr, M. Lalmas, S. Malik, Z. Szlávik (Eds.), Advances in XML Information Retrieval. XI, 438 pages. 2005.

Vol. 3492: P. Blache, E. Stabler, J. Busquets, R. Moot (Eds.), Logical Aspects of Computational Linguistics. X, 363 pages. 2005. (Subseries LNAI).

Vol. 3489: G.T. Heineman, I. Crnkovic, H.W. Schmidt, J.A. Stafford, C. Szyperski, K. Wallnau (Eds.), Component-Based Software Engineering. XI, 358 pages. 2005.

Vol. 3488: M.-S. Hacid, N.V. Murray, Z.W. Raś, S. Tsumoto (Eds.), Foundations of Intelligent Systems. XIII, 700 pages. 2005. (Subseries LNAI).

Vol. 3486: T. Helleseth, D. Sarwate, H.-Y. Song, K. Yang (Eds.), Sequences and Their Applications - SETA 2004. XII, 451 pages. 2005.

Vol. 3483: O. Gervasi, M.L. Gavrilova, V. Kumar, A. Laganà, H.P. Lee, Y. Mun, D. Taniar, C.J.K. Tan (Eds.), Computational Science and Its Applications – ICCSA 2005, Part IV. XXVII, 1362 pages. 2005.

Vol. 3482: O. Gervasi, M.L. Gavrilova, V. Kumar, A. Laganà, H.P. Lee, Y. Mun, D. Taniar, C.J.K. Tan (Eds.), Computational Science and Its Applications – ICCSA 2005, Part III. LXVI, 1340 pages. 2005.

Vol. 3481: O. Gervasi, M.L. Gavrilova, V. Kumar, A. Laganà, H.P. Lee, Y. Mun, D. Taniar, C.J.K. Tan (Eds.), Computational Science and Its Applications – ICCSA 2005, Part II. LXIV, 1316 pages. 2005.

Vol. 3480: O. Gervasi, M.L. Gavrilova, V. Kumar, A. Laganà, H.P. Lee, Y. Mun, D. Taniar, C.J.K. Tan (Eds.), Computational Science and Its Applications – ICCSA 2005, Part I. LXV, 1234 pages. 2005.

Vol. 3479: T. Strang, C. Linnhoff-Popien (Eds.), Location- and Context-Awareness. XII, 378 pages. 2005.

Vol. 3478: C. Jermann, A. Neumaier, D. Sam (Eds.), Global Optimization and Constraint Satisfaction. XIII, 193 pages. 2005.

Vol. 3477: P. Herrmann, V. Issarny, S. Shiu (Eds.), Trust Management. XII, 426 pages. 2005.

Vol. 3475: N. Guelfi (Ed.), Rapid Integration of Software Engineering Techniques. X, 145 pages. 2005.

Vol. 3474: C. Grelck, F. Huch, G.J. Michaelson, P. Trinder (Eds.), Implementation and Application of Functional Languages. X, 227 pages. 2005.

Vol. 3468: H.W. Gellersen, R. Want, A. Schmidt (Eds.), Pervasive Computing. XIII, 347 pages. 2005.

Vol. 3467: J. Giesl (Ed.), Term Rewriting and Applications. XIII, 517 pages. 2005.

Vol. 3465: M. Bernardo, A. Bogliolo (Eds.), Formal Methods for Mobile Computing. VII, 271 pages. 2005.

Vol. 3464: S.A. Brueckner, G.D.M. Serugendo, A. Karageorgos, R. Nagpal (Eds.), Engineering Self-Organising Systems. XIII, 299 pages. 2005. (Subseries LNAI).

Vol. 3463: M. Dal Cin, M. Kaâniche, A. Pataricza (Eds.), Dependable Computing - EDCC 2005. XVI, 472 pages. 2005.

Vol. 3462: R. Boutaba, K.C. Almeroth, R. Puigjaner, S. Shen, J.P. Black (Eds.), NETWORKING 2005. XXX, 1483 pages. 2005.

Vol. 3461: P. Urzyczyn (Ed.), Typed Lambda Calculi and Applications. XI, 433 pages. 2005.

Vol. 3460: Ö. Babaoglu, M. Jelasity, A. Montresor, C. Fetzer, S. Leonardi, A. van Moorsel, M. van Steen (Eds.), Self-star Properties in Complex Information Systems. IX, 447 pages. 2005.

Vol. 3459: R. Kimmel, N.A. Sochen, J. Weickert (Eds.), Scale Space and PDE Methods in Computer Vision. XI, 634 pages. 2005.

Vol. 3458: P. Herrero, M.S. Pérez, V. Robles (Eds.), Scientific Applications of Grid Computing. X, 208 pages. 2005.

Vol. 3456: H. Rust, Operational Semantics for Timed Systems. XII, 223 pages. 2005.

Vol. 3455: H. Treharne, S. King, M. Henson, S. Schneider (Eds.), ZB 2005: Formal Specification and Development in Z and B. XV, 493 pages. 2005.

Vol. 3454: J.-M. Jacquet, G.P. Picco (Eds.), Coordination Models and Languages. X, 299 pages. 2005.

Vol. 3453: L. Zhou, B.C. Ooi, X. Meng (Eds.), Database Systems for Advanced Applications. XXVII, 929 pages. 2005.

Vol. 3452: F. Baader, A. Voronkov (Eds.), Logic for Programming, Artificial Intelligence, and Reasoning. XI, 562 pages. 2005. (Subseries LNAI).

Vol. 3450: D. Hutter, M. Ullmann (Eds.), Security in Pervasive Computing. XI, 239 pages. 2005.

Vol. 3449: F. Rothlauf, J. Branke, S. Cagnoni, D.W. Corne, R. Drechsler, Y. Jin, P. Machado, E. Marchiori, J. Romero, G.D. Smith, G. Squillero (Eds.), Applications of Evolutionary Computing. XX, 631 pages. 2005.

Vol. 3448: G.R. Raidl, J. Gottlieb (Eds.), Evolutionary Computation in Combinatorial Optimization. XI, 271 pages. 2005.

Vol. 3447: M. Keijzer, A. Tettamanzi, P. Collet, J.v. Hemert, M. Tomassini (Eds.), Genetic Programming. XIII, 382 pages. 2005.

Vol. 3444: M. Sagiv (Ed.), Programming Languages and Systems. XIII, 439 pages. 2005.

Vol. 3443: R. Bodik (Ed.), Compiler Construction. XI, 305 pages. 2005.

Vol. 3442: M. Cerioli (Ed.), Fundamental Approaches to Software Engineering. XIII, 373 pages. 2005.

Vol. 3441: V. Sassone (Ed.), Foundations of Software Science and Computational Structures. XVIII, 521 pages. 2005.

Vol. 3440: N. Halbwachs, L.D. Zuck (Eds.), Tools and Algorithms for the Construction and Analysis of Systems. XVII, 588 pages. 2005.

Vol. 3439: R.H. Deng, F. Bao, H. Pang, J. Zhou (Eds.), Information Security Practice and Experience. XII, 424 pages. 2005.

Vol. 3438: H. Christiansen, P.R. Skadhauge, J. Villadsen (Eds.), Constraint Solving and Language Processing. VIII, 205 pages. 2005. (Subseries LNAI).

Vol. 3437: T. Gschwind, C. Mascolo (Eds.), Software Engineering and Middleware. X, 245 pages. 2005.